Introduction to Computational Fluid Dynamics

Introduction to Computational Fluid Dynamics

P. Niyogi

S. K. Chakrabartty

M. K. Laha

Pearson

ISBN 978-81-7758-764-7

First Impression, 2006

Published by Pearson India Education Services Pvt. Ltd, CIN: U72200TN2005PTC057128.

Head Office: 1st Floor, Berger Tower, Plot No. C-001A/2, Sector 16B, Noida – 201 301, Uttar Pradesh, India

Registered Office: 7th Floor, SDB2, ODC 7, 8 & 9, Survey No.01 ELCOT IT/ ITES SEZ, Sholinganallur, Chennai – 600119, Tamil Nadu, India.
Website: in.pearson.com, Email: companysecretary.india@pearson.com

Digitally printed in India by Manipal Technologies Limited. in the year of 2022

To our wives

Supriya Niyogi,

Shikha Chakrabartty,

Jayasree Laha

About the Authors

Pradip Niyogi obtained his M.Sc. (1955) in Applied Mathematics from Calcutta University. He was a Research Associate of late Prof K. Os-watitsch at the then DVL Institute of Theoretical Gasdynamics, Aachen (1961–65), Germany. Under his supervision he carried out research work and received a doctorate degree from Technische Hochschule, Aachen (1965) with a thesis on Wing Theory. He served in Jadavpur University, Kolkata from 1958 to 1979 and then joined IIT, Kharagpur, as a Professor of Mathematics and held the position till 1997.

An author of several books and monographs, Dr Niyogi's major contributions are in Integral Equations, High Speed Gasdynamics and Computational Fluid Dynamics (CFD), Numerical Weather Prediction and Database Systems.

S. K. Chakrabartty obtained his M.Sc. (1970) in Applied Mathematics and Ph.D. (1976) from Jadavpur University, Kolkata. He worked as a Senior Research Fellow (1975–77) at the Mathematics Department, IISc, Bangalore, and a Research Associate in the Department of Aerospace Engineering, IIT, Kharagpur, for a year. Since 1978 Dr Chakrabartty has been working as a scientist at National Aerospace Laboratories (NAL), Bangalore, in CFD, contributing consistently to the development of software for solving Transonic Small Perturbation (TSP), and Full Potential (FP), Euler and the Reynolds Averaged Navier-Stokes (RANS) equations governing fluid flows past aerospace vehicles.

Dr Chakrabartty has made a significant contribution in the area of CFD especially in compressible viscous flow computation, and the CFD codes developed by him and his team have contributed to aircraft design and development activities at NAL. In 1986–87, he spent about a year at German Aerospace Center (DLR) in Braunschweig, Germany. He has several research papers in national and international journals to his credit.

M. K. Laha obtained his B.Tech. (1979) and M.Tech. (1981) degrees in Aeronautical Engineering, and Ph.D. (1993) in unsteady aerodynamics from IIT, Kharagpur. Currently Associate Professor, he has been a member of the faculty of the Department of Aerospace Engineering at his alma mater since 1987. Apart from teaching a variety of subjects related to aerospace engineering, he has been involved in collaborative research with the aircraft industry in the area of unsteady aerodynamics for the development of the Light Combat Aircraft (LCA). His interests include aerodynamics and computers.

Preface

Computational Fluid Dynamics (CFD) is an introductory level text book for undergraduate and postgraduate students of engineering and postgraduate students of mathematics and physics. It is an outcome of lectures delivered for courses at different levels to the students of various disciplines in science and engineering and the active involvement in research and development of the subject by the authors.

CFD complements experimental and theoretical fluid dynamics by providing an efficient means of simulating fluid flows of practical interest. With the rapid development of computer technology and advancement in numerical analysis and algorithms over the past few decades, CFD has found application in a variety of fields including aerospace, naval and surface transport engineering, physiological fluid dynamics, oceanography, meteorology and astrophysics. Growing interest in the subject among the students of mechanical, aerospace, civil, and chemical engineering and also among students of physics and applied mathematics has made a basic text book on the subject necessary.

CFD is a rapidly growing area and a large number of new concepts have emerged. Thus, it is necessary that a student of CFD becomes familiar with these concepts and assimilates them. For this, he needs some knowledge of partial differential equations, numerical analysis and algorithms and fluid dynamics. Books on mathematics usually do not deal with these diverse topics in a single book. In view of this, as far as practicable, a full mathematical treatment of the required topics has been presented. This book is self-contained and divided into two parts. The first part of the book, consisting of the first six chapters, is essentially devoted to this endeavor. The subject has been introduced in a self-contained manner, where the basic fundamentals in fluid dynamic equations and the step-by-step development of numerical algorithms along with questions of numerical stability, convergence and accuracy have been discussed in detail. Also, the authors have tried to anticipate and answer questions that students usually ask when first learning the subject.

The theory of partial differential equations and finite difference and finite volume methods for their numerical solution is introduced first. This mathematical foundation aims to provide answers to basic questions and equip the student with the concepts and tools needed. Chapter 2 deals with finite difference and finite volume methods. Computational methods for solving partial differential equations together with initial and/or boundary conditions differ depending on the type of the equation. The more important properties of the computational methods for the three main types, namely, parabolic, hyperbolic and elliptic type equations have been discussed in Chapters 3, 4 and 5.

Chapter 6 is devoted to problems governed by equations of mixed elliptic–hyperbolic type. Such problems are of great importance in the study of transonic flows—a field which eluded a solution even for the simplest model, namely, the transonic small perturbation (TSP) model, for several decades. To keep matters within bounds, we have omitted any discussion on the finite-element method. Interested readers may look in the bibliography for introductory level text books on this topic.

The second part (Chapters 7–13) starts with the description of the fundamental equations in fluid dynamics, their hierarchy at different levels of approximation and the boundary conditions suitable for practical problems. Chapter 8 deals with numerical methods to generate curvilinear body-fitted grids, which is the first and one of the most important steps in the use of CFD for practical applications. Mathematical details of each and every step required have been explained. The aim is that a student should be able to write his/her own computer program stepwise and finally get a grid surrounding any two-dimensional object. Depending on the complexity of the governing equations, the remaining chapters have been arranged, starting from the Laplace equation to the compressible Reynolds Averaged Navier–Stokes (RANS) equations with turbulence modeling. A mathematical model describing the inviscid incompressible flow governed by the Laplace equation has been described in Chapter 9. It introduces the panel method, which is highly suitable for preliminary design and quick analysis of flow past complex configurations and is widely used in the aerospace industry. Computation of inviscid compressible flows governed by the transonic small perturbation, transonic full-potential and Euler equations is considered next and described in detail in Chapter 10. Computations of boundary layer flows and incompressible and compressible viscous flows are explained thoroughly in Chapters 11, 12 and 13, respectively, along with suitable examples.

In order to provide motivation to the student, each chapter begins with a brief introduction that explains the importance of the chapter and indicates how it is connected to the previous and subsequent chapters. Important concepts are explained through simple worked-out examples. Summary, keywords and exercises are provided. The more important technical terms have been explained in the glossary. The bibliography should help the student continue further studies on selected topics. For some problems, computer programs in FORTRAN and C have been provided. These will help students to write their own programs for solving problems.

Some topics of interest in CFD, such as the formulation of the governing equations in non-inertial frames of reference, internal flows, and TVD schemes, are not discussed since these are beyond the scope of this book.

P. Niyogi

S. K. Chakrabartty

M. K. Laha

Acknowledgements

We have received constant support and technical help from many people at the National Aerospace Laboratories (NAL), Bangalore. We are specially thankful to K. Dhanalaksmi who read all the chapters of the manuscript carefully and computed solutions of some of the examples. Sincere thanks are also due to J. S. Mathur and V. Ramesh for their efforts. Discussions with scientists at the CTFD Division of NAL at different stages of preparation of the manuscript are gratefully acknowledged.

The help and suggestions unstintingly given by N. Singh of the Department of Aerospace Engineering, IIT, Kharagpur, is gratefully acknowledged as well.

It is a pleasure to acknowledge the help received from all our students, in particular, we are thankful to G. C. Layek, S. B. Hazra, S. Ghosh, T. R. Mahapatra and T. K. Maikap.

In teaching the subject we have often consulted several valuable books listed in the bibliography, whose influence may be seen in many parts of the book. We express our sincere thanks and indebtedness to all of them.

Last but not the least, we gratefully acknowledge the interest, encouragement and support received from our families during the preparation of the manuscript.

P. Niyogi

S. K. Chakrabartty

M. K. Laha

Contents

Part I
Finite Difference Method for Partial Differential Equations

1

Introduction and Mathematical Preliminaries

Computational fluid dynamics, (abbreviated CFD) studies problems of fluid dynamics using numerical methods on a computer. Although the subject fluid dynamics is about three centuries old, CFD is a relatively new subject. It is hardly four decades old. Remarkable success has been achieved in CFD during this time, particularly, in handling nonlinear problems and problems involving complex geometries. Computational methods developed in CFD are being increasingly applied to other branches of science and engineering. The problems of study in CFD are often formulated in terms of a partial differential equation or in terms of a system of such equations, together with boundary and/or initial conditions. This chapter recapitulates the basic concepts of partial differential equations which are useful in the subsequent chapters.

1.1 INTRODUCTION

Computational fluid dynamics is an extension of classical fluid dynamics, originating from the amalgamation of classical fluid dynamics and numerical analysis supported by powerful electronic digital computers. Thousands of research workers, mathematicians, physicists and engineers all over the world contributed towards its growth and development and the progress is remarkable. It has created an atmosphere of confidence, even to the extent that almost no problem of fluid dynamics is now considered unsolvable. The zeal that has been generated in the minds of research workers and students of fluid dynamics is unparallel.

The year 1755 was a decisive year in the development of classical fluid dynamics when Leonhard Euler established the momentum–flux conservation equations for an inviscid fluid, now known as *Euler's equations*. The Navier–Stokes equations for a viscous fluid were developed in 1840. Through the development of boundary layer concept of Prandtl (1904), it became possible to unite inviscid and viscous flow studies and obtain satisfactory explanation of a large number of flow phenomena. Approximate methods were employed and the results obtained were found to be qualitatively correct and often showed excellent agreement with experimental results. Frequently, expansion procedures in terms of certain small parameters were used to simplify the governing equations, which were then solved approximately. No general procedures for solving nonlinear problems were available, and in the language of von Karman (1941) "engineer grapples with nonlinear problems". The scene totally changed through the advent of electronic digital computers when powerful numerical methods like finite difference, finite element and their variants were developed which could deal with nonlinear problems effectively.

CFD, is a subject developed recently, almost simultaneously with electronic digital computers. Foundations of finite difference method for partial differential equations was laid by Courant et al. (1928) in a celebrated paper for studying existence and uniqueness of solutions of partial differential equations. Although, early works on CFD may be traced back to Richardson (1910) and Emmons (1944), the systematic study of CFD started in the sixties. During the past 35 years, CFD made spectacular progress, and several powerful computational methods have come up during this period, the most prominent among them being the finite difference method, the finite element method, the finite volume method and the spectral method. These methods depend heavily on the use of powerful electronic digital computers.

The basic equations of inviscid fluid dynamics are the Euler equations, which constitute a system of first-order nonlinear partial differential equations. Along with a particular problem of fluid dynamics, certain initial and boundary conditions are associated, which are also often nonlinear. For more than two centuries, mathematicians and engineers introduced various kinds of approximations leading to linearization of the basic equations. Often, the boundary conditions were also linearized in order to make the problems tractable. Even for such linearized problems, recourse to further approximation had to be taken while solving the linearized problems because most frequently no exact analytical method might be available.

A primary difference between CFD methods and the approximate methods of classical fluid dynamics may be noted. Once a model has been chosen, the model equations and boundary and/or initial conditions are solved in CFD by *numerically exact methods*, without introducing any further approximation. The only kind of error introduced are the numerical errors for which often numerical error estimates or some kind of error bounds are available. By sufficient refinement of the mesh system, the computed solutions are expected to approach the exact solution. This is not true about the above-mentioned approximate methods which introduce approximations to the model equations and boundary conditions which are then solved by approximate methods.

During the fifties and sixties various computational methods were developed, as already mentioned, which aided by an electronic digital computer could successfully treat nonlinear problems. All these methods discretize the problem so that a solvable system of algebraic equations are obtained, the number of equations being equal to the number of discrete points in the domain. The resulting algebraic system is linear if the original problem is linear, otherwise a system of nonlinear equations is obtained, which is then solved by a standard numerical method on a digital computer.

A computer plays a vital role in view of the huge number of arithmetic operations involved in a typical fluid flow problem and our desire is to solve such systems in a reasonable amount of time. To form an idea of the amount of labour involved, we consider in the following, a simple problem involving two-dimensional Laplace equation.

Example 1.1

Consider the two-dimensional Laplace equation

$$\frac{\partial^2 u}{\partial x^2} + \frac{\partial^2 u}{\partial y^2} = 0 \tag{1.1}$$

in the unit square OABC in the first quadrant with Dirichlet boundary condition, by finite difference method as shown in Fig. 1.1. The domain OABC is subdivided into a network by drawing straight lines $x = j\Delta x$, $y = k\Delta y$, $j = 1, 2 \ldots, M, M+1$, $k = 1, 2 \ldots, N, N+1$, parallel to the coordinate axes. The points of intersection of these lines are known as *mesh points*.

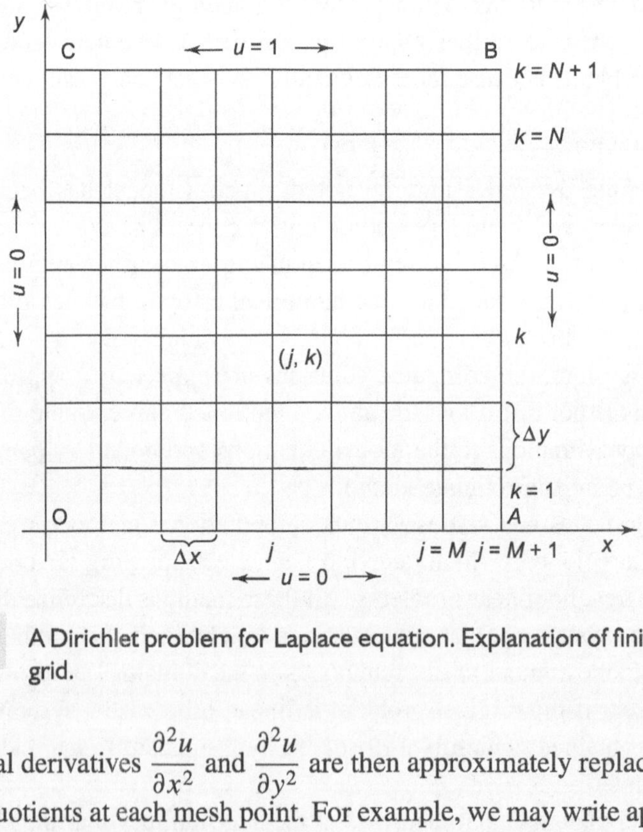

Figure 1.1 A Dirichlet problem for Laplace equation. Explanation of finite difference grid.

The partial derivatives $\frac{\partial^2 u}{\partial x^2}$ and $\frac{\partial^2 u}{\partial y^2}$ are then approximately replaced by partial difference quotients at each mesh point. For example, we may write approximately

$$\left.\frac{\partial^2 u}{\partial x^2}\right]_{(x,y)} = \frac{u(x+\Delta x, y) - 2u(x,y) + u(x-\Delta x, y)}{\Delta x^2} + O(\Delta x^2) \quad (1.2)$$

In fact, assuming $u(x, y)$ to be sufficiently *smooth* (that is, sufficient number of times continuously differentiable in the considered domain), it is possible to write, using Taylor's expansion about the point (x, y)

$$\frac{1}{\Delta x^2}\Big[u(x+\Delta x, y) - 2u(x,y) + u(x-\Delta x, y)\Big]$$

$$= \frac{1}{\Delta x^2}\Bigg[\bigg\{u(x,y) + \Delta x u_x + \frac{\Delta x^2}{2}u_{xx} + \frac{\Delta x^3}{6}u_{xxx}$$

$$+\frac{\Delta x^4}{24}u_{xxxx}|_{(\xi_1,y)}\bigg\} - 2u(x,y) + \bigg\{u(x,y) - \Delta x u_x + \frac{\Delta x^2}{2}u_{xx}$$

$$-\frac{\Delta x^3}{6}u_{xxx} + \frac{\Delta x^4}{24}u_{xxxx}|_{(\xi_2,y)}\bigg\}\Bigg]$$

$$= u_{xx} + \frac{\Delta x^2}{24}\Big[u_{xxxx}|_{(\xi_1,y)} + u_{xxxx}|_{(\xi_2,y)}\Big] \quad (1.3)$$

where ξ_1, ξ_2 are points in the intervals $(x, x+\Delta x)$ and in $(x-\Delta x, x)$ respectively. In view of the smoothness of $u(x, y)$, the derivatives of u are bounded, and the term in the square bracket on the right side of Eq. (1.3), known as the *truncation error term*, may be rewritten as $O(\Delta x^2)$, representing *a quantity of the order* of Δx^2; this means that it is a quantity which when divided by Δx^2, tends to a finite nonzero limit as $\Delta x \to 0$. In finite difference notation, it is customary to write

$$u(x, y) = u(j\Delta x, k\Delta y) = u_{j,k}.$$

Then $u(x+\Delta x, y) = u(j\Delta x + \Delta x, k\Delta y) = u_{j+1,k}$
and $u(x-\Delta x, y) = u(j\Delta x - \Delta x, k\Delta y) = u_{j-1,k}$.
So, Eq. (1.2) may be rewritten as

$$u_{xx}|_{j,k} = \frac{1}{\Delta x^2}\left[u_{j+1,k} - 2u_{j,k} + u_{j-1,k}\right] + O(\Delta x^2) \quad (1.4)$$

This is *central difference* representation for the second derivative. Since the values of u at three abscissa points $j+1, j, j-1$ are involved, it is called *three-point central difference formula*. Similarly, it is possible to express u_{yy} as

$$u_{yy}|_{j,k} = \frac{1}{\Delta y^2}\left[u_{j,k+1} - 2u_{j,k} + u_{j,k-1}\right] + O(\Delta y^2) \quad (1.5)$$

If the second derivatives in Eqs. (1.4) and (1.5) be replaced by the corresponding first terms on the right, a *truncation error* is committed which is of the second order, that is, proportional to the square of the mesh lengths Δx or Δy. It is called truncation error because it arises through the truncation of an infinitesimal or limiting process by a finite one.

Substituting Eqs. (1.4) and (1.5) in Eq. (1.1) and neglecting the truncation error terms for the present, the following linear algebraic equation is obtained for the mesh point $(j,\ k)$:

$$\frac{1}{\Delta x^2}\left[u_{j+1,k} - 2u_{j,k} + u_{j-1,k}\right] + \frac{1}{\Delta y^2}\left[u_{j,k+1} - 2u_{j,k} + u_{j,k-1}\right] = 0 \quad (1.6)$$

One such equation is obtained for each of the internal mesh points $j = 1, 2, \ldots, M$ and $k = 1, 2, \ldots, N$ so that a system of MN equations are obtained for equal number of unknowns. The values of the unknown $u(x,\ y)$ on the boundaries may be obtained from the prescribed boundary conditions. Under the boundary conditions shown in Fig. 1.1,

$$u_{o,k} = 0, \quad 0 \le k \le N+1, \quad u_{j,o} = 0, \quad 0 \le j \le M, \quad (1.7)$$

$$u_{M+1,k} = 0, \quad 0 \le k \le N+1, \quad u_{j,N+1} = 1, \quad 1 \le j \le M \quad (1.8)$$

So a system of $n = MN$ number of linear algebraic equations is to be solved for n-unknowns. It is not difficult to prove that the coefficient determinant of the system of equations (1.6) does not vanish.

Now, suppose we wish to obtain results correct to three-decimal places and wish to know, what should be the mesh sizes Δx, Δy so that this could be achieved. For the sake of simplicity, assume $\Delta x = \Delta y$; then we should have the neglected truncation error

$$\Delta x^2 \approx \frac{1}{2} \times 10^{-3} \quad \Rightarrow \Delta x \sim (5 \times 10^{-4})^{1/2} = 2.36 \times 10^{-2}.$$

Also, $n = N \times M \approx 1600$, so that $n \approx 1600$ number of mesh points. Noting that one of the most economical elimination methods for solving linear systems requires approximately $\frac{n^3}{3}$ number of multiplications and divisions, it follows that for solving the system of Eq. (1.6), the number of multiplication and division needed is at least

$$\frac{1}{3} \times (1600)^3 \sim 1.4 \times 10^9,$$

which is a huge number, essentially requiring evaluation with the help of a powerful digital computer. These days in CFD, it is common to demand accuracy greater than three decimal places and often six-decimal place accuracy is required for many design problems and applications. This would mean significantly greater number of operations requiring more powerful computers, in order that we may be able to solve a typical CFD problem in a reasonable amount of time.

1.2 TYPICAL PARTIAL DIFFERENTIAL EQUATIONS IN FLUID DYNAMICS

The principles of conservation of mass, momentum and energy flux constitute the basic principles of fluid flow and deliver the requisite number of partial differential equations. If the fluid is assumed to be inviscid, then Euler equations are the governing equations. In vector notation, assuming no external body forces, heat or energy supply, the Euler equations are (Landau and Lifshitz, 1989; Oswatitsch 1956; Niyogi, 1977):

$$\text{the continuity equation, } \frac{\partial \rho}{\partial t} + \nabla.(\rho\vec{q}) = 0, \tag{1.9}$$

$$\text{momentum–flux equations , } \frac{\partial \vec{q}}{\partial t} + (\vec{q}.\nabla)\vec{q} = -\frac{1}{\rho}\nabla p, \tag{1.10}$$

$$\text{energy–flux equation, } \frac{\partial s}{\partial t} + (\vec{q}\cdot\nabla)s = 0 \tag{1.11}$$

and the vector operator

$$\nabla \equiv i\frac{\partial}{\partial x} + j\frac{\partial}{\partial y} + k\frac{\partial}{\partial z}, \tag{1.12}$$

i, j, k being unit vectors along the Cartesian coordinate directions x, y and z.

Here, $\vec{q} = (u, v, w)$ is the velocity vector, with Cartesian components u, v and w, p is the pressure, ρ the density and s is the specific entropy of the fluid medium. Also, it may be noted that although the momentum–flux equations (1.10) were derived by Euler, the set of Eqs. (1.9)–(1.11) are often referred to as Euler equations. The thermodynamic variables of state are connected by an equation of state

$$p = p(\rho, s) \tag{1.13}$$

Most gases at normal temperature and pressure obey the equation of state

$$p = \frac{R}{m}\rho T \tag{1.14}$$

where R is a universal constant, T is the absolute temparature and m is the molar mass of the fluid medium. At normal temperature and pressure

$$R = 1.986\ \text{cal/g°C} = 8.31\ \text{Joule/g°C},$$

For oxygen $m = 32$, for nitrogen $m = 28$ and for air $m = 29.0$.

At very high speeds (hypersonic flow), the temperature can become very high and dissociation and ionization take place. In such cases also, one can work with an equation of the form (1.14), although with a smaller value of m. Equations (1.9)–(1.11), together with Eq. (1.13) or Eq. (1.14) constitute the governing equations for *inviscid fluid flow.* In three dimensions, these constitute a system of five first-order nonlinear partial differential equations and one thermodynamic relation for the six unknowns, namely, the three velocity components (for Cartesian coordinates these are usually denoted by u, v and w), the pressure p, density ρ and the specific entropy s. For a specific fluid flow problem, appropriate initial and boundary conditions must be prescribed. For example, at a solid body, it is required that the normal velocity must vanish relative to the body.

At a boundary between two immiscible fluids the pressure and the velocity component normal to the surface of separation must be the same for the two fluids. Moreover, each of these velocity components must be equal to the corresponding component of the velocity of the surface.

A detailed discussion on the basic equations of fluid dynamics, together with different kinds of boundary and/or initial conditions may be found in Chapter 7.

If further, the fluid is assumed to be *incompressible* then the density ρ is constant throughout the fluid. The Euler equations may then be simplified to

$$\nabla \cdot \vec{q} = 0, \quad \text{continuity,} \tag{1.15}$$

$$\frac{\partial \vec{q}}{\partial t} + (\vec{q} \cdot \nabla)\vec{q} = -\frac{1}{\rho}\nabla p, \quad \text{momentum flux.} \tag{1.16}$$

The equations are particularly simple for *irrotational flow* for which

$$\nabla \times \vec{q} = 0 \tag{1.17}$$

everywhere in the flow field. A *velocity potential* ϕ exists then, such that

$$\vec{q} = \nabla\phi \tag{1.18}$$

and the continuity equation (1.15) reduces to

$$\nabla^2\phi = 0, \tag{1.19}$$

where ∇^2 denotes the *Laplacian operator.*

For steady inviscid isentropic irrotational flow the governing equation is the gasdynamic equation (see Chapter 7)

$$\nabla^2\phi = \frac{1}{c^2}\nabla\phi \cdot \nabla\left(\frac{1}{2}q^2\right) \tag{1.20}$$

where the local sound speed c is given by the *Bernoulli equation*

$$q^2 + \frac{2}{\gamma - 1}c^2 = q_\infty^2 + \frac{2}{\gamma - 1}c_\infty^2, \tag{1.21}$$

suffix ∞ denoting free-stream condition, γ being the ratio of the specific heats $\gamma = C_p/C_v$.

Thus we see that the exact equations of *Euler model* are a system of first order nonlinear partial differential equations. As discussed in Chapter 7, those of Navier–Stokes model are a system of second-order nonlinear partial differential equations. Under various simplifying assumptions, the governing equations may be approximated by a linear or quasilinear second-order partial differential equation. We begin by recapitulating some well-known results from theory of partial differential equations, which would be helpful in developing numerical solution and assessing their correctness. For a thorough discussion of the topic, books of Courant and Hilbert (1953), Hellwig (1964) and Prasad and Ravindran (1985) may be consulted.

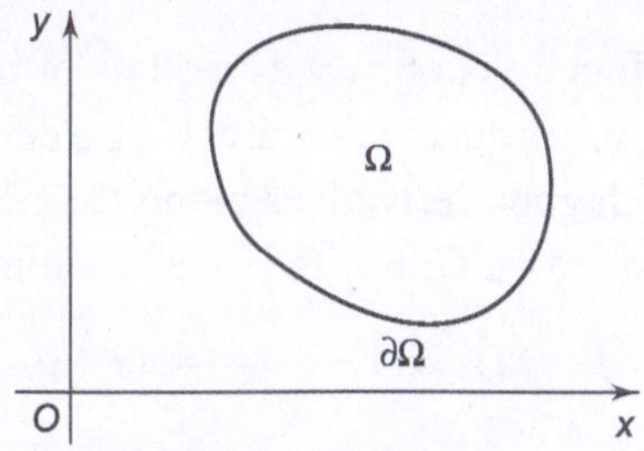

Figure 1.2 A domain Ω with boundary ∂Ω for a partial differential equation.

1.3 TYPES OF SECOND-ORDER EQUATIONS

We consider partial differential equations of the form

$$au_{xx} + 2bu_{xy} + cu_{yy} + g(x, y, u, u_x, u_y) = 0 \tag{1.22}$$

where suffixes denote differentiation with respect to the variables x and y in a certain domain Ω of the x, y-plane bounded by the curve $\partial\Omega$ (Fig. 1.2).

The unknown $u = u(x, y)$ is a function of the independent variables x and y, and is assumed to be sufficient number of times continuously differentiable in Ω. Equation (1.22) is of second order, since the order of the highest derivative u_{xx}, u_{xy} or u_{yy} is second. If the quantities a, b and c be functions of x and y only and g be a linear function in u, u_x and u_y then (1.22) is a *linear* partial differential equation. Otherwise, it is nonlinear. If a, b and c be functions of x, y as well as of u, u_x, u_y then (1.22) is said to be *quasilinear*. Quasilinear partial differential equations are a special class of nonlinear equations which occur frequently in problems of fluid dynamics.

At a point $P(x, y)$ in the domain Ω, Eq. (1.22) is classified into three different types depending on the values of a, b, c, which may vary from point to point in Ω. It is said to be of elliptic, parabolic and hyperbolic type at a point according as:
elliptic type if $ac - b^2 > 0$,
parabolic type if $ac - b^2 = 0$,
and hyperbolic type if $ac - b^2 < 0$.

1.3.1 Characteristics of Second-Order Equations

Consider a second-order partial differential equation

$$au_{xx} + 2bu_{xy} + cu_{yy} + g = 0, \tag{1.23}$$

where a, b, c and g are functions of x, y as well as of u, u_x, u_y only, in a certain domain Ω, so that Eq. (1.23) is quasilinear. Let C be a curve in the domain such that u, x, y together with the higher derivatives along C satisfy Eq. (1.23). The initial conditions are not prescribed on C. For the sake of simplicity, let us put

$$u_x = p, \quad u_y = q, \quad u_{xx} = r, \quad u_{xy} = s, \quad u_{yy} = t.$$

Then along C,

$$dp = \frac{\partial p}{\partial x} dx + \frac{\partial p}{\partial y} dy = r\,dx + s\,dy,$$

$$dq = \frac{\partial q}{\partial x} dx + \frac{\partial q}{\partial y} dy = s\,dx + t\,dy,$$

$$ar + 2bs + ct + g = o, \tag{1.24}$$

Let r and t be eliminated from the third equation of (1.24) with the help of the first two. We get

$$\frac{a}{dx}(dp - sdy) + 2bs + \frac{c}{dy}(dq - sdx) + g = 0,$$

which simplifies to

$$s\left[-a(\frac{dy}{dx})^2 + 2b\frac{dy}{dx} - c\right] + a\frac{dp}{dx}\frac{dy}{dx} + c\frac{dq}{dx} + g\frac{dy}{dx} = 0 \tag{1.25}$$

If now, the coefficient of s be chosen to vanish , that is, if we choose

$$a\left(\frac{dy}{dx}\right)^2 - 2b\frac{dy}{dx} + c = 0 \tag{1.26}$$

then from Eq. (1.25) we necessarily have

$$a\frac{dp}{dx}\frac{dy}{dx} + c\frac{dq}{dx} + g\frac{dy}{dx} = 0 \tag{1.27}$$

Equation (1.26) gives two values for the slope $\frac{dy}{dx}$ of the required curve C, corresponding to which the higher derivatives of u become indeterminate . These directions are known as *the characteristic directions*. Then Eq. (1.27) delivers the *compatibility condition*, which must be satisfied along the characteristics (see Chapter 4, Section 4.9).

Solving Eq. (1.26) as a quadratic equation in $\frac{dy}{dx}$ we get

$$\frac{dy}{dx} = \frac{1}{a}\left[b \pm \sqrt{b^2 - ac}\right] \tag{1.28}$$

This shows that a pair of real characteristic directions exist for $b^2 > ac$, that is, for equations of hyperbolic type, while parabolic type equations have only one real characteristic, and elliptic type equations have no real characteristics.

Example 1.2

Consider Laplace equation in two dimensions

$$\frac{\partial^2 u}{\partial x^2} + \frac{\partial^2 u}{\partial y^2} = 0 \tag{1.29}$$

Here $a = 1 = c$, $b = 0$. So, $ac - b^2 = 1 > 0$, at all points of the domain. Hence the Laplace equation (1.29) is of elliptic type. In this equation, x and y are independent space directions. Since two space dimensions are involved, it is said to be two-dimensional Laplace equation.

Laplace equation occurs abundantly in science and engineering. Steady-state potential problems are governed by Laplace equation.

Example 1.3

Consider the equation

$$\frac{\partial u}{\partial t} = K\frac{\partial^2 u}{\partial x^2}, \tag{1.30}$$

where K is a positive constant and t represents the physical quantity time. Comparing with Eq. (1.23) we see that here, $a = K, b = 0, c = 0$, so that $ac - b^2 = 0$. Consequently, Eq. (1.30) is of parabolic type. It is known as the *heat conduction equation*.

If in an equation of the form (1.30) one of the variables x or y represents the physical quantity time, then instead of x or y the symbol t is used explicitly.

Equation (1.30) represents the propagation of heat in a thin long insulated bar of unit length and uniform cross-section (Fig. 1.3). The initial temperature distribution is prescribed.

$$\text{At} \quad t = 0, \quad u(x, t) = f(x), \quad 0 < x < 1 \tag{1.31}$$

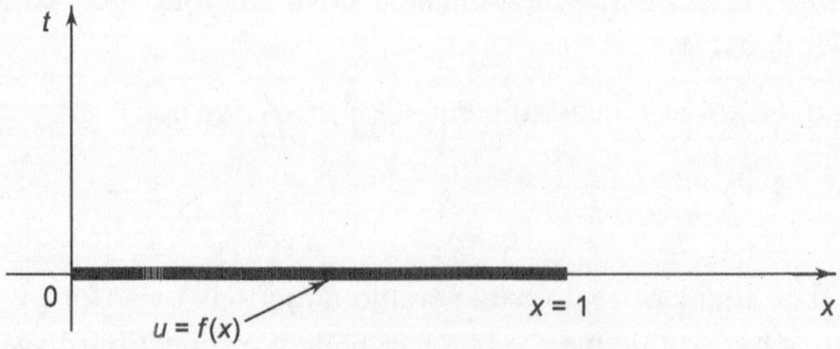

Figure 1.3 Heat conduction in a thin long bar.

This is the *initial condition.* The ends $x = 0$ and $x = 1$ are maintained at suitable temperatures, which are then the *boundary conditions*. For example,

$$\text{at} \quad x = 0, \quad u(x, t) = 0$$
$$\text{and at} \quad x = 1, \quad u(x, t) = 1, \tag{1.32}$$

where the temperatures and lengths have been suitably made dimensionless. The conditions (1.32) hold at the ends of the space domain $0 < x < 1$, for all time $t \geq 0$.

Example 1.4

Consider the equation

$$\frac{\partial^2 u}{\partial t^2} = c^2 \frac{\partial^2 u}{\partial x^2}, \quad c = \text{const.} > 0, \tag{1.33}$$

known as the one-dimensional wave equation. Here, $a = c^2, b = 0, c = -1$, $ac - b^2 = -c^2 < 0$, so that Eq. (1.33) is of hyperbolic type.

The vibrations of a thin uniform elastic string, of dimensionless length unity, fixed for example, at both the ends are governed by the wave equation (1.33).

Suppose, initially at time $t = 0$, the string is slightly displaced and let loose from this displaced position (Fig. 1.4). To find the shape of the string $u(x, t)$ at any later time $t > 0$. Let the initial shape be

$$u(x, t) = f(x), \qquad 0 < x < 1. \tag{1.34}$$

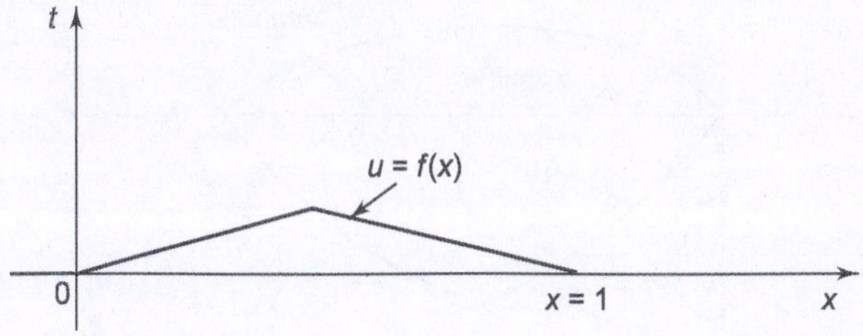

Figure 1.4 Initial position of a thin vibrating string.

Also, since initially it is let loose, its velocity is zero, that is,

$$\left.\frac{\partial u(x,t)}{\partial t}\right|_{t=0} = 0 \tag{1.35}$$

These are the initial conditions. The string is fixed at both the ends, so that

$$u(o,t) = 0, \quad u(1,t) = 0, \text{ for all time.} \tag{1.36}$$

Equations (1.36) are the *boundary conditions, which hold at the ends of the space domain* $0 < x < 1$, *for all time* $t \geq 0$.

Example 1.5

Consider the equation

$$yu_{xx} + u_{yy} = 0 \tag{1.37}$$

Here, $a = y, b = 0, c = 1, ac - b^2 = y$. Hence, Eq. (1.37) is of elliptic type for $y > 0$, parabolic type for $y = 0$ and hyperbolic type for $y < 0$. It is an equation of *mixed type* (Fig. 1.5).

1.4 WELL-POSED PROBLEMS

We are often interested in finding solution of special problems for partial differential equations which satisfy certain additional conditions, as we see in the above examples. This is contrary to the usual approach for ordinary differential equations where we first look for general solutions and then find special solutions satisfying certain initial conditions and/or boundary conditions. Additional conditions like initial and/or boundary conditions are associated with partial differential equations. The set of initial and/or boundary conditions together with the partial differential equation constitute a problem. It is important to note that these additional conditions

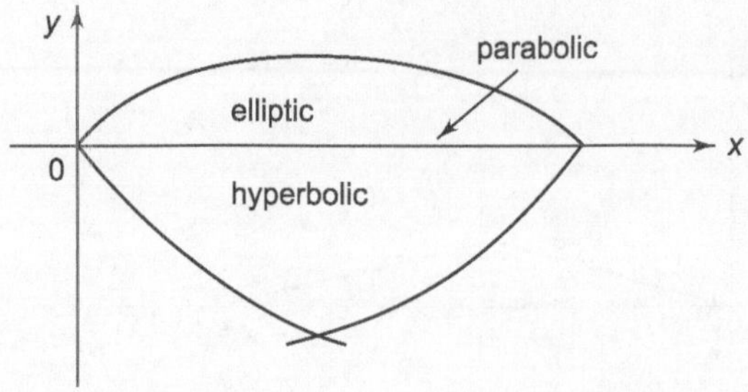

Figure 1.5 Domain of a mixed type equation.

cannot be prescribed arbitrarily, because then a solution may not exist or the solution may not be unique. In order that a problem may be mathematically meaningful, it must satisfy both the following conditions:

(i) *Existence*: at least one solution of the problem exists.

(ii) *Uniqueness*: at most one solution of the problem exists.

In order that a problem may be physically meaningful, the solution must satisfy a third condition, over and above (i) and (ii).

(iii) *Requirement of continuous dependence on boundary and/or initial data*: This is a physical requirement which demands that, if the boundary and/or initial conditions are changed by a small amount, the solution must change by at most a small amount.

The condition of continuous dependence on boundary and/or initial data is particularly important in CFD because some amount of small error, like round-off error is almost always present in the evaluation of the boundary and/or initial data. It would be a very severe difficulty if as a result of the small errors the solution were to change by a large amount. If a problem satisfies all the three of the above conditions, then the problem is said to be *well-posed in the sense of Hadamard*. In the present work, we are interested in well-posed problems.

It is of much help if we know beforehand, if the problem under investigation is well-posed or not. For second-order linear partial differential equations well-known results are available in the literature. We quote here some examples of well-posed problems without proof. Proofs may be found in the standard literature on partial differential equations, Hellwig (1964), Courant and Hilbert (1953), Prasad and Ravindran (1985).

1.4.1 Examples of Well-Posed Problems

1. Dirichlet problem for the Laplace equation is well-posed. We state the result formally, without proof, using the notation that $u \in C^0$ means that u is continuous in the given domain and $u \in C^j$ means that u is j-times continuously differentiable in the domain.

   We state here the Poisson theorem, that gives a general result.

   Poisson's Theorem: If $g(x) \in C^0$ on $|x| = R$, then for $n \geq 2$

$$u(x) = \begin{cases} \dfrac{R^2 - |x|^2}{R\omega_n} \displaystyle\int_{|y|=R} \dfrac{g(y)}{|x-y|^n}\, ds, & \text{for } |x| < R, \\ g(x) & \text{for } |x| = R \end{cases} \tag{1.38}$$

   belongs to C^0 in $|x| \leq R$, to C^2 in $|x| < R$ and $u(x)$ is a solution of the problem

$$\begin{aligned} \Delta_n u &= 0, && \text{for } |x| < R, \\ u &= g(x), && \text{for } |x| = R, \end{aligned} \tag{1.39}$$

   where ω_n is the solid angle in n dimensions

$$\omega_n = \frac{2\pi^{n/2}}{\Gamma(n/2)},$$

   and x and y are the n-component vectors $x = (x_1, x_2, \ldots, x_n)$, $y = (y_1, y_2, \ldots, y_n)$, Δ_n denoting the Laplacian in n dimensions

$$\Delta_n = \frac{\partial^2}{\partial x_1^2} + \frac{\partial^2}{\partial x_2^2} + \cdots + \frac{\partial^2}{\partial x_n^2}.$$

   The problem is well-posed.

2. For the heat conduction problem, the result may be stated as follows :

   Let $g(x) \in C^0$ in R_1 and $|f(x)| \leq M \exp(Kx^2)$, with constants $M, K \geq 0$. Then

$$u(x,t) = \begin{cases} \dfrac{1}{\sqrt{4\pi t}} \displaystyle\int_{-\infty}^{\infty} e^{-(x-y)^2/4t} g(y)\, dy, & \text{for } 0 < t \leq T, \\ g(x), & \text{for } t = 0, \end{cases} \tag{1.40}$$

   where $T < \frac{1}{4K}$, is a solution of the heat conduction problem

$$\frac{\partial u}{\partial t} = \frac{\partial^2 u}{\partial x^2}, \quad \text{in} \quad -\infty < x < \infty, \quad 0 < t \leq T, \tag{1.41}$$

   with initial condition

$$u(x,0) = \phi(x) \quad \text{in} \quad -\infty < x < \infty. \tag{1.42}$$

The problem is well-posed.

3. Cauchy problem for the wave equation is well-posed.

If $f(x) \in C^\circ$ and $g(x) \in C^1$ in $-\infty < x < \infty$, then the function

$$u(x,t) = \frac{1}{2c}\left[f(x+ct) + f(x-ct) + \int_{x-ct}^{x+ct} g(\xi)\,d\xi \right], \tag{1.43}$$

belongs to C^2 on $-\infty < x, t < \infty$ and is a solution of the *Cauchy problem* for the wave equation

$$\frac{\partial^2 u}{\partial x^2} = c^2 \frac{\partial^2 u}{\partial x^2},\; c = \text{const.} > 0,$$

$$u(x,0) = f(x), \qquad \frac{\partial u(x,0)}{\partial t} = g(x) \tag{1.44}$$

1.4.2 An Ill-Posed Problem

Example 1.6

Cauchy problem for Laplace equation is not well-posed. Consider the problem

$$u_{xx} + u_{yy} = 0, \tag{1.45}$$

with initial conditions on $y = 0$

$$u(x,0) = e^{-\sqrt{m}} \sin mx, \quad m = \text{ a positive integer},$$

$$\frac{\partial u(x,0)}{\partial y} = 0. \tag{1.46}$$

It is easy to verify that the solution of the above problem is

$$u(x,y) = \frac{1}{2} e^{-\sqrt{m}} \left[e^{my} + e^{-my} \right] \sin mx. \tag{1.47}$$

As $m \to \infty$, the initial data tend to zero with all its derivatives , while $u(x, y)$ diverges rapidly for $y \neq 0$.

1.5 PROPERTIES OF LINEAR AND QUASILINEAR EQUATIONS

The theory of linear and quasilinear partial differential equations is a well-developed discipline of mathematical analysis. In this section, we state without proof some basic qualitative properties of second order linear and quasilinear partial differential equations. These properties are useful in our subsequent studies of CFD. For a detailed discussion and proofs of the properties the references Courant and Hilbert (1953), Hellwig (1964), DuChateau and Zachmann (1986), Prasad and Ravindran (1985), may be consulted.

1.5.1 Qualitative Properties of Partial Differential Equations

1. Let us consider a second-order linear partial differential equation

$$L[u] + d\,u_x + e\,u_y + f\,u = g$$

where
$$L[u] = a\,u_{xx} + 2b\,u_{xy} + c\,u_{yy} \tag{1.48}$$

in a given domain Ω. If Eq. (1.48) be of elliptic type and the coefficients be analytic functions of x and y in the given domain Ω then the solution must also be analytic.

If on the other hand, the coefficients satisfy a Hölder condition (or have derivatives of order k satisfying Hölder conditions) then the solution has second derivatives satisfying a Hölder condition (or have derivatives up to order $k+2$ satisfying a Hölder condition). It may be noted that a function f is said to satisfy a Hölder condition with constant λ and exponent α, $0 < \alpha < 1$, if

$$|f(P) - f(Q)| \le \lambda \overline{PQ}^{\,\alpha} \tag{1.49}$$

It may be mentioned that if, for example, the coefficients are continuous but not Hölder continuous, no twice continuously differentiable solution need exist.

Similar results hold for quasilinear equations.

2. A basic property of a linear elliptic type partial differential equation is the *maximum–minimum principle*. It states that, if $f \le 0$ and $g \ge 0$, a solution of Eq. (1.48) defined in a domain Ω, cannot have a positive maximum (or a negative minimum) at an interior point unless it is a constant. A consequence of the maximum principle is that the Dirichlet problem for an elliptic type equation (1.48), with $f \le 0$, has at most one solution. The maximum principle also holds for quasilinear equations.
3. The solution of an equation of *hyperbolic type* with analytic coefficients need not be analytic. Discontinuities in the higher derivatives of the solution are propagated along the *characteristic curves*. For quasilinear equations of hyperbolic type, solutions with stronger discontinuity in the form of *shocks* may exist, across which the unknown function experiences a jump discontinuity.
4. For parabolic type equations also the maximum–minimum principle hold. Heat conduction equation and other parabolic type equations with smooth coefficients satisfy a *maximum–minimum principle.* Let Ω be a bounded region in R^3 (three-dimensional Euclidian space) with a smooth closed surface S as boundary. Assuming $u(x, y, z, t)$ continuous in the domain $\overline{\Omega}$, consisting of Ω and its boundary and let $0 \le t \le T$, where T is some fixed value. Let m_s denote the minimum of u on the surface S for $0 \le t \le T$, m_o the minimum

in Ω at the initial time $t = 0$, and further $m = \min(m_s, m_o)$. Similarly, let M_s and M_o denote the corresponding maxima of u and $M = \max(M_s, M_o)$. Then the maximum–minimum principle states that a solution $u(x, y, z, t)$ of the 3-D heat conduction equation in the domain consisting of the Cartesian product $\Omega \times (0, T)$

$$u_t - \nabla^2 u = 0 \tag{1.50}$$

satisfies the condition

$$m \le u \le M \tag{1.51}$$

The wave equation or hyperbolic type equations do not exhibit any kind of maximum–minimum principle. On the other hand, the more important qualitative properties of hyperbolic type equations originate from the existence of real characteristics. These properties are discussed in the subsequent sections.

1.6 PHYSICAL CHARACTER OF SUBSONIC AND SUPERSONIC FLOWS

As previously pointed out, the physical character of a flow field is closely related to the type of the governing partial differential equations. At sufficiently low speeds a fluid medium irrespective of whether it is a liquid or a gas, behaves as an incompressible fluid, for which the density may be taken to be constant. The continuity equation for irrotational flow then reduces to the Laplace equation, which is of elliptic type. This means that small pressure variations are instantaneously propagated in all directions with infinite speed. In other words, the concept of incompressible fluid does not take into account the fact that pressure variations propagate in the fluid with a finite speed.

Let us consider flows produced by small disturbances in an inviscid fluid. Since a small pressure change is propagated with the speed of sound, it is evident that the effect of pressure changes produced, for example, in the air by a body moving faster than the speed of sound cannot reach points ahead of the body. It may be said that the body is unable to send signals ahead. There is a fundamental difference between subsonic and supersonic motion. We consider in the following a well known example due to von Karman (1941).

Consider the case of subsonic motion, for example, the uniform level flight of an aircraft. In this case a pressure signal travels ahead at sound speed minus the flight speed relative to the aircraft whereas a signal travels backward at a speed equal to the sum of the flight and sound speeds. Thus, although the mechanism of pressure propagation is no longer symmetric, every point in space is reached by a signal, provided of course that the flight started at a sufficiently distant point. This is not so in the case of a supersonic flight. To see this, let us take the simplest case

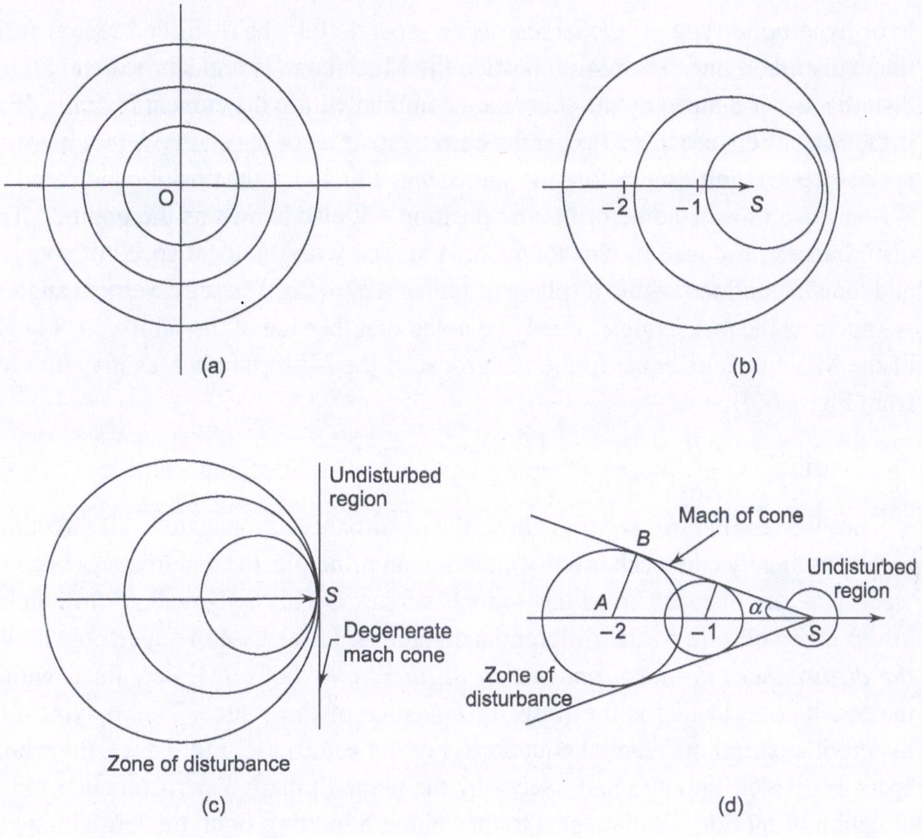

Figure 1.6 Disturbances produced by a point source in different speed ranges. (a) Source stationary at O. (b) Source S moving right at subsonic speed. (c) Source S moving right at sonic speed. (d) Source S moving right at supersonic speed.

of motion of a point source. Figure 1.6(a) shows the spherical surfaces formed by the pressure disturbance in equal time intervals in the case of a point source at rest, which are concentric spheres. Figure 1.6(b) shows the same surfaces relative to the point source moving to the right with subsonic speed. Figure 1.6(c) shows the same surfaces in the case of a point source moving to the right with sonic speed, while Fig. 1.6 (d) represents the case of a source moving with supersonic speed. It is to be noted that in the last case all the disturbances are restricted to the interior of a cone that includes all the spherical disturbances emitted by the source before the instant considered. This cone is known as the *Mach cone*. The generators of the cone are the characteristics of the governing partial differential equation, which is of hyperbolic type. The characteristics separate out the disturbed region from the undisturbed one. The region outside the Mach cone is undisturbed and all the disturbances produced by the source are confined within the conical region, which is the zone of disturbance. Taking the current position of the source S moving with speed $U(> c)$, unit time before the source was situated at the position indicated by -1 and two units time before at the position -2 also shown as the point A. The disturbances produced by the source at A spread with the local speed of sound c and remain confined within a sphere of radius $AB = 2c$. The semi-vertical angle α is known as the *Mach angle*. It may be noted that the sine of the semi-vertex angle of the Mach cone is equal to the reciprocal of the Mach number, as may be seen from Fig. 1.6(d).

$$\sin\alpha = \frac{AB}{SA} = \frac{2c}{2U} = \frac{1}{M} \tag{1.52}$$

Thus we see that in a subsonic flow, the disturbances propagate in all directions and theoretically can reach infinity although, in principle, the disturbances become weaker as the distance from the source increases. This is typical of flow fields governed by elliptic partial differential equations. In the case of supersonic flows, *the disturbances do not propagate in all directions* but remain confined within the Mach cone, which is the *range of influence* of the source. This is typical of hyperbolic partial differential equations. For the sonic case, Fig. 1.6(c), the whole space is divided into two half-spaces by the plane through S perpendicular to the direction of motion. With respect to the source S moving right, the left half-space is referred to as the upstream region, while that to the right of S, that is ahead of S, is the downstream region. The disturbances remain confined within the upstream half-space. Region ahead of the source S, remains undisturbed, while the upstream half-space is the range of influence. Flows governed by parabolic partial differential equations exhibit such a behaviour.

When constructing finite difference representation of partial differential equations, it is helpful to note the differences among the above three types of influences of a disturbance. It is desirable that the finite difference model for a partial differential equation should reflect the type of influence corresponding to the type of the partial differential equation.

1.7 SECOND-ORDER WAVE EQUATION

The standard linear second order partial differential equation of hyperbolic type in two independent variables, namely, the physical space x and the time t is the wave equation

$$u_{tt} = c^2 u_{xx},\ c = \text{const.} > 0,$$

In view of the presence of only one space direction, it is said to be one-dimensional, often abbreviated as 1-D. For hyperbolic type equations a pair of real *characteristics* exist along which discontinuities in the first (or higher) derivatives propagate. Along the characteristics, the partial differential equation may be considerably simplified. For example, for two independent variables it reduces to an ordinary differential equation in each direction. This is the first step for the method of characteristics, which is a successful computational method for hyperbolic type quasi-linear partial differential equations (Oswatitsch, 1956; Niyogi, 1977), originating in the precomputer days and important even now. The domains for hyperbolic type equations are generally open in the time or time-like direction. The numerical schemes are usually *marching schemes,* that is, schemes that march forward in the time–direction, showing *directional bias*. The question of stability is particularly important for such schemes.

1.7.1 Cauchy Problem for the Wave Equation

Let us consider the Cauchy problem for the one-dimensional wave equation

$$u_{tt} = c^2 u_{xx}, \quad c = \text{ const.} > 0 \tag{1.53}$$

with initial conditions

$$u(x,0) = f(x), \quad u_t(x,0) = g(x), \quad -\infty < x < \infty \tag{1.54}$$

$f(x)$, $g(x)$ being continuous functions.

To solve this problem, let us put

$$\xi = x + ct, \text{ and } \eta = x - ct, \tag{1.55}$$

so that $x = \frac{1}{2}(\xi + \eta)$, and $t = \frac{1}{2c}(\xi - \eta)$. Then,

$$u_x = u_\xi \xi_x + u_\eta \eta_x = u_\xi + u_\eta,$$

$$u_{xx} = u_{\xi\xi} + 2u_{\xi\eta} + u_{\eta\eta},$$

and similarly, $u_{tt} = c^2 \left[u_{\xi\xi} - 2u_{\xi\eta} + u_{\eta\eta}\right]$. Substituting these values in Eq. (1.53) yields on simplification

$$4c^2 u_{\xi\eta} = 0.$$

On integration, it follows that

$$u = F(\xi) + G(\eta)$$

where F and G are arbitrary, twice continuously differentiable functions. We thus obtain the general solution, known as *D'Alembert's solution* of the wave equation

$$u(x,t) = F(x+ct) + G(x-ct). \tag{1.56}$$

It may be noted that if F and G be functions which are not twice differentiable, then it is possible to extend the meaning of a solution and talk of *generalized solutions* or *weak solutions*. Roughly speaking, these are different from *genuine solutions (that is, solutions satisfying differentiability requirement)* in that the derivatives of u or the function u itself may experience jump discontinuity. The concept is particularly important for computing solutions with a *shock discontinuity* or *contact discontinuity*.

Now, to determine the functions F and G, the initial conditions (1.54) are used, to yield

$$F(x)+G(x)=f(x)$$
$$\text{and}\quad cF'(x)-cG'(x)=g(x),$$

where the prime denotes differentiation with respect to the argument. Integrating the second equation we get

$$F(x)-G(x)=\frac{1}{c}\int_0^x g(\tau)\,d\tau+c_1,\qquad c_1=\text{const.}$$

Solving for $F(x)$ and $G(x)$,

$$F(x)=\frac{1}{2}\left[f(x)+\frac{1}{c}\int_0^x g(\tau)\,d\tau\right]+\frac{c_1}{2},$$
$$G(x)=\frac{1}{2}\left[f(x)-\frac{1}{c}\int_0^x g(\tau)\,d\tau\right]+\frac{c_2}{2},$$

so that from Eq. (1.56)

$$u(x,t)=\frac{1}{2}\left[f(x+ct)+f(x-ct)+\frac{1}{2c}\int_{x-ct}^{x+ct} g(\tau)\,d\tau\right]+c_2,$$
$$c_2=\text{ arbitrary const.}$$

For $t=0$, this reduces to

$$f(x)=f(x)+c_2,\qquad\Longrightarrow\qquad c_2=0.$$

Hence the solution of the Cauchy problem is

$$u(x,t)=\frac{1}{2}\left[f(x+ct)+f(x-ct)\right]+\frac{1}{2c}\int_{x-ct}^{x+ct} g(\tau)\,d\tau \tag{1.57}$$

The straight lines $\xi=$ constant and $\eta=$ constant are the characteristics of the wave equation (1.53). Thus, for different values of the constants two different families of characteristics of the wave equation (1.53) are given by

$$\xi=x+ct=\text{constant},\qquad\text{and}\qquad \eta=x-ct=\text{constant}, \tag{1.58}$$

The straight lines $x + ct =$ const., for different values of the constant represent a family of parallel straight lines which are inclined to the left if we look in the t direction, Fig. 1.7(a). Similarly, the straight lines $x - ct =$ const., represent a family of parallel straight lines inclined to the right, if we look in the t direction, Fig. 1.7(b). So, the lines $\xi =$ const. and $\eta =$ const. are called respectively *left- and right-running characteristics* of the wave equation.

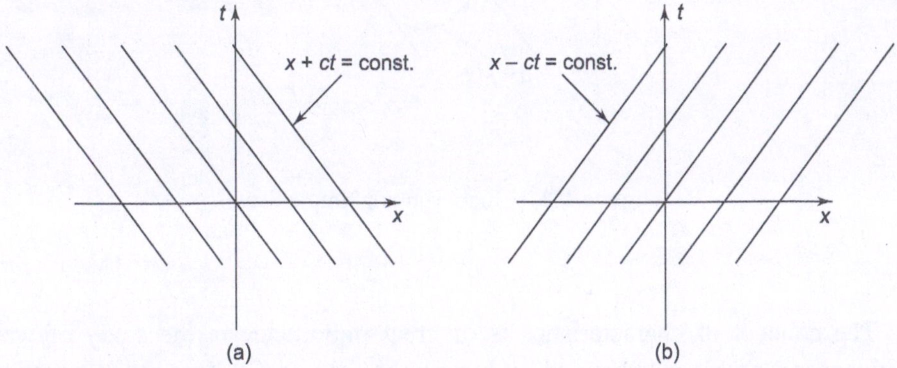

Figure 1.7 Characteristics of the wave equation (a) left running and (b) right running.

A solution

$$u(x, t) = F(x - ct) \tag{1.59}$$

represents in the x, t-plane the propagation of a wave form which at initial time $t = 0$, had the shape $u = F(x)$ (Fig. 1.8). A remarkable feature is that the wave propagates in a single direction with constant speed and that the wave-form does not suffer any distortion with time. Such a wave is known as a *simple wave*. Thus, we see that the D'Alembert's solution, Eq. (1.56) represents the superposition of right running and left running simple waves.

It may be noted that real characteristics exist for any hyperbolic type partial differential equation. For linear equations the characteristics are straight lines and are known explicitly while for quasilinear partial differential equations they are curved lines. Their shapes are unknown, and may be determined only as a part of the solution. Further, it may be noted that weak discontinuities of the solution are propagated along the characteristics, while stronger discontinuities are propagated along the *shock waves, which are surfaces in the fluid medium across which the physical flow variables, like the velocity components, the pressure, density and entropy experience jump.*

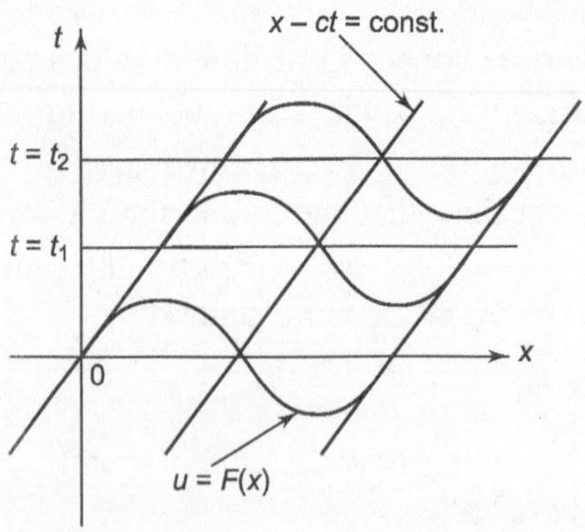

Figure 1.8 Right running simple wave.

The concept of characteristics is of great importance in the study of wave propagation problems governed by hyperbolic type partial differential equations as well as in CFD. A rich literature exists on hyperbolic type partial differential equations and their applications in gasdynamics, with particular reference to steady supersonic flow and flow problems governed by unsteady Euler equations. We discuss here some basic concepts like the *domain of dependence* and the *range of influence*. For a detailed study references Courant and Friedrichs (1948), Courant and Hilbert (1953), and Oswatitsch (1956) may be consulted.

1.7.2 Domain of Dependence and Range of Influence

Consider solution of the Cauchy problem for the 1-D wave equation (1.53) and Eq. (1.54) in a certain domain. Let us consider the solution $u(x_0, t_0)$ at any point $P(x_0, t_0)$ of the domain. Through P, the left and right running characteristics PB and PA are drawn to intersect the initial line $t = 0$ (x-axis) at A and B (Fig. 1.9). Then A is $(x_0 - ct_0, 0)$ and B is $(x_0 + ct_0, 0)$. The solution of the Cauchy problem, Eq. (1.57) shows that the solution at P depends on the initial values f at the points A and B and that on g between A and B only, that is, on the initial values between the segment AB and on no other points. The segment AB is known as the *domain of dependence* of the solution at the point P. It may be noted that the solution at P does not depend on initial values at points to the left of A or on values at those points to the right of B.

Again if we draw through A the right running characteristic AR and the left running characteristic AL (Fig. 1.10), then we see that the solution at any point in the triangular region LAR will depend on the initial values at the point A on the initial line. The domain LAR (unbounded) is known as the *range of influence* of the point A.

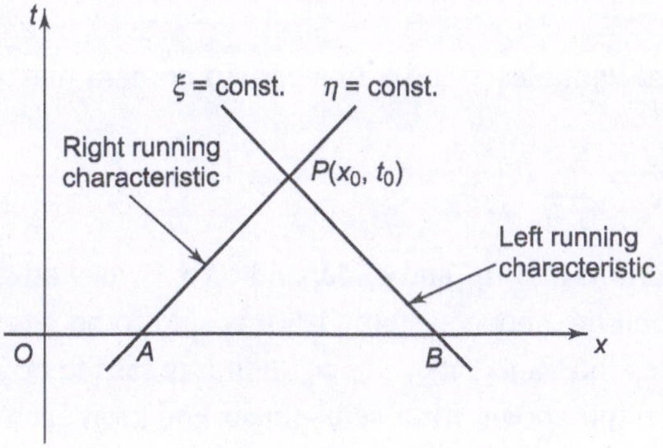

Figure 1.9 The domain of dependence AB of the point P.

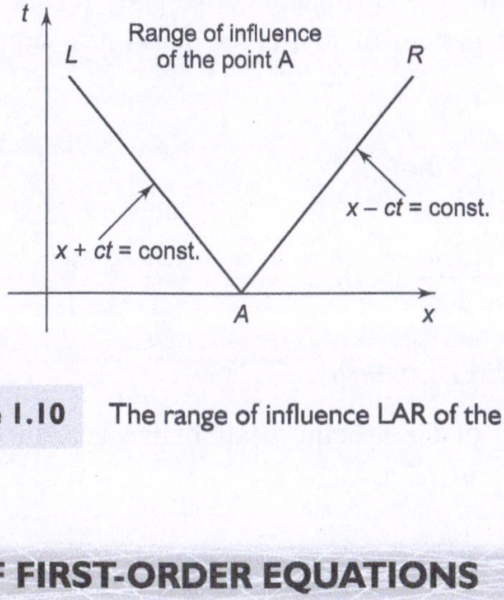

Figure 1.10 The range of influence LAR of the point A.

1.8 SYSTEM OF FIRST-ORDER EQUATIONS

As observed earlier, the basic conservation laws of mass, momentum and energy–flux for an inviscid compressible fluid yield the Euler equations which are a system of first-order partial differential equations. It is also possible to express the Navier–Stokes equations in the form of a first-order system. These facts show the importance of understanding the theory of such systems. A great amount of research has been done on hyperbolic systems of conservation laws and their numerical solution. In the present section, preliminary theoretical background is provided and numerical solution of hyperbolic systems is discussed in Chapter 4.

1.8.1 Classification and Types of First-Order Systems

A system of n first-order partial differential equations in n-functions

$$u_1, u_2, \ldots, u_n$$

of two independent variables x and y in a certain domain may be written as

$$\sum_{j=1}^{n} a_{ij} \frac{\partial u_j}{\partial x} + \sum_{j=1}^{n} b_{ij} \frac{\partial u_j}{\partial y} = c_i, \quad i = 1, 2, \cdots, n \tag{1.60}$$

If each of the functions a_{ij}, b_{ij} and c_i depend on x, y as well as on the unknowns $u_1, u_2, \ldots, u_n$, then the above system (1.60) is said to be *quasi-linear*. However, if a_{ij}, b_{ij} do not depend on $u_1, u_2, \ldots, u_n$ then it is said to be *semi-linear*. On the other hand, if over and above, for a semi-linear equation each of the functions c_i depends linearly on $u_1, u_2, \ldots, u_n$, the system (1.60) is said to be *linear*.

Example 1.7

The Euler equations for inviscid compressible 1-D flow of a perfect gas with velocity u, pressure p and density ρ, without external body forces, heat or energy supply may be written as a system of first-order quasilinear equations (Niyogi, 1977; Oswatitsch, 1956)

$$\begin{aligned}
&\frac{\partial \rho}{\partial t} + u\frac{\partial \rho}{\partial x} + \rho\frac{\partial u}{\partial x} = 0, \\
&\frac{\partial u}{\partial t} + u\frac{\partial u}{\partial x} + \frac{1}{\rho}\frac{\partial p}{\partial x} = 0, \\
&\frac{\partial p}{\partial t} + u\frac{\partial p}{\partial x} + \gamma p\frac{\partial u}{\partial x} = 0,
\end{aligned} \tag{1.61}$$

γ denoting the ratio of the specific heats of the gas. In matrix notation, these equations may be expressed as

$$\frac{\partial U}{\partial t} + A\frac{\partial U}{\partial x} = 0 \tag{1.62}$$

where U is the vector $U = (\rho,\ u,\ p)^{\mathrm{T}}$, the superscript T denoting the transpose of the vector and A is the 3×3 matrix

$$A = \begin{pmatrix} u & \rho & 0 \\ 0 & u & \frac{1}{\rho} \\ 0 & \gamma p & u \end{pmatrix} \tag{1.63}$$

If $A = (a_{ij})$, $B = (b_{ij})$ denote $n \times n$ matrices and U and C denote respectively the column vectors $U = (u_1,\ u_2, \ldots, u_n)^{\mathrm{T}}$ and $C = (c_1,\ c_2, \ldots, c_n)^{\mathrm{T}}$, then we note that the general form of first-order system, Eq. (1.60) may be expressed as

$$A\frac{\partial U}{\partial x} + B\frac{\partial U}{\partial y} = C. \tag{1.64}$$

1.8.2 Conservation Form and Conservation-Law Form

Often, systems like (1.64) are said to be in non-conservative or non-conservation form as against equations in *divergence form* or *conservation form.* System of equations expressed in divergence form, as

$$\frac{\partial F(U)}{\partial x} + \frac{\partial G(U)}{\partial y} = 0 \tag{1.65}$$

are said to be in *conservation form or conservative form*, while equations of the form (1.64) are said to be in non-conservative form or non-conservation form. The terminology draws analogy from the definition of divergence of a vector function $\vec{H}$ with components F and G, $\vec{H} = (F, G)$, so that the divergence of $\vec{H}$ is given by

$$\nabla \cdot \vec{H} = \frac{\partial F}{\partial x} + \frac{\partial G}{\partial y}.$$

An important particular case of equations in divergence form occurs for $G(U) = U$. Then Eq. (1.65) reduces to

$$\frac{\partial U}{\partial t} + \frac{\partial F(U)}{\partial x} = 0, \tag{1.66}$$

where we have written the independent variable y as t. Such systems are called *consevation-law form.* Frequently in such equations the physical quantity time, denoted by t, appears explicitly, which is taken as the second independent variable in place of y. Note that Eq. (1.66) may be rewritten in the form Eq. (1.62) where A is the matrix $A = \dfrac{\partial F}{\partial U}$.

Example 1.8

The system of Equations (1.62) may be expressed as (Niyogi, 1977; Oswatitsch, 1956)

$$\begin{aligned}
\frac{\partial \rho}{\partial t} + \frac{\partial(\rho u)}{\partial x} &= 0,\\
\frac{\partial(\rho u)}{\partial t} + \frac{\partial(p + \rho u^2)}{\partial x} &= 0,\\
\frac{\partial(\frac{1}{2}\rho u^2 + \frac{p}{\gamma-1})}{\partial t} + \frac{\partial}{\partial x}\left[u\left(\frac{\gamma p}{\gamma - 1} + \frac{1}{2}\rho u^2\right)\right] &= 0
\end{aligned} \tag{1.67}$$

We may rewrite this system in matrix notation as a conservation-law system

$$\frac{\partial U}{\partial t} + \frac{\partial F(U)}{\partial x} = 0,$$

where the column vectors U and $F(U)$ are

$$U = \begin{pmatrix} \rho \\ \rho u \\ E \end{pmatrix} \text{ and } F(U) = \begin{pmatrix} \rho u \\ \rho u^2 + p \\ (E+p)u \end{pmatrix} \tag{1.68}$$

Here, E denotes the total energy, expressed as the sum of kinetic energy and the specific internal energy e per unit mass

$$E = \frac{p}{\gamma - 1} + \frac{1}{2}\rho u^2 = \rho\left(\frac{1}{\gamma - 1}\frac{p}{\rho} + \frac{1}{2}u^2\right) = \rho\left(e + \frac{1}{2}u^2\right) \tag{1.69}$$

System (1.64) may be classified into elliptic, hyperbolic or parabolic types depending on the zeros of the generalized characteristic polynomial $P_n(\lambda) = \det(A - \lambda B)$, where *det* denotes the determinant of the matrix.

It may be recalled that for an $n \times n$ matrix A if a nonnull n component vector V and a scalar λ exist such that $AV = \lambda V$, then λ is an eigenvalue and V is the corresponding eigenvector of the matrix A and the polynomial $\det(A - \lambda I)$, where I denotes the unit matrix, is known as the *characteristic polynomial of the matrix* A.

The system is said to be of *elliptic type* if $P_n(\lambda)$ possesses no real zeros. Alternatively, it is of hyperbolic type if $P_n(\lambda)$ has n real zeros with at least one repeated root and n linearly independent corresponding eigenvectors. It is of *parabolic type* if $P_n(\lambda)$ has n real zeros, at least one of which is repeated and smaller than n linearly independent eigenvectors.

In CFD, most frequently we encounter first order systems like Eq. (1.62), commonly known as conservation laws. If the matrix A has n real and distinct eigenvalues or else if the eigenvalues of the matrix A be real with n linearly independent eigenvectors then system (1.62) is of *hyperbolic type*.

More generally, first order systems may be put to the form

$$\frac{\partial U}{\partial t} + A\frac{\partial U}{\partial x} + C = 0 \tag{1.70}$$

where U is the unknown vector function, A and C are matrix and column vector functions of x, t and U respectively. This is in non-conservation form. The corresponding conservation-law form

$$\frac{\partial U}{\partial t} + \frac{\partial F(U)}{\partial x} + C = 0 \tag{1.71}$$

is often said to be a *weak-conservation form*, as against Eq. (1.66) which is called *strong conservation law*. The condition of hyperbolicity of Eq. (1.71) is the same as that of the strong conservation law form Eq. (1.66). Note that in most cases of practical interest, F is a nonlinear function of U and the corresponding systems Eqs. (1.62), (1.66) or (1.70) are nonlinear.

Example 1.9

Let us consider the Cauchy–Riemann equations $u_x = v_y$, and $u_y = -v_x$, which may be put to the form

$$\begin{pmatrix} 1 & 0 \\ 0 & 1 \end{pmatrix}\begin{pmatrix} u_x \\ v_x \end{pmatrix} + \begin{pmatrix} 0 & -1 \\ 1 & 0 \end{pmatrix}\begin{pmatrix} u_y \\ v_y \end{pmatrix} = 0$$

It is of the form Eq. (1.64) and the characteristic polynomial

$$P_n(\lambda) = \det(A - \lambda B) = \begin{vmatrix} 1 & \lambda \\ -\lambda & 1 \end{vmatrix} = 1 + \lambda^2.$$

The characteristic polynomial has no real zeros and the system is of elliptic type.

Example 1.10

Let us consider the system

$$u_x = v_y \quad \text{and} \quad u_y = v$$

Then with the matrices

$$A = \begin{pmatrix} 1 & 0 \\ 0 & 0 \end{pmatrix}$$

and

$$B = \begin{pmatrix} 0 & -1 \\ 1 & 0 \end{pmatrix}$$

this system is of the form Eq. (1.64). So, the characteristic polynomial $P_n(\lambda) = \lambda^2$. All the eigenvectors corresponding to the real repeated zero $\lambda = 0$ are multiples of the vector $(0, 1)^{\mathrm{T}}$. Thus there is only one linearly independent eigenvector, implying that the system is of parabolic type. Note that eliminating v, the given system is $u_x = u_{yy}$ which is the parabolic type heat conduction equation.

Example 1.11

Consider the system

$$\frac{\partial U}{\partial t} + A\frac{\partial U}{\partial x} = 0 \tag{1.72}$$

where $U = \begin{pmatrix} v \\ w \end{pmatrix}$ and A is the matrix $A = \begin{pmatrix} 0 & -c \\ -c & 0 \end{pmatrix}$. Then

$$P_n(\lambda) = \begin{vmatrix} 1 & 0 \\ 0 & 1 \end{vmatrix} - \lambda \begin{vmatrix} 0 & -c \\ -c & 0 \end{vmatrix} = \begin{vmatrix} 1 & \lambda c \\ \lambda c & 1 \end{vmatrix} = 1 - \lambda^2 c^2.$$

The zeros are $\lambda = \frac{1}{c}, -\frac{1}{c}$ which are real and distinct. It is of hyperbolic type. Eliminating, for example w, we get back the 1-D second-order wave equation

$$v_{tt} = c^2 v_{xx}$$

1.9 WEAK SOLUTIONS

As mentioned earlier, solutions of hyperbolic type equations with smooth initial data may develop discontinuities. For the 1-D wave equation (1.53), the arbitrary functions F and G in the D'Alembert's solution (1.56) should be twice differentiable functions. Such solutions with requisite smoothness are called *genuine solutions*. We often encounter in CFD, on the other hand, problems with curves in the interior of the domain across which the unknown function may not be differentiable or even may not be continuous. It is possible to interpret such solutions as *weak solutions*. We begin with a formal definition of a weak solution (Courant and Hilbert, 1953; Prasad and Ravindran, 1985; DuChateau and Zachman, 1986; Lax, 1954; Richtmyer and Morton, 1967).

The vector function U is called a *weak solution* of the conservation law (1.71) with initial value Φ if the integral relation

$$\int\int [W_t U + W_x F - WC]\,\mathrm{d}x\,\mathrm{d}t + \int W(x,0)\Phi(x)\,\mathrm{d}x = 0, \qquad (1.73)$$

holds for every *test vector* W which has continuous first derivatives and which vanishes outside of some bounded set. The integral relation has been obtained by multiplying Eq. (1.71) by W on the left, integrating it over the domain and then integrating the result by parts.

We note that Eq. (1.73) does not require any continuity of U. So it is possible for weak solutions of Eq. (1.71) to have discontinuities. Further, we note that according to the definition (1.73), a genuine solution is a weak solution and conversely, a weak solution with continuous first derivatives is a genuine solution. Moreover, weak solutions need not be differentiable. Let U_1 and U_2 be two genuine solutions of Eq. (1.71) on the two sides of a smooth curve Γ dividing the domain into two parts D_1 and D_2 of the x, t-plane (Fig. 1.11). Then the solutions U_1, U_2 taken together constitute a weak solution, if and only if the slope τ of the curve Γ satisfies the relation (Lax, 1954; DuChateau and Zachmann, 1986; Richtmyer and Morton, 1967).

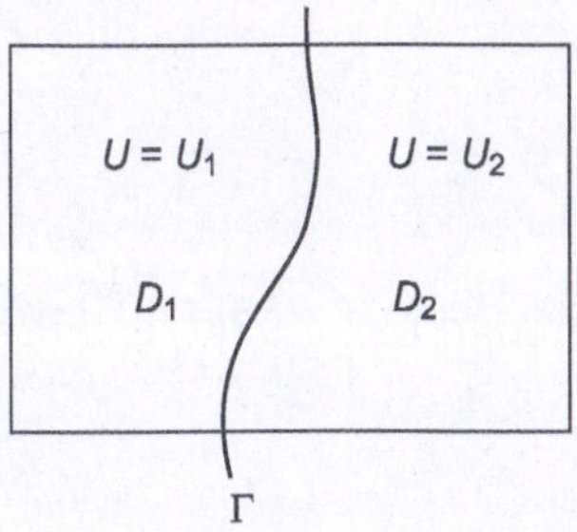

Figure 1.11 Explaining weak solution. The smooth curve Γ divides the domain into two parts D_1 and D_2.

$$\frac{1}{\tau}(U_1 - U_2) = F(U_1) - F(U_2) \tag{1.74}$$

For conservation laws of mass, momentum and energy flux, Eq. (1.74) represents the Rankine–Hugoniot shock conditions.

Some important properties of weak solutions have been discussed by Lax (1954) which are briefly noted here :

- Weak solutions cannot be obtained as limits of genuine solutions.
- The initial values do not, in general, determine a unique weak solution.

It may be mentioned that uniqueness requires a supplementary condition, called *entropy condition* (Lax, 1972; Harten, 1983). We discuss this point further in Chapter 10, in connection with shock computation.

Example 1.12

The functions

$$u(x,t) = \begin{cases} 0, & x < 0 \\ x/t, & 0 < x < t \\ 1, & t < x \end{cases} \tag{1.75}$$

and

$$u(x,t) = \begin{cases} 0, & 2x < t \\ 1, & 2x > t \end{cases} \tag{1.76}$$

are both weak solutions of the inviscid Burger's equation

$$u_t + \left(\frac{1}{2}u^2\right)_x = 0, \tag{1.77}$$

with initial values

$$\phi(x) = \begin{cases} 0, & x < 0 \\ 1, & x > 0 \end{cases}$$

For details the reader is referred to Lax (1974), and Prasad and Ravindran (1985).

1.10 SUMMARY

Computational fluid dynamics solves fluid flow problems using numerical methods. For this purpose fluid flow problems are formulated and stated in terms of ordinary or partial differential equations. This introductory chapter recapitulates classification and mathematical and physical properties of partial differential equations (abbreviated PDE) of interest in computational fluid dynamics. In particular, the following concepts have been discussed.

1. Characteristics and types of PDE.
2. Well-posed problems.
3. Important properties of linear and quasi-linear equations of different types.
4. Domain of dependence and range of influence.
5. First-order systems and conservation-law form.
6. Genuine and weak solutions.

1.11 KEY TERMS

Boundary condition
Cauchy problem
Conservation-law form
D'Alembert solution
Dirichlet problem
Domain of dependence
Elliptic type equation
First order system
Genuine solution
Hyperbolic type equation
Initial condition
Laplace equation
Mach angle
Mach cone
Maximum principle
Mixed type equation
Neumann problem
Parabolic type equation
Poisson equation
Quasilinear equation
Range of influence
Simple wave
Subsonic flow
Supersonic flow
Transonic flow
Wave equation
Weak solution
Well-posed problem

2

Finite Difference and Finite Volume Discretisations

2.1 INTRODUCTION

Partial differential equations occur in almost every branch of science and engineering. Exact analytical solutions may be obtained only for very few simple cases among them. In most cases one has to adopt some numerical method to solve them. Among the numerical methods, the finite difference, finite volume and finite element methods are the most popular. They are applicable to linear as well as nonlinear partial differential equations and the methods can produce results of arbitrarily high accuracy, by refining the grid sizes. In this sense the results delivered are said to be *numerically exact*. Further, the methods are applicable to higher dimensions also.

Computational methods like finite difference, finite volume, finite element or their variants, that are most frequently used in CFD, all begin by subdividing the domain of integration into a large but finite number of small elements. The elements are so small that the flow quantities like the pressure, velocity or density and others may be assumed to be constant over the element, equal to the value, say, at the centroid of the element. Thus a finite set of points is associated with a given domain. Instead of the centroid, for example, the vertices of the elements may also be considered. The concept of approximating the flow quantities by a discrete model that is composed of a set of piecewise continuous functions which are defined over a finite number of elements is the basis of *the finite element* method. The readers interested in finite element methods may refer any text book on the subject like Segerlind (1976). For the finite difference method, a two-dimensional domain is preferably subdivided into rectangular elements, although other types of discretisation are possible. If the flow domain is of irregular shape, then suitable coordinate transformations in terms of curvilinear coordinates are applied to transform the irregular domain into a regular domain like a rectangle in the *computational plane*. This rectangle is then subdivided into a network by drawing suitable straight lines parallel to the edges of the rectangle, taken as coordinate axes. The points of intersection of the straight lines are called *mesh points* or *grid points*. This is known as *grid generation*. The grids have to satisfy a number of criteria, and a large number of methods to generate suitable grids are now available, discussed in detail in Chapter 8 on grid generation. It may be mentioned that the finite element or the finite volume methods do not require any such coordinate transformation of the governing partial differential equations and can handle quite satisfactorily domains of irregular shape or complex geometry. However, grid generation is essential in these methods also. The present work emphasises mainly the *finite difference* and *finite volume* methods for computational solution of partial differential equations and their applications to fluid dynamic problems. The procedures are applicable to higher order linear and nonlinear partial differential equations, as well as to nonlinear systems of partial differential equations.

2.2 FINITE DIFFERENCE DISCRETISATION

Let us consider first the finite difference method and assume that necessary grids have been generated using transformation to a suitable curvilinear coordinate system. The transformed plane, known as the *computational plane,* is assumed to be rectangular. The governing partial differential equations are also transformed to the computational plane. The partial derivatives in the partial differential equations as well as those occurring in the boundary and/or initial conditions are then approximated by partial difference quotients of the unknown function at each mesh point. In practice, the partial derivatives of the unknown function at a given mesh point are approximated by difference quotients involving a finite number of ordinates (usually two, three or five). The term "finite" in "finite difference" is used in contrast with the term "infinitesimal" used in "infinitesimal calculus" implying small quantities approaching zero in the limit. In methods like the finite difference, finite volume (or finite element), the concerned elements involved are not infinitesimal but finite.

So, for each mesh point of the domain, an algebraic equation is obtained. Thus, we obtain a system of algebraic equations, the number of unknowns being equal to the number of equations, each being equal to the number of mesh points. If the partial differential equation as well as the boundary and/or initial conditions are linear, then the resulting system of algebraic equations is also linear; otherwise the system is nonlinear. This algebraic system is then solved on a digital computer using standard numerical procedures. It may be noted that the finite difference method (and other computational methods) deliver the solution only at a finite number of discrete mesh points of the domain. If a solution is desired at any other point of the domain, an interpolation would be necessary.

2.3 DISCRETISATION OF DERIVATIVES

The process of discretisation requires approximate representation of the partial derivatives occurring in the partial differential equations or in the boundary and/or initial conditions. Consider a domain D of the computational plane subdivided into a network by straight lines $x = j\Delta x$, $y = k\Delta y$, $j = 1, 2, \dots, N$, $k = 1, 2, \dots, M$, parallel to the rectangular coordinate axes, as shown in Fig. 2.1.

Here Δx and Δy denote the mesh spacings in the x- and y-directions, respectively. The mesh point (x, y) is indicated by the pair (j, k), the mesh spacings Δx, Δy being understood.

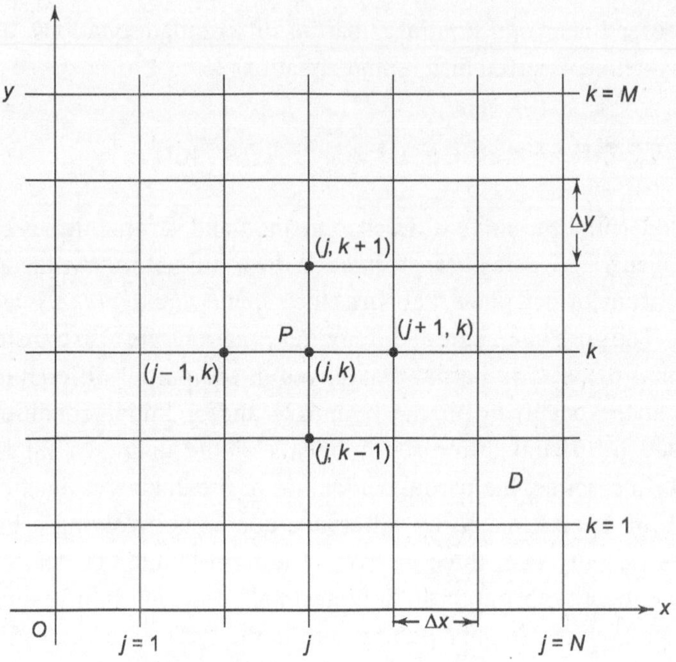

Figure 2.1 Finite difference discretisation.

Let the unknown function be $u(x, y)$ and its value at the mesh point (x, y) be $u(j\Delta x, k\Delta y)$, denoted by $u_{j,k}$. Often, the variable t, denoting the physical quantity time, is taken as $t = n\Delta t$, n being used as a superscript, for example, $v(x, t) = v(j\Delta x, n\Delta t) = v_j^n$. The approximation to the derivatives may be obtained by Taylor's expansion. Assuming the function $u(x, y)$ is continuously differentiable sufficient number of times in the given domain, we may write, for example,

$$u(x + \Delta x, y) = u(x, y) + \Delta x u_x|_{(x,y)} + \frac{\Delta x^2}{2!} u_{xx}|_{(\xi_1,y)}$$

where ξ_1 is a point in the interval $(x, x + \Delta x)$. That is

$$u_x|_{(x,y)} = \frac{u(x + \Delta x, y) - u(x, y)}{\Delta x} - \frac{\Delta x}{2}(u_{xx})|_{(\xi_1,y)}.$$

It is to be noted, that in view of the assumption of existence of the derivatives up to requisite orders, the second derivative is finite. So, the second term on the right when divided by Δx, approaches a finite nonzero limit as $\Delta x \to 0$.

A quantity $f(x)$ is said to be *of the order of* $h^m (m > 0)$ denoted by $f(x) \sim O(h^m)$, if $\lim_{h \to 0} \frac{f(x)}{h^m} = l$, where l is a finite non-zero quantity. In other words, for sufficiently small h, there exists a constant K, such that

$$|f(x)| \leq Kh^m \tag{2.1}$$

In finite difference notation, the above equation may be rewritten as

$$u_x|_{j,k} = \frac{1}{\Delta x}\left[u_{j+1,k} - u_{j,k}\right] + O(\Delta x) \tag{2.2}$$

Equation (2.2) is the *forward difference* representation of the derivative u_x at the point (j, k), expressed in terms of the value of u at the forward mesh point $(j + 1, k)$ and at (j, k). If $u_x|_{j,k}$ is approximated by

$$u_x|_{j,k} = \frac{1}{\Delta x}\left[u_{j+1,k} - u_{j,k}\right],$$

an error $O(\Delta x)$ is committed known as the *truncation error* (abbreviated as T.E.). Similarly, the *backward difference (also called upwind difference)* is obtained by expanding $u(x - \Delta x, y)$ about the point (x, y)

$$u_x|_{j,k} = \frac{1}{\Delta x}\left[u_{j,k} - u_{j-1,k}\right] + O(\Delta x) \tag{2.3}$$

The *central difference* representation for the first derivative u_x is given by

$$u_x|_{j,k} = \frac{1}{2\Delta x}\left[u_{j+1,k} - u_{j-1,k}\right] + O(\Delta x^2) \tag{2.4}$$

It is called central difference because here the first term on the right is expressed in terms of the values of u at the two equally spaced neighbouring mesh points $(j + 1, k)$ and $(j - 1, k)$ on the two sides of the point (j, k) and that the resulting truncation error is $O(\Delta x^2)$, that is, second order. This is more accurate than the forward or backward difference representations having first order truncation error.

Similarly, by Taylor's expansion the central difference representation for the second derivatives u_{xx}, u_{yy} may be obtained as

$$u_{xx}|_{j,k} = \frac{1}{\Delta x^2}\left[u_{j+1,k} - 2u_{j,k} + u_{j-1,k}\right] + O(\Delta x^2) \tag{2.5}$$

$$u_{yy}|_{j,k} = \frac{1}{\Delta y^2}\left[u_{j,k+1} - 2u_{j,k} + u_{j,k-1}\right] + O(\Delta y^2) \tag{2.6}$$

The central difference representation is the most accurate among forward, backward and central difference representations. However, it does not necessarily follow that the best result would always be obtained by using the central difference representation, and some caution is necessary. For example, it is known that for elliptic type equations it is advantageous to use central difference representation, while for parabolic and hyperbolic type equations which have integration domains open in the time or time-like direction, one sided differences like the forward or backward difference representation may be more appropriate. Indiscriminate use of central difference in such cases might lead to stability problems, discussed later.

An alternative and convenient way of deriving the finite difference representations for the derivatives at (j, k) is to express the derivatives as a linear combination of neighbouring pivotal values. For example, let us consider a function of a single variable $u = u(x)$, and $x = j\Delta x$. Let us assume a three-point formula

$$u_x|_j = a\,u_{j-1} + b\,u_j + c\,u_{j+1} + \text{T.E.}, \tag{2.7}$$

where a, b, c are constants to be determined, and the last term indicates the truncation error.

By Taylor's expansion about the point j we get

$$\begin{aligned} u_x|_j = {} & a\left[u_j - \Delta x(u_x)_j + \frac{\Delta x^2}{2}(u_{xx})_j - \frac{\Delta x^3}{6}(u_{xxx})_j + \cdots\right] + b\,u_j \\ & + c\left[u_j + \Delta x(u_x)_j + \frac{\Delta x^2}{2}(u_{xx})_j + \frac{\Delta x^3}{6}(u_{xxx})_j + \cdots\right] + \text{T.E.} \end{aligned} \tag{2.8}$$

In order to solve for the three unknowns a, b and c, three equations are required. These are obtained by comparing coefficients of u_j, $(u_x)_j$ and $(u_{xx})_j$ on both sides. The following equations are obtained :
$a + b + c = 0$, $(-a + c)\Delta x = 1$ and $(a + c)\frac{\Delta x^2}{2} = 0$.

Solving these, we get $c = \frac{1}{2\Delta x} = -a$ and $b = 0$. Substituting the values of a, b, c in Eq. (2.7), we get

$$u_x\big|_j = \frac{1}{2\Delta x}\left(u_{j+1} - u_{j-1}\right) + \text{T.E.} \tag{2.9}$$

The leading term of the truncation error follows on simplification from Eq. (2.8) as

$$\text{T.E.} = -\frac{\Delta x^2}{6}u_{xxx} + \cdots \tag{2.10}$$

This is a quantity $O(\Delta x^2)$, which means that for sufficiently small Δx, there exists a constant K such that

$$|\text{T.E.}| \le K\Delta x^2$$

It is interesting to note that with three pivotal values, the highest order of the truncation error that may be achieved is $O(\Delta x^2)$.

As another example, we consider the three-point forward difference representation for the first derivative,

$$u_x|_j = au_j + bu_{j+1} + cu_{j+2} + \text{T.E.},$$

a, b, c being constants to be determined. Then by Taylor's expansion

$$u_x\big|_j = au_j + b\left[u_j + \Delta x u_x|_j + \frac{\Delta x^2}{2}u_{xx}|_j + \frac{\Delta x^3}{6}u_{xxx}|_j + \cdots\right]$$

$$+ c\left[u_j + 2\Delta x u_x|_j + 2\Delta x^2 u_{xx}|_j + \frac{8\Delta x^3}{6}u_{xxx}|_j + \cdots\right] + \text{T.E.} \quad (2.11)$$

Equating coefficients, yields

$$a + b + c = 0, \quad 1 = b\Delta x + 2c\Delta x, \quad b\frac{\Delta x^2}{2} + c.2\Delta x^2 = 0.$$

Solving these equations we get $a = -\frac{3}{2\Delta x}$, $b = \frac{2}{\Delta x}$, $c = -\frac{1}{2\Delta x}$. Consequently,

$$u_x\big|_j = \frac{1}{2\Delta x}\left(-3u_j + 4u_{j+1} - u_{j+2}\right) \quad (2.12)$$

The corresponding leading term in the truncation error is obtained from Eq. (2.11) on simplification as

$$\text{T.E.} = -(b + 8c)\frac{\Delta x^3}{6}u_{xxx}|_j + \cdots$$

$$= \frac{\Delta x^2}{3}(u_{xxx})_j + \cdots \quad (2.13)$$

Hence,

$$u_x|_j = \frac{1}{2\Delta x}\left(-3u_j + 4u_{j+1} - u_{j+2}\right) + O(\Delta x^2) \quad (2.14)$$

which is the *three-point forward difference formula for the first derivative.* Similar formulae may be derived for other derivatives also. For ready reference, some of the more useful formulae, together with leading truncation error terms evaluated at the pivotal point j, are given below:

three-point backward $$\frac{\partial u}{\partial x}\bigg|_j = \frac{1}{2\Delta x}\left[3u_j - 4u_{j-1} + u_{j-2}\right] - \frac{\Delta x^2}{3}u_{xxx}\bigg|_j \quad (2.15)$$

five-point central $$\frac{\partial u}{\partial x}\Big|_j = \frac{1}{12\Delta x}\left(u_{j-2} - 8u_{j-1} + 8u_{j+1} - u_{j+2}\right) + \frac{\Delta x^4}{30}u_{xxxxx} \quad (2.16)$$

three-point forward $$\frac{\partial^2 u}{\partial x^2}\Big|_j = \frac{1}{\Delta x^2}\left[u_j - 2u_{j+1} + u_{j+2}\right] - \Delta x u_{xxx} \quad (2.17)$$

$$\text{five-point central } \frac{\partial^2 u}{\partial x^2}\Big|_j = \frac{1}{12\Delta x^2}\left[-u_{j-2} + 16u_{j-1} - 30u_j + 16u_{j+1} - u_{j+2}\right] + \frac{\Delta x^4}{90} u_{xxxxxx} \quad (2.18)$$

2.4 CONSISTENCY, CONVERGENCE, AND STABILITY

The first two steps in any finite difference method are: (i) discretisation of the problem and (ii) numerical solution of the discretised system of equations. Once a problem is discretised and the solution of the discretised system has been obtained, the following question arises: what relation does the computed finite difference solution have with the solution of the original problem with which we started? The concepts of *consistency, convergence and stability* are intimately related to the answer to this question.

The discretised version of the problem is called a *finite difference scheme* (also called *finite difference analogue*). If the truncation error term of the finite difference scheme approaches zero, as the mesh spacings approach zero, the procedure is said to be *consistent*. If again, the solution of the finite difference scheme approaches the exact solution as the mesh spacings approach zero, the procedure is said to be *convergent*. Further, we wish to emphasise that in the finite difference method, since the discretised version of a problem is solved numerically, numerical *round-off errors* are committed with each of the arithmetic operations, addition, subtraction, multiplication and division. If the cumulative effect of the round-off errors committed at all the mesh points remains bounded then the finite difference scheme is said to be *stable*.

It is difficult to study stability based on such a definition. A simple procedure was put forward by von Neumann (1950) in which he introduced the concept of studying the effect of the finite difference scheme on a row of small errors. If, at every step, the effect of these errors remain bounded, it is said to be *stepwise stable*; otherwise it is *unstable*. The stability analysis method of von Neumann is, strictly speaking, applicable to linear partial differential equations only. In this method, the influence of the boundary conditions is ignored.

Parabolic and hyperbolic type equations have generally open integration domains in the time direction or time-like direction and the numerical solution process is a *marching process* in that direction. The question of stability is particularly important for problems associated with such equations.

The concepts of consistency, stability and convergence are closely connected with each other. The celebrated *Lax Equivalence Theorem* (1954), explains the interconnection and guarantees that *for a well-posed, consistent initial value problem governed by linear partial differential equations, stability is necessary and sufficient for convergence.* The proof of convergence of a finite difference scheme, is generally difficult and in CFD this result is widely used, though it is not strictly true for nonlinear partial differential equations.

From the above discussions, it may be observed that convergence is directly associated with the concept of truncation error, while stability is associated with the concept of round-off error. In the subsequent sections, we shall make detailed study of consistency, stability and convergence for the model linear equations of different types. Such a study will bring out the qualitative properties of the various finite difference schemes.

2.5 FINITE VOLUME DISCRETISATION

In the finite volume formulation, computations are carried out in the physical flow domain. The computational domain is divided into a network of finite volumes/cells. The generation of a body-fitted grid using curvilinear co-ordinates and the solution process are decoupled since no global transformation is used. The required data concerning the grid are only the Cartesian co-ordinates of the vertices of every cell in the given mesh. Elementary volumes are formed by joining the vertices by straight lines. The main advantage of the finite volume method is its flexibility in treating arbitrary geometries efficiently. Nowadays, it has become very popular for two and three-dimensional flow computation. In this approach the governing equations are considered in their integral form. The derivatives are not approximated by the difference quotients as in the finite difference method. Instead, the divergence theorem of Gauss or the Green's theorem is used over a control volume to get the divergence of a vector field. If V is the volume bounded by a closed surface S and $\vec{A}$ is a vector function of position with continuous derivatives, then

$$\iiint_V \vec{\nabla} \cdot \vec{A} \, dV = \iint_S \vec{A} \cdot \vec{n} \, ds = \oint_S \vec{A} \cdot d\vec{s}, \tag{2.19}$$

where, $\vec{n}$ is the outward drawn normal to S. Let div$\vec{A}$ be the divergence of a vector field $\vec{A}$ at a point P. Imagine P to be enclosed by a surface ΔS of volume ΔV such that if $\Delta V \to 0$, ΔS shrinks to the point P. Then by the mean-value theorem, the left hand-side of (2.19) may be written as

$$\overline{\text{div}\vec{A}} \iiint_V dV = \overline{\text{div}\vec{A}} \, \Delta V. \tag{2.20}$$

$\overline{\text{div}\vec{A}}$ denotes some representative value of div$\vec{A}$ between its maximum and minimum in ΔV. Then

$$\overline{\text{div}\vec{A}} = \frac{\iint\limits_{\Delta S} \vec{A} \cdot \vec{n}\, \mathrm{d}s}{\Delta V} \tag{2.21}$$

In the limit $\Delta V \to 0$ and P being interior to ΔV, $\overline{\text{div}\vec{A}} \to \text{div}\vec{A}$ at the point P; hence

$$\text{div}\vec{A} = \lim_{\Delta V \to 0} \frac{\iint\limits_{\Delta S} \vec{A} \cdot \vec{n}\, \mathrm{d}s}{\Delta V}, \tag{2.22}$$

which defines the quantity div$\vec{A}$. Physically, $\dfrac{\iint\limits_{\Delta S} \vec{A} \cdot \vec{n}\, \mathrm{d}s}{\Delta V}$ implies the flux or net outflow per unit volume of the vector $\vec{A}$ through the surface ΔS. In other words, it is the density of sources per unit volume.

The semi-discretisation method due to Jameson et al. (1981a), which completely separates the discretisation of space and time derivatives will be followed here. After discretising the spatial derivatives the resulting system of ordinary differential equations in time can be solved using explicit time stepping scheme and many techniques to accelerate the convergence can be used to reach the steady state. There exists a variety of finite volume methods to discretise the governing equations in space depending on the choice of the control volume and the position where the flow variables are defined. The method of discretisation is called *either cell-centred or cell-vertex finite volume* depending on whether the flow variables are stored at the centre or at the vertices of the cell respectively. In the cell-centred formulation, the flow quantites are associated with the centre of a cell in the computational mesh and the fluxes across the cell boundaries are calculated using arithmetic means of values in the adjacent cells. In the nodal-point or cell-vertex discretisation the flow quantities are ascribed to the vertices of the cell. This arrangement can give better accuracy for the highly stretched and skewed grids (Rossow, 1987; Chakrabartty, 1989, 1990a, 1990b) that are necessary for viscous flow computations. This is because in the cell-vertex scheme the surface boundary conditions can be satisfied exactly at the vertices along the body surface, and the pressure on the wall can be computed directly, whereas an extrapolation is necessary for these if one uses the cell-centred scheme. For simple geometries, with nearly uniform grids, both the methods give almost the same results. Finite volume methods are normally applied to cells defined by the primary grid (formed by joining the vertices), so that certain cell faces will coincide with the flow boundary (body-fitted grid). One can also apply these methods to arbitrary secondary cells formed by using the basic grid data, in which case the boundary cells are not full cells. In any case, Eq. (2.19) will be used to get the derivatives at a point inside the control volume. The net flux out of the control volume has to be evaluated by algebraic sum of the fluxes through each face of it. The staggered arrangement of the spatial positions, namely that, the conserved variables are to be stored at a point inside the cell (Fig. 2.2) and the flux quantities are to be evaluated at the faces of the cell, is one of the main features of the finite volume concept.

2.5.1 Cell-Centred Scheme

Let us consider the two-dimensional Euler equations representing the conservation laws (mass, momentum and energy) in integral form suitable for finite volume discretisation as (see Chapter 7),

$$\frac{\partial}{\partial t}\iiint_V \vec{W}\mathrm{d}V + \iint_S \overline{\overline{F}}\cdot\vec{n}\mathrm{d}s = 0, \tag{2.23}$$

where $\vec{W}$, the conservative variables/unit volume and the flux $\overline{\overline{F}}$, /unit area/unit time, are defined as

$$\vec{W} = \begin{bmatrix}\rho\\ \rho u\\ \rho v\\ \rho E\end{bmatrix}, \quad \overline{\overline{F}} = \begin{bmatrix}\rho\vec{q}\\ \rho u\vec{q} + p\vec{e}_x\\ \rho v\vec{q} + p\vec{e}_y\\ \rho H\vec{q}\end{bmatrix}, \tag{2.24}$$

$$\vec{q} = u\vec{e}_x + v\vec{e}_y \tag{2.25}$$

$$H = E + p/\rho \tag{2.26}$$

$$E = \frac{RT}{\gamma - 1} + \frac{1}{2}\left(u^2 + v^2\right). \tag{2.27}$$

$\vec{e}_x$ and $\vec{e}_y$ are the unit vectors along x and y directions respectively. E and H are the total internal energy and the total enthalpy respectively. If W is a scalar then $\vec{F}$ is a vector and if $\vec{W}$ is a vector $\overline{\overline{F}}$ is a tensor. The second term in Eq. (2.23) is the net flux out of the control volume and has to be evaluated by algebraic sum of the fluxes through each face of the control volume. In the context of finite volume, a face is a vector represented by its area (length) and the normal direction associated to it.

For the cell-centred finite volume spatial discretisation in two dimensions, the computational domain is sub-divided into quadrilateral cells by joining the cell vertices by straight lines as shown in Fig. 2.2. Conservative variables are assigned at the centre of the cell and fluxes are computed on the boundary of the cell.

Consider a particular cell $\Omega_{i,j}$ defined by the four vertices a, b, c and d in the i–j plane by (i, j), $(i+1, j)$, $(i, j+1)$ and $(i+1, j+1)$ respectively. Since (2.23) is valid for any control volume, it also holds locally for each cell. Hence,

$$\frac{\mathrm{d}}{\mathrm{d}t}\int_{\Omega_{i,j}} \vec{W}\mathrm{d}\Omega + \int_{\partial\Omega_{i,j}} \bar{\bar{F}} \cdot \vec{n}\mathrm{d}s = 0. \tag{2.28}$$

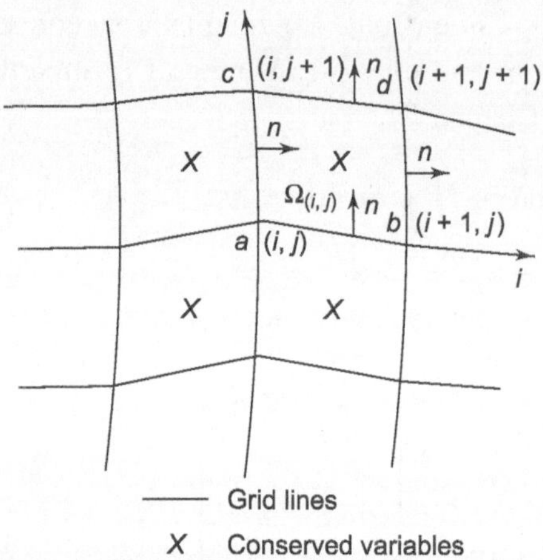

Figure 2.2 Spatial discretisation in cell-centred scheme.

where the boundary $\partial\Omega_{i,j}$ is given by the four faces ab, bd, dc and ca. Let $\bar{\bar{F}}$ have two components $(\vec{F}_x, \vec{F}_y)$ and the vector $\vec{n}\mathrm{d}s$ represent the face vector $\overline{ab}$ having two components $-\mathrm{d}y$ and $\mathrm{d}x$ in Cartesian x and y directions respectively (Fig. 2.3). So,

$$\bar{\bar{F}} = \vec{F}_x\vec{e}_x + \vec{F}_y\vec{e}_y \text{ and } \overline{ab} = \vec{n}ds = -dy\vec{e}_x + dx\vec{e}_y.$$

Then

$$\bar{\bar{F}} \cdot \vec{n}\mathrm{d}s = -\vec{F}_x\mathrm{d}y + \vec{F}_y dx. \tag{2.29}$$

where dx and $-dy$ are the increments in x and y respectively from the point a to b.

Flow quantities $\vec{W}$ are taken to be the volume averaged values located at the centre of the cell. Hence,

$$\vec{W}_{i,j} = \frac{1}{V_{i,j}}\int_{\Omega_{i,j}} \vec{W}d\Omega, \tag{2.30}$$

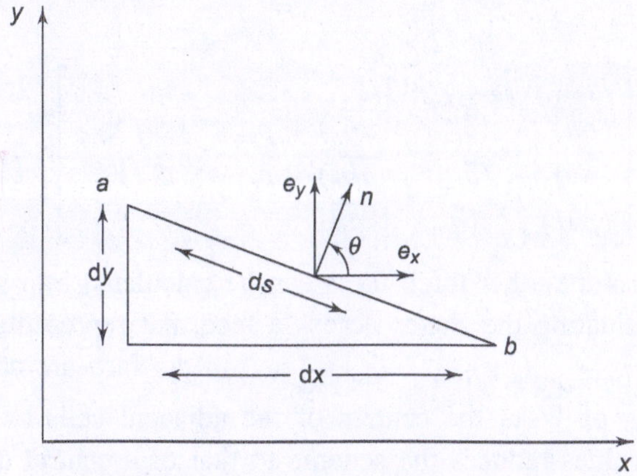

Figure 2.3 Components of a face vector.

where $V_{i,j}$ is the area of the cell $\Omega_{i,j}$ and is given by

$$V_{i,j} = 0.5\{(x_{i+1,j+1} - x_{i,j})(y_{i,j+1} - y_{i+1,j}) - (y_{i+1,j+1} - y_{i,j})(x_{i,j+1} - x_{i+1,j})\} \quad (2.31)$$

The discrete analog of Eq. (2.28) can be written as

$$V_{i,j}\left(\frac{d}{dt}\vec{W}_{i,j}\right) + Q_{i,j} = 0, \quad (2.32)$$

where $Q_{i,j}$ represents the net flux out of a cell and is balanced by the rate of change of $\vec{W}_{i,j}$. The flux $Q_{i,j}$ across the cell $\Omega_{i,j}$ bounded by the faces $\overline{ab}$, $\overline{bd}$, $\overline{dc}$ and $\overline{ca}$ can be calculated as

$$Q_{i,j} = \overline{\overline{F}} \cdot \overline{ab}_n + \overline{\overline{F}} \cdot \overline{bd}_n + \overline{\overline{F}} \cdot \overline{dc}_n + \overline{\overline{F}} \cdot \overline{ca}_n, \quad (2.33)$$

where the suffix n denotes the area vectors of the faces $\overline{ab}$, $\overline{bd}$ etc. Following (2.29), the terms on the right hand side of (2.33) can be calculated in similar way as for the first term,

$$\overline{\overline{F}} \cdot \overline{ab}_n = -\vec{F}_x \Delta y + \vec{F}_y \Delta x. \quad (2.34)$$

The normal direction of $\overline{ab}_n$ is along the increasing j-direction. Let SJX and SJY denote the x and y components of the surface vector respectively whose normal is along the j-direction. For the face $\overline{ab}$, $SJX = -\Delta y = -(y_b - y_a)$ and $SJY = \Delta x = (x_b - x_a)$, where the coordinates of the points a and b are (x_a, y_a) and (x_b, y_b) respectively. Derivation of these formulae for face area in two dimensions and cell volume in three dimensions are given in the next section. The flux across the side $\overline{ab}$ can be calculated in the following way.

$$\overline{\overline{F}} \cdot \overline{ab_n} = \begin{bmatrix} \rho_{i+\frac{1}{2},j}(\vec{q}_{i+\frac{1}{2},j} \cdot \overline{ab_n}) \\ (\rho u)_{i+\frac{1}{2},j}(\vec{q}_{i+\frac{1}{2},j} \cdot \overline{ab_n}) + p_{i+\frac{1}{2},j} \cdot SJX_{i,j} \\ (\rho v)_{i+\frac{1}{2},j}(\vec{q}_{i+\frac{1}{2},j} \cdot \overline{ab_n}) + p_{i+\frac{1}{2},j} \cdot SJY_{i,j} \\ (\rho H)_{i+\frac{1}{2},j}(\vec{q}_{i+\frac{1}{2},j} \cdot \overline{ab_n}) \end{bmatrix}.$$

The fluxes across the other three faces can be calculated in a similar way. In the process of evaluating the fluxes across a face, the representative values of the variables like $\rho_{i+\frac{1}{2},j}$, $(\rho u)_{i+\frac{1}{2},j}$, $\vec{q}_{i+\frac{1}{2},j}$ etc. on the face are calculated using the average values of $\vec{W}$ at the centres of the adjacent cells of the face. Use of these average values reduces the scheme to that of a central difference on a smooth Cartesian grid and makes it second order accurate. It can be shown that cell-centred finite volume discretisation is equivalent to central differencing when the conservation equations are written in general curvilinear co-ordinate system. It should be noted here, that taking arithmetic mean of the values on either side of a face is an approximate solution of the *Riemann problem*. Since locally, from the two adjacent cells, where the representative values of the flow variables are given, we are searching for a value on the connecting face, which is the famous Riemann problem. Many schemes are available in literature for better approximation of the solution of the non-linear Riemann problem, particularly when the constant states on either side are connected either by a contact or a rarefaction wave. This problem can be avoided by nodal-point/cell-vertex scheme. Details of spatial discretisation and the calculation of the fluxes using nodal-point/cell-vertex scheme are discussed in Chapter 10 for Euler equations and in Chapter 13 for Navier–Stokes equations.

2.6 FACE AREA AND CELL VOLUME

Computation of volume in the physical space is essential for the finite volume method. We will follow here the method discussed by Vinokur (1986), and Kordulla and Vinokur (1983). In two dimensions, it is essentially the area of a cell described by either three or four corner points for triangular or quadrilateral grids respectively. If the cell is closed, then it follows from the divergence theorem (2.19) applied to a constant vector that

$$\oint_S \vec{n}\, dS = 0, \tag{2.35}$$

which means that any open surface whose boundary is a given closed curve has a unique surface vector independent of the shape of the surface. The surface area vector is only a function of the shape of the edges. It reveals further that there may

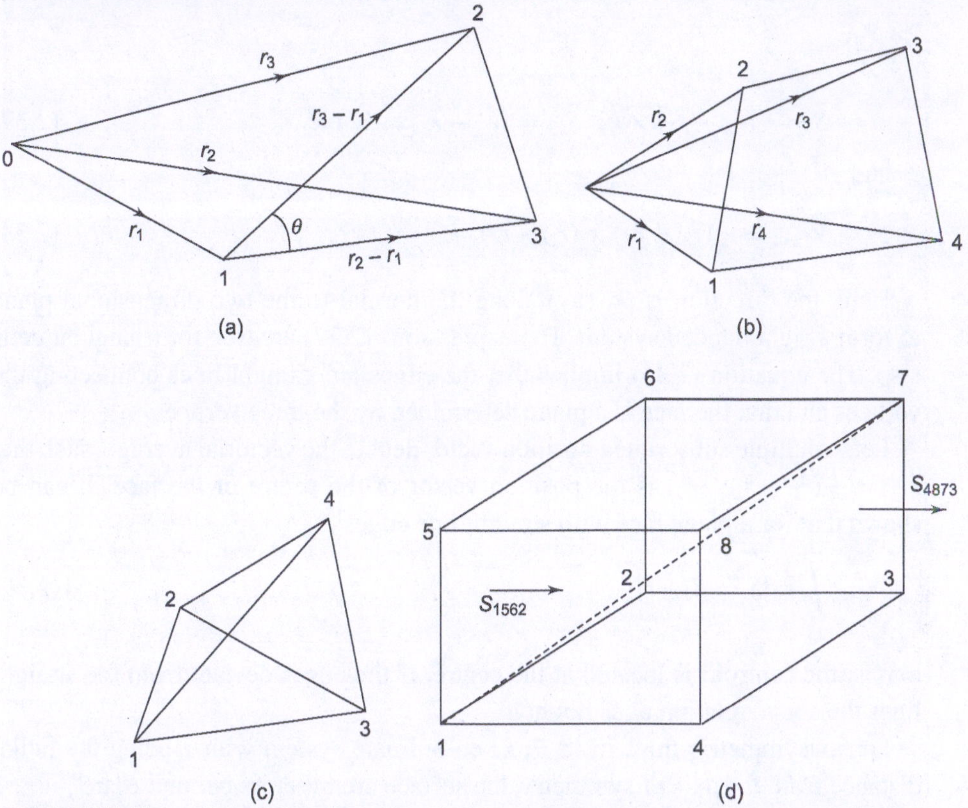

Figure 2.4 Face area and volume of a cell. (a) Triangular face. (b) Quadrilateral face. (c) Tetrahedral cell. (d) Hexahedral cell.

exist an infinite number of edge-shapes connecting two vertices that will give the same surface area vectors as that obtained for the straight line edges. These shapes can be called *equivalent straight line shapes.*

For a triangular cell, let the vertices be 1, 2 and 3, whose position vectors with respect to the origin O are $\vec{r}_1, \vec{r}_2$ and $\vec{r}_3$ respectively (Fig. 2.4a). Then the area (two-dimensional equivalent of the volume) vector of the cell is

$$\vec{S}_{123} = S_{123}\,\vec{n}_{123} = \int\limits_{123} \vec{n}\,dS = \frac{1}{2}(\vec{r}_2 - \vec{r}_1) \times (\vec{r}_3 - \vec{r}_1). \tag{2.36}$$

The implied direction $\vec{k}$ (say), is the normal to the plane containing the two line vectors $\overline{12}$ and $\overline{13}$ forming a right handed system. For a quadrilateral cell (Figs. 2.4(a) and 2.4(b)) defined by the co-planar vertices $\vec{r}_1, \vec{r}_2, \vec{r}_3$ and $\vec{r}_4$, the face area (means line in two dimensions) vectors and the volume (means area in two dimensions) take the form:

$$\vec{S}_{12} = (\vec{r}_2 - \vec{r}_1) \times \vec{k},\ \ \vec{S}_{23} = (\vec{r}_3 - \vec{r}_2) \times \vec{k},\ \text{etc.,} \tag{2.37}$$

and

$$V_{1234} = \frac{1}{2}(\vec{r}_3 - \vec{r}_1) \times (\vec{r}_2 - \vec{r}_4) \cdot \vec{k}. \tag{2.38}$$

Here, the direction is ($\vec{k}$, say) along the normal to the two dimensional plane to form a right handed system. The expressions (2.37) are true for triangular cells also. The equation (2.36) implies that the edges are straight lines connecting the vertices and that the face is a plane determined by the three vertices.

Let a multiple subscripted position vector denote the vectorial average, such that $\vec{r}_{123} = \frac{1}{3}(\vec{r}_1 + \vec{r}_2 + \vec{r}_3)$ is the position vector of the centre of the face. It can be shown that for a plane face with straight line edges

$$\int_{123} \vec{r}\, dS = S_{123}\,\vec{r}_{123}, \tag{2.39}$$

that is, the centroid is located at the centre. If the edges deviate from the straight lines the above relation does not hold.

For axisymmetric flow, in (z, θ, x) co-ordinate system with z being the radial distance from the axis of symmetry, the surface area vectors per unit θ are

$$\vec{S}_{12} = z_{12}(\vec{r}_2 - \vec{r}_1) \times \vec{k},\ \ \vec{S}_{23} = z_{23}(\vec{r}_3 - \vec{r}_2) \times \vec{k}. \tag{2.40}$$

The lateral surface area vector becomes

$$\vec{S}_{1234} = \frac{1}{2}(\vec{r}_3 - \vec{r}_1) \times (\vec{r}_2 - \vec{r}_4) \tag{2.41}$$

where $\vec{r}_i,\ i = 1, \cdots, 4$ are the position vectors of the quadrilateral cell with vertices 1, 2, 3 and 4 at $\theta =$ const. plane. Similarly, the volume per unit θ can be obtained from

$$V_{1234} = \frac{1}{2}[z_{132}(\vec{r}_3 - \vec{r}_1) \times (\vec{r}_2 - \vec{r}_1) + z_{134}(\vec{r}_4 - \vec{r}_1) \times (\vec{r}_3 - \vec{r}_1)] \cdot \vec{k}. \tag{2.42}$$

For a tetrahedral cell with vertices 1, 2, 3 and 4 (see Figs. 2.4(c)) having position vectors $\vec{r}_1, \vec{r}_2, \vec{r}_3, \vec{r}_4$, defined by plane faces and straight edges, the volume is given by the one-sixth of the triple product of the three vectors emanating from one of the vertices and ordered according to right handed system. So, the volume of the tetrahedron V_{1234} is

$$V_{1234} = \frac{1}{6}(\vec{r}_2 - \vec{r}_1) \times (\vec{r}_3 - \vec{r}_1) \cdot (\vec{r}_4 - \vec{r}_1) = \frac{1}{3}\vec{S}_{123} \cdot (\vec{r}_4 - \vec{r}_1). \tag{2.43}$$

These formulae can be used to make geometric calculations for an arbitrary cell with straight line edges. For a polygonal face, it can be divided into plane triangular facets and the total volume can be calculated as the sum of all the tetrahedra. The surface area vectors and their moments for each face are unique, but the total volume will depend on the method of subdivision, since the diagonals of four non-planar points do not intersect.

Let a regular hexahedral cell (Fig. 2.4(d)) be defined by eight arbitrary vertices numbered 1, 2, … , 8 with edges 14, 12, 15 directed in the positive ξ, η and ζ directions, respectively. The simplest way to define a shape whose volume can be precisely calculated is to partition each face into two planar triangles. Then the total volume will depend on the orientation of partitioning, since the diagonals of four non-planar points do not intersect. Same partitioning is necessary for neighbouring cells to be contiguous. Surface area vectors in positive ξ direction are $\vec{S}_{1562}$ and $\vec{S}_{4873}$, as shown in the figure. The expression for the first can be written using (2.41) as:

$$\vec{S}_{1562} = \frac{1}{2}(\vec{r}_6 - \vec{r}_1) \times (\vec{r}_5 - \vec{r}_2) = (\vec{r}_{56} - \vec{r}_{12}) \times (\vec{r}_{15} - \vec{r}_{26}). \tag{2.44}$$

Similarly, the surface area vectors of the other faces can be calculated. It is to be noted that the first formula is obtained as the vector product of the two diagonals, showing that each diagonal is perpendicular to the surface normal. The second formula is in terms of two vectors joining opposite edge midpoints. Since these vectors intersect at the centre of the face, it follows that the midpoints of the four edges and the face centre all lie in a plane in between the planes containing the two diagonals. Kordulla and Vinokur (1983), observed that if one vertex of a main diagonal is chosen as the common apex and the other vertex as the intersection of three equivalent plane faces, then the number of pyramids reduces to three sharing the main diagonal as a common edge. They arrived at the following expression for the volume of the hexahedron;

$$V_{12345678} = \frac{1}{3}(\vec{S}_{1485} + \vec{S}_{1234} + \vec{S}_{1562}) \cdot (\vec{r}_7 - \vec{r}_1). \tag{2.45}$$

This reflects the formula for calculating the volume of a pyramid: the complete cell consists of three topological pyramids with the three base faces $\vec{S}_{1485}$, $\vec{S}_{1234}$ and $\vec{S}_{1562}$ intersecting at the vertex-1 and $(\vec{r}_7 - \vec{r}_1)$ determines the height with respect to each base. The volume of each pyramid is one-third of the corresponding height times the base area. Three other similar expressions can be derived based on the other three choices for main diagonal, each yielding a different but nearly the same value for the volume.

2.6.1 Equivalence Between Finite Difference and Finite Volume Methods

Let us consider a two-dimensional flow. Governing equations in differential form can be obtained as a limiting case of $V \to 0$ from (2.23) and can be written as:

$$\frac{\partial \vec{W}}{\partial t} + \frac{\partial \vec{F}}{\partial x} + \frac{\partial \vec{G}}{\partial y} = 0, \tag{2.46}$$

where

$$\vec{W} = \begin{bmatrix} \rho \\ \rho u \\ \rho v \\ \rho E \end{bmatrix}, \quad \vec{F} = \begin{bmatrix} \rho u \\ \rho u^2 + p \\ \rho u v \\ \rho u H \end{bmatrix}, \quad \vec{G} = \begin{bmatrix} \rho v \\ \rho u v \\ \rho v^2 + p \\ \rho v H \end{bmatrix}.$$

Change of co-ordinate system from physical (x, y) to curvilinear co-ordinate system (ξ, η) in two dimensions has been discussed in detail in Chapter 8. Its extension to three-dimensional case is straight forward. Equations (2.46) can be writen in (ξ, η) co-ordinate system using chain rule as,

$$\frac{\partial \vec{W}}{\partial t} + (\xi_x \vec{F}_\xi + \xi_y \vec{G}_\xi) + (\eta_x \vec{F}_\eta + \eta_y \vec{G}_\eta) = 0. \tag{2.47}$$

Following a similar way as explained in Chapter 8, Eq. (2.47) can be written in conservative form as,

$$\frac{\partial \overline{\vec{W}}}{\partial t} + \frac{\partial}{\partial \xi}\overline{\vec{F}} + \frac{\partial}{\partial \eta}\overline{\vec{G}} = 0, \tag{2.48}$$

where,

$$\overline{\vec{W}} = \frac{\vec{W}}{J}, \tag{2.49}$$

$$\overline{\vec{F}} = \frac{\vec{F}\xi_x + \vec{G}\xi_y}{J}, \tag{2.50}$$

$$\overline{\vec{G}} = \frac{\vec{F}\eta_x + \vec{G}\eta_y}{J} \tag{2.51}$$

and

$$J = \xi_x \eta_y - \eta_x \xi_y. \tag{2.52}$$

Let the directions i and j be along the curvilinear co-ordinates ξ and η respectively, and for simplicity let us assume $\Delta\xi = \Delta\eta = 1$. After discretisation in finite difference, Eq. (2.48) becomes

$$\frac{\partial \overline{W}}{\partial t} + \vec{\bar{F}}_{i+\frac{1}{2},j} - \vec{\bar{F}}_{i-\frac{1}{2},j} + \vec{\bar{G}}_{i,j+\frac{1}{2}} - \vec{\bar{G}}_{i,j-\frac{1}{2}} = 0. \tag{2.53}$$

The term $\vec{\bar{F}}_{i+\frac{1}{2},j}$ in the above equation implies

$$\vec{\bar{F}}_{i+\frac{1}{2},j} = (\vec{F}\frac{\xi_x}{J} + \vec{G}\frac{\xi_y}{J})_{i+\frac{1}{2},j} \tag{2.54}$$

Since $\frac{\xi_x}{J}$ and $\frac{\xi_y}{J}$ are the x- and y-components of the area vector of the face whose normal is along i-direction at $(i + \frac{1}{2}, j)$,

$$\vec{\bar{F}}_{i+\frac{1}{2},j} = \bar{\bar{F}}_{i+\frac{1}{2},j} \cdot \vec{S}_{i+\frac{1}{2},j} \tag{2.55}$$

is the flux across this face. Similarly, the other terms in (2.53) represent the flux across other three faces of the control volume and Eq. (2.53) is equivalent to Eqs. (2.32) and (2.33) derived for the finite volume discretisation. The area vectors $\vec{S}_{i+\frac{1}{2},j}$, etc. are equivalent to the area vectors $\overline{ab}_n$.

2.7 SUMMARY

The basic concepts of finite difference and finite volume spatial discretisation schemes have been introduced. Representation of derivatives of a function by finite difference quotients and by using Green's theorem has been discussed and their equivalence is also shown. Use of Taylor's series to get the order of accuracy of discretisation schemes has been explained. Concepts of consistency, convergence and stabilty of difference schemes have been introduced. Fundamental concept of finite volume approach has been introduced in a concise way. Use of cell-centred finite volume method has been explained in detail for two-dimensional Euler equations governing the inviscid flow. Computation of face area and cell volumes are very important in finite volume approach since they are treated like vectors. Detailed formulations of these quantities are given.

2.8 KEY TERMS

Cell centred scheme
Cell vertex scheme
Cell volume
Consistency
Convergence
Grid generation
Grid points
Lax equivalence theorem
Marching process
Mesh points

Face area
Finite difference
Finite element
Finite volume
Riemann problem
Round-off error
Stability
Truncation error.

2.9 EXERCISE 2

2.1 Use Taylor's expansion for $u(x+\Delta x, y)$ and $u(x-\Delta x, y)$ about the point (x, y) and verify that the central difference approximation of u_x, often called three-point central-difference formula is:

$$u_x = \frac{u(x+\Delta x, y) - u(x-\Delta x, y)}{2\Delta x} + O(\Delta x^2).$$

2.2 Use Taylor's expansion and verify Eqs. (2.5) and (2.6) for the second derivative of the function $u(x, y)$.

(a) $u_{xx}|_{j,k} = \dfrac{1}{\Delta x^2}\left[u_{j+1,k} - 2u_{j,k} + u_{j-1,k}\right] + O(\Delta x^2)$

(b) $u_{yy}|_{j,k} = \dfrac{1}{\Delta y^2}\left[u_{j,k+1} - 2u_{j,k} + u_{j,k-1}\right] + O(\Delta y^2)$.

2.3 Verify the last terms in Eqs.(2.15)–(2.18) using Taylor's expansion.

(a) $\left.\dfrac{\partial u}{\partial x}\right|_j = \dfrac{1}{2\Delta x}\left[3u_j - 4u_{j-1} + u_{j-2}\right] - \dfrac{\Delta x^2}{3}u_{xxx}$

(b) $\left.\dfrac{\partial u}{\partial x}\right|_j = \dfrac{1}{12\Delta x}\left(u_{j-2} - 8u_{j-1} + 8u_{j+1} - u_{j+2}\right) + \dfrac{\Delta x^4}{30}u_{xxxxx}$

(c) $\dfrac{\partial^2 u}{\partial x^2}\Big|_j = \dfrac{1}{\Delta x^2}\left[u_j - 2u_{j+1} + u_{j+2}\right] - \Delta x u_{xxx}$

(d) $\dfrac{\partial^2 u}{\partial x^2}\Big|_j = \dfrac{1}{12\Delta x^2}\left[-u_{j-2} + 16u_{j-1} - 30u_j + 16u_{j+1} - u_{j+2}\right] + \dfrac{\Delta x^4}{90}u_{xxxxxx}$

the derivatives being evaluated at the point j.

2.4 Use Taylor's expansion to establish the following representations:

(a) $u_x|_j = \dfrac{1}{4\Delta x}\left[-5u_{j-1} + 4u_j + u_{j+1}\right] + O(\Delta x^2)$.

(b) $u_{xy} = \dfrac{1}{4\Delta x \Delta y}\left[u_{j+1,k+1} - u_{j+1,k-1} - u_{j-1,k+1} + u_{j-1,k-1}\right] +$
$O(\Delta x^2, \Delta y^2)$.

2.5 Establish the following approximations for the mixed partial derivatives:

(a) $\dfrac{\partial^2 u}{\partial x \partial y}\Big|_{j,k} = \dfrac{1}{\Delta x}\left[\dfrac{u_{j+1,k} - u_{j+1,k-1}}{\Delta y} - \dfrac{u_{j,k} - u_{j,k-1}}{\Delta y}\right] + O(\Delta x, \Delta y)$

(b) $\dfrac{\partial^2 u}{\partial x \partial y}\Big|_{j,k} = \dfrac{1}{\Delta x}\left[\dfrac{u_{j,k+1} - u_{j,k}}{\Delta y} - \dfrac{u_{j-1,k+1} - u_{j-1,k}}{\Delta y}\right] + O(\Delta x, \Delta y)$

(c) $\dfrac{\partial^2 u}{\partial x \partial y}\Big|_{j,k} = \dfrac{1}{\Delta x}\left[\dfrac{u_{j+1,k} - u_{j+1,k}}{\Delta y} - \dfrac{u_{j,k+1} - u_{j,k}}{\Delta y}\right] + O(\Delta x, \Delta y)$

(d) $\dfrac{\partial^2 u}{\partial x \partial y}\Big|_{j,k} = \dfrac{1}{\Delta x}\left[\dfrac{u_{j,k} - u_{j,k-1}}{\Delta y} - \dfrac{u_{j-1,k} - u_{j-1,k-1}}{\Delta y}\right] + O(\Delta x, \Delta y)$

2.6 Establish the following approximations.

(a) $\dfrac{\partial^2 u}{\partial x^2}\Big|_j = \dfrac{1}{\Delta x^2}\left[-u_{j+3} + 4u_{j+2} - 5u_{j+1} + 2u_j\right] + O(\Delta x^2)$

(b) $\dfrac{\partial^3 u}{\partial x^3}\Big|_j = \dfrac{1}{2\Delta x^3}\left[u_{j+2} - 2u_{j+1} + 2u_{j-1} - u_{j-2}\right] + O(\Delta x^2)$

2.7 Show that the truncation error in the representation

$$\frac{du}{dx}\Big|_j = \frac{1}{2\Delta x}\left[(1-\alpha)u_{j+1} + 2\alpha u_j - (1+\alpha)u_{j+1}\right],$$

is

$$-\frac{\alpha \Delta x}{2}\frac{d^2 u}{dx^2} - \frac{\Delta x^2}{6}\frac{d^3 u}{dx^3} + O(\Delta x^3).$$

2.8 If the function $\varphi(x, y)$ is a component of a vector, verify the following relation to calculate $\dfrac{\partial \varphi}{\partial x}$ using the Green's theorem.

$$\iiint_V \frac{\partial \varphi}{\partial x} dV = \iint_S \varphi \cos(\vec{n}, x) ds. \tag{2.56}$$

2.9 Using Taylor's expansion verify that in one dimension cell-centred finite volume scheme is equivalent to central-difference scheme.

2.10 Using Taylor's expansion verify that the cell-centred finite volume discretization gives a second order accurate partial derivative at a point in two dimensional space for a uniform grid.

3

Equations of Parabolic Type

Finite difference and finite volume discretizations have been introduced in the previous chapter. In the present one and in the subsequent three chapters basic finite difference method have been explained using model parabolic, hyperbolic, elliptic and mixed elliptic–hyperbolic equations. These methods have been applied to fluid dynamic problems in the Chapters 9–13. The finite volume method have been applied to study numerical solution of inviscid and viscous compressible flow problems governed by Euler and Navier–Stokes equations in Chapters 10 and 13.

3.1 INTRODUCTION

As already pointed out in Chapter 1, the methods for solving problems governed by partial differential equations, together with appropriate boundary and initial conditions, change depending on the type of the governing equations. Parabolic type equations are associated with domains, usually open in the time or time-like direction. Computational schemes for such problems are schemes that march forward step by step in this direction. The question of stability is important for such schemes. These schemes may be explicit or implicit. Analysing the truncation error term, one finds certain quasi-physical effects associated with them, which in turn could provide valuable information about designing schemes with desirable physical effects. Methods for analysing the stability using von Neumann method have been discussed in detail. Convergence of the schemes have been discussed. All these basic concepts have been introduced with reference to the model 1-D and 2-D (one and two-dimensional) heat conduction problems. The Crank–Nicholson implicit scheme (Section 3.3) and the ADI schemes (Section 3.8.1) lead to tridiagonal system of algebraic equations which may be solved very efficiently by means of the Thomas algorithm, discussed in Section 3.4.1.

Subsonic and supersonic regions in steady transonic flow field are separated by a line called sonic line. Governing equations are of elliptic, hyperbolic and parabolic type in subsonic, supersonic and on the sonic line respectively. So, the governing differential equations for transonic flow field change type depending on the local flow speed. Parabolic equations are connected with some kind of diffusion. Momentum equations for unsteady viscous flows contain diffusion terms and they are of parabolic type. Another interesting case is the Schrödinger equation, which appears in quantum mechanics. Most commonly used parabolic-type differential equation is the heat conduction equation and the finite difference analysis of this equation will follow in the next sections.

3.2 FINITE DIFFERENCE SCHEME FOR HEAT CONDUCTION EQUATION

We illustrate the finite difference method by considering a simple problem, namely the problem of heat conduction in a thin long insulated bar of uniform cross-section described by the equation

$$u_t = Ku_{xx}, K = \text{const.} > 0 \tag{3.1}$$

Here, the temperature distribution at any lengthwise position x at time t is denoted by $u(x, t)$. K denotes the coefficient of thermal diffusion, the suffixes denoting differentiation. The bar is situated along the x-axis, between $x = 0$ and $x = 1$, where the lengths have been made dimensionless with the length of the bar. The temperature distribution at the initial time $t = 0$ is given as

$$u(x, 0) = f(x), \quad 0 < x < 1, \tag{3.2}$$

$f(x)$ being a known continuous function. At the ends of the bar, a constant dimensionless temperature $u = 0$ is maintained for all time, so that the boundary conditions are

$$u(0, t) = 0, \text{ and } u(1, t) = 0, \text{ for all time } t > 0. \tag{3.3}$$

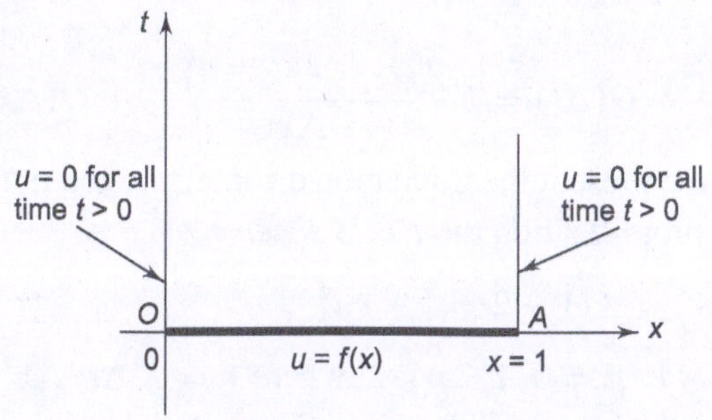

Figure 3.1 Heat conduction problem in a thin long bar *OA* of unit length.

The boundary and initial value problem stated above is shown in Fig. 3.1.

In order to solve this problem by finite difference, the domain $0 \le x \le 1, t > 0$, is subdivided into a network by drawing straight lines

$$x = j\Delta x, \quad t = n\Delta t, \quad j = 1, 2, \ldots, N-1, \quad n = 1, 2, \ldots$$

parallel to the coordinate axes, with $N\Delta x = 1$, Δx and Δt being the mesh spacings in x and t directions, as shown in Fig. 3.2.

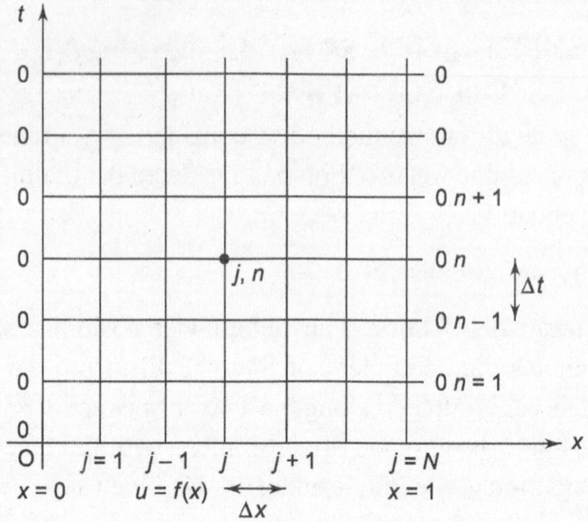

Figure 3.2 Finite difference representation of the heat conduction problem in Fig. 3.1.

The time derivative in Eq. (3.1) is now replaced by forward difference representation and the space derivative by three-point central difference, and using the notation $u(x, t) = u(j\Delta x,\ n\Delta t) = u_j^n$, yields

$$\frac{u_j^{n+1} - u_j^n}{\Delta t} + O(\Delta t) = K\frac{u_{j+1}^n - 2u_j^n + u_{j-1}^n}{\Delta x^2} + O(\Delta x^2) \tag{3.4}$$

Neglecting for the present the truncation error terms, and multiplying both sides by Δt, we get on simplification the *FTCS scheme*

$$u_j^{n+1} = ru_{j+1}^n + (1 - 2r)u_j^n + ru_{j-1}^n \tag{3.5}$$

$j = 1, 2, \cdots, N - 1,\ n = 0, 1, 2, \cdots,$ where $r = K\Delta t/\Delta x^2$. The above equations are valid for all internal mesh points. The initial condition is discretised as

$$u(j\Delta x, 0) = f(j\Delta x), \text{ or } u_j^0 = f_j, \quad j = 1, 2, \cdots, N - 1. \tag{3.6}$$

The discretised form of the boundary conditions are

$$u(0, n\Delta t) = 0, \quad \text{or,} \quad u_0^n = 0, \text{ and } u(N\Delta x, n\Delta t) = 0, \text{ or } u_N^n = 0 \tag{3.7}$$

for all time $t > 0$. Equations (3.5)–(3.7) constitute *a finite difference analog or a finite difference scheme* for solving the problem stated by Eqs. (3.1)–(3.3).

Note that the right hand side of Eq. (3.5) involves quantities at level n only, so that the solution at the time-level $(n + 1)$ is obtained directly or *explicitly* in terms of those at level n. Such a scheme is called an *explicit scheme*, as against *implicit schemes* where a system of algebraic equations is required to be solved for finding the solution at the time-level $n+1$. Due to the type of finite differencing used, the above scheme is known as the *forward time central space* scheme, abbreviated as *FTCS* scheme. The mesh points involved in a particular scheme may be represented schematically in the form of a *computational molecule.*

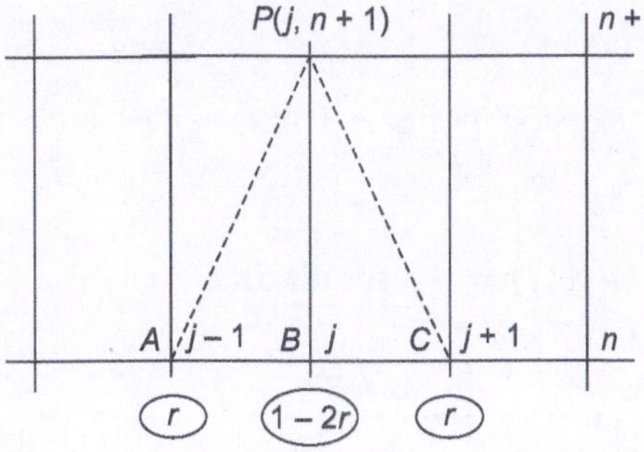

Figure 3.3 Computational molecule for the FTCS scheme.

The computational molecule for the FTCS scheme is shown in Fig. 3.3. It shows that the solution at the point $P(j, n+1)$ depends on the values only at three mesh points of the previous time-step, namely, those at $A(j-1)$, $B(j)$ and at $C(j+1)$ and at no other mesh points of the time-level n. Consequently, the solution at the point P is influenced by the values on the segment ABC of the previous time step and that it is not influenced by the values at the mesh points to the left of the point A or to the right of C. The region of influence is triangular, much like that of a hyperbolic-type equation. This implies that FTCS scheme models the parabolic type heat conduction equation by a hyperbolic type difference scheme, which is considered to be a drawback of the scheme.

As mentioned in Section 1.4, if as Δt and $\Delta x \to 0$ the finite difference solution u_j^n approaches the exact solution $u(x,t)$ then the method is said to be *convergent*. In computing the finite difference solution round-off errors are associated with each arithmetic operation. If the cumulative effect of the round-off errors remain bounded, the method is stable. As mentioned earlier, von Neumann introduced the concept of Fourier stability, in which the effect of a row of round-off errors in any step on the next time-step of computed solution is studied. If the solution remains bounded in the next step, it is said to be *stepwise stable*. Convergence and stability for the FTCS scheme is studied in the subsequent sections.

The FTCS solution is sensitive to the value of the ratio $r = K\dfrac{\Delta t}{\Delta x^2}$, involving the mesh lengths. It can be proved that if $0 < r \le \frac{1}{2}$, the FTCS scheme is *both convergent and stable*. For $r > \frac{1}{2}$, it is neither convergent nor stable. A formal proof appears in the next section. We consider next some illustrative examples.

Example 3.1

Consider the initial boundary value problem

$$\frac{\partial u}{\partial t} = K\frac{\partial^2 u}{\partial x^2}, \quad K = \text{const}, \quad t > 0,$$

with initial condition

$$u(x, 0) = \sin \pi x, \text{ for } 0 < x < 1$$

and boundary conditions

$$u(0, t) = 0, \quad u(1, t) = 0 \quad \text{for all time} \quad t \geq 0.$$

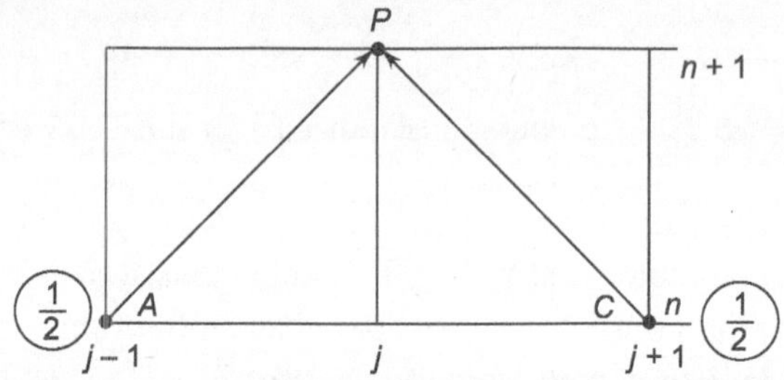

Figure 3.4 Computational molecule for the FTCS scheme with $r = K\Delta t/\Delta x^2 = \frac{1}{2}$.

Case 1: For the sake of simplicity, let us choose $r = K\frac{\Delta t}{\Delta x^2} = \frac{1}{2}$. The initial values are $\sin j\pi \Delta x$, $j = 1, 2, 3$. Correct to two decimals $\sin \frac{\pi}{4} = 0.71 = \sin \frac{3\pi}{4}$. From Eq. (3.5) follows

$$u_j^{n+1} = \frac{1}{2}\left(u_{j+1}^n + u_{j-1}^n\right) \tag{3.8}$$

The computational molecule is shown in Fig. 3.4. The solution at any point $P(j, n+1)$ is just the average of the two adjacent mesh points $A(j-1, n)$ and $C(j+1, n)$ of the previous time step, on the two sides of the point P.

The computed solution for a few time steps are shown in Fig. 3.5(a). We note that there is a steady decay of the solution in the t-direction. This is in accordance with the known exact solutions of heat conduction problem. Further the computed solution lies between 0 and 1, obeying the *maximum principle*, according to which, under appropriate smoothness assumptions, the solution cannot attain a positive maximum or a negative minimum at an interior point, that is the interior values are bounded by the boundary and/or initial values. Suppose further, that a small round-off error ϵ is committed at the mesh point (2, 1) and that no other round-off errors are committed. (This assumption is quite hypothetical, and in practice round-off errors would be committed at each and every mesh point.) The effect of the single round-off error on the computed solution is shown in Fig. 3.5(b), indicating a steady decay of the round-off error with increasing time step. We have no difficulty in believing that the solution is both convergent and stable in this case.

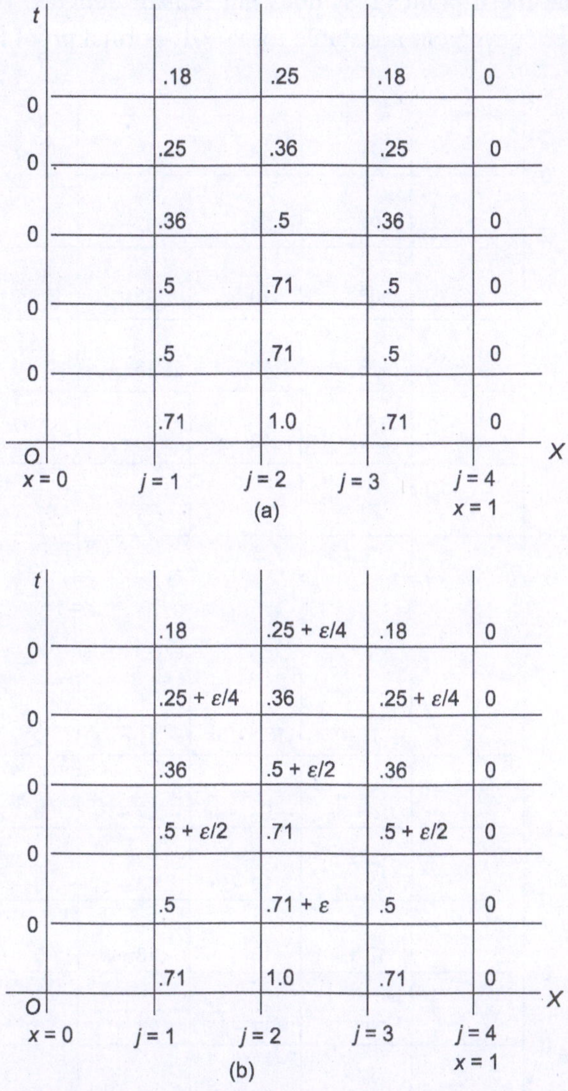

Figure 3.5 (a) Computed FTCS solution for $r = K\Delta t/\Delta x^2 = \frac{1}{2}$. (b) Effect of a single round-off error ϵ for $r = \frac{1}{2}$.

Case 2: Let us consider next the case $r = 1$, all other conditions remaining the same. Then from Eq. (3.5), we get

$$u_j^{n+1} = u_{j-1}^n + u_{j+1}^n - u_j^n \tag{3.9}$$

The computed solution is shown in Fig. 3.6(a).

The magnitude of the solution u_j^{n+1} increases with increasing time step, and violates the maximum principle. Also, the effect of a small round-off error ϵ committed at the mesh point (2, 1) does not remain bounded (Fig. 3.6(b)). The scheme is neither convergent nor stable for $r = 1$. Formal proof follows.

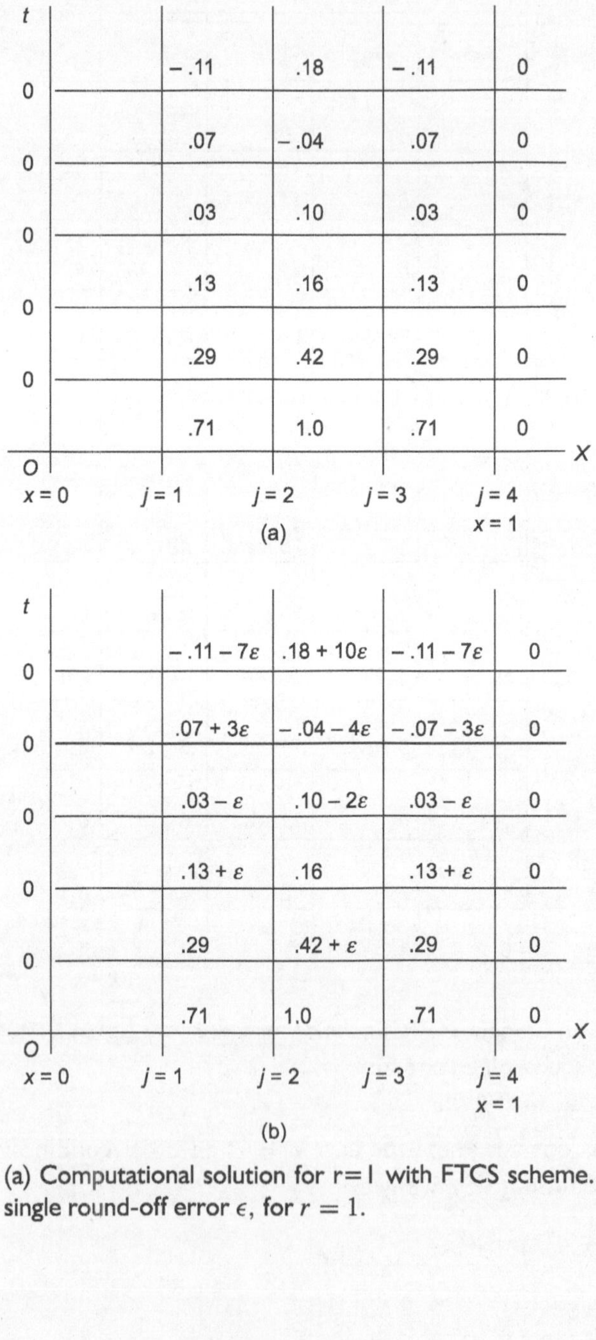

Figure 3.6 (a) Computational solution for r=1 with FTCS scheme. (b) Effect of a single round-off error ϵ, for $r = 1$.

3.2.1 FTCS Scheme: Truncation Error and Consistency

Let the function $u(x, t)$ to be sufficient number of times continuously differentiable in the considered domain. By Taylor's expansion about the mesh point (j, n) follows

$$\begin{aligned}\frac{1}{\Delta t}\left(u_j^{n+1}-u_j^n\right) &= \frac{1}{\Delta t}[u(x,t+\Delta t)-u(x,t)]\\ &= \frac{1}{\Delta t}\left[\{u_j^n+\Delta t(u_t)_j^n+\frac{\Delta t^2}{2}(u_{tt})_j^n+\cdots\}-u_j^n\right]\\ &= (u_t)_j^n+\frac{\Delta t}{2}(u_{tt})_j^n+O(\Delta t^2),\end{aligned} \tag{3.10}$$

suffix t denoting differentiation with respect to t. Also, by Taylor's expansion

$$\frac{1}{\Delta x^2}\left(u_{j+1}^n+u_{j-1}^n-2u_j^n\right) = (u_{xx})_j^n+\frac{\Delta x^2}{12}(u_{xxxx})_j^n+O(\Delta x^4) \tag{3.11}$$

The truncation error T.E. of the FTCS scheme is

$$\begin{aligned}\text{T.E.} &= \text{True value} - \text{Approximate value}\\ &= (u_t - Ku_{xx}) - \left\{\frac{u_j^{n+1}-u_j^n}{\Delta t} - K\frac{u_{j+1}^n-2u_j^n+u_{j-1}^n}{\Delta x^2}\right\}\end{aligned}$$

Substituting from Eqs. (3.10) and (3.11) in it, we get the leading term in the T.E., for any general mesh point (x, t) omitting the subscripts and the superscripts, as

$$\text{T.E.} = -\frac{\Delta t}{2}u_{tt} + K\frac{\Delta x^2}{12}u_{xxxx} + O(\Delta t^2) + O(\Delta x^4) \tag{3.12}$$

Equation (3.12) shows that the FTCS scheme is first order accurate in time and second order accurate in space. The scheme is often termed as first order, equal to the order of the lower of the accuracies in time and space. Moreover, differentiating Eq. (3.1), we see that $u_{tt} = K^2u_{xxxx}$, so that for $r = K\Delta t/\Delta x^2 = 1/6$, the first two terms in the right hand side of Eq. (3.12) cancel and the truncation error is second order in time and fourth order in space.

In view of the assumption of smoothness, the derivatives $u_{tt}, u_{xx}, \ldots,$ are bounded quantities. Hence, taking the limit $\Delta t \to 0$ and $\Delta x \to 0$, it follows from Eq. (3.12), that the truncation error approaches zero. Hence the FTCS scheme is *consistent*.

Note that the FTCS scheme was obtained through discretisation of partial differential equation (3.1). Consistency of the scheme means that, in this case, the original equation may be recovered from the discretised equation. The process is reversible in the case of consistency.

3.2.2 Modified Equation

While solving the discretised equations numerically, a question that naturally comes to our mind is "which equation are we actually solving?" In order to answer this question, we go back to Eqs. (3.10) and (3.11) obtained by Taylor's expansion of the different terms in the FTCS scheme

$$\frac{1}{\Delta t}\left(u_j^{n+1}-u_j^n\right)=\frac{K}{\Delta x^2}\left[u_{j+1}^n-2u_j^n+u_{j-1}^n\right] \tag{3.13}$$

In fact, substituting the representations Eqs. (3.10) and (3.11) in Eq. (3.13) and omitting the subscript j and the superscript n for any general point (x,t) we obtain

$$u_t+\frac{\Delta t}{2}u_{tt}+O(\Delta t^2)=Ku_{xx}+\frac{K\Delta x^2}{12}u_{xxxx}+O(\Delta x^4)$$

which we rewrite as

$$u_t-Ku_{xx}=K\frac{\Delta x^2}{12}u_{xxxx}-\frac{\Delta t}{2}u_{tt}+O(\Delta t^2)+O(\Delta x^2) \tag{3.14}$$

Neglecting higher order terms, we see that the equation that is computationally solved is really

$$\frac{\partial u}{\partial t}-K\frac{\partial^2 u}{\partial x^2}=K\frac{\Delta x^2}{12}\frac{\partial^4 u}{\partial x^4}-\frac{\Delta t}{2}\frac{\partial^2 u}{\partial t^2} \tag{3.15}$$

Noting that $\Delta x^2/12$ is a small quantity compared to the other terms, the presence of $\frac{\partial^2 u}{\partial t^2}$ in Eq. (3.15) shows that the modified equation (3.15) is hyperbolic in character.

Strictly speaking, Eq. (3.15) is also not the equation that is actually solved numerically by finite difference because in the derivation we have eliminated the time derivatives using the differential equation. The time derivatives ought to have been eliminated using only the difference equations (Warming and Hyett, 1974) as is done in Chapter 4, Sections 4.2 or 4.3.

3.2.3 FTCS Scheme: Convergence

Let us denote the exact solution of the FTCS scheme as $U(x,t)=U_j^n$ and let u_j^n denote the solution of the finite difference equations

$$u_j^{n+1}=ru_{j+1}^n+(1-2r)u_j^n+ru_{j-1}^n$$

Let us write the exact solution of Eq. (3.1) also in the same form as

$$U_j^{n+1}=rU_{j+1}^n+(1-2r)U_j^n+rU_{j-1}^n+\tau \tag{3.16}$$

where $\tau=\tau_1+\tau_2$, denotes the truncation error term. By straightforward Taylor expansion it may be seen that

$$\tau_1 = O(\Delta t^2) \le M_1 \Delta t^2 \quad \text{and} \quad \tau_2 = O(\Delta t.\Delta x^2) \le M_2 \Delta t \Delta x^2 \tag{3.17}$$

where M_1 and M_2 are constants independent of Δt and Δx.

Let e_j^n denote the error at the mesh point (j, n), so that $e_j^n =$ true value − approximate value $= U_j^n - u_j^n$.

Then subtracting Eq. (3.5) from Eq. (3.16) we get

$$e_j^{n+1} = re_{j+1}^n + (1-2r)e_j^n + re_{j-1}^n + \tau_1 + \tau_2 \tag{3.18}$$

For $r \le \frac{1}{2}$ the middle term on the right is ≥ 0. Then, taking the absolute value of both sides, it follows that

$$|e_j^{n+1}| \le r|e_{j+1}^n| + (1-2r)|e_j^n| + r|e_{j-1}^n| + |\tau_1| + |\tau_2| \tag{3.19}$$

which shows that the error at any mesh point j at time level $n+1$ is bounded by the weighted average error at three mesh points of the previous time level, increased by $|\tau_1|$ and $|\tau_2|$. Since the average of three quantities cannot be greater than their maximum, it follows that

$$|e_j^{n+1}| \le \max|e_j^n| + |\tau_1| + |\tau_2|$$

Since this result is true for any mesh point j, in time level $n+1$ it is also true for the maximum, and

$$\max|e_j^{n+1}| \le \max|e_j^n| + |\tau_1| + |\tau_2| \tag{3.20}$$

Replacing n by $n-1$, we see that

$$\max|e_j^n| \le \max|e_j^{n-1}| + |\tau_1| + |\tau_2|$$

Substituting this in Eq. (3.20) we get

$$\max|e_j^{n+1}| \le \max|e_j^{n-1}| + 2(|\tau_1| + |\tau_2|).$$

Replacing n by $n-1$ follows

$$\max|e_j^n| \le \max|e_j^{n-2}| + 2(|\tau_1| + |\tau_2|).$$

Repeating the process, we get

$$\max|e_j^n| \le \max|e_j^o| + n(|\tau_1| + |\tau_2|) \tag{3.21}$$

Let the solution be computed for $0 < t \le T$ where the time T may be large but finite. Since $t = n\Delta t$, the maximum time step n is $T/\Delta t$. Consequently, Eq. (3.17) yields

$$n(|\tau_1| + |\tau_2|) \le \frac{T}{\Delta t}\left[M_1 \Delta t^2 + M_2 \Delta t \Delta x^2\right] = TM_1 \Delta t + TM_2 \Delta x^2$$

which approaches zero as $\Delta t \to 0$ and $\Delta x \to 0$.

Assuming that there is no prescribed initial error, max $|e_j^o| = 0$, so that from Eq. (3.21) it follows that the maximum absolute error in the nth step approaches zero as $\Delta t \to 0$, $\Delta x \to 0$. This completes the convergence proof.

If $r > \frac{1}{2}$, then $(1 - 2r)$ is negative and the inequality (3.19) will not follow from Eq. (3.18).

3.2.4 FTCS Scheme: Stability

We noted that, the question of convergence is associated with the concept of truncation error. The question of stability is closely connected to the growth or decay of round-off errors, which are committed at each mesh point while solving the discretised equations numerically. There are mainly two methods for studying stability of a finite difference scheme, namely, the *matrix stability analysis* and the *von Neumann stability analysis*. Both the methods are applicable to linear partial differential equations only. The matrix stability analysis is the more accurate of the two, which takes into account the boundary conditions of the problem. Stability of a numerical scheme depends on the eigenvalues of an associated matrix. On the other hand, von Neumann stability analysis, also known as Fourier stability analysis, effectively ignores the boundary conditions and is easier to apply. In fact, it is the most popular method of stability analysis. The method studies the stepwise growth or decay of an initial row of round-off errors, which may be expanded in a finite Fourier series for linear equations with constant coefficients. Because of linearity it is enough to consider just one term of the series. If the stepwise growth of this term remains bounded, it is said to be *stable*; otherwise it is *unstable*. Such an analysis is applicable only to linear equations. However, in CFD, local linear stability of nonlinear equations are analysed taking the coefficients in the equations to be locally frozen and assuming them to be known constants.

Let U_j^n denote the exact solution of the finite difference FTCS equations

$$u_j^{n+1} = ru_{j+1}^n + (1 - 2r)u_j^n + ru_{j-1}^n \tag{3.22}$$

Since round-off errors are committed with each arithmetic operation, u_j^n is contaminated with round-off errors. U_j^n is the solution that would have been obtained if exact arithmetic could be performed, without committing any round-off error.

We define error ϵ_j^n as

$$\epsilon_j^n = \text{true value} - \text{approximate value} = U_j^n - u_j^n.$$

Substituting for u_j^n in Eq. (3.22) we get

$$U_j^{n+1} - \epsilon_j^{n+1} = r(U_{j+1}^n - \epsilon_{j+1}^n) + (1 - 2r)(U_j^n - \epsilon_j^n) + r(U_{j-1}^n - \epsilon_{j-1}^n).$$

But the exact solution U_j^n must satisfy Eq. (3.22) so that

$$\epsilon_j^{n+1} = r\epsilon_{j+1}^n + (1-2r)\epsilon_j^n + r\epsilon_{j-1}^n \tag{3.23}$$

This shows that the propagation of error ϵ_j^n is governed by the same difference equation as that satisfied by the unknown solution u_j^n. Thus, we see that both the round-off error and the numerical solution have the same growth or decay behaviour with respect to time.

In the *von Neumann method,* a distribution of errors at time level $t = 0$ is introduced, which propagates with increasing time t in accordance with the governing Eq. (3.23). For difference equations with constant coefficients and assuming that the problem has a periodic boundary condition, the error may be expanded in a finite Fourier series, in the interval of interest L. We write it as

$$\epsilon_j^o = \sum_j A_{jo}\exp(ik_j j\Delta x), \quad i = \sqrt{-1} \tag{3.24}$$

where the wave number $k_j = \dfrac{j\pi}{L}$, $j = 0, 1, 2, \cdots, M$. Here, M denotes the number of intervals of length Δx contained in L, so that $L = M\Delta x$. The frequencies are $f_j = k_j/2\pi = j/2L$. For $j = 0$, the value f_0 corresponds to a steady state. The number of terms in Eq. (3.24) is equal to the number of mesh points on the line $t = 0$. Let us seek solution of Eq. (3.23) such that it reduces to Eq. (3.24) on the initial line $t = 0$, that is on $n = 0$. With this motivation, let us assume

$$\epsilon_j^n = \sum A_j Z^n(j)\exp(ik_j j\Delta x) \tag{3.25}$$

where Z is an arbitrary real or complex number.
Since the difference equation is linear and homogeneous, superposition of solutions is also a solution and it is enough to consider the growth of error of a typical term

$$\epsilon_j^n \sim AZ^n \exp(ik_j j\Delta x) \tag{3.26}$$

where A is an arbitrary constant. This term reduces to $\epsilon_j^o \sim A\exp(ik_j j\Delta x)$, when $t = 0$ (that is, on $n = 0$.) The necessary and sufficient condition for the error Eq. (3.26) to remain bounded is that

$$|Z| \le 1 \tag{3.27}$$

It is to be noted that the quantity Z, is the ratio of the error at two consecutive time steps

$$Z = \frac{\epsilon_j^{n+1}}{\epsilon_j^n} \tag{3.28}$$

and is known as the *amplification factor* of the scheme.

Let us choose $Z = \exp(\alpha\,\Delta t)$, as a trial solution, where α may be complex. Then, omitting the arbitrary constant A, which has no effect on the stability of the scheme, we have

$$\epsilon_j^n = \exp(\alpha n \Delta t)\exp(ik_j j\Delta x) = \exp(\alpha n \Delta t)\exp(i\beta j) \tag{3.29}$$

where $\beta = k_j \Delta x$. The quantity β is real. Substituting (3.29) in Eq. (3.23), we obtain

$$\exp(\alpha(n+1)\Delta t)\exp(i\beta j) = (1-2r)\exp(\alpha n \Delta t)\exp(i\beta j)$$
$$+r\exp(\alpha n \Delta t)\left[\exp(i\beta(j+1)) + \exp(i\beta(j-1)) - 2\exp(i\beta j)\right]$$

Cancelling the factor $\exp(\alpha n\Delta t)\exp(i\beta j)$ from both sides and simplifying we get for the FTCS scheme

$$\begin{aligned}\exp(\alpha\,\Delta t) &= 1 + r\left[\exp(i\beta) + \exp(-i\beta) - 2\right] \\ &= 1 - 4r\sin^2\frac{\beta}{2}\end{aligned} \tag{3.30}$$

The quantity on the right of Eq. (3.30) represents the amplification factor for the FTCS scheme. According to Eq. (3.27), for stability $|\exp(\alpha\,\Delta t)| \le 1$, so that $|1-4r\sin^2\frac{\beta}{2}| \le 1$. Therefore $-1 \le 1-4r\sin^2\frac{\beta}{2} \le 1$. Since β is real, the right hand inequality is automatically satisfied. The left inequality delivers

$$r \le \frac{1}{2\sin^2\dfrac{\beta}{2}}. \tag{3.31}$$

Eq. (3.31) should be satisfied for all values of β, so that r must be less than or equal to the least value of the right hand side. Consequently,

$$r \le \frac{1}{2} \tag{3.32}$$

for stability.

3.2.5 Derivative Boundary Conditions

Let us consider solution of the one-dimensional (1-D) heat conduction Eq. (3.1), with initial condition

$$u(x,0) = f(x), \quad 0 \le x \le 1, \tag{3.33}$$

and with boundary conditions expressed in terms of the derivatives of the unknown, for example, as

$$\frac{\partial u}{\partial x} = u, \quad \text{at } x = 0, \tag{3.34}$$

and

$$\frac{\partial u}{\partial x} = -u \quad \text{at } x = 1. \tag{3.35}$$

In such cases, if we represent the derivative by central difference, then Eqs. (3.34) and (3.35) yield

$$\frac{u^n_{j+1} - u^n_{j-1}}{2\Delta x} = u^n_j, \quad \text{for j} = 0, t \geq 0, \text{ and} \tag{3.36}$$

$$\frac{u^n_{j+1} - u^n_{j-1}}{2\Delta x} = -u^n_j, \quad \text{for } j = N, t \geq 0. \tag{3.37}$$

At $j = 0$ and at $j = N$, these equations require the knowledge of u^n_{-1} and u^n_{N+1} which are values at points outside the domain. In such cases, we assume that the differential equation is valid at the so-called *false boundaries* $j = -1$ and $j = N + 1$, and use the differential equation to eliminate the values on the false boundaries. The following example explains this.

Example 3.2

Here, we choose $r = K\Delta t/\Delta x^2 = \frac{1}{2}$, $\Delta x = \frac{1}{4}$. The false boundaries are shown by dotted vertical lines in Fig. 3.7. FTCS scheme has been used.

For the explicit FTCS scheme with $r = \frac{1}{2}$,

$$u^{n+1}_j = \frac{1}{2}\left(u^n_{j+1} + u^n_{j-1}\right), \quad j = 1, 2, 3, 4 \tag{3.38}$$

Putting $j = 0$ in Eq. (3.36) and $j = 4$ in Eq. (3.37), we get respectively

$$u^n_1 - u^n_{-1} = \frac{1}{2}u^n_0,$$

$$\text{and } u^n_5 = u^n_3 - \frac{1}{2}u^n_4 \tag{3.39}$$

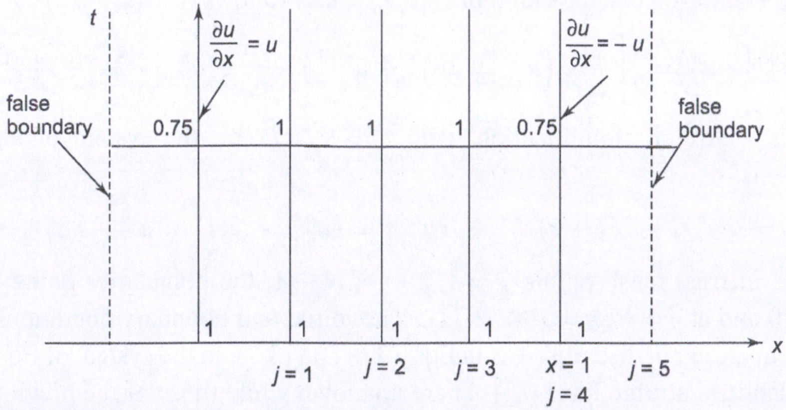

Figure 3.7 Derivative boundary conditions and false boundaries (dotted lines).

From Eq. (3.38) we get for $j = 0$ and $j = 4$

$$u_0^{n+1} = \frac{1}{2}(u_1^n + u_{-1}^n) \text{ and } u_4^{n+1} = \frac{1}{2}(u_5^n + u_3^n) \tag{3.40}$$

Eliminating now u_{-1}^n and u_5^n between Eqs. (3.39) and (3.40) results in

$$u_0^{n+1} = u_1^n - \frac{1}{4}u_0^n, \text{ and } u_4^{n+1} = u_3^n - \frac{1}{4}u_4^n$$

The first time-step solutions are shown in Fig. 3.7.

3.3 CRANK–NICHOLSON IMPLICIT SCHEME

The FTCS explicit scheme for the 1-D heat conduction problem, discussed in the previous section, is only conditionally stable. This is particularly inconvenient for nonlinear partial differential equations for which the mesh ratio $r = K\Delta t/\Delta x^2$ may involve some function of the unknown u. In order that the stability condition be satisfied for some chosen value of the mesh spacing Δx, an unusually small corresponding value of Δt might have to be chosen. This would lead to a large increase in labour. Also, although the accuracy of the scheme is second order in space, it is only first order in time. A serious drawback of the scheme is that it models a parabolic-type equation with a difference scheme having hyperbolic character. In the Crank–Nicholson implicit scheme all these drawbacks are avoided at a moderate cost.

In this scheme, as before, the time derivative is represented by forward difference while the space derivative is represented by the average central difference at the present and the new time step. Neglecting truncation errors for the present, the Crank–Nicholson discretisation of Eq. (3.1) is as follows:

$$\frac{1}{\Delta t}\left(u_j^{n+1} - u_j^n\right) = \frac{K}{2}\left[\frac{1}{\Delta x^2}\left(u_{j+1}^{n+1} - 2u_j^{n+1} + u_{j-1}^{n+1}\right) + \frac{1}{\Delta x^2}\left(u_{j+1}^n - 2u_j^n + u_{j-1}^n\right)\right] \tag{3.41}$$

This yields on simplification, with $r = K\Delta t/\Delta x^2$, the system of algebraic equations

$$-ru_{j-1}^{n+1} + 2(1+r)u_j^{n+1} - ru_{j+1}^{n+1} = ru_{j+1}^n + 2(1-r)u_j^n + ru_{j-1}^n \tag{3.42}$$

for the internal mesh points $j = 1, 2, \cdots, N-1$, the boundaries being located at $j=0$ and at $j=N$, $n = 0, 1, 2, \cdots$. The initial and boundary conditions of the problem are again discretised as in Eqs. (3.6) and (3.7). In the system of Eq. (3.42), the quantities at time level $(n+1)$ are unknown while those of the previous time level n are known quantities. Since a system of equations is to be solved for each time step, the Crank–Nicholson method is called an *implicit method*. Each of the equations consists of three unknowns except the first and the last, that is, those corresponding to $j = 1$ and $j = N - 1$, which contain only two unknowns each. This is a *tridiagonal system.* In the next subsection, we describe an elimination algorithm known as the *Thomas algorithm,* which is very efficient and requires only $(5M - 4)$ number of multiplications and divisions for a system of M equations. The computational molecule for the Crank–Nicholson scheme is shown in Fig. 3.8.

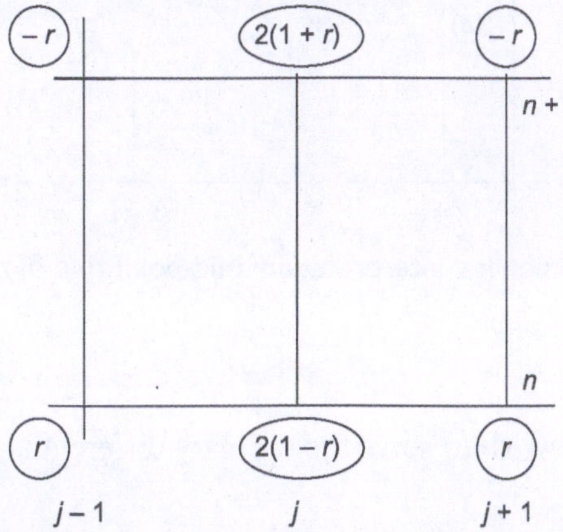

Figure 3.8 Computational mesh and influence of mesh points for the Crank–Nicholson scheme.

We see from Fig. 3.8 that all points in the semi-infinite region on the $(n+1)^{\text{th}}$ and previous time levels influence the solution u_j^{n+1}. This is typical of parabolic-type equations.

As shown in the subsequent subsections, it is second order accurate in both time and space and that it is unconditionally stable. These are some of the features which make the scheme *highly attractive.*

3.4 ANALOGY WITH SCHEMES FOR ORDINARY DIFFERENTIAL EQUATIONS

Many of the well-known finite difference schemes for partial differential equations are nothing but direct application of low-order numerical integration formulae like the mid-point rule or the trapezoidal rule.

We recall that the mid-point rule for integrating $f(x)$ from a to b is

$$\int_a^b f(x)\,dx = (b-a)\cdot f\left(\frac{a+b}{2}\right) \tag{3.43}$$

That is, it approximates the area under the curve between $x = a$, $x = b$ and the x-axis by the area of the rectangle formed by the ordinate $f\left(\dfrac{a+b}{2}\right)$ at the mid-point of the interval and the length of the interval $(b - a)$, Fig. 3.9.

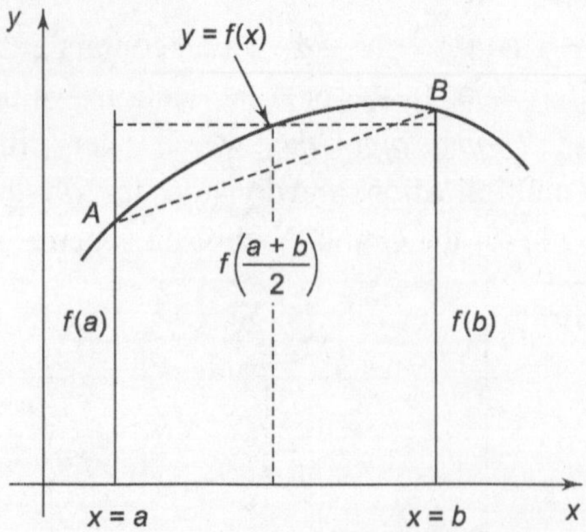

Figure 3.9 Geometrical interpretation: Mid-point rule of numerical integration.

For example, if we wish to solve the ordinary differential equation

$$\frac{dy}{dx} = f(x, y) \tag{3.44}$$

in the subinterval (x_j, x_{j+1}), then by integration using mid-point rule, we get

$$y_{j+1} - y_j = \int_{x_j}^{x_{j+1}} f(x, y) \quad dx$$

$$= hf\left(x_{j+\frac{1}{2}}, y_{j+\frac{1}{2}}\right)$$

$x_{j+\frac{1}{2}}$ denoting the abscissa of the mid-point of the interval, and $h = x_{j+1} - x_j$ = length of the subinterval. The point $x_{j+\frac{1}{2}}$ is not a pivotal point, which requires a method to express it in terms of pivotal values. If instead, we integrate $f(x, y)$ in the interval (x_{j-1}, x_{j+1}) and use mid-point rule we get

$$y_{j+1} - y_{j-1} = 2h \cdot f(x_j, y_j) \tag{3.45}$$

which is the well-known *leap-frog* scheme.

Recalling that the trapezoidal rule is

$$\int_a^b f(x)\,dx = \frac{b-a}{2}[f(b) + f(a)] \tag{3.46}$$

by applying the trapezoidal rule to Eq. (3.44) in (x_j, x_{j+1}), it follows that

$$y_{j+1} - y_j = \frac{h}{2}\left[f(x_j, y_j) + f(x_{j+1}, y_{j+1})\right] \tag{3.47}$$

Comparing with Eq. (3.41) it may be noted that the Crank–Nicholson scheme is a direct application of the trapezoidal rule of integration. Further, it may be noted that both mid-point rule and trapezoidal rule are second order, the truncation errors being

$-\frac{h^2}{24}f'''(\xi_1)$ and $\frac{-h^2}{12}f'''(\xi_2)$ respectively, ξ_1, ξ_2 being points inside the interval of integration.

The simplest of the numerical integration formulae is the *Euler explicit or Euler forward formula*, which for Eq. (3.44) in $(x_j,\ x_{j+1})$ may be written as

$$y_{j+1} - y_j = hf(x_j, y_j) \tag{3.48}$$

It is only first order accurate. A direct application of this to the 1-D heat conduction Eq. (3.1) yields the explicit FTCS scheme Eq. (3.5). It may be noted that the more accurate Simpson's $\frac{1}{3}$-rule often leads to instability and is not of much use.

3.4.1 Thomas Algorithm for Tridiagonal Systems

The linear system of Eq. (3.42) is of the form

$$\begin{aligned} b_1 f_1 + c_1 f_2 &= d_1, \\ a_i f_{i-1} + b_i f_i + c_i f_{i+1} &= d_i, \qquad i = 2, 3, \cdots, M-1, \\ a_M f_{M-1} + b_M f_M &= d_M, \end{aligned} \tag{3.49}$$

where $f_1, f_2, \cdots, f_M$ are unknown. The coefficients a_i, b_i, c_i, d_i are prescribed quantities. The coefficient matrix of this system is

$$\begin{pmatrix} b_1 & c_1 & 0 & 0 & \cdots & 0 \\ a_2 & b_2 & c_2 & 0 & \cdots & 0 \\ \cdots & \cdots & \cdots & \cdots & \cdots & \cdots \\ 0 & \cdots & 0 & a_{M-1} & b_{M-1} & c_{M-1} \\ 0 & 0 & \cdots & 0 & a_M & b_M \end{pmatrix} \tag{3.50}$$

In this matrix, elements on the leading diagonal and on the two diagonals on the two sides of it only are nonzero. Such matrices are known as *tridiagonal matrices*. It is known that the coefficient determinant of the system is nonzero, so that the system (3.49) is uniquely solvable.

For solving it, the first equation is used to eliminate f_1 from the second. The new second equation is used to eliminate f_2 from the third equation, and in general, the new i-th equation is used to eliminate f_i from the $(i+1)^{\text{th}}$-equation. Repeating the process of elimination $(M-1)$-times, we are left with an equation containing only the last unknown f_M. Using this value, the unknowns $f_{M-1}, f_{M-2}, \cdots, f_1$ are determined recursively.

In fact, dividing both sides of the first equation in (3.49) and transposing, we have $f_1 = p_1 + q_1 f_2$, where $p_1 = \frac{d_1}{b_1}$, $q_1 = -\frac{c_1}{b_1}$, assuming $b_1 \neq 0$. Substituting it in the second equation, we get $a_2(p_1 + q_1 f_2) + b_2 f_2 + c_2 f_3 = d_2$, which on simplification may be put to the form $f_2 = p_2 + q_2 f_3$, where $p_2 = \dfrac{d_2 - p_1 a_2}{b_2 + q_1 a_2}$, $q_2 = \dfrac{-c_2}{b_2 + q_1 a_2}$, assuming that the denominator does not vanish.

Repeating this process, yields the recurrence relations

$$f_M = p_M, \quad p_M = \frac{d_M - p_{M-1} a_M}{b_M + q_{M-1} a_M}, \tag{3.51}$$

$$f_i = p_i + q_i f_{i+1}, \quad i = M-1, M-2, \cdots, 1 \tag{3.52}$$

where

$$p_i = \frac{d_i - p_{i-1} a_i}{b_i + q_{i-1} a_i}, \quad q_i = \frac{-c_i}{b_i + q_{i-1} a_i}, \quad i = 2, 3, \cdots, M-1 \tag{3.53}$$

assuming that none of the denominators vanish. If however, some of the denominators vanish, then also the above scheme may be used with minor modifications.

In fact, noting that if the denominator of p_i vanishes, then the $i^{-\text{th}}$ equation determines f_{i+1} uniquely and the system breaks-up into two tridiagonal systems. The procedure explained above for obtaining the solution of a tridiagonal system of equations is known as the *Thomas algorithm.*

The tridiagonal system belongs to the category of *sparse systems*, in view of the relatively large number of zero elements in the coefficient matrix. In CFD, we frequently encounter large sparse systems.

Operational Count for Thomas Algorithm: A small number of arithmetic operations are required for the Thomas algorithm. Let us count only the number of multiplications and divisions and neglect the additions and subtractions.

We note that according to Eqs. (3.52) and (3.53), computation of p_i and q_i taken together need four operations and f_i requires one operation, so that five operations are needed for computing each of the (M–2)-number of f_i's. The first and the last unknowns f_1 and f_M each require three operations, so that the total for Thomas algorithm is

$$5(M-2) + 3 + 3 = 5M - 4 \text{ operations.} \tag{3.54}$$

It may be noted that straightforward application of Gaussian elimination subroutine needs $O(\frac{1}{3}M^3)$ number of multiplications and divisions, which is very large compared to the count Eq. (3.54). This is due to the fact that zero operations have been avoided altogether. Also, the zero elements are not stored, saving storage space.

It can be shown (Richtmeyer and Morton, 1967) that if the coefficient matrix is *diagonally dominant*, that is, if $b_i| \geq |a_i| + |c_i|$, then round-off errors remain bounded and the Thomas algorithm is stable.

Example 3.3

We shall solve the problem in Example 3.1 by Crank–Nicholson scheme, with $r = K\Delta t/\Delta x^2 = 1$, and $\Delta x = \frac{1}{4}$. Then Eq. (3.42) becomes

$$-u_{j-1}^{n+1} + 4u_j^{n+1} - u_{j+1}^{n+1} = u_{j+1}^n + u_{j-1}^n, \quad j = 1, 2, 3 \tag{3.55}$$

The computational molecule is shown in Fig. 3.10(a) and the initial boundary-value problem in Fig. 3.10(b). Putting $n = 0$ in Eq. (3.55), the resulting system is, omitting the superscript,

$$4u_1 - u_2 = 1$$
$$-u_1 + 4u_2 - u_3 = 1.42$$
$$-u_2 + 4u_3 = 1,$$

which yield on solution by Thomas algorithm the values
$u_1^1 = u_3^1 = 0.387$ and $u_2^1 = 0.548$, correct to three decimal places.

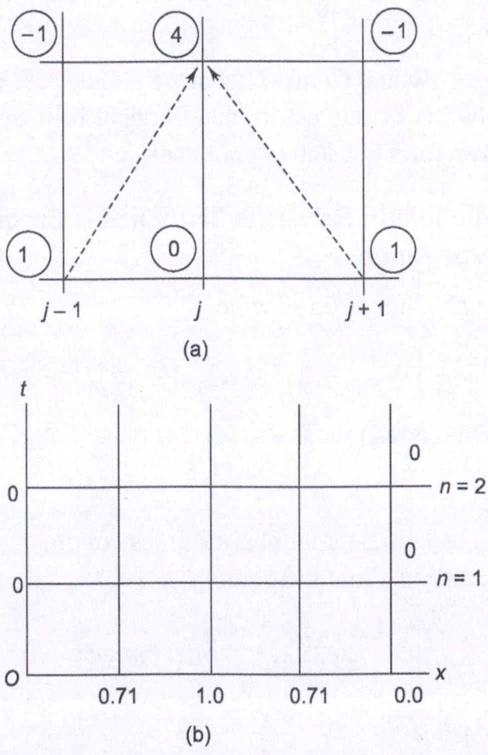

Figure 3.10 Crank–Nicholson scheme with r = 1. (a) Computational molecule, (b) Boundary and initial conditions.

Exact analytical solution of the problem may be found by separation of variables as

$$u(x,t) = \exp(-\pi^2 t)\sin(n\pi x) \tag{3.56}$$

At the point $x = \frac{1}{2}$, $t = \frac{1}{16}$, we have approximately from the above exact solution

$$u(\frac{1}{2}, \frac{1}{16}) \approx \exp(-\frac{\pi^2}{16}) \approx 0.5396 = 0.540$$

For the FTCS explicit method with $r = \frac{1}{2}$, $\Delta x = \frac{1}{4}$, $K = 1$, $\Delta t = \frac{1}{2} \cdot \frac{1}{16} = \frac{1}{32}$. Consequently, $t = \frac{1}{16}$ for $j = 2$. The FTCS solution is

$$u\left(\frac{1}{2}, \frac{1}{16}\right)\bigg|_{\text{FTCS}} = 0.50$$

and the corresponding value with Crank–Nicholson scheme as calculated above is

$$u\left(\frac{1}{2}, \frac{1}{16}\right)\bigg|_{\text{C–N}} = 0.548.$$

Thus, FTCS indicates 8% and Crank–Nicholson about 1.5% error.

The above problem is of interest in metallurgical heat treatment studies, as it models the time taken for a hot slab to cool itself.

3.4.2 Crank–Nicholson Scheme: Truncation Error, Consistency, and Convergence

We consider the Crank–Nicholson scheme

$$\frac{u_j^{n+1} - u_j^n}{\Delta t} = \frac{K}{2} \cdot \frac{1}{\Delta x^2}\left[\{u_{j+1}^n - 2u_j^n + u_{j-1}^n\} + \{u_{j+1}^{n+1} - 2u_j^{n+1} + u_{j-1}^{n+1}\}\right] \tag{3.57}$$

for the heat-conduction equation

$$u_t = K u_{xx} \tag{3.58}$$

Assuming $u(x,t)$ to be sufficient number of times continuously differentiable, we find by Taylor's expansion about the point $(j, n + \frac{1}{2})$

$$\frac{u_j^{n+1} - u_j^n}{\Delta t} = u_t\bigg|_j^{n+\frac{1}{2}} + \frac{\Delta t^2}{24} u_{ttt}\bigg|_j^{n+\frac{1}{2}} + O(\Delta t^4) \tag{3.59}$$

Also, by Taylor's expansion

$$\frac{u_{j+1}^n - 2u_j^n + u_{j-1}^n}{\Delta x^2} = u_{xx}\bigg|_j^n + \frac{\Delta x^2}{12} u_{xxxx}\bigg|_j^n + O(\Delta x^4) \tag{3.60}$$

and

$$\frac{u_{j+1}^{n+1} - 2u_j^{n+1} + u_{j-1}^{n+1}}{\Delta x^2} = u_{xx}\Big|_j^{n+1} + \frac{\Delta x^2}{12} u_{xxxx}\Big|_j^{n+1} + O(\Delta x^4) \tag{3.61}$$

Moreover, it is to be noted that for any smooth function $\phi(t)$,

$$\frac{1}{2}\left(\phi^{n+1} + \phi^n\right) = \phi^{n+\frac{1}{2}} + \frac{\Delta t^2}{8}\phi_{tt}\Big|^{n+\frac{1}{2}} + O(\Delta t^3) \tag{3.62}$$

Substituting Eq. (3.59)–(3.62) in Eq. (3.57), we obtain, after some simplification,

$$\begin{aligned}
u_t\Big|_j^{n+\frac{1}{2}} + \frac{\Delta t^2}{24} u_{ttt}\Big|_j^{n+\frac{1}{2}} + O(\Delta t^4) &= \\
K\left\{\frac{u_{xx}^n + u_{xx}^{n+1}}{2} + \frac{\Delta x^2}{12}\frac{u_{xxxx}^n + u_{xxxx}^{n+1}}{2}\right\} + O(\Delta x^4) &= \\
K\left[u_{xx}\Big|_j^{n+\frac{1}{2}} + \frac{\Delta t^2}{8}u_{xxtt}\Big|_j^{n+\frac{1}{2}} + 0(\Delta t^3)\right] + K\frac{\Delta x^2}{12}\Bigg[u_{xxxx}\Big|_j^{n+\frac{1}{2}} + \frac{\Delta t^2}{8}u_{xxxxtt}\Big|_j^{n+\frac{1}{2}} & \\
+ O(\Delta t^3)\Bigg] + O(\Delta x^4) &= \\
Ku_{xx}\Big|_j^{n+\frac{1}{2}} + K\frac{\Delta t^2}{8}u_{xxtt}\Big|_j^{n+\frac{1}{2}} + K\frac{\Delta x^2}{12}u_{xxxx}\Big|_j^{n+\frac{1}{2}} + K\frac{\Delta x^2\Delta t^2}{96}u_{xxxxtt}\Big|_j^{n+\frac{1}{2}} & \\
+ O(\Delta x^2\Delta t^3) + O(\Delta x^6) &
\end{aligned}$$

which yields the leading terms of the truncation error at the point $(j, n + 1/2)$

$$u_t - Ku_{xx} = -\frac{\Delta t^2}{24}u_{ttt} + K\frac{\Delta t^2}{8}u_{xxtt} + K\frac{\Delta x^2}{12}u_{xxxx} \tag{3.63}$$

Equation (3.63) shows that the truncation error of the Crank–Nicholson scheme is a quantity $O(\Delta t^2) + O(\Delta x^2)$, that is, it is second order in both time and space. Further, the truncation error approaches zero as $\Delta t \to 0$ and $\Delta x \to 0$, showing that the Crank–Nicholson scheme is consistent.

The time derivatives may be eliminated from the right side of Eq. (3.63) using the differential equation (3.58). For, differentiating Eq. (3.58) with respect to t, we get for u_{tt} and u_{ttt}

$$u_{tt} = K\frac{\partial}{\partial t}(u_{xx}) = K\frac{\partial^2}{\partial x^2}(u_t) = K^2 u_{xxxx} \tag{3.64}$$

$$u_{ttt} = \frac{\partial}{\partial t}(K^2 u_{xxxx}) = K^2\frac{\partial^4}{\partial x^4}(u_t) = K^3 u_{xxxxxx} \tag{3.65}$$

Further

$$u_{xxtt} = \frac{\partial^2}{\partial x^2}(u_{tt}) = \frac{\partial^2}{\partial x^2}(K^2 u_{xxxx}) = K^2 u_{xxxxxx} \tag{3.66}$$

So, using Eqs. (3.65)–(3.66) in Eq. (3.63) we get, omitting the subscripts and superscripts,

$$\begin{aligned} u_t - K u_{xx} &= -\frac{\Delta t^2 K^3}{24} u_{xxxxxx} + K \frac{\Delta t^2}{8} K^2 u_{xxxxxx} + K \frac{\Delta x^2}{12} u_{xxxx} \\ &= K^3 \frac{\Delta t^2}{12} u_{xxxxxx} + K \frac{\Delta x^2}{12} u_{xxxx} \end{aligned} \tag{3.67}$$

If, instead of the differential equation, the difference equation is used to eliminate the time derivatives, we obtain after some simple but somewhat lengthy calculations

$$u_t - K u_{xx} = K \frac{\Delta x^2}{12} u_{xxxx} + \left[\frac{K^3}{12} \Delta t^2 + \frac{1}{360} \Delta x^4 \right] u_{xxxxxx} + \ldots \tag{3.68}$$

which is the *modified equation* for the Crank–Nicholson scheme. It may be noted that *only even-order space-derivatives appear in the modified equation.*

Proceeding as in the case of FTCS scheme (Section 3.2.3) it is possible to prove the convergence of the Crank–Nicholson scheme. Formal proofs may be found, for example, in Jain (1984).

3.4.3 Dissipative and Dispersive Errors

We note that in the truncation error terms of the FTCS scheme, from which the time derivatives have been eliminated, Eq. (3.15), and in the modified equation (3.68) of the Crank–Nicholson scheme, no odd-order space derivatives appear. Certain quasi-physical effects are associated with the nature of the lowest order truncation error terms. The lowest order term containing odd-order space-derivatives are associated with the presence of *dispersive errors* while *even-order* derivatives are the source of *dissipative errors*.

Comparing with the 1-D Navier–Stokes equation

$$\frac{\partial u}{\partial t} + u \frac{\partial u}{\partial x} = \nu \frac{\partial^2 u}{\partial x^2} + \ldots \tag{3.69}$$

assuming a constant coefficient of kinematic viscosity ν of small magnitude, it is expected that a term containing u_{xx} would behave like a physical viscosity.

In fact, it is found from numerical experiments, that the presence of such even-order derivative terms would have the effect of damping out any steep gradients in the flow field. A truncation error term of this type is said to introduce an *artificial viscosity* into the solution.

This is said to be an *implicit artificial viscosity* as against *explicit artificial viscosity* often added to a difference scheme in order to obtain a desired effect. So, truncation errors with leading even-order space-derivative terms are often called dissipative terms and the corresponding scheme is referred to as a *dissipative scheme*.

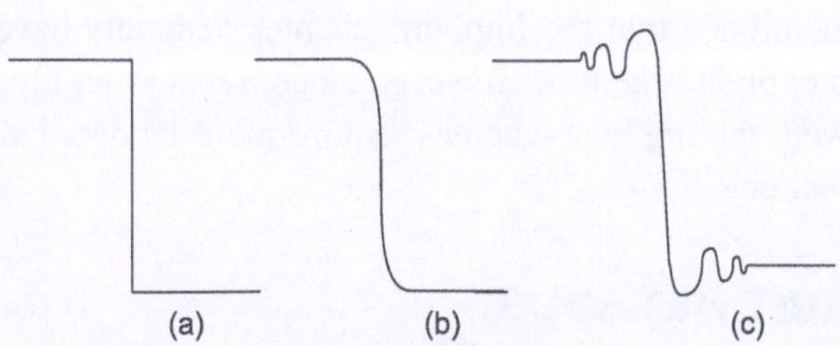

Figure 3.11 Effect of dissipation and dispersion. (a) Exact solution of a pressure jump. (b) Damping because of dissipation (typical of first order schemes). (c) Distortion mainly due to dispersive error (typical of second-order schemes).

The presence of an odd-order space-derivative in the lowest order truncation error term is responsible for dispersion, in which phase relations between the various Fourier components of the solution are distorted, as shown in Fig. 3.11. Such errors are called *dispersive errors* and the scheme is said to be a *dispersive scheme*. The combined effect of dispersion and dissipation is sometimes called *diffusion*, which appears to spread out sharp dividing lines that may appear in the computational domain.

3.4.4 Stability of the Crank–Nicholson Scheme

Let us investigate the stability of the Crank–Nicholson scheme, Eq. (3.42) by the von Neumann method. As in the case of FTCS scheme, the equation being linear, the error satisfies the difference equation and it is enough to study the growth of a typical term

$$u_j^n = e^{\alpha t} e^{i k_j x} = e^{\alpha n \Delta t} e^{i\beta j}, \tag{3.70}$$

where $\beta = k_j \Delta x$ is real, $i = \sqrt{-1}$, and $Z = e^{\alpha \Delta t}$ is the amplification factor, where α may be complex. Substituting Eq. (3.70) in the Crank–Nicholson scheme (3.42) and canceling throughout the factor $e^{\alpha n \Delta t} e^{i\beta j}$, yields

$$e^{\alpha \Delta t}\left\{-\frac{r}{2}(e^{i\beta} + e^{-i\beta}) + (1+r)\right\} = \frac{r}{2}\left(e^{i\beta} + e^{-i\beta}\right) + (1-r)$$

which gives

$$Z = e^{\alpha \Delta t} = \frac{r\cos\beta + 1 - r}{1 + r - r\cos\beta} = \frac{1 - 2r\sin^2\dfrac{\beta}{2}}{1 + 2r\sin^2\dfrac{\beta}{2}} \tag{3.71}$$

This shows that for all values of r, $r > 0$, $|Z| < 1$. Hence the Crank–Nicholson scheme is *unconditionally stable*. Since the stability does not depend on the value of r, it is said to be *unconditionally stable*.

It may be mentioned that the implicit schemes generally have better stability behaviour than explicit schemes. However, since a system of algebraic equations have to be solved, the implicit schemes require more labour. This is particularly true for nonlinear equations.

3.5 A NOTE ON IMPLICIT METHODS

Consider a generalization of the Crank–Nicholson scheme for the 1-D heat-conduction equation (3.1),

$$\frac{\Delta u_j^n}{\Delta t} = \frac{K}{\Delta x^2}\left\{\theta\delta_{xx}u_j^{n+1} + (1-\theta)\delta_{xx}u_j^n\right\}, \tag{3.72}$$

where θ is a parameter $0 \leq \theta \leq 1$, and δ denotes the central difference operator (explained in detail in Section 3.7) and $\Delta u_j^n = u_j^{n+1} - u_j^n$, so that Δu_j^n denotes the correction to be added to u_j^n in order to obtain u_j^{n+1}. Note that for $\theta = 0$, Eq. (3.72) represents the explicit FTCS scheme while for $\theta = 1$, the scheme is fully implicit. For $\theta = \frac{1}{2}$ we regain the trapezoidal rule, that is the Crank–Nicholson scheme, which is implicit. For studying the stability of the scheme let us consider, as before, a trial solution $u_j^n \sim e^{\alpha n\Delta t}.e^{i\beta j}$, where β is real, α may be complex. Then noting that

$$\delta_{xx}u_j^n = u_{j+1}^n - 2u_j^n + u_{j-1}^n = e^{\alpha n\Delta t}e^{i\beta j}\left(-4\sin^2\frac{\beta}{2}\right)$$

equation (3.72) delivers, upon cancelling the common factor $e^{\alpha n\Delta t}e^{i\beta j}$ and simplifying

$$e^{\alpha\Delta t} = \frac{1 - 4r(1-\theta)\sin^2\dfrac{\beta}{2}}{1 + 4r\theta\cdot\sin^2\dfrac{\beta}{2}} \tag{3.73}$$

showing that it is unconditionally stable for $\frac{1}{2} \leq \theta \leq 1$. For $0 \leq \theta < \frac{1}{2}$, it is stable if

$$-1 \leq \frac{1 - 4r(1-\theta)\sin^2\dfrac{\beta}{2}}{1 + 4r\theta\cdot\sin^2\dfrac{\beta}{2}} \leq 1$$

The right inequality is automatically satisfied. The other inequality is satisfied if

$$4(1-2\theta)r\sin^2\frac{\beta}{2} \leq 2$$

implying that

$$r \leq \frac{1}{2(1-2\theta)}, \tag{3.74}$$

which is the condition of stability.

It may be noted that the Crank–Nicholson scheme is on the boundary of the unconditionally stable region. We shall see in Chapter 5, that for the solution of steady state problems (elliptic type problems) it may be more efficient to solve an equivalent transient problem and continue marching in the time (or time-like) direction till the solution does not change up to a prescribed tolerance. However, it is often found that the solution in the different parts of the computational domain approaches the steady state at significantly different rates. A differential equation showing such a behaviour is said to be *stiff*. Experience shows that, the Crank–Nicholson scheme often produces an oscillatory solution in this situation. Although the scheme remains stable, the steady state is approached at a slow rate. In such cases, certain three time-level schemes behave better than the Crank–Nicholson scheme.

A *generalised three time-level* scheme may be defined for the heat conduction equation (3.1) as

$$(1+\xi)\frac{\Delta u_j^n}{\Delta t} - \xi\frac{\Delta u_j^{n-1}}{\Delta t} = \frac{K}{\Delta x^2}\left[(1-\theta)\delta_{xx}u_j^n + \theta\delta_{xx}u_j^{n+1}\right] \tag{3.75}$$

where ξ and θ are parameters.

The inclusion of the extra time level demands additional storage and execution time. A particularly effective three-level scheme is obtained by choosing $\xi = \frac{1}{2}, \theta = 1$. This scheme, known as *3LFI (3–level fully implicit) scheme,* has truncation error $O(\Delta t^2, \Delta x^2)$ and is unconditionally stable. This scheme damps out spurious oscillations for stiff problems. The resulting system of linear algebraic equations may be solved by the Thomas algorithm.

3.6 LEAP–FROG AND DUFORT–FRANKEL SCHEMES

As we have seen in the previous section, the Crank–Nicholson scheme has better stability and accuracy properties compared to the FTCS method which is explicit. If both the time and space derivatives in the 1-D heat conduction equation (3.1) be represented by central difference, we obtain

$$\frac{1}{2\Delta t}(u_j^{n+1} - u_j^{n-1}) + O(\Delta t^2) = K\frac{u_{j+1}^n - 2u_j^n + u_{j-1}^n}{\Delta x^2} + O(\Delta x^2) \tag{3.76}$$

which is known as the *leap–frog scheme*. It is second-order accurate in both time and space. However, this scheme is of no practical use, because it is *unconditionally unstable*, as may be easily established by the von Neumann stability analysis. It was observed by DuFort and Frankel (1953) that if in the above equation the term u_j^n be approximated by the average value of u_j at $n-1$ and at $n+1$ time levels, the stability behaviour improves significantly. The resulting scheme, neglecting truncation error terms,

$$\frac{1}{2\Delta t}(u_j^{n+1} - u_j^{n-1}) = \frac{K}{\Delta x^2}[u_{j+1}^n - (u_j^{n+1} + u_j^{n-1}) + u_{j-1}^n] \tag{3.77}$$

is known as *the DuFort and Frankel scheme*. Introducing $r = K\Delta t/\Delta x^2$, we get on simplification

$$(1+2r)u_j^{n+1} = (1-2r)u_j^{n-1} + 2r(u_{j+1}^n + u_{j-1}^n) \tag{3.78}$$

The DuFort–Frankel scheme Eq. (3.78) indicates certain surprising features. The quantities on the right side of Eq. (3.78) are at time-level n and $n-1$, so that they are known quantities when one computes the solution at the time-level $(n+1)$. Consequently, it is an explicit scheme. From our experience with the explicit FTCS scheme, we would expect it to be at best, conditionally stable. However, it turns out to be unconditionally stable, as we prove in Section 3.6.2. Another surprising fact is that *under certain conditions it is not consistent,* as shown in the following subsections.

3.6.1 Truncation Error of the DuFort–Frankel Scheme

Assuming the quantities to be sufficient number of times continuously differentiable, we obtain by Taylor expansion about the point (j, n), the following equations,

$$u_j^{n+1} - u_j^{n-1} = 2\Delta t(u_t)_j^n + \frac{\Delta t^3}{3}(u_{ttt})_j^n + \frac{\Delta t^5}{60}(u_{ttttt})_j^n + O(\Delta t^6),$$

$$2r(u_j^{n+1} + u_j^{n-1}) = 2r[2u_j^n + \Delta t^2(u_{tt})_j^n + \frac{\Delta t^4}{12}(u_{tttt})_j^n] + O(\Delta t^6),$$

$$u_{j+1}^n + u_{j-1}^n = 2u_j^n + \Delta x^2(u_{xx})_j^n + \frac{\Delta x^4}{12}(u_{xxxx})_j^n + O(\Delta x^6).$$

Substituting these values in Eq. (3.77) yields, dropping the subscripts j and superscripts n

$$2\Delta t(u_t)+\frac{\Delta t^3}{3}(u_{ttt})+\frac{\Delta t^5}{60}(u_{ttttt})+O(\Delta t^6)=4ru_j^n+2r\Delta x^2(u_{xx})+r\frac{\Delta x^4}{6}(u_{xxxx})$$
$$+O(\Delta x^6) - 2r[2u_j^n + \Delta t^2(u_{tt}) + \frac{\Delta t^4}{12}(u_{tttt}) + O(\Delta t^6)]$$

Dividing both sides by $2\Delta t$ yields using $r = K\Delta t/\Delta x^2$,

$$u_t - Ku_{xx} = -\frac{\Delta t^2}{6}u_{ttt} - K\frac{\Delta t^2}{\Delta x^2}u_{tt} + K\frac{\Delta x^2}{12}u_{xxxx} - \frac{K}{12}\frac{\Delta t^4}{\Delta x^2}u_{tttt} + \cdots \quad (3.79)$$

The quantities on the right hand side of Eq. (3.79) represent the truncation error term of the DuFort–Frankel scheme. If as $\Delta t \to 0$, $\Delta x \to 0$, the ratio $\frac{\Delta t}{\Delta x} \to c$, where c is some nonzero constant, then we see from Eq. (3.79) that the DuFort–Frankel scheme approaches

$$u_t - Ku_{xx} + Kc^2u_{tt} = 0 \quad (3.80)$$

which is different from the 1-D heat conduction equation (3.1) implying that in this case it is inconsistent. On the other hand, if $c = 0$, the scheme is consistent.

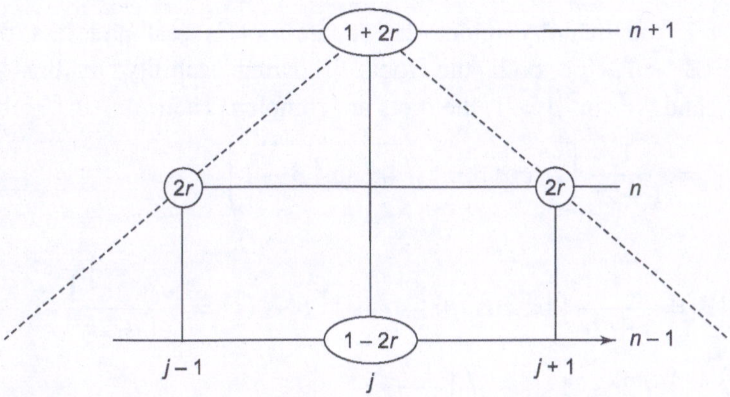

Figure 3.12 Computational molecule for the DuFort–Frankel Scheme (dotted lines enclose zone of influence).

It may be noted that Eq. (3.80) is of hyperbolic type. This could have been surmised from the computational molecule for the scheme, shown in Fig. 3.12.

In spite of the above mentioned drawbacks, the DuFort–Frankel scheme was quite popular, mainly due to the fact that it may be applied to problems of two or three space dimensions with only a little more effort than is required for that in one dimension. This feature is contrary to that of the Crank–Nicholson scheme which requires prohibitive amount of labour in more than one dimensions. The DuFort–Frankel scheme was used for quite some years, till it gave way to the alternating direction implicit (ADI) method.

3.6.2 Stability of DuFort–Frankel Scheme

For studying stability by the von Neumann method, let us take a trial solution

$$u_j^n = e^{\alpha n \Delta t} e^{i\beta j},$$

where α is a complex and β a real constant and $i = \sqrt{-1}$.

Substituting in Eq. (3.78) and cancelling the common factor $e^{\alpha n \Delta t} e^{i\beta j}$ from both sides, we get

$$(1 + 2r)e^{\alpha \Delta t} = (1 - 2r)e^{-\alpha \Delta t} + 2r\left(e^{i\beta} + e^{-i\beta}\right)$$

Solving it as a quadratic equation in $Z = \exp(\alpha \Delta t)$ we get for the amplification factor Z

$$Z = \frac{1}{1 + 2r}\left[2r \cos \beta \pm \sqrt{1 - 4r^2 \sin^2 \beta}\right] \tag{3.81}$$

For $2r < 1$, the quantity under the square-root is real and less than unity, so that $|Z| < 1$, for both the roots, showing stability in this case. For $2r > 1$, and $4r^2 \sin^2 \beta > 1$, the roots are complex. Then

$$Z = \frac{1}{1 + 2r}\left[2r \cos \beta \pm i\sqrt{4r^2 \sin^2 \beta - 1}\right]$$

so that

$$|Z| = \frac{1}{1 + 2r}\left[4r^2 \cos^2 \beta + 4r^2 \sin^2 \beta - 1\right]^{\frac{1}{2}} = \frac{\sqrt{4r^2 - 1}}{1 + 2r}$$

$$= \left(\frac{2r - 1}{2r + 1}\right)^{\frac{1}{2}} = \left(\frac{1 - \frac{1}{2r}}{1 + \frac{1}{2r}}\right)^{\frac{1}{2}}.$$

Since $0 < \frac{1}{2r} < 1$, it follows that $|Z| < 1$, showing that the scheme is stable. For the case $2r > 1$ and $4r^2 \sin^2 \beta < 1$, the scheme may also be shown to be stable. Note that for $2r = 1$, it follows from Eq. (3.81) that

$$Z = \begin{cases} \cos \beta, & \text{for the upper sign} \\ 0, & \text{for the lower sign} \end{cases}$$

so that $|Z| \leq 1$ in all the cases, irrespective of the value of r. Hence, DuFort–Frankel scheme is unconditionally stable.

3.7 OPERATOR NOTATION

In the previous sections, we have considered the 1-D heat-conduction equation. For more than one space dimensions, it appears convenient to introduce operator notations. We introduce the forward, backward (or upwind) and central-difference operators denoted respectively by $\vec{\delta}$, $\overline{\delta}$ and δ, the space coordinates being indicated by the appropriate suffixes. We define

$$\begin{aligned} \text{forward difference} \quad & \vec{\delta}_x u_j^n = u_{j+1}^n - u_j^n, \\ \text{backward difference} \quad & \overline{\delta}_x u_j^n = u_j^n - u_{j-1}^n, \\ \text{central difference} \quad & \delta_x u_j^n = u_{j+1}^n - u_{j-1}^n \end{aligned} \tag{3.82}$$

Higher order differences are represented by repetition of the operators and suffixes, for example,

$$\begin{aligned} \delta_{xx} u_j^n &= \text{second central difference of } u_j^n \text{ along } x \\ &= \overline{\delta}_x(\vec{\delta}_x) u_j^n = \vec{\delta}_x(\overline{\delta}_x) u_j^n = u_{j+1}^n - 2u_j^n + u_{j-1}^n. \end{aligned} \tag{3.83}$$

Similarly,

$$\overline{\delta}_{xx} u_j^n = u_j^n - 2u_{j-1}^n + u_{j-2}^n. \tag{3.84}$$

Observe that

$$\delta_x(\delta_x u_j^n) \neq \delta_{xx} u_j^n. \tag{3.85}$$

Using operator notations, we may rewrite the FTCS scheme, Eq. (3.5) in the form

$$\frac{u_j^{n+1} - u_j^n}{\Delta t} = \frac{K}{\Delta x^2} \delta_{xx} u_j^n$$

which may be simplified, using $r = K\Delta t/\Delta x^2$ as

$$u_j^{n+1} = u_j^n + r\delta_{xx} u_j^n = (I + r\delta_{xx}) u_j^n, \tag{3.86}$$

the quantity I denoting the identity operator. We may further rewrite Eq. (3.86) as

$$u_j^{n+1} = L_{\mathrm{E}}\, u_j^n,$$

where L_{E} denotes the explicit operator $L_{\mathrm{E}} = I + r\delta_{xx}$. Similarly, the 1-D Crank–Nicholson scheme Eq. (3.42) may be expressed as

$$u_j^{n+1} - u_j^n = \frac{r}{2}\delta_{xx} u_j^n + \frac{r}{2}\delta_{xx} u_j^{n+1},$$

which may be further rewritten in the form

$$(I - \frac{r}{2}\delta_{xx}) u_j^{n+1} = (I + \frac{r}{2}\delta_{xx}) u_j^n \tag{3.87}$$

It may be noted that the operator $I - \frac{r}{2}\delta_{xx}$ yields a tridiagonal coefficient matrix.

In the next section we consider the heat conduction equation in two-space dimensions, where the operator notation may be conveniently used.

3.8 THE ALTERNATING DIRECTION IMPLICIT (ADI) METHOD

Let us consider the two-dimensional heat conduction equation

$$\frac{\partial u}{\partial t} = K\left(\frac{\partial^2 u}{\partial x^2} + \frac{\partial^2 u}{\partial y^2}\right), \qquad K = \text{const.} > 0 \tag{3.88}$$

together with appropriate initial and boundary conditions, to be stated in due course.

The Crank–Nicholson scheme for this equation may be written in operator notation, with $u(x, y, t) = u(j\Delta x, k\Delta y, n\Delta t) = u^n_{j,k}$ as

$$\frac{1}{\Delta t}\left[u^{n+1}_{j,k} - u^n_{j,k}\right] = \frac{K}{2}\left[\frac{\delta_{xx}}{\Delta x^2}(u^{n+1}_{j,k} + u^n_{j,k}) + \frac{\delta_{yy}}{\Delta y^2}(u^{n+1}_{j,k} + u^n_{j,k})\right] \tag{3.89}$$

Let us put

$$K\frac{\Delta t}{\Delta x^2} = r_1 \quad \text{and} \quad K\frac{\Delta t}{\Delta y^2} = r_2$$

Then Eq. (3.89) may be simplified as

$$u^{n+1}_{j,k} - u^n_{j,k} = \frac{r_1}{2}\delta_{xx}(u^{n+1}_{j,k} + u^n_{j,k}) + \frac{r_2}{2}\delta_{yy}(u^{n+1}_{j,k} + u^n_{j,k}),$$

which is rewritten in operator notation as

$$\left[I - \frac{r_1}{2}\delta_{xx} - \frac{r_2}{2}\delta_{yy}\right]u^{n+1}_{j,k} = (I + \frac{r_1}{2}\delta_{xx} + \frac{r_2}{2}\delta_{yy})u^n_{j,k} \tag{3.90}$$

Noting the central difference representations

$$\delta_{xx}u^{n+1}_{j,k} = u^{n+1}_{j+1,k} - 2u^{n+1}_{j,k} + u^{n+1}_{j-1,k}, \tag{3.91}$$

$$\delta_{yy}u^{n+1}_{j,k} = u^{n+1}_{j,k+1} - 2u^{n+1}_{j,k} + u^{n+1}_{j,k-1}, \tag{3.92}$$

the left hand side of Eq. (3.90) simplifies to the expression

$$-\frac{r_1}{2}u^{n+1}_{j+1,k} + \{1 + (r_1 + r_2)\}u^{n+1}_{j,k} - \frac{r_1}{2}u^{n+1}_{j-1,k} - \frac{r_2}{2}u^{n+1}_{j,k+1} - \frac{r_2}{2}u^{n+1}_{j,k-1} \tag{3.93}$$

This yields a *pentadiagonal coefficient matrix*, of which two diagonals with nonzero elements are on the two sides of the leading diagonal and two others are respectively at horizontal and vertical distances M rows or M columns away from the left hand top corner, $M \times M$ denoting the number of unknowns. We can see this clearly if we write out the complete set of equations, as in the next example 3.4.

Example 3.4

Let us consider discretization of a domain which is a unit square in the first quadrant. Let us take $\Delta x = \Delta y = \frac{1}{5}$, so that the number of internal mesh points in the x and y directions are $M = N = 5 - 1 = 4$ each. The internal mesh points correspond to the suffixes $j = 1, 2, 3, 4$ and $k = 1, 2, 3, 4$, while $j = 0$, $j = 5$, $k = 0$ and $k = 5$ represent mesh points on the boundaries of the domain.

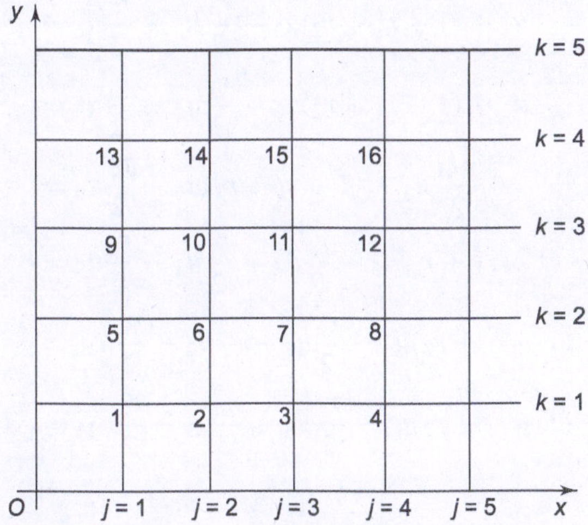

Figure 3.13 Renumbering of the mesh points (Example 3.4).

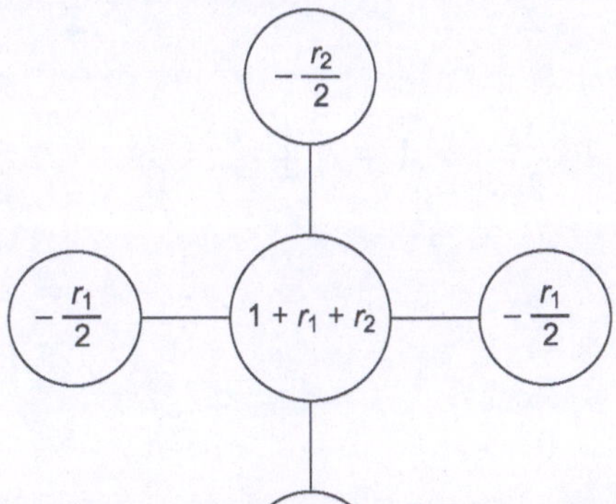

Figure 3.14 Computational molecule for the Crank–Nicholson scheme in 2-D.

The $16 (= 4 \times 4)$ internal mesh points are renumbered as $1, 2, \cdots, 16$, the unknowns being denoted as $u_1, u_2, \cdots, u_{16}$ as shown in Fig. 3.13. The computational molecule for the 2-D Crank–Nicholson scheme is shown in Fig. 3.14.

We get the equations

$$
\begin{aligned}
(1+r_1+r_2)u_1 - \frac{r_1}{2}u_2 - \frac{r_2}{2}u_5 &= \cdots \\
\frac{-r_1}{2}u_1 + (1+r_1+r_2)u_2 - \frac{r_1}{2}u_3 - \frac{r_2}{2}u_6 &= \cdots \\
\frac{-r_1}{2}u_2 + (1+r_1+r_2)u_3 - \frac{r_1}{2}u_4 - \frac{r_2}{2}u_7 &= \cdots \\
\frac{-r_1}{2}u_3 + (1+r_1+r_2)u_4 - \frac{r_2}{2}u_8 &= \cdots \\
(1+r_1+r_2)u_5 - \frac{r_1}{2}u_6 - \frac{r_2}{2}u_1 - \frac{r_2}{2}u_9 &= \cdots \\
\frac{-r_1}{2}u_5 + (1+r_1+r_2)u_6 - \frac{r_1}{2}u_7 - \frac{r_2}{2}u_2 - \frac{r_2}{2}u_{10} &= \cdots \\
\frac{-r_1}{2}u_6 - (1+r_1+r_2)u_7 - \frac{r_1}{2}u_8 - \frac{r_2}{2}u_3 - \frac{r_2}{2}u_{11} &= \cdots \\
&\vdots
\end{aligned}
$$

The right hand side quantities, most of which are zeros, contain known boundary values. The resulting coefficient matrix is pentadiagonal, as follows

$$
\begin{pmatrix}
1+r_1+r_2 & \frac{-r_1}{2} & \cdots & \cdots & \frac{-r_2}{2} & \cdots & \cdots & \cdots \\
\frac{-r_1}{2} & 1+r_1+r_2 & \frac{-r_1}{2} & \cdots & \cdots & \frac{-r_2}{2} & \cdots & \cdots \\
0 & \frac{-r_1}{2} & 1+r_1+r_2 & \cdots & \cdots & \cdots & \frac{-r_2}{2} & \cdots \\
\cdots & \cdots & \cdots & \cdots & \cdots & \cdots & \cdots & \cdots \\
\frac{-r_2}{2} & \cdots & \cdots & \cdots & \cdots & \cdots & \cdots & \cdots \\
0 & \frac{-r_2}{2} & \cdots & \cdots & \cdots & \cdots & \cdots & \cdots \\
0 & 0 & \cdots & \cdots & \cdots & \cdots & \cdots & \cdots
\end{pmatrix}
\tag{3.94}
$$

To solve such a system of MN equations by direct elimination methods one needs at least $MN \cdot M^2 = NM^3$ arithmatic operations where M is the bandwidth. For large values of M and N this is quite large and we look for other methods, of solving the set of linear algebraic equations with coefficient matrix Eq. (3.94), particularly iterative methods.

3.8.1 ADI Scheme

An alternative way to solve the above set of MN equations, was proposed by Peaceman and Rachford (1955). In this scheme each time step is subdivided into two half steps. In the first half time step, one space direction, say y, is treated implicitly while the other space direction x is treated explicitly. In the next half step, the roles are reversed and the y-direction is treated explicitly while x-direction is treated implicitly. In operator notation, the steps from n^{th} time level to $(n+1)^{\text{th}}$ time level may be expressed, assuming that the values at level n have been already computed, as

$$\frac{2}{\Delta t}\left(u_{j,k}^{n+\frac{1}{2}} - u_{j,k}^{n}\right) = K\left[\frac{\delta_{xx}}{\Delta x^2}u_{j,k}^{n} + \frac{\delta_{yy}}{\Delta y^2}u_{j,k}^{n+\frac{1}{2}}\right] \tag{3.95}$$

$$\frac{2}{\Delta t}\left(u_{j,k}^{n+1} - u_{j,k}^{n+\frac{1}{2}}\right) = K\left[\frac{\delta_{xx}}{\Delta x^2}u_{j,k}^{n+1} + \frac{\delta_{yy}}{\Delta y^2}u_{j,k}^{n+\frac{1}{2}}\right] \quad j=1, 2, \cdots, M, k=1, 2, \cdots, N \tag{3.96}$$

Use of explicit discretisation suggests such a scheme only to be conditionally stable. But this scheme, known as the *alternating direction implicit* or *ADI scheme*, shows certain unexpected features. The scheme is *unconditionally stable* as we shall prove presently. Further, the scheme is *second order in both time and space*, although the individual component equations are only first order in time and second order in space. Moreover, an unexpectedly low amount of computational labour is required for each time step. These are all highly desirable features. We seldom find them to be present all at the same time in a scheme. For these reasons, it is often the first choice of the user.

Adding Eqs. (3.95) and (3.96) we get

$$u_{j,k}^{n+1} - u_{j,k}^{n} = \frac{K\Delta t}{2}\left[2\frac{\delta_{yy}}{\Delta y^2}u_{j,k}^{n+\frac{1}{2}} + \frac{\delta_{xx}}{\Delta x^2}\left(u_{j,k}^{n} + u_{j,k}^{n+1}\right)\right] \tag{3.97}$$

$$= \frac{K\Delta t}{\Delta y^2}\delta_{yy}u_{j,k}^{n+\frac{1}{2}} + \frac{K\Delta t}{\Delta x^2}\frac{1}{2}\delta_{xx}(u_{j,k}^{n} + u_{j,k}^{n+1})$$

The first term on the right, approximates the time integration by the *mid-point rule* while the second term is the result of *trapezoidal rule* of integration, both of which yield second order accuracy. Since three-point central difference has been used for the space derivatives, these are also second order.

We note that each of the half-time steps Eqs. (3.95) and (3.96) consists of tridiagonal systems, so that one full time step require the solution of two tridiagonal systems. This requires for each k, $(5M-4)$ multiplications and divisions and there are N such k's, so that the total is proportional to MN, that is proportional to the number of internal mesh points. This is quite a small number, which makes the scheme highly attractive.

3.8.2 Splitting and Approximate Factorisation

Both splitting and approximate factorisation reduce the operators involved to factors. When an implicit operator is reduced to factors, it is called *approximate factorisation* while factorisation of explicit operators is known as *splitting*.

Motivation behind approximate factorisation or splitting is to achieve certain advantages, for example, reduction in operational cost or improvement in stability. On the other hand, the order of the error arising on account of splitting or approximate factorisation should not be larger than that of the truncation error of the scheme. The ADI scheme is the earliest of the approximate factorisation schemes. Many other similar factorised schemes have been studied since then, like the AF1, AF2 and AF3 schemes (Baker, 1984; Steger and Baldwin, 1972). In order to explain these schemes better, let us look back to the 2-D implicit Crank–Nicholson scheme in operator notation, Eq. (3.90) which is

$$\left(I - \frac{r_1}{2}\delta_{xx} - \frac{r_2}{2}\delta_{yy}\right) u_{j,k}^{n+1} = \left(I + \frac{r_1}{2}\delta_{xx} + \frac{r_2}{2}\delta_{yy}\right) u_{j,k}^{n} \tag{3.98}$$

δ_{xx}, δ_{yy} representing central difference representation of the second derivatives respectively along x and y directions. On the other hand, the ADI equations (3.95) and (3.96) may be put respectively as

$$(I - \frac{r_2}{2}\delta_{yy})u_{j,k}^{n+\frac{1}{2}} = (I + \frac{r_1}{2}\delta_{xx})u_{j,k}^{n} \tag{3.99}$$

$$(I - \frac{r_1}{2}\delta_{xx})u_{j,k}^{n+1} = (I + \frac{r_2}{2}\delta_{yy})u_{j,k}^{n+\frac{1}{2}} \tag{3.100}$$

The quantities at the half-step with superscript $n + \frac{1}{2}$ may be eliminated, assuming that all the operators commute, yielding

$$\left(I - \frac{r_1}{2}\delta_{xx}\right)\left(I - \frac{r_2}{2}\delta_{yy}\right) u_{j,k}^{n+1} = \left(I + \frac{r_2}{2}\delta_{yy}\right)\left(I + \frac{r_1}{2}\delta_{xx}\right) u_{j,k}^{n} \tag{3.101}$$

We note that

$$\begin{aligned}\left(I - \frac{r_1}{2}\delta_{xx}\right)\left(I - \frac{r_2}{2}\delta_{yy}\right) &= I - \frac{r_1}{2}\delta_{xx} - \frac{r_2}{2}\delta_{yy} + \frac{r_1 r_2}{2}\delta_{xx}\delta_{yy} \\ &= I - \frac{r_1}{2}\delta_{xx} - \frac{r_2}{2}\delta_{yy} + \frac{K^2}{2}\Delta t^2 \frac{\delta_{xx}}{\Delta x^2}\frac{\delta_{yy}}{\Delta y^2}\end{aligned}$$

so that Eq. (3.101) is the same as Eq. (3.98) except for the extra term

$$\frac{K^2}{2}\Delta t^2 \frac{\delta_{xx}}{\Delta x^2}\frac{\delta_{yy}}{\Delta y^2}(u_{j,k}^{n+1} - u_{j,k}^{n}) \approx \frac{K^2}{2}\Delta t^3 \frac{\delta_{xx}}{\Delta x^2}\frac{\delta_{yy}}{\Delta y^2}\frac{\partial u}{\partial t}. \tag{3.102}$$

Since $\dfrac{\partial u}{\partial t}$ is a finite quantity, this term is a third-order quantity, and is of smaller order than the truncation error of the Crank–Nicholson scheme, which is second order. This leads to the idea, successfully used by many authors, that implicit operators which are difficult to solve may be factored into two or more simpler factors approximately, that may be solved much more easily.

3.8.3 Stability of the ADI Scheme

Let us assume trial solutions of the difference equations (3.95) and (3.96) for u^n and $u^{n+\frac{1}{2}}$ in the form

$$u^n = Z^n e^{i k_1 x} e^{i k_2 y} \tag{3.103}$$

$$\text{and} \quad u^{n+\frac{1}{2}} = \xi Z^n e^{i k_1 x} e^{i k_2 y}, \quad i = \sqrt{-1} \tag{3.104}$$

where ξ is a constant, and $Z = e^{\alpha \Delta t}$, the amplification factor. The subscripts corresponding to x and y have been suppressed. Then $x = j\Delta x$, $y = k\Delta y$, and we rewrite u^n as

$$u^n = \rho^n e^{i \beta_1 j} e^{i \beta_2 k}, \text{ with } \beta_1 = k_1 \Delta x \; \beta_2 = k_2 \Delta y, \; \beta_1, \beta_2 \text{ real},$$

α may be a complex quantity. First, we observe that

$$\delta_{xx} u^n = u^n_{j+1,k} - 2u^n_{j,k} + u^n_{j-1,k},$$

delivers

$$\begin{aligned} \delta_{xx} u^n &= Z^n e^{i \beta_2 k} \left\{ e^{i \beta_1 (j+1)} - 2e^{i \beta_1 j} + e^{i \beta_1 (j-1)} \right\} \\ &= Z^n e^{i \beta_2 k} e^{i \beta_1 j}\, 2(\cos \beta_1 - 1) \end{aligned} \tag{3.105}$$

Substituting this in Eq. (3.99) we get

$$\xi Z^n e^{i \beta_1 j} e^{i \beta_2 k} - Z^n e^{i \beta_1 j} e^{i \beta_2 k} =$$

$$K \frac{\Delta t}{2} \left[\frac{\xi}{\Delta x^2} Z^n e^{i \beta_1 j} e^{i \beta_2 k} 2(\cos \beta_1 - 1) + \frac{1}{\Delta y^2} \rho^n e^{i \beta_1 j} e^{i \beta_2 k}\, 2(\cos \beta_2 - 1) \right]$$

which gives on simplification

$$\xi - 1 = \frac{r_1}{2} \xi [2(\cos \beta_1 - 1)] + \frac{r_2}{2} [2(\cos \beta_2 - 1)] \tag{3.106}$$

Further, substituting Eqs. (3.103)–(3.105) in Eq. (3.100) we obtain on simplification,

$$Z - \xi = \frac{r_1}{2} \xi [2(\cos \beta_1 - 1)] + \frac{r_2}{2} Z [2(\cos \beta_2 - 1)] \tag{3.107}$$

Using the abbreviations $\lambda_1 = r_1(\cos \beta_1 - 1)$ and $\lambda_2 = r_2(\cos \beta_2 - 1)$ in Eqs. (3.106)– (3.107), we get the following two equations for the unknowns ξ and Z

$$\xi - 1 = \lambda_1 \xi + \lambda_2,$$

$$\rho - \xi = \lambda_1 \xi + Z \lambda_2$$

Solving these equations simultaneously, follows

$$\xi = \frac{1 + \lambda_2}{1 - \lambda_1}$$

and the amplification factor Z is given by

$$Z = \frac{1+\lambda_1}{1-\lambda_2}\xi = \left(\frac{1+\lambda_1}{1-\lambda_2}\right)\left(\frac{1+\lambda_2}{1-\lambda_1}\right). \tag{3.108}$$

Hence, noting that $\lambda_1 < 0, \quad \lambda_2 < 0$, it follows that $|Z| \leq 1$ and consequently the *ADI scheme is unconditionally stable*. Thus, we see that the ADI scheme has all the desirable features that we expect from a numerical scheme, namely

1. it is an implicit scheme,
2. it is second order accurate in both time and space,
3. it is unconditionally stable, and
4. it has low computational cost, the number of multiplications and divisions required for each time step being proportional to the number of mesh points.

All these features make the ADI scheme a very attractive one, and the scheme has found wide application in CFD (Anderson et al, 1984; Mitchell and Griffiths, 1980).

3.8.4 Program 3.1: adi.f

The following program solves the 2-D heat conduction equation by the ADI-scheme.

```
c......Program 3.1: adi.f.
c..... It solves by ADI-scheme 2-D heat-conduction Eq. u_t=K( u_xx+ u_yy),
c..... initial condition  u(x,y,o)= sin( pi x)sin(pi y), 0<= x,y <= 1,
c......boundary conditions u=0, on all the boundaries.
c......Known exact solution u(x,y,t) = e^(-pi^2 t)*sin(pi x) sin(pi y).
c..... Take delta x= .2=delta y, r_1=1=r_2,KK=1.
c......dth is half time- step. nm=nmax, jm=jmax,km=kmax
c.........................................................
parameter( ix=50, it=50)
dimension u(0:ix,0:ix,0:it),x(0:ix), y(0:ix),t(0:it)
dimension cc(ix), dd(ix),ue(0:ix,0:ix,0:it),aa(ix),bb(ix)
open(unit=12, file='adi.out')
pi=3.1415926
r1=1.0
r2=1.0
kk=1.0
dx=0.125
dy=0.125
dt=r1*dx*dx
dth=0.5*dt
km=8
jm=8
```

```
nm=8
jmm=jm-1
kmm=km-1
do 10 i=0,jm
x(i)=i*dx
y(i)=i*dy
10  continue
c......Initial condition evaluation ......................
do 20 j=0,jm
do 21 k=0,km
u(j,k,0)= sin(pi*x(j))*sin(pi*y(k))
21 continue
20 continue
c.......Main iteration loop for two- half  time-steps.  .....
c.......Formation of the coefficient matrices................
do 40 n=0,nm,2
nh=n+1
do 61 k=1,kmm
do 30 j=1,jmm
cc(j)=-0.5*r1
aa(j)=-0.5*r1
bb(j)=1.+r1
dd(j)=0.5*r2*(u(j,k-1,n)+u(j,k+1,n))+ (1.0 - r2)*u(j,k,n)
30 continue
call strid(aa,bb,cc,dd,jmm)
do 31 j=1,jmm
u(j,k,nh)=dd(j)
31 continue
61 continue
do 62 j=1,jmm
do 35 k=1,kmm
bb(k)=1.+r2
cc(k)=-0.5*r2
aa(k)=-0.5*r2
dd(k)=0.5*r1*(u(j-1,k,nh)+u(j+1,k,nh))+(1.0-r1)*u(j,k,nh)
35 continue
call strid(aa,bb,cc,dd,kmm)
do 32 k=1,kmm
u(j,k,nh+1)=dd(k)
32 continue
62 continue
40 continue
8 format(1x,2i4,2f10.5)
9 format(1x,f10.5)
c......Exact solution computation................
tt=nm*dt
```

```
fac=exp(-pi*pi*tt)
do 110 j=1,jmm
do 110 k=1,kmm
ue(j,k,nm)=fac *sin(pi*x(j))*sin(pi*y(k))
110 continue
write(12,*)'--------------------------------------------------'
write(12,*) 'Output of program adi.f. '
write(12,*)'--------------------------------------------------'
write(12,*) 'dx= ', dx, ' dy= ', dy, ' r1= ', r1,  ' r2= ',r2
write(12,*)'jm= ',jm,' km= ',km,' nm = ',nm,' tt= ',tt,' fac= ',fac
write(12,*)'--------------------------------------------------'
write(12,*) 'Comparison with the exact solution'
write(12,*)'..................................................'
write(12,*) '  j ', '   k ', '  u(j,k,nm) ', ' ue(j,k,nm) '
write(12,*)'..................................................'
do 111 j=1,jmm,2
do 111 k=1,kmm,2
write(12,8) j, k, u(j,k,nm), ue(j,k,nm)
111 continue
j=16
do 112 k=1,kmm
112 continue
close(12)
stop
end

  subroutine strid(a,b,c,d,m)
parameter(ix=50)
dimension a(ix),b(ix),c(ix),d(ix),p(ix)
C  Forward elimination..........................
p(1) = c(1)/b(1)
d(1)=d(1)/b(1)
do 1 j=2,m
jm=j-1
factor= 1./(b(j)-a(jm)*p(jm))
p(j)= c(j)*factor
1 d(j)=(d(j)-a(jm)*d(jm))*factor
C Back-substitution sweep......................
do 2 j=m-1,1,-1
jp=j+1
2 d(j)=d(j)-p(j)*d(jp)
return
end
```

```
Output of program adi.f.
-----------------------------------------------
dx=    0.125 dy=    0.125 r1=    1. r2=    1.
jm=  8 km=  8 nm =  8 tt=   0.125 fac=   0.291212976
-----------------------------------------------
Comparison with the exact solution
...............................................
   j    k   u(j,k,nm)  ue(j,k,nm)
...............................................
    1   1   0.04322    0.04265
    1   3   0.10435    0.10296
    1   5   0.10435    0.10296
    1   7   0.04322    0.04265
    3   1   0.10435    0.10296
    3   3   0.25192    0.24857
    3   5   0.25192    0.24857
    3   7   0.10435    0.10296
    5   1   0.10435    0.10296
    5   3   0.25192    0.24857
    5   5   0.25192    0.24857
    5   7   0.10435    0.10296
    7   1   0.04322    0.04265
    7   3   0.10435    0.10296
    7   5   0.10435    0.10296
    7   7   0.04322    0.04265
-----------------------------------
```

3.9 SUMMARY

Finite difference discretisation of parabolic-type differential equations have been discussed in detail, using a simple model problem of one dimensional heat conduction. Convergence, stability and consistency of the explicit FTCS (forward time and central space) and implicit Crank–Nicholson schemes have been studied. Leap-Frog scheme and DuFort–Frankel schemes are also studied through the same equation. von Neumann stability analysis method has been introduced and used for explicit and implicit finite difference schemes for the simple heat conduction equation. The resulting tri-diagonal system of algebraic equations, solved by Thomas algorithm. It is found that the FTCS scheme is conditionally stable and conditionally convergent, Crank–Nicholson implicit and alternative direction implicit (ADI) schemes are unconditionally stable. In particular, the following items have been discussed.

1. 1-D heat conduction problem.
2. Explicit FTCS (forward time central space) scheme.
3. Consistency, convergence.
4. von Neumann stability analysis.
5. Crank–Nicholson implicit scheme.
6. Consistency, convergence and unconditional stability.
7. Leap–Frog scheme, unconditionally unstable.
8. DuFort–Frankel scheme is explicit and unconditionally stable, although not consistent under certain conditions.
9. ADI method for 2-D equations.
10. Splitting and approximate factorisation.
11. Unconditional stability.

The following program in Fortran (C program in Appendix) with test results presented.

1. adi.f: solves by the ADI scheme 2-D heat conduction equation.

3.10 KEY TERMS

ADI (alternating direction implicit scheme)
Amplification factor
Artificial viscosity,
Boundary condition
Convergence of FTCS
Crank–Nicholson scheme
Derivative boundary condition
Diffusion
Dispersive error
Dissipative error
Explit scheme
DuFort–Frankel scheme
Elliptic-type equation
Finite difference grid
FTCS (forward time central space) scheme
Fourier stability analysis
Implicit scheme
Leap–Frog scheme
Matrix stability analysis
Maximum principle
Mesh size
Midpoint rule
Three-level fully implicit scheme
Sparse system
Stiff differential equation
Thomas algorithm
Tridiagonal matrix
Trapezoidal rule
Truncation error
Unconditional stabilty
von Neumann stability analysis

3.11 EXERCISE 3

3.1. Establish the following forward difference representation for the first derivative:

(a) $\left.\frac{\partial u}{\partial x}\right|_j = \frac{1}{2\Delta x}\left[-3u_j + 4u_{j+1} - u_{j+2}\right] + \frac{1}{3}\Delta x^2 u_{xxx}.$

(b) Establish the representation

$$\left.\frac{\partial^2 u}{\partial x^2}\right|_j = \frac{1}{12\Delta x}\left[u_{j-2} - 8u_{j-1} + 8u_{j+1} - u_{j+2}\right] + \frac{\Delta x^4}{30}u_{xxxxx}.$$

3.2. Establish the following operator relations for the forward difference $\vec{\delta}_x$, backward difference (upwind) $\overline{\delta}_x$ and central difference δ_x:

$$\text{(a) } \vec{\delta}_x + \overline{\delta}_x = \delta_x;\ \text{(b) } \vec{\delta}_x - \overline{\delta}_x = \Delta x \delta_{xx};\ \text{(c) } \vec{\delta}_x \overline{\delta}_x = \delta_{xx}.$$

3.3. Solve the heat-conduction problem: $u_t = Ku_{xx}$, $K = \text{const.} > 0$. under initial condition

$$u(x,0) = \begin{cases} x, 0 \le x \le 0.5, \\ (1-x), 0.5 \le x \le 1. \end{cases}$$

and boundary conditions

$$u(0,x) = 0 = u(1,x), \text{ for all time.}$$

Take (a) $K = 1, r = \frac{1}{2}, \Delta x = \frac{1}{4}$; (b) $K = 1, r = \frac{1}{6}, \Delta x = \frac{1}{4}$; (c) $K = 1, r = \frac{1}{6}, \Delta x = \frac{1}{10}$. Compute by FTCS scheme, the solution for the first four-time steps and compare the final step results with that of the exact solution.

3.4. Do the same problem as in the previous one, by FTCS scheme, with initial condition $u(x,0) = 0.8 \sin \pi x, 0 \le x \le 1$. Other conditions remain the same.

3.5. Do the same problem as in the previous one, by FTCS scheme, with initial condition $u(x,0) = \frac{1}{1+x^2}$, $0 \le x \le 1$. Compute only the first two steps for $r = 0.5$, and $r = \frac{1}{6}$, $\Delta x = 0.20$. Other conditions remain the same.

3.6. Do the same problem as in the previous one, by FTCS scheme, with initial condition $u(x,0) = \frac{2}{1+x}$ $0 \le x \le 1$. Other conditions remain the same.

3.7. Solve the heat-conduction problems 3.3 and 3.4, by FTCS scheme, with derivative boundary conditions:

(a) $u_x|_{x=0} = 0, \quad u_x|_{x=1} = 0.$

(b) $u_x|_{x=0} = u, \quad u_x|_{x=1} = -u.$

Compute the solution for the first two steps only.

3.8. Write a subroutine subprogram to solve a tridiagonal system of linear algebraic equations.

3.9. Solve by Crank–Nicholson scheme the heat conduction equation, respectively with initial conditions

(a) $u(x,0) = \begin{cases} x, 0 \le x \le 0.5, \\ (1-x), 0.5 \le x \le 1. \end{cases}$

(b) $u(x,0) = 0.8 \sin \pi x, \quad 0 \le x \le 1,$

and boundary conditions

$$u(0,t) = 0 = u(1,t), \text{ for all time.}$$

Taking $K = 1$, $r = 1$, $\Delta x = \frac{1}{5}$, compute the solution for the first two time steps.

3.10. Taking boundary condition on $x = 1$ as $u(1,t) = 1$ for all time, compute the solution of the previous problem by Crank–Nicholson scheme. Other conditions remain unaltered.

3.11. Do problems 3.5 and 3.6 by Crank–Nicholson scheme with $r = 1$, and $r = \frac{1}{6}$, respectively. Use the subroutine developed earlier, to solve these problems. Compute solution for the first four steps.

3.12. Show that the *leap–Frog* scheme (central-time-central-space)

$$\frac{1}{2\Delta t}\left[u_j^{n+1} - u_j^{n-1}\right] = \frac{1}{\Delta x^2}\left[u_{j-1}^n - 2u_j^n + u_{j+1}^n\right], \tag{3.109}$$

is second-order accurate in both time and space. Show further, by von Neumann stability analysis, that this scheme is unconditionally unstable; in other words the scheme is practically useless.

3.13. Convert the subprogram strid in Program 3.1 adi.f, into a FORTRAN program STRID.f for use as a tridiagonal system solver and verify the results of the worked out Example 3.3.

3.14. If the differential equation

$$u_t = \alpha\left(u_{xx} + u_{yy}\right), \alpha = \text{const.} > 0$$

be discretised by the FTCS scheme, in operator notation,

$$u_{j,k}^{n+1} = u_{j,k}^n + \alpha\Delta t\left(\frac{\delta_{xx}}{\Delta x^2} + \frac{\delta_{yy}}{\Delta y^2}\right)u_{j,k}^n,$$

show that for stability

$$\alpha\Delta t\left(\frac{1}{\Delta x^2} + \frac{1}{\Delta y^2}\right) \le \frac{1}{2}.$$

4

Equations of Hyperbolic Type

Finite difference method for model partial differential equations of parabolic type have been discussed in the previous chapter. In the present chapter we discuss finite difference methods for model equations of hyperbolic type. We begin our discussion with the simplest equation of hyperbolic type, namely, the *linear 1-D wave equation,* also known as *the convection equation.* Hyperbolic system of conservation laws, which occur frequently in the study of Euler equations of fluid dynamics is discussed next. The convection–diffusion equation is an equation of much importance in fluid dynamics. Although it is of parabolic type, it has features different from parabolic type heat-conduction equation and knowledge of methods for convection equation is useful in its study. So, we also briefly discuss in this chapter, the convection–diffusion equation.

4.1 INTRODUCTION

Linear and nonlinear wave propagation problems are governed by partial differential equations of hyperbolic type. One frequently encounters equations of hyperbolic type in CFD. The unsteady Eulers equations governing flow of an inviscid fluid are a system of first order nonlinear hyperbolic type equations. The well-known linear one-dimensional wave equation

$$u_{tt} = c^2 u_{xx}, \; c = \text{ const.} > 0, \tag{4.1}$$

is the governing equation for sound wave propagation. The model equation of hyperbolic type is the first-order equation

$$\frac{\partial u}{\partial t} + c\frac{\partial u}{\partial x} = 0, \; c = \text{ const.} > 0, \tag{4.2}$$

commonly known as the *first-order linear wave equation or the convection equation*. It is also known as the *advection equation* in meteorology. It represents convection of u with constant speed c. The importance of this simple equation may be seen from the fact that almost all problems in fluid dynamics contain this operator, although with nonuniform velocity.

Differentiating Eq. (4.2) with respect to t we get, assuming $u(x, t)$ to be sufficiently smooth,

$$\begin{aligned}\frac{\partial^2 u}{\partial t^2} &= -c\frac{\partial}{\partial t}\left(\frac{\partial u}{\partial x}\right) = -c\frac{\partial}{\partial x}\left(\frac{\partial u}{\partial t}\right)\\ &= -c\frac{\partial}{\partial x}\left(-c\frac{\partial u}{\partial x}\right) = c^2\frac{\partial^2 u}{\partial x^2},\end{aligned}$$

which is the second order wave equation (4.1). The convection equation (4.2) is taken as the model equation of hyperbolic type in view of its simplicity. We study the finite difference methods for computational solution of it in the present chapter.

Exact solution of Eq. (4.2) with initial condition

$$u(x,0) = F(x), \quad \text{in} \quad -\infty < x < \infty,$$

where F is a differentiable function, is given by

$$u(x,t) = F(x - ct) \tag{4.3}$$

As pointed out in Chapter 1, it represents a wave-form $F(x)$ propagating with constant speed c in a single direction. The slope of the direction is given by

$$\frac{\mathrm{d}x}{\mathrm{d}t} = c. \tag{4.4}$$

The characteristics of Eq. (4.2) are the straight lines $x - ct = \text{const.}$ which are right running, if we look in the positive t-direction.

We begin our study of finite difference schemes for hyperbolic equations with some of the simplest explicit schemes.

4.2 EXPLICIT SCHEMES

4.2.1 FTCS Scheme

Let us consider first the forward-time central-space (FTCS) scheme for the first-order wave equation (4.2). Using the notation $x = j\Delta x, t = n\Delta t, u(x,t) = u_j^n$, the FTCS scheme is

$$\frac{1}{\Delta t}\left(u_j^{n+1} - u_j^n\right) + O(\Delta t) = -\frac{c}{2\Delta x}\left(u_{j+1}^n - u_{j-1}^n\right) + O(\Delta x^2). \tag{4.5}$$

The truncation error of the scheme is $O(\Delta t,\ \Delta x^2)$. Because of the first-order accuracy in time, it is considered to be a first-order scheme, equal to the order of the lower accuracy between time and space truncation errors. Let us study its stability by von Neumann method. For the heat conduction equation in Chapter 3, the FTCS scheme was found to be conditionally stable.

In order to study the stability, let us take

$$u_j^n = \exp(\alpha n\Delta t)\exp(ik_m j\Delta x) = \exp(\alpha n\Delta t)\exp(i\beta j), \quad i = \sqrt{-1}, \tag{4.6}$$

where k_m denotes the wave number, $\beta = k_m\Delta x$, β is real, α may be complex. Substituting Eq. (4.6) in Eq. (4.5), omitting the truncation error terms and simplifying we get

$$G = \exp(\alpha\Delta t) = \text{ amplification factor } = 1 - i\nu\sin\beta \tag{4.7}$$

where

$$\nu = \frac{c\Delta t}{\Delta x}. \tag{4.8}$$

So, $|G|^2 = 1 + \nu^2 \sin^2 \beta > 1$ for all ν. Hence the scheme is unconditionally unstable. Thus the FTCS method is of no use in practice. The quantity ν is known as the *Courant number.*

4.2.2 FTFS Scheme

We consider next the forward-time forward-space (FTFS) scheme. Omitting the truncation error terms, it is

$$\frac{u_j^{n+1} - u_j^n}{\Delta t} = -c\frac{u_{j+1}^n - u_j^n}{\Delta x},$$

which may be simplified, using $\nu = c\Delta t/\Delta x$, to

$$u_j^{n+1} = (1+\nu)u_j^n - \nu u_{j+1}^n. \tag{4.9}$$

This is an explicit scheme with truncation error $O(\Delta t,\ \Delta x)$, that is, it is a first-order explicit scheme. For studying von Neumann stability, let us substitute a trial solution Eq. (4.6) in Eq. (4.9) to yield on simplification, the value of the amplification factor $G = \exp(\alpha\Delta t)$ as

$$G = \exp(\alpha\Delta t) = (1 + \nu - \nu\cos\beta) - i\nu\sin\beta \tag{4.10}$$

It follows then

$$|G|^2 = 1 + 2\nu(1+\nu)(1-\cos\beta) > 1, \quad \nu > 0.$$

Hence, the FTFS scheme is also unconditionally unstable and is of no use.

4.2.3 Upwind Scheme: First Order

Among the first-order explicit schemes, we consider next the forward-time backward-space (FTBS) or the first-order upwind scheme which is of considerable interest. Neglecting truncation error terms, it is given by

$$\frac{u_j^{n+1} - u_j^n}{\Delta t} = -c\frac{u_j^n - u_{j-1}^n}{\Delta x}, \quad c = \text{const.} > 0, \tag{4.11}$$

which simplifies, using $\nu = c\Delta t/\Delta x$, to

$$u_j^{n+1} = u_j^n - \nu(u_j^n - u_{j-1}^n) = (1-\nu)u_j^n + \nu u_{j-1}^n \tag{4.12}$$

This is again an explicit scheme with truncation error $O(\Delta t,\ \Delta x)$, that is, it is a first-order scheme.

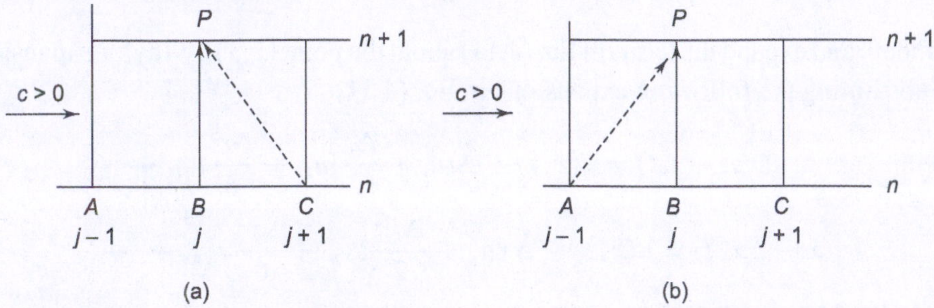

Figure 4.1 Influence regions of different schemes; (a) FTFS, (b) Upwind scheme

Using the trial solution Eq. (4.6) in Eq. (4.12) we get on simplification, the amplification factor $G = \exp(\alpha \Delta t)$ as

$$G = 1 - \nu + \nu \exp(-i\beta) = (1 - \nu + \nu \cos\beta) - i \sin\beta \tag{4.13}$$

So, on simplification,

$$|G|^2 = 1 - 2\nu(1 - \nu)(1 - \cos\beta)$$

Noting that $1 - \cos\beta \geq 0$ and that $2\nu(1 - \nu) \geq 0$ if and only if $0 \leq \nu \leq 1$, it follows that the upwind scheme Eq. (4.12) is stable if and only if

$$0 \leq c\frac{\Delta t}{\Delta x} \leq 1 \tag{4.14}$$

The quantity $c\dfrac{\Delta t}{\Delta x}$, is known as the *Courant number,* and the condition (4.14) the *Courant– Friedrichs–Lewy (or CFL) condition.* The CFL condition Eq. (4.14) means that the propagating wave or signal should not travel more than one mesh length Δx in time Δt, in order to maintain stability.

It may be noted that the upwind scheme Eq. (4.12) expresses the solution at the mesh point $P(j, n + 1)$ in terms of values at the upwind mesh point $A(j - 1, n)$ of the previous time level (Fig. 4.1(b)). In other words, a forward propagating wave, (that is, $c > 0$,) carries information from an upwind point A to the desired point P (Fig. 4.1(b)), which is quite natural to understand. For the FTFS scheme Eq. (4.9) it is just the reverse. It is required to carry information from the downstream point $C(j + 1, n)$ to the point P, (Fig. 4.1(a)). This goes against the nature of the convection equation and leads to unconditional instability.

4.2.4 Upwind Scheme: Modified Equation

It is quite illuminating to study the truncation error term and to derive the modified equation for the upwind scheme. For this, let us assume $u(x, t)$ to be sufficiently smooth and expand the terms in Eq. (4.11) about the point (x, t) by Taylor expansion. Substituting the following expansions in Eq. (4.11)

$$u(x, t + \Delta t) = u(x, t) + \Delta t u_t + \frac{\Delta t^2}{2} u_{tt} + \frac{\Delta t^3}{6} u_{ttt} + \cdots$$

$$u(x - \Delta x, t) = u(x, t) - \Delta x u_x + \frac{\Delta x^2}{2} u_{xx} - \frac{\Delta x^3}{6} u_{xxx} + \cdots,$$

it follows on simplification

$$u_t + c u_x = -\frac{\Delta t}{2} u_{tt} - \frac{\Delta t^2}{6} u_{ttt} + \frac{c\Delta x}{2} u_{xx} - \frac{c\Delta x^2}{6} u_{xxx} + \cdots \tag{4.15}$$

Noting that,
Truncation Error = True Value – Approximate Value,
it follows that the truncation error (abbreviated T.E.) is

$$\begin{aligned} \text{T.E.} &= (u_t + c u_x) - \left[\frac{1}{\Delta t}(u_j^{n+1} - u_j^n) + \frac{c}{\Delta x}(u_j^n - u_{j-1}^n) \right] \\ &= -\frac{\Delta t}{2} u_{tt} - \frac{\Delta t^2}{6} u_{ttt} + \frac{c\Delta x}{2} u_{xx} - \frac{c\Delta x^2}{6} u_{xxx} + \cdots \end{aligned}$$

This shows that the righthand side of Eq. (4.15) gives the T.E., which is $O(\Delta t, \Delta x)$, as already stated. The modified equation is derived by eliminating the time-derivatives in Eq. (4.15) using Taylor's expansion (the differential equation should not be used for this purpose). For a detailed derivation and discussion, the book by Anderson et.al. (1984) and the paper by Warming and Hyett (1974) may be consulted. Differentiating both sides of Eq. (4.15) with respect to time t, and again with respect to x, we get

$$u_{tt} = -c u_{xt} - \frac{\Delta t}{2} u_{ttt} - \frac{\Delta t^2}{6} u_{tttx} + \frac{c\Delta x}{2} u_{xxt} - \frac{c\Delta x^2}{6} u_{xxxt} + \cdots \tag{4.16}$$

$$u_{tx} = -c u_{xx} - \frac{\Delta t}{2} u_{ttx} - \frac{\Delta t^2}{6} u_{tttx} + \frac{c\Delta x}{2} u_{xxx} - \frac{c\Delta x^2}{6} u_{xxxx} + \cdots \tag{4.17}$$

Multiplying Eq. (4.17) by $-c$ and adding to Eq. (4.16), gives

$$u_{tt} = c^2 u_{xx} + \Delta t \left(-\frac{u_{ttt}}{2} + \frac{c}{2} u_{ttx} + O(\Delta t) \right) + \Delta x \left(\frac{c}{2} u_{xxt} - \frac{c^2}{2} u_{xxx} + O(\Delta x) \right) \tag{4.18}$$

Further, differentiating Eq. (4.18) respectively with respect to x and t and Eq. (4.17) with respect to x, yields the following relations

$$u_{ttt} = -c^3 u_{xxx} + O(\Delta t, \Delta x), \quad u_{ttx} = c^2 u_{xxx} + O(\Delta t, \Delta x),$$
$$u_{xxt} = -c u_{xxx} + O(\Delta t, \Delta x).$$

Substituting these relations in Eq. (4.15) delivers on simplification *the modified equation*

$$u_t + cu_x = \frac{c}{2}(1-\nu)\Delta x u_{xx} - \frac{c}{6}\Delta x^2(2\nu^2 - 3\nu + 1)u_{xxx} + \cdots \tag{4.19}$$

The right hand side of equation (4.19) represents the truncation error term of the first-order upwind scheme. It is emphasised that *this is the equation that is actually solved by numerical means.*

For the Courant number $\nu = c\dfrac{\Delta t}{\Delta x} = 1$, the truncation error vanishes and from Eq. (4.12) we see that the scheme reduces to

$$u_j^{n+1} = u_{j-1}^n$$

which is equivalent to solving the first-order wave equation exactly using the method of characteristics. For $0 < \nu < 1$, the leading term in the T.E., on the right of Eq. (4.19) is αu_{xx}, where

$$\alpha = \frac{c}{2}(1-\nu)\Delta x > 0, \tag{4.20}$$

is a small positive quantity. This term acts essentially like the viscous term $\dfrac{\mu}{\rho}\dfrac{\partial^2 u}{\partial x^2}$ in the right hand side of the 1-D unsteady Navier–Stokes equation

$$\frac{\partial u}{\partial t} + u\frac{\partial u}{\partial x} = \frac{\mu}{\rho}\frac{\partial^2 u}{\partial x^2} + \cdots,$$

μ being the dynamic viscosity and ρ the density of the fluid. It is called *artificial viscosity, whose main action is to reduce sharp gradients in a flow field.* Since this term is implicitly present in the truncation error term, it is called *implicit artificial viscosity* as against other schemes where such artificial viscosity terms are explicitly added to capture appropriate physical effects. For example, in order to damp out certain undesirable effects such terms may be used. The presence of the odd-order derivative term u_{xxx} in the truncation error is a *dispersive error*, responsible for distortion of wave shapes.

To gain more insight into the stepwise growth of round-off errors, let us look into the amplification factor G given by Eq. (4.13) and express the complex quantity G in polar form as

$$G = |G|\exp(i\theta)$$

where $|G|$ denotes the amplitude and θ denotes the phase. Then, as already seen in Eq. (4.13), the amplitude and phase of the first-order upwind scheme is given by

$$|G| = \{1 - 2\nu(1-\nu)(1-\cos\beta)\}^{\frac{1}{2}} \tag{4.21}$$

and

$$\theta = \tan^{-1}\left(\frac{-\nu \sin\beta}{1-\nu+\nu\cos\beta}\right) \tag{4.22}$$

both of which depend on the CFL number ν and the wave number β. By comparing $|G|$ and θ, with that of the exact solution we may determine the amplitude and phase errors. For this, the amplification factor of the exact solution should be determined. Let $\bar{u}(x, t)$ denote the exact solution of the wave equation (4.2).

Then using, as before, a trial solution

$$\bar{u}(x, t) = e^{\alpha t} e^{i\,k_m x}, \quad k_m = \text{wave number}, \quad i = \sqrt{-1},$$

in the first-order wave equation (4.2) we get

$$\alpha e^{\alpha t} e^{i\,k_m x} + c.e^{\alpha t} i\,k_m\, e^{i\,k_m x} = 0$$

which may be simplified to

$$\alpha + c\,i\,k_m = 0 \quad \Longrightarrow \quad \alpha = -i\,c\,k_m.$$

Thus, the exact solution is

$$\bar{u}(x, t) = e^{i\,k_m(x-ct)}.$$

The exact amplification factor is

$$\bar{G} = \frac{\bar{u}(x, t+\Delta t)}{\bar{u}(x, t)} = \frac{e^{i\,k_m(x-c(t+\Delta t))}}{e^{i\,k_m(x-ct)}} = e^{-i\,k_m\,c\,\Delta t} \tag{4.23}$$

If we rewrite the amplification factor $\bar{G}$ in the polar form

$$\bar{G} = |\bar{G}|e^{i\,\bar{\theta}},$$

then the amplitude of the exact amplification factor is $|\bar{G}| = 1$. The exact phase

$$\bar{\theta} = -c\,k_m\,\Delta t = -c\frac{\Delta t}{\Delta x}\cdot k_m\,\Delta x = -\nu\beta, \tag{4.24}$$

where $\nu = c\dfrac{\Delta t}{\Delta x}$ and $\beta = k_m \Delta x$, is proportional to the wave number k_m.

The error in the amplitude of the amplification factor compared to that of the exact solution is often called *amplitude error*. This is also known as *dissipative error*. On the other hand the relative error in the phase $\theta/\bar{\theta}$ is called *phase error*, known as *dispersive error*. If, for a particular value of the wave number β, the relative phase error $\theta/\bar{\theta} > 1$, then the numerical solution is said to have a *leading phase error*. The opposite case, in which $\theta/\bar{\theta} < 1$ is one of *lagging phase error*.

The relative amplitude and phase error of the first-order upwind scheme is shown in Figs. 4.2(a) and 4.2(b). It may be seen from Fig. 4.2(b), that the first-order upwind scheme has a leading phase error for $0.5 < \nu < 1$ and a lagging phase error for $\nu < 0.5$.

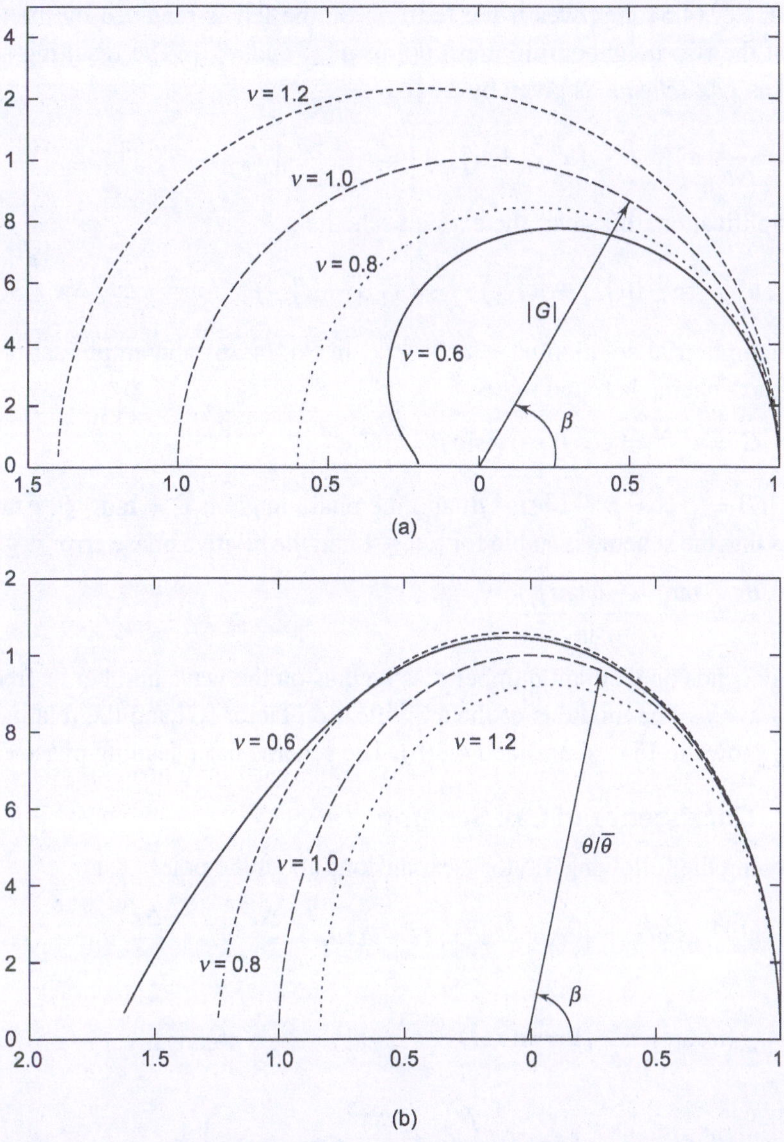

Figure 4.2 Upwind scheme (a) relative amplitude errror and (b) relative phase error for different values of ν.

4.2.5 The Lax Scheme

It was observed by Lax (1954) that the stability behaviour of the explicit FTCS scheme, Eq. (4.5) improves if the term u_j^n on the left is replaced by the average value at the two neighbouring mesh points u_{j-1}^n and u_{j+1}^n. The resulting scheme, known as *Lax scheme*, is given by

$$\frac{1}{\Delta t}\left[u_j^{n+1} - \frac{1}{2}\left(u_{j-1}^n + u_{j+1}^n\right)\right] = -\frac{c}{2\Delta x}\left[u_{j+1}^n - u_{j-1}^n\right] \tag{4.25}$$

On simplification, this gives the explicit scheme

$$u_j^{n+1} = \frac{1}{2}\left(u_{j-1}^n + u_{j+1}^n\right) - \frac{\nu}{2}\left(u_{j+1}^n - u_{j-1}^n\right), \quad \nu = c\Delta t/\Delta x \tag{4.26}$$

Introducing a trial solution $u_j^n = e^{\alpha n \Delta t} e^{i\beta j}$, in Eq. (4.26), the amplification factor of the Lax scheme is found to be

$$G = e^{\alpha \Delta t} = \cos\beta - i\nu\sin\beta = |G|e^{i\theta} \tag{4.27}$$

where $|G| = \sqrt{\cos^2\beta + \nu^2\sin^2\beta}$, and the phase angle is $\theta = \tan^{-1}(-\nu\tan\beta)$. It follows that the scheme is stable for $|\nu| \leq 1$ and the relative phase error is given by

$$\frac{\theta}{\overline{\overline{\theta}}} = \frac{\tan^{-1}(-\nu\tan\beta)}{-\beta\nu} \tag{4.28}$$

which depends on Courant number ν as well as on the wave number or frequency of the wave β. The modulus of the amplification factor $|G|$ and the relative phase error is shown in Figs. 4.3(a) and 4.3(b). The scheme has a leading phase error.

4.2.6 Consistency of Lax Scheme

Introducing the following Taylor's expansions about the point (x, t)

$$u_j^{n+1} = u(x, t+\Delta t) = u(x,t) + \Delta t u_t + \frac{\Delta t^2}{2}u_{tt} + \frac{\Delta t^3}{6}u_{ttt} + \cdots, \tag{4.29}$$

$$\frac{1}{2}\left(u_{j+1}^n + u_{j-1}^n\right) = u(x,t) + \frac{\Delta x^2}{2}u_{xx} + \frac{\Delta x^4}{24}u_{xxxx} + \cdots, \tag{4.30}$$

$$\frac{\nu}{2}\left(u_{j+1}^n - u_{j-1}^n\right) = c\Delta t\left(u_x + \frac{\Delta x^2}{6}u_{xxx} + \cdots\right) \tag{4.31}$$

in Eq. (4.26), we obtain on simplification

$$u_t + cu_x = -\frac{\Delta t}{2}u_{tt} + \frac{\Delta x^2}{2\Delta t}u_{xx} - \frac{\Delta t^2}{6}u_{ttt} - \frac{c\Delta x^2}{6}u_{xxx} + \cdots \tag{4.32}$$

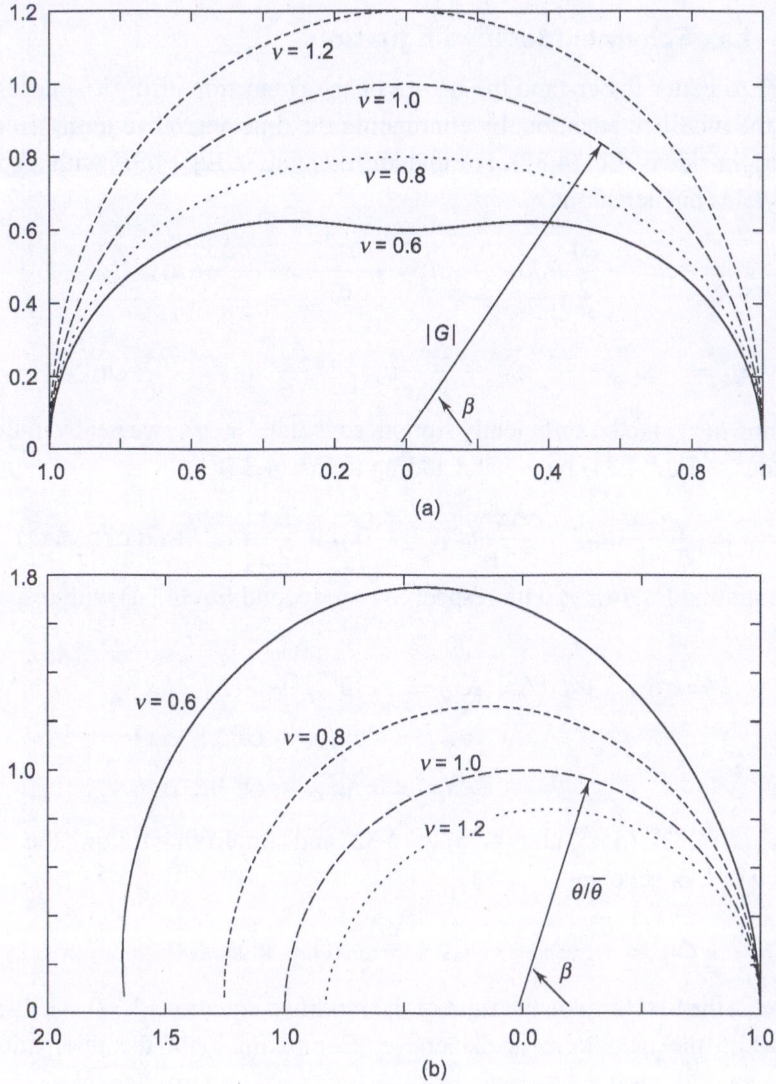

Figure 4.3 Lax scheme (a) modulus of amplification (b) relative phase error.

The quantity on the right side gives the truncation error of the Lax Scheme. It shows that the truncation error is $O(\Delta t, \frac{\Delta x^2}{\Delta t})$. As $\Delta x \to 0,\ \Delta t \to 0,\ \dfrac{\Delta x^2}{\Delta t}$ may not tend to zero, so that the scheme is not uniformly consistent. However, for a fixed value of the Courant number $\nu = c\Delta t/\Delta x$, the scheme is consistent.

4.2.7 Lax Scheme: Modified Equation

In order to better understand the nature of the truncation error, it is necessary to derive the modified equation, by eliminating the time derivative terms u_{tt} and u_{ttt} on the right side of Eq. (4.32). For this, differentiating Eq. (4.32) with respect to t and x yields the equations

$$u_{tt} = -cu_{xt} - \frac{\Delta t}{2} u_{ttt} + \frac{\Delta x^2}{2\Delta t} u_{xxt} - \frac{\Delta t^2}{6} u_{ttt} - \frac{c\Delta x^2}{6} u_{xxxt} + \cdots, \tag{4.33}$$

$$u_{tx} = -cu_{xx} - \frac{\Delta t}{2} u_{ttx} + \frac{\Delta x^2}{2\Delta t} u_{xxx} - \frac{\Delta t^2}{6} u_{ttx} - \frac{c\Delta x^2}{6} u_{xxxx} + \cdots \tag{4.34}$$

Assuming $u(x, t)$ to be sufficiently smooth so that $u_{xt} = u_{tx}$, we get by multiplying both sides of Eq. (4.34) by $-c$ and adding to Eq. (4.33)

$$u_{tt} = c^2 u_{xx} + \frac{c\Delta t}{2} u_{ttx} - \frac{c\Delta x^2}{2\Delta t} u_{xxx} - \frac{\Delta t}{2} u_{ttt} + \frac{\Delta x^2}{2\Delta t} u_{xxt} + O(\Delta t^2, \Delta x^2) \tag{4.35}$$

Differentiating Eq. (4.35) with respect to t, and x and Eq. (4.34) with respect to x, we get

$$\begin{aligned} u_{ttt} &= c^2 u_{xxt} + O(\Delta t, \Delta x) = -c^3 u_{xxx} + O(\Delta t, \Delta x) \\ u_{txx} &= -cu_{xxx} + O(\Delta t, \Delta x) \\ u_{ttx} &= c^2 u_{xxx} + O(\Delta t, \Delta x) \end{aligned} \tag{4.36}$$

Using Eqs. (4.33), (4.35) and (4.36) in (4.32) and simplifying follows the modified equation of Lax scheme

$$u_t + cu_x = \frac{c\Delta x}{2} \frac{1 - \nu^2}{\nu} u_{xx} + \frac{c\Delta x^2}{3} (1 - \nu^2) u_{xxx} + \cdots \tag{4.37}$$

For $\nu < 1$, the first term on the right of the modified equation (4.37) is a dissipative error while the next term is dispersive. Comparing with the dissipation term $\frac{c}{2}(1 - \nu)\Delta x\, u_{xx}$ of the first-order upwind scheme Eq. (4.19), it may be noted that Lax scheme introduces relatively large dissipation error.

4.2.8 The Leap-Frog Scheme

The leap-frog scheme is obtained by representing both the time and space derivatives in Eq. (4.2) by central difference, which yields

$$\frac{u_j^{n+1} - u_j^{n-1}}{2\Delta t} = -c \frac{u_{j+1}^n - u_{j-1}^n}{2\Delta x} \tag{4.38}$$

Using the Courant number $\nu = c\Delta t/\Delta x$, this equation simplifies to

$$u_j^{n+1} = u_j^{n-1} - \nu \left(u_{j+1}^n - u_{j-1}^n \right) \tag{4.39}$$

This is an explicit, three-time-level (namely, involving time-levels $n-1$, n and $n+1$) scheme. It is *second order accurate* in both time and space. To study its von Neumann stability we use, as before a trial solution

$$u_j^n = e^{\alpha n \Delta t} e^{i \beta j}, \quad i = \sqrt{-1}, \quad \beta \text{ real}, \ \alpha \text{ complex},$$

in Eq. (4.38), which yields the amplification factor $G = e^{\alpha \Delta t}$ as solution of the quadratic equation

$$\left(e^{\alpha \Delta t}\right)^2 + 2 i \nu \sin \beta e^{\alpha \Delta t} - 1 = 0$$

Solving this quadratic equation, we get

$$G = e^{\alpha \Delta t} = -i \nu \sin \beta \pm \sqrt{1 - \nu^2 \sin^2 \beta} \tag{4.40}$$

The amplification factor depends on the Courant number ν and the wave number or frequency β. Further, the modulus of the amplification factor is given by

$$|G|^2 = \nu^2 \sin^2 \beta + 1 - \nu^2 \sin^2 \beta = 1$$

so that the scheme is *unconditionally stable*. This is a little surprising in view of our experience with the parabolic type heat-conduction equation for which it is unconditionally unstable. A little calculation shows that the modified equation for the leap-frog scheme is

$$u_t + c u_x = -\frac{c \Delta x^2 (1 - \nu^2)}{6} u_{xxx} - \frac{c \Delta x^4}{120} (9\nu^4 - 10\nu^2 + 1) u_{xxxxx} + \cdots \tag{4.41}$$

which shows that the scheme is second order accurate. It is worth noting that only odd-order derivatives appear in the truncation error terms on the right of Eq. (4.41) indicating that the scheme is only dispersive and not dissipative. As a consequence, small errors for example in boundary and/or initial conditions would not be damped out. Extra artificial viscosity may be required to be added explicitly. Further, it is a little inconvenient that the solution requires prescription of initial conditions at two time-levels. This difficulty is circumvented by using a suitable two-time-level scheme to start with. Moreover, note that u_j^{n+1} does not depend on the value of u_j^n, but depends on u_j^{n-1}. This leads to two sets of solutions developing independently, which is highly undesirable. The modulus of the amplification factor and relative phase errors are shown in Figs. 4.4(a) and 4.4(b).

4.3 LAX–WENDROFF SCHEME AND VARIANTS

In this section, we consider a few second-order schemes which have been widely studied. Among these, the Lax–Wendroff scheme, Lax and Wendroff (1960), is a basic work. We derive it here for the first-order linear wave equation.

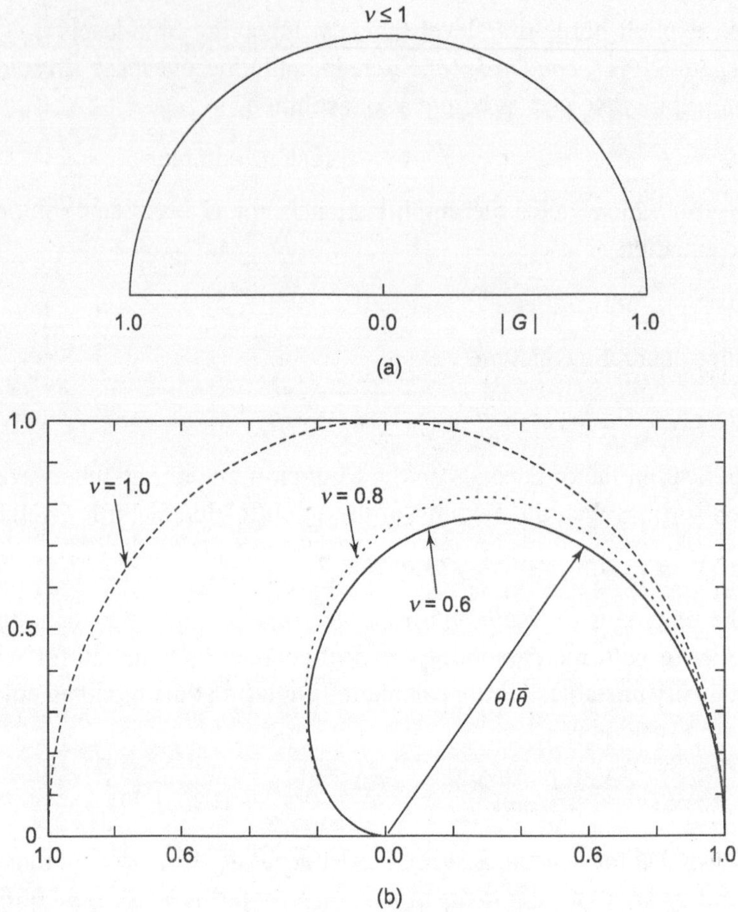

Figure 4.4 Leap-frog scheme (a) modulus of amplification (b) relative phase error.

Let $u(x, t)$ be sufficiently smooth. Then by Taylor's expansion about the point (x, t), we have keeping terms up to the second order

$$u(x, t + \Delta t) = u(x, t) + \Delta t u_t + \frac{\Delta t^2}{2} u_{tt} \tag{4.42}$$

The time-derivatives u_t and u_{tt} are replaced by space-derivatives using the partial differential equation. Noting that

$$u_t = -cu_x,$$

and

$$u_{tt} = \frac{\partial}{\partial t}(u_t) = \frac{\partial}{\partial t}(-c\frac{\partial u}{\partial x}) = -c\frac{\partial}{\partial x}(\frac{\partial u}{\partial t}) = -c\frac{\partial}{\partial x}(-c\frac{\partial u}{\partial x}) = c^2 u_{xx},$$

we have from Eq. (4.42)

$$u(x, t + \Delta t) = u(x, t) - c\Delta t u_x + c^2 \frac{\Delta t^2}{2} u_{xx}.$$

Using finite difference notation and representing u_x and u_{xx} by the corresponding second-order central difference, *the Lax–Wendroff scheme* is obtained

$$u_j^{n+1} = u_j^n - c\Delta t \frac{u_{j+1}^n - u_{j-1}^n}{2\Delta x} + \frac{c^2 \Delta t^2}{2} \frac{u_{j+1}^n - 2u_j^n + u_{j-1}^n}{\Delta x^2}. \tag{4.43}$$

Introducing the Courant number $\nu = c\Delta t / \Delta x$, Eq. (4.43) may be expressed as

$$u_j^{n+1} = u_j^n - \frac{\nu}{2}(u_{j+1}^n - u_{j-1}^n) + \frac{\nu^2}{2}(u_{j+1}^n - 2u_j^n + u_{j-1}^n), \tag{4.44}$$

which is an explicit scheme.

For studying stability, as before let us take a trial solution $u_j^n = e^{\alpha n \Delta t} e^{i\beta j}$. Substituting it in the Lax–Wendroff scheme, Eq. (4.44) we get on simplification the amplification factor $G = e^{\alpha \Delta t}$ as

$$G = e^{\alpha \Delta t} = 1 - \nu^2 + \nu^2 \cos\beta - i\nu \sin\beta \tag{4.45}$$

A little calculation and trigonometric simplification shows that

$$\begin{aligned} |G|^2 &= (1 - \nu^2 + \nu^2 \cos\beta)^2 + \nu^2 \sin^2\beta \\ &= 1 - 4\nu^2(1 - \nu^2)\sin^4\phi, \end{aligned}$$

where $\phi = \dfrac{\beta}{2}$. Hence the scheme is stable for $|\nu| \leq 1$, that is

$$\left|\frac{c\Delta t}{\Delta x}\right| \leq 1 \tag{4.46}$$

4.3.1 Lax–Wendroff Scheme: Modified Equation

Assuming $u(x, t)$ to be sufficient number of times continuously differentiable, the following formulae may be easily derived by Taylor's expansion :

$$\frac{1}{2\Delta x}\left[u_{j+1}^n - u_{j-1}^n\right] = u_x + \frac{\Delta x^2}{6} u_{xxx} + O(\Delta x^4)$$

$$\frac{1}{\Delta t}\left(u_j^{n+1} - u_j^n\right) = u_t + \frac{\Delta t}{2} u_{tt} + \frac{\Delta t^2}{6} u_{ttt} + O(\Delta t^4) \tag{4.47}$$

$$\frac{1}{\Delta x^2}\left(u_{j+1}^n - 2u_j^n + u_{j-1}^n\right) = u_{xx} + \frac{\Delta x^2}{12} u_{xxxx} + O(\Delta x^4)$$

Substituting these in the Lax–Wendroff scheme, Eq. (4.44) we get on simplification

$$u_t+cu_x=-\frac{\Delta t}{2}u_{tt}+\frac{c^2\Delta t}{2}u_{xx}-\frac{\Delta t^2}{6}u_{ttt}-\frac{c\Delta x^2}{6}u_{xxx}+c^2\Delta t.\frac{\Delta x^2}{24}u_{xxxx}+\cdots \quad (4.48)$$

Differentiating both sides of Eq. (4.48) with respect to t and x and eliminating u_{xt}, yields

$$u_{tt}=c^2u_{xx}-\frac{\Delta t}{2}u_{ttt}+\frac{c\Delta t}{2}u_{ttx}+\frac{c^2\Delta t}{2}u_{xxt}-\frac{c^3\Delta t}{2}u_{xxx}+O(\Delta t^2,\ \Delta x^2) \quad (4.49)$$

Differentiating Eq. (4.49) with respect to t and x and proceeding as in the previous cases one finds

$$u_{xtt}=c^2u_{xxx}+O(\Delta t), \quad u_{txx}=-cu_{xxx}+O(\Delta t)$$

and

$$u_{ttt}=-c^3u_{xxx}+O(\Delta t).$$

Substituting these relations and for u_{tt} from Eq. (4.49) in Eq. (4.48) follows on simplification the modified equation for the Lax–Wendroff scheme

$$u_t+cu_x=-\frac{c\Delta x^2}{6}(1-\nu^2)u_{xxx}-\frac{c\Delta x^3}{8}\nu(1-\nu^3)u_{xxxx}+\cdots \quad (4.50)$$

For $\nu \neq 1$, the scheme is predominantly dispersive.

The modulus of the amplification factor and the relative phase error are shown in Figs. 4.5(a) and 4.5(b), which show that the scheme has mainly a lagging phase error except for relatively large wave numbers in the range $\sqrt{0.5} < \nu \leq 1$.

4.3.2 Two-Step Lax–Wendroff Scheme

This is a variant of the Lax–Wendroff scheme, which is much easier to compute, particularly for nonlinear equations or systems. The scheme may be looked upon as a kind of predictor–corrector formula, popular in numerical analysis. Here two half time-steps constitute one full time-step. The scheme may be expressed as

$$\text{Predictor}: \quad \frac{1}{\Delta t/2}\left[u_{j+\frac{1}{2}}^{n+\frac{1}{2}}-\frac{1}{2}\left(u_{j+1}^n+u_j^n\right)\right]=-\frac{c}{\Delta x}\left(u_{j+1}^n-u_j^n\right) \quad (4.51)$$

$$\text{Corrector}: \quad \frac{1}{\Delta t}\left[u_j^{n+1}-u_j^n\right]=-\frac{c}{\Delta x}\left(u_{j+\frac{1}{2}}^{n+\frac{1}{2}}-u_{j-\frac{1}{2}}^{n+\frac{1}{2}}\right) \quad (4.52)$$

The predictor step predicts values at the half time-level $n+\frac{1}{2}$ by applying the Lax scheme, given by Eq. (4.25) at the point $j+\frac{1}{2}$. The corrector step is a half-step leap-frog scheme Eq. (4.38). The truncation error of the scheme is $O(\Delta t^2,\ \Delta x^2)$. For linear equations, the scheme is equivalent to the one-step Lax–Wendroff scheme.

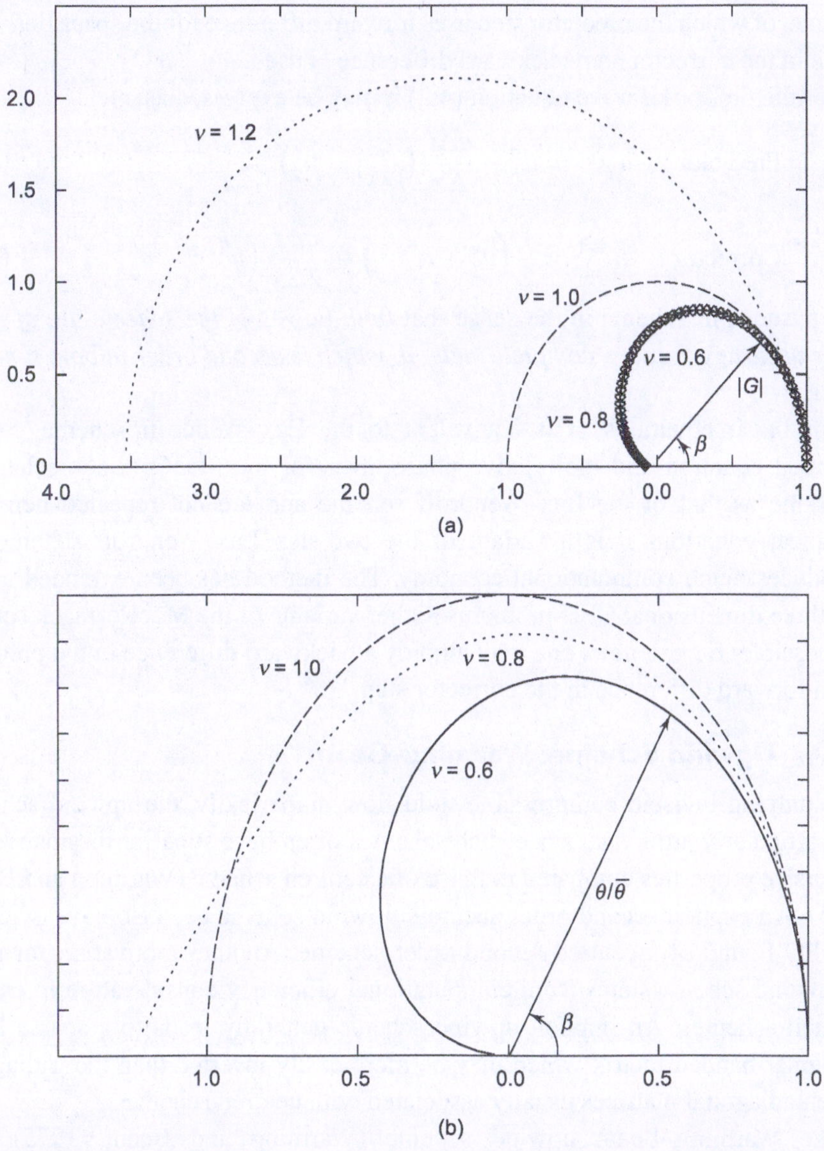

Figure 4.5 Lax–Wendroff scheme (a) modulus of amplification (b) relative phase error.

4.3.3 The MacCormack Scheme

The MacCormack (1969) scheme, which became very popular in CFD during the seventies, is a variant of the Lax–Wendroff scheme. It may be expressed as a split scheme, of which the predictor step uses forward difference for the space derivative while in the corrector step backward difference is used.

For the first-order wave equation (4.2) it may be expressed as

$$\text{Predictor}: \quad u_j^{n+\frac{1}{2}} = u_j^n - \frac{c\Delta t}{\Delta x}\left(u_{j+1}^n - u_j^n\right) \tag{4.53}$$

$$\text{Corrector}: \quad u_j^{n+1} = \frac{1}{2}\left(u_j^n + u_j^{n+\frac{1}{2}}\right) - \frac{c\Delta t}{2\Delta x}\left(u_j^{n+\frac{1}{2}} - u_{j-1}^{n+\frac{1}{2}}\right) \tag{4.54}$$

It is a true split scheme in the sense that *both halves of the scheme are of lower order accuracy than the complete method, which is second order in both time and space.*

For linear equations, it is equivalent to the Lax–Wendroff scheme. So the modified equation and the relative phase error for the MacCormack scheme is the same as that of the Lax–Wendroff scheme and are not repeated here. For nonlinear equations it is a variant of the two-step Lax–Wendroff scheme and introduces much computational economy. The method has been extended to two and three dimensional CFD problems. Other variants of the MacCormack scheme are possible, for example, one may employ a backward difference in the predictor and a forward difference in the corrector step.

4.3.4 Upwind Scheme: Warming–Beam

For studying inviscid compressible fluid flow numerically, the upwind schemes are particularly attractive, since such schemes often have superior dissipative and dispersive properties compared to those of a centred scheme (Warming and Beam, 1975). An explicit second order accurate upwind scheme can also have twice the stability bound of a centred second-order scheme. Another motivation for using an upwind scheme stems from computational efficiency consideration in case of implicit schemes. An implicit upwind scheme generally leads to a sparse lower triangular banded matrix which may be more easily inverted than the tridiagonal or pentadiagonal matrices usually associated with centred schemes.

The Warming–Beam upwind scheme (Warming and Beam, 1975) is a second-order scheme and can be considered a variant of the MacCormack scheme (4.53)–(4.54).

Both the predictor and the corrector steps are now backward differenced as

$$\text{Predictor}: u_j^{n+\frac{1}{2}} = u_j^n - \frac{c\Delta t}{\Delta x}\left(u_j^n - u_{j-1}^n\right)$$

$$\text{Corrector}: u_j^{n+1} = \frac{1}{2}\left[u_j^n + u_j^{n+\frac{1}{2}} - \frac{c\Delta t}{\Delta x}\left(u_j^{n+\frac{1}{2}} - u_{j-1}^{n+\frac{1}{2}}\right) - \frac{c\Delta t}{\Delta x}\left(u_j^n - 2u_{j-1}^n + u_{j-2}^n\right)\right] \quad (4.55)$$

Eliminating the half-step values or the predicted values, the scheme may be expressed as a one-step method

$$u_j^{n+1} = u_j^n - \nu(u_j^n - u_{j-1}^n) + \frac{\nu(\nu-1)}{2}\left(u_j^n - 2u_{j-1}^n + u_{j-2}^n\right), \quad (4.56)$$

where $\nu = c\Delta t/\Delta x$ is the Courant number.

A little calculation following von Neumann stability analysis shows that the amplification factor of the scheme is (Warming and Beam, 1975)

$$G = 1 - 2\nu\left[\nu + 2(1-\nu)\sin^2\frac{\beta}{2}\right]\sin^2\frac{\beta}{2} - i\nu\sin\beta\left[1 + 2(1-\nu)\sin^2\frac{\beta}{2}\right] \quad (4.57)$$

and

$$|G| = 1 - 4\nu(1-\nu)^2(2-\nu)\sin^4\frac{\beta}{2} \quad (4.58)$$

From Eq. (4.58), it may be seen that the Warming–Beam upwind scheme is stable for $0 \le \nu \le 2$.

Fig. 4.6(a) shows the magnitude of the amplification factor $|G|$ and Fig. 4.6(b) shows the relative phase error for $0 \le \beta \le \pi$, for several Courant numbers in $0 < \nu < 1$. For a given value of β, the deviation of the curves from the unit circle indicates the relative error per time step for the wave number $k_m = \beta/\Delta x$. Further, it may be noted that the upwind scheme has more damping for small Courant numbers compared to the MacCormack scheme while both the schemes have large dispersion. In the range $0 < \nu < 1$, the MacCormack scheme has a predominantly lagging phase error (i.e. $\phi/\phi_e < 1$) and the upwind scheme has a leading phase error (i.e. $\phi/\phi_e > 1$). Thus, it can be surmised that considerable reduction in phase error would occur if the schemes were alternated on successive time steps. This is the basis of Fromm's method of zero average phase error (Fromm, 1971).

Example 4.1

Compute the solution of the equation $u_t + cu_x = 0, \quad c = \text{const.} > 0$ for the first two-steps, using the upwind scheme, with initial condition

$$u(x,0) = \begin{cases} x - x^2, & 0 \le x \le 1, \\ 0, & x > 1, \end{cases}$$

and boundary condition $u(0,t) = 0$ for all t, taking $\Delta x = \frac{1}{4}$, $\nu = c\Delta t/\Delta x = \frac{1}{2}$.

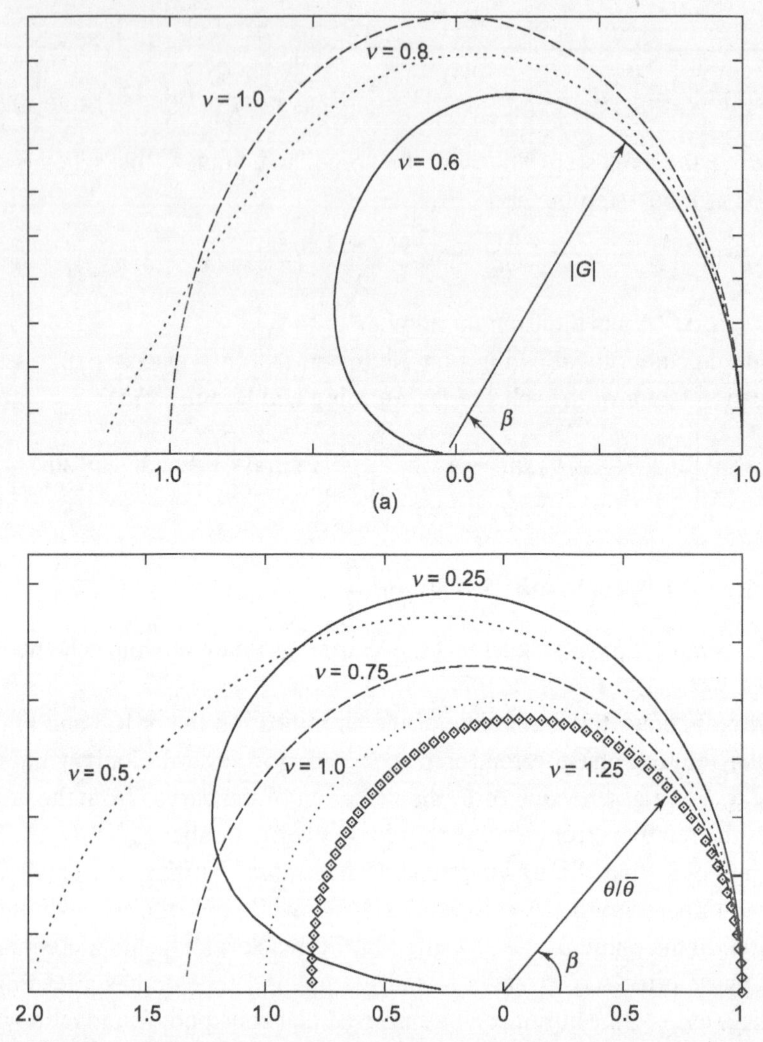

Figure 4.6 Warming-Beam upwind scheme (a) modulus of amplification (b) relative phase error.

Here $u_j^0 = j\Delta x - (j\Delta x)^2 = \frac{j}{4} - \frac{1}{16}j^2$, $j = 0, 1, \ldots, 5$. The first-order upwind scheme for $\nu = \frac{1}{2}$ is

$$u_j^{n+1} = \frac{1}{2}\left[u_j^n + u_{j-1}^n\right]$$

The first two-step solutions are shown in Fig. 4.7.

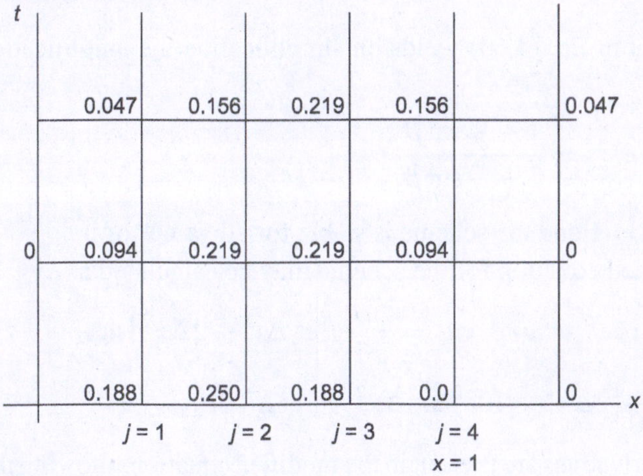

Figure 4.7 First two-step solutions by the upwind scheme.

4.4 IMPLICIT SCHEMES

Crank–Nicholson type of implicit scheme may be obtained by applying the trapezoidal rule of numerical integration Eq. (3.47) between time levels n and $(n+1)$ to the linear wave equation (4.2). We get,

$$u_j^{n+1} - u_j^n = -\int_{t=n\Delta t}^{(n+1)\Delta t} c\frac{\partial u}{\partial x}\,\mathrm{d}t = -\frac{c\Delta t}{2}\left[\left(\frac{\partial u}{\partial x}\right)_j^n + \left(\frac{\partial u}{\partial x}\right)_j^{n+1}\right]$$

Replacing $\frac{\partial u}{\partial x}$ by central difference at j, yields

$$u_j^{n+1} = u_j^n - \frac{\nu}{4}\left[\left(u_{j+1}^n - u_{j-1}^n\right) + \left(u_{j+1}^{n+1} - u_{j-1}^{n+1}\right)\right], \tag{4.59}$$

where $\nu = c\Delta t/\Delta x$, the Courant number.

The method is second-order in both time and space, which may be seen from the accuracy of the formulae, namely, the trapezoidal rule and central difference, used in deriving Eq. (4.59). This equation may be rearranged as

$$-\frac{\nu}{4}u_{j-1}^{n+1} + u_j^{n+1} + \frac{\nu}{4}u_{j+1}^{n+1} = u_j^n - \frac{\nu}{4}\left(u_{j+1}^n - u_{j-1}^n\right) \tag{4.60}$$

showing that the scheme is implicit and a tridiagonal system is to be solved. The Thomas algorithm may conveniently be used for this.

In order to study stability of the scheme by von Neumann method, let us choose as before, a trial solution

$$u_j^n = e^{\alpha n\Delta t}e^{i\beta j}, \qquad \text{with } \beta \text{ real, } \alpha \text{ complex.}$$

Substituting it in Eq. (4.60) yields on simplification the amplification factor $G = e^{\alpha \Delta t}$ as

$$G = e^{\alpha \Delta t} = \frac{1 - \frac{i\nu}{2}\sin\beta}{1 + \frac{i\nu}{2}\sin\beta} \tag{4.61}$$

so that $|e^{\alpha \Delta t}| = 1$ and the scheme is stable for all values of ν.

The modified equation for the scheme may be calculated as

$$u_t + cu_x = -\frac{c}{12}\left[c^2\,\Delta t^2 + 2\Delta x^2\right] u_{xxx}$$

$$-\frac{c}{240}\left[2\Delta x^4 + 10c^2 \Delta t^2 \Delta x^2 + 3c^3 \Delta t^4\right] u_{xxxxx} + \cdots \tag{4.62}$$

Odd-order derivatives are present in the modified equation, showing that the scheme is essentially dispersive and is devoid of any implicit artificial viscosity. Often, some explicit artificial viscosity may have to be added in order to prevent blowing-up of the numerical solution for nonlinear equations of fluid dynamics.

4.5 MORE ON UPWIND SCHEMES

For hyperbolic type equations, the first and higher order upwind schemes have been studied, quite thoroughly, in the literature (Beam and Warming, 1976; Warming and Hyett, 1974; Yee 1985). The first-order upwind scheme and its extension to second-order for the scalar linear convection Eq. (4.2) have been studied in the previous sections for the case $c =$ const. > 0. von Neumann stability analysis showed the forward-time backward-space (FTBS) first-order scheme to be stable. On the other hand, if the convection speed c be negative, then the FTBS is no longer stable, but the forward-time forward-space (FTFS) scheme is stable. Thus if the problem be such that the convection speed c is a constant, but may be of either sign, then both the cases may be combined as follows (Yee, 1985).

We consider the upwind scheme for the scalar convection equation

$$u_t + cu_x = 0, \quad c = \text{const.}, \tag{4.63}$$

given by

$$\frac{1}{\Delta t}\left(u_j^{n+1} - u_j^n\right) = \begin{cases} -\frac{c}{\Delta x}\left(u_j^n - u_{j-1}^n\right), & c > 0 \\ -\frac{c}{\Delta x}\left(u_{j+1}^n - u_j^n\right), & c < 0 \end{cases}$$

Substituting $r = \frac{\Delta t}{\Delta x}$, this may be simplified to

$$u_j^{n+1} = \begin{cases} u_j^n - rc\left(u_j^n - u_{j-1}^n\right), & c > 0 \\ u_j^n - rc\left(u_{j+1}^n - u_j^n\right), & c < 0 \end{cases} \tag{4.64}$$

If we use the substitutions

$$c^+ = \frac{1}{2}(c + |c|) \quad \text{and} \quad c^- = \frac{1}{2}(c - |c|),$$

then c^+ and c^- denote the value of c respectively in the cases $c > 0$ and $c < 0$. Then, both the cases of Eq. (4.64) may be combined to yield

$$u_j^{n+1} = u_j^n - r\left[c^+(u_j^n - u_{j-1}^n) + c^-(u_{j+1}^n - u_j^n)\right], \tag{4.65}$$

which may be rewritten as

$$u_j^{n+1} = u_j^n - \frac{r}{2}c(u_{j+1}^n - u_{j-1}^n) + \frac{r}{2}|c|(u_{j+1}^n - 2u_j^n + u_{j-1}^n) \tag{4.66}$$

When the upwind scheme is extended to nonlinear equations and systems, the form of Eq. (4.66) is more compact and efficient in terms of operations count, as pointed out in Yee (1985).

Equation (4.66) may be put to the form

$$u_j^{n+1} = u_j^n - r\left(h_{j+\frac{1}{2}}^n - h_{j-\frac{1}{2}}^n\right) \tag{4.67}$$

where $r = \Delta t/\Delta x$ and $h_{j+\frac{1}{2}}$, called the *numerical flux function* is defined by

$$h_{j+\frac{1}{2}} = \frac{1}{2}\left[c\left(u_{j+1} + u_j\right) - |c|\left(u_{j+1} - u_j\right)\right] \tag{4.68}$$

It may be verified that the scheme is stable for

$$0 < c\frac{\Delta t}{\Delta x} \le 1, \tag{4.69}$$

and that this is first order and explicit. It is to be noted that in Eq. (4.67) the space derivative appears to be centrally differenced. Such schemes are called *conservative schemes*, to be defined shortly.

Several other spatially centred second order accurate schemes for the scalar convection equation with constant convection speed, Eq. (4.67), may be found in Yee (1989). We only mention that the Lax–Wendroff scheme (4.44) may also be put to the form

$$u_j^{n+1} = u_j^n - r\left(h_{j+\frac{1}{2}}^n - h_{j-\frac{1}{2}}^n\right), \quad r = \frac{\Delta t}{\Delta x},$$

with the numerical flux function

$$h_{j+\frac{1}{2}} = \frac{1}{2}\left[c(u_{j+1} + u_j) - rc^2(u_{j+1} - u_j)\right] \tag{4.70}$$

Such schemes may be extended to first-order hyperbolic systems with constant coefficient matrix case, similar to the upwind methods.

4.6 SCALAR CONSERVATION LAW: LAX–WENDROFF AND RELATED SCHEMES

The Lax–Wendroff scheme, Eq. (4.43) for the scalar hyperbolic equation (4.2) may be readily extended to the case of hyperbolic system of conservation laws. To see this, we consider first the case of a *scalar conservation law*, consisting of *one equation in only one unknown* $u(x,t)$

$$\frac{\partial u}{\partial t} + \frac{\partial F}{\partial x} = 0, \tag{4.71}$$

where $F = F(u)$, may be a *nonlinear function of* u. Then by Taylor's expansion of the function $u(x,\ t+\ \Delta t)$ about the point (x,t) we have, in finite difference notation

$$u_j^{n+1} = u_j^n + \Delta t\, u_t|_j^n + \frac{\Delta t^2}{2}\, u_{tt}|_j^n + \cdots \tag{4.72}$$

Noting that $u_t = -F_x$, $u_{tt} = (-F_x)_t = -(F_t)_x$, and that $F_t = F_u\, u_t = A(u)(-F_x)$, where the derivative $F_u = A(u)$, it follows from Eq. (4.72)

$$u_j^{n+1} = u_j^n + \Delta t\,(-F_x)_j^n + \frac{\Delta t^2}{2}\frac{\partial}{\partial x}\,[A(u)\,F_x]_j^n$$

keeping terms up to second order. Replacing the x-derivatives by the corresponding central difference representations, given by

$$\frac{\partial}{\partial x}\,[A(u)\,F_x]_j^n = \frac{1}{\Delta x}\left[(A\,F_x)_{j+\frac{1}{2}}^n - (A\,F_x)_{j-\frac{1}{2}}^n\right]$$

$$= \frac{1}{\Delta x}\left[A_{j+\frac{1}{2}}^n\,\frac{F_{j+1}^n - F_j^n}{\Delta x} - A_{j-\frac{1}{2}}^n\,\frac{(F_j^n - F_{j-1}^n)}{\Delta x}\right],$$

the *Lax–Wendroff scheme* (applicable to nonlinear equations as well) is obtained as

$$u_j^{n+1} = u_j^n - \frac{\Delta t}{2\Delta x}(F_{j+1}^n - F_{j-1}^n) + \frac{\Delta t^2}{2\Delta x^2}\left[A_{j+\frac{1}{2}}^n(F_{j+1}^n - F_j^n) - A_{j-\frac{1}{2}}^n(F_j^n - F_{j-1}^n)\right] \tag{4.73}$$

Here, we have used the notations

$$F_j^n = F(u_j^n),\quad A_j^n = A(u_j^n),\quad A_{j+\frac{1}{2}}^n = \frac{1}{2}\,(A_j^n + A_{j+1}^n)$$

It is not difficult to see that the corresponding *MacCormack scheme* (see Eqs. (4.53) and (4.54)) may be put to the form

$$\text{Predictor: } \overline{u_j^{n+1}} = u_j^n - \frac{\Delta t}{\Delta x}\,(F_j^n - F_{j-1}^n) \tag{4.74}$$

$$\text{Corrector: } u_j^{n+1} = \frac{1}{2}\left(\overline{u_j^{n+1}} + u_j^n\right) - \frac{\Delta t}{2\Delta x}\left[\overline{F_{j+1}^{n+1}} - \overline{F_j^{n+1}}\right], \quad \overline{F_j^{n+1}} = F(\overline{u_j^{n+1}}).$$

It is not difficult to verify that both the Lax–Wendroff scheme (4.73) and the MacCormack scheme (4.74) are conservative schemes.

Example 4.2

Solve using the Lax–Wendroff scheme the first time-step solution of inviscid Burger's equation $u_t + (\frac{1}{2}u^2)_x = 0$, with initial condition $u(x, 0) = \sqrt{x}, \quad 0 \le x \le 1$ and boundary condition $u(0, t) = 0$ for all time. Compare with the exact solution $u(x, t) = \frac{1}{2}[-t + \sqrt{t^2 + 4x}]$.

Take $\Delta x = 0.2$ and $r = \Delta t/\Delta x = 0.5$. Then, $\Delta t = 0.1$. Here $F(u) = \frac{u^2}{2}$, so that $A(u) = \dfrac{dF}{du} = u$. Also,

$$u_{j+\frac{1}{2}}^n = \frac{1}{2}\left(u_j^n + u_{j+1}^n\right)$$

and

$$u_{j-\frac{1}{2}}^n = \frac{1}{2}\left(u_j^n + u_{j-1}^n\right).$$

Consequently, Lax–Wendroff scheme (4.73) becomes

$$u_j^{n+1} = u_j^n - \frac{1}{8}\left[\left(u_{j+1}^n\right)^2 - \left(u_{j-1}^n\right)^2\right] \tag{4.75}$$
$$+\frac{1}{32}\left[\left(u_{j+1}^n + u_j^n\right)\left\{\left(u_{j+1}^n\right)^2 - \left\{u_j^n\right)^2\right\} - \left(u_j^n + u_{j-1}^n\right)\left\{\left(u_j^n\right)^2 - \left(u_{j-1}^n\right)^2\right\}\right]$$

The initial values are $u_j^0 = \sqrt{j\Delta x} = \sqrt{0.2j}, \quad 0 \le j \le 5$. Boundary values : $u_0^n = 0$, for all $n \ge 0$. This gives,

$$u_0^0 = 0, \quad u_1^0 = 0.447, \quad u_2^0 = 0.633, \quad u_3^0 = 0.775, \quad u_4^0 = 0.894, \quad u_5^0 = 1.000.$$

The first step solutions are:

$$u_0^1 = 0, \quad u_1^1 = 0.401, \quad u_2^1 = 0.585, \quad u_3^1 = 0.726, \quad u_4^1 = 0.846,$$

the corresponding exact solution for the first step being (0.400, 0.584, 0.726, 0.846).

Example 4.3

Compute the first two steps of the numerical solution of the inviscid Burger's equation in $0 \le x \le 1, t > 0$,

$$\frac{\partial u}{\partial t} + \frac{\partial}{\partial x}\left(\frac{1}{2}u^2\right) = 0, \qquad x > 0, \qquad t > 0 \tag{4.76}$$

subject to the initial and boundary conditions

$$u(x,0) = x, \quad x > 0, \quad \text{and} \quad u(0,t) = 0, \quad t > 0,$$

using the MacCormack scheme. Compare the numerical solution with the exact solution $u = x/(1+t)$, taking $\Delta x = 0.2$ and $r = \dfrac{\Delta t}{\Delta x} = 0.5$.

Here, $x = j\Delta x = 0.2j$. The MacCormack scheme for $r = \frac{1}{2}$, and $F(u) = \frac{1}{2}u^2$ is

$$\text{Predictor} : \overline{u_j^{n+1}} = u_j^n - \frac{1}{4}\left[(u_j^n)^2 - (u_{j-1}^n)^2\right],$$

$$\text{Corrector} : u_j^{n+1} = \frac{1}{2}\left(u_j^n + \overline{u_j^{n+1}}\right) - \frac{1}{8}\left[\overline{(u_{j+1})^2} - \overline{\left(u_j^{n+1}\right)^2}\right]$$

The initial values are $u_j^0 = j\Delta x = 0.2j$, $0 \le j \le 5$. Boundary values : $u_0^n = 0$, for all $n \ge 0$. This gives,

$$u_0^0 = 0, \quad u_1^0 = 0.2, \quad u_2^0 = 0.4, \quad u_3^0 = 0.6, \quad u_4^0 = 0.8, \quad u_5^0 = 1.000.$$

Then,

$$\overline{u_1^1} = u_1^0 - \frac{1}{4}[(u_1^0)^2 - (u_0^0)^2] = 0.397, \overline{u_2^1} = 0.582, \overline{u_3^1} = 0.725, \overline{u_4^1} = 0.844, \overline{u_5^1} = 0.950$$

Using the corrector formula, the values obtained are $u_1^2 = 0.167$, $u_2^2 = 0.334$, $u_3^2 = 0.500$, $u_4^2 = 0.668$. However, u_5^1 cannot be calculated, since the corrector formula requires the value $\overline{u_6^1}$, which in turn requires u_6^0 which is outside the given domain and unknown. The corresponding *exact values* are : $u_1^2 = 0.167$, $u_2^2 = 0.333$, $u_3^2 = 0.500$, $u_4^2 = 0.668$. The agreement is very good.

Note that for computing the first two-step values at the boundary $j = 5$, requires the initial value at $j = 6$ and $j = 7$ which are given by the initial condition as $u_6^0 = 1.2$, $u_7^0 = 1.4$. For example,

$$u_5^1 = 1.0 - \frac{1}{8}\left\{(1.2)^2 - (0.8)^2\right\} + \frac{1}{32}\left[(1.2 + 1.0)\left\{(1.2)^2 - (1.0)^2\right\} - (1.0 + 0.8)\left\{(1.0)^2 - (0.8)^2\right\}\right] = 0.907.$$

For computing u_5^2, the value of u_6^1 is required, which is

$$u_6^1 = 1.2 - \frac{1}{8}\left\{(1.4)^2 - (1.0)^2\right\} + \frac{1}{32}\left[(1.4 + 1.2)\left\{(1.4)^2 - (1.2)^2\right\} - (1.2 + 1.0)\left\{(1.2)^2 - (1.0)^2\right\}\right] = 1.092.$$

A small computer program lxmc.f is written to compute the solution for 10–time steps with $\Delta t = 0.1$, $\Delta x = 0.2$. Table 4.1, shows the solution by the Lax–Wendroff and MacCormack schemes compared with the exact solution, as output of Program 4.1, at time $t = 0.1$ and $t = 0.5$. For small time the agreement is very good. With increasing time the errors in the computed solutions increase.

4.6.1 Program 4.1: lxmc.f

The following program computes solution of inviscid Burger's equation by Lax–Wendroff and MacCormack schemes and compares with the exact solution.

```
c Program 4.1: lxmc.f
c       Lax-Wendroff and MacCormack solutions of Inviscid
c Burger's equation, u(x,0)=x, x>= 0, u(0,t)= 0.
c Exact solution: u=x/(1+t).
c Take dx=.2, r= dt/dx=0.5, 0 <=x <=2. 10 time steps.
c......................................................
parameter ( im=20, in=10 )
dimension x(0:im), ulw(0:im, 0:in), ux(0:im,0:in),t(0:in)
dimension utm(0:im, 0:in),umc(0:im,0:in)
open(unit=8, file='lxmc.out')
dx =0.2
dt = 0.1
r= 0.5
imt=im/2
x(0)=0.0
ulw(0,0)=0.0
umc(0,0)=0.0
utm(0,0)=0.0
21 format(1x,2f10.4)
c.......Initial condition...............................
do 10 j=1,im
x(j)= x(j-1)+ dx
ulw(j,0)= x(j)
umc(j,0)=x(j)
utm(j,0)=x(j)
10 continue
c.......Boundary condition...............................
do 23 n=1,in
ulw(0,n)=0.0
umc(0,n)=0.0
utm(0,n)=0.0
23 continue
c...... Lax-Wendorf solution computation  ......................
do 30 n=0,in-1
np=n+1
ijm=im-1
do 20 j=1, ijm
uj=ulw(j,n)
```

```
ujp= ulw(j+1,n)
ujm=ulw(j-1,n)
ujph= (ujp+uj)
ujmh=(uj+ujm)
t1 = (ujp*ujp-ujm*ujm)/8.
ulw(j,np)=uj-t1+(ujph*(ujp*ujp- uj*uj)-ujmh*(uj*uj-ujm*ujm))/32.
20 continue
30 continue
c...... Computation of Exact solution ux....................
do 40 n=0,in
t(n)= n* dt
do 50 j=0,im
ux(j,n)= x(j)/(1.+ t(n))
50 continue
40 continue
c......Computation by MacCormack scheme .....................
do 80 n=0,in-1
np= n+1
ijm=im-1
do 81 j=1,ijm
utj= utm(j,n)
utjm = utm(j-1,n)
utm(j,np)=utj- .5*r*(utj*utj- utjm*utjm)
81 continue
do 83 j=1,ijm
unp=utm(j,np)
utmp=utm(j+1,np)
umj=umc(j,n)
umc(j,np)= .5*(unp+ umj) - .25*r*(utmp*utmp - unp*unp)
83 continue
80 continue
c.............write output...................................
write(8,*)'Output of L-W and McCormack Sol.ofInviscid Burgers Eq'
write(8,*)       ' dx= ', dx, '   dt= ', dt
write(8,*)'n ',' x ',' t ','ulw(j,n) ','umc(j,n) ','ux (j,n)'
do 60 n=1,in
do 70 j=1,imt
write (8,13) n, x(j), t(n) , ulw(j,n), umc(j,n) , ux(j,n)
13  format(1x,  I4, f8.3, 4f10.4 )
70 continue
60 continue
close(8)
stop
end

%-----------------------------------------------------------------------
```

Table 4.1 Solution of Burger's equation by Lax–Wendroff and MacCormack schemes at 1st and 5th-time steps. Output of Program 4.1

```
Output of Program 4.1: lxmc.f
----------------------------------------------------

Inviscid Burger solution dx=.2, dt=.1, r=.5
   Lax-Wendorf(ulw), MacCormack(umc), exact(ux)
 dx=    0.200000003     dt=    0.100000001
    n      x         t        ulw(j,n)  umc(j,n)   ux (j,n)
    1   0.200     0.1000     0.1820    0.1824    0.1818
    1   0.400     0.1000     0.3640    0.3643    0.3636
    1   0.600     0.1000     0.5460    0.5462    0.5455
    1   0.800     0.1000     0.7280    0.7281    0.7273
    1   1.000     0.1000     0.9100    0.9100    0.9091
    5   0.200     0.5000     0.1337    0.1499    0.1333
    5   0.400     0.5000     0.2674    0.2811    0.2667
    5   0.600     0.5000     0.4011    0.4122    0.4000
    5   0.800     0.5000     0.5347    0.5432    0.5333
    5   1.000     0.5000     0.6684    0.6742    0.6667
-----------------------------------------------------------
```

4.6.2 Implicit Schemes for Scalar Conservation Law

The method of deriving the implicit scheme in Section 4.4 may be readily extended to the case of a scalar conservation law and a system of conservation laws. Such schemes have been derived, among others, by Briley and McDonald (1973) and Beam and Warming (1976).

Application of the trapezoidal rule of numerical integration Eq. (3.47) between time levels n and $n+1$ to the scalar conservation law Eq. (4.71) yields

$$u_j^{n+1} - u_j^n = -\int_{t=n\Delta t}^{(n+1)\Delta t} \frac{\partial F}{\partial x}\,dt = -\frac{1}{2}\Delta t\left[\left(\frac{\partial F}{\partial x}\right)_j^n + \left(\frac{\partial F}{\partial x}\right)_j^{n+1}\right] \quad (4.77)$$

where $F = F(u)$ may be a non-linear function of u. It may be noted that system of conservation laws may also be treated in the same way as described presently, by taking u and F as vector quantities. The presence of the quantity $(\frac{\partial F}{\partial x})$ at the time-level $(n+1)$ leads to a system of non-linear algebraic equations and some kind of linearisation is highly desirable. The kind of linearisation presented here, has been suggested by Briley and McDonald (1973) as well as by Beam and Warming (1976).

By Taylor's expansion of $\left(\frac{\partial F}{\partial x}\right)_j^{n+1}$ about the time level n and neglecting terms with powers higher than the first in Δt, we get

$$\left.\frac{\partial F}{\partial x}\right]_j^{n+1} = \left.\frac{\partial F}{\partial x}\right]_j^n + \Delta t \frac{\partial}{\partial t}\left(\frac{\partial F}{\partial x}\right)_j^n + O(\Delta t^2).$$

Interchanging the order of differentiation in the second term on the right, this yields

$$\left.\frac{\partial F}{\partial x}\right]_j^{n+1} = \left.\frac{\partial F}{\partial x}\right]_j^n + \Delta t \frac{\partial}{\partial x}\left(\frac{\partial F}{\partial t}\right)_j^n + O(\Delta t^2). \tag{4.78}$$

Here, $\dfrac{\partial F}{\partial t} = A\dfrac{\partial u}{\partial t}$, where $A = \dfrac{\partial F}{\partial u}$. Substituting Eq. (4.78) in Eq. (4.77) we get

$$u_j^{n+1} = u_j^n - \frac{\Delta t}{2}\left[2\left(\frac{\partial F}{\partial x}\right)_j^n + \Delta t \frac{\partial}{\partial x}\left(A\frac{\partial u}{\partial t}\right)_j^n\right]. \tag{4.79}$$

Replacing here the time derivative by the forward difference and the space derivative by the three-point central difference, follows

$$u_j^{n+1} = u_j^n - \frac{\Delta t}{2}\left[\frac{1}{\Delta x}(F_{j+1}^n - F_{j-1}^n)\right]$$
$$-\frac{\Delta t}{4\Delta x}\left[A_{j+1}^n\left(u_{j+1}^{n+1} - u_{j+1}^n\right) - A_{j-1}^n(u_{j-1}^{n+1} - u_{j-1}^n)\right].$$

This yields on simplification, the tridiagonal system

$$-\frac{r}{4}A_{j-1}^n u_{j-1}^{n+1} + u_j^{n+1} + \frac{r}{4}A_{j+1}^n u_{j+1}^{n+1} = u_j^n - \frac{r}{2}\left(F_{j+1}^n - F_{j-1}^n\right)$$
$$-\frac{r}{4}A_{j-1}^n u_{j-1}^n + \frac{r}{4}A_{j+1}^n u_{j+1}^n \tag{4.80}$$

where $r = \dfrac{\Delta t}{\Delta x}$ and $F_j^n = F(u_j^n)$. This is a tridiagonal system of linear algebraic equations and may be solved by the Thomas algorithm. The scheme is second-order accurate in both time and space and unconditionally stable. However, as in the case of the linear wave equation, the modified equation (see for example, Eq. (4.62)) contains no even-order derivative term and the scheme is dispersive. In general, some artificial smoothing term is necessary for the computation of a solution containing a shock. Beam and Warming (1976) suggest the use of the following fourth difference term for this purpose,

$$-\frac{\omega}{8}\left[u_{j+2}^n - 4u_{j+1}^n + 6u_j^n - 4u_{j-1}^n + u_{j-2}^n\right], \tag{4.81}$$

which does not affect the overall accuracy of the scheme and is stable for values of the parameter ω in $0 < \omega \leq 1$.

Instead of solving Eq. (4.80) for the unknowns u_j^{n+1}, Beam and Warming (1976) recommend the use of the differences Δu_j, defined by

$$\Delta u_j = u_j^{n+1} - u_j^n. \tag{4.82}$$

Then Δu_j is called *the correction*, which is to be added to u_j^n in order to obtain u_j^{n+1}. Rewritten in terms of the corrections, Eq. (4.80) takes the form

$$-\frac{r}{4}A_{j-1}^n\Delta u_{j-1} + \Delta u_j + \frac{r}{4}A_{j+1}^n\Delta u_{j+1} = -\frac{r}{2}\left(F_{j+1}^n - F_{j-1}^n\right) \tag{4.83}$$

which is again a tridiagonal system for the unknowns Δu_j. It may be noted that the corrections Δu_j are small quantities compared to u_j, which are particularly advantageous for non-linear systems leading to economy in labour. For steady-state computations also, the unknown corrections are better behaved than the unknowns u_j^n and asymptotically approach zero as convergence is reached, leading to computational economy.

Apart from the above implicit scheme, Beam and Warming (1978) have developed a factored scheme, which they use in their implicit scheme.

4.7 HYPERBOLIC SYSTEM OF CONSERVATION LAWS

Hyperbolic system of conservation laws in two independent variables t and x, may be expressed in the form

$$\frac{\partial U}{\partial t} + \frac{\partial F(U)}{\partial x} = 0 \tag{4.84}$$

where U and $F(U)$ are m-component vectors, $U = (U_1, U_2, \cdots, U_m)^{\mathrm{T}}$ and

$$F = (F_1, F_2, \cdots, F_m)^{\mathrm{T}}.$$

In terms of the Jacobian matrix $A(U) = \dfrac{\partial F(U)}{\partial U}$ the above system may also be expressed in the *quasi-linear form*

$$\frac{\partial U}{\partial t} + A(U)\frac{\partial U}{\partial x} = 0 \tag{4.85}$$

The system (4.84) is assumed to be hyperbolic. Consequently, all the eigenvalues $\lambda_1, \lambda_2, \cdots, \lambda_m$ of the matrix A are real and a similarity transformation exists such that

$$Q^{-1}A\,Q = \Lambda = \mathrm{diag}(\lambda_1, \lambda_2, \cdots, \lambda_m) \tag{4.86}$$

Equations in hyperbolic conservation law form are particularly important in computation of discontinuous solutions like those containing *shock or contact discontinuities, generally referred to as weak solutions* (see Section 1.9). The numerical methods for computing weak solutions belong to the categories of *shock-fitting* and *shock-capturing*.

A shock-fitting method almost always needs some previous knowledge or detection of the shock location, where explicitly the Rankine–Hugoniot shock conditions (Niyogi, 1977; Oswatitsch, 1956) are satisfied. On the other hand, the shock-capturing methods require no special technique for detection of shocks. The shocks are automatically captured by a shock-capturing scheme. The integral equation method of Oswatitsch (1950), Niyogi (1982), Oswatitsch (1956, 1977) is an example of one of the earliest shock-fitting methods used for computing solutions of the transonic small perturbation flow past thin symmetric airfoils at zero incidence. The method of Lax (1954) defined by Eq. (4.25) is one of the earliest shock-capturing schemes. It was shown by Lax and Wendroff (1960) that discontinuous solutions can be computed without any special treatment of the discontinuity if the differential equation is in a *conservative form* (also called *conservation form*) (4.84) and if the *scheme is conservative*.

A two-level explicit scheme for Eq. (4.84) is said to be *conservative* if it is of the general form

$$u_j^{n+1} = u_j^n - \frac{\Delta t}{\Delta x}\left[H\left(u_{j+k}^n, \cdots, u_{j-k+1}^n\right) - H\left(u_{j+k-1}^n, \cdots, u_{j-k}^n\right)\right] \tag{4.87}$$

where H is a function of $2k$-arguments, which must satisfy the *consistency requirement*

$$H(U, U, \cdots, U) = F(U) \text{ for any } U. \tag{4.88}$$

Note that, for the particular case $k = 1$, Eq. (4.87) is of the form

$$U_j^{n+1} = U_j^n - \frac{\Delta t}{\Delta x}\left[H\left(U_{j+1}^n, U_j^n\right) - H\left(U_j^n, U_{j-1}^n\right)\right] \tag{4.89}$$

4.7.1 System of Conservation Laws

We next consider the case of the system of *conservation laws*

$$\begin{aligned}
&\frac{\partial U_1}{\partial t} + \frac{\partial}{\partial x} F_1(U_1, U_2, \ldots, U_m) = 0, \\
&\frac{\partial U_2}{\partial t} + \frac{\partial}{\partial x} F_2(U_1, U_2, \ldots, U_m) = 0, \\
&\vdots \qquad\qquad\qquad\qquad \vdots \\
&\frac{\partial U_m}{\partial t} + \frac{\partial}{\partial x} F_m(U_1, U_2, \ldots, U_m) = 0,
\end{aligned} \tag{4.90}$$

for the unknowns $U_1, U_2, \ldots, U_m$. In general the U_i, $i = 1, \ldots, n$, are functions of x and t, and the functions F_i, $i = 1, \ldots, n$ may be any nonlinear functions. Introducing the column vectors $U = (U_1, U_2, \ldots, U_m)^{\mathrm{T}}$ and $F = (F_1, F_2, \ldots, F_m)^{\mathrm{T}}$, we note that the system of equations (4.90) may be put to the form (4.84). In order to derive the Lax–Wendroff scheme for this case, let us consider the ith equation

$$\frac{\partial U_i}{\partial t} + \frac{\partial}{\partial x} F_i(U_1, U_2, \ldots, U_m) = 0,$$

where each U_i are twice differentiable functions of x and t in the considered domain. By Taylor's expansion about the point (x, t), keeping terms up to the second order in Δt, we have

$$U_i(x, t + \Delta t) = U_i(x, t) + \Delta t (U_i)_t + \frac{\Delta t^2}{2}(U_i)_{tt} \tag{4.91}$$

We note that the time derivatives may be replaced by the space derivatives

$$(U_i)_t = -(F_i)_x$$

and

$$\frac{\partial^2 U_i}{\partial t^2} = \frac{\partial}{\partial t}\left(-\frac{\partial F_i}{\partial x}\right) = -\frac{\partial}{\partial x}\left(\frac{\partial F_i}{\partial t}\right)$$
$$= -\frac{\partial}{\partial x}\left[\frac{\partial F_i}{\partial U_1}\frac{\partial U_1}{\partial t} + \frac{\partial F_i}{\partial U_2}\frac{\partial U_2}{\partial t} + \cdots + \frac{\partial F_i}{\partial U_m}\frac{\partial U_m}{\partial t}\right]$$
$$= -\frac{\partial}{\partial x}\left[\sum_{j=1}^{m} \frac{\partial F_i}{\partial U_j}\frac{\partial U_j}{\partial t}\right]$$

Substituting in Eq. (4.91), and replacing the space derivatives by the corresponding central difference representations as in the earlier cases, follows the *Lax–Wendroff scheme for systems* in vector notation

$$U_j^{n+1} = U_j^n - \frac{\Delta t}{2\Delta x}\left(F_{j+1}^n - F_{j-1}^n\right) + \frac{\Delta t^2}{2\Delta x^2}\left[A_{j+\frac{1}{2}}^n\left(F_{j+1}^n - F_j^n\right) - A_{j-\frac{1}{2}}^n\left(F_j^n - F_{j-1}^n\right)\right] \tag{4.92}$$

where $A = A(U)$ denotes the Jacobian matrix with elements $\left(\frac{\partial F_i}{\partial U_j}\right)$ and we have used the notations

$$A_j^n = A(U_j^n),\ A_{j+\frac{1}{2}}^n = \frac{1}{2}\left[A(U_j^n) + A(U_{j+1}^n)\right],$$
$$A_{j-\frac{1}{2}}^n = \frac{1}{2}\left[A(U_j^n) + A(U_{j-1}^n)\right],\ F_j^n = F(U_j^n).$$

Eqs. (4.73) and (4.92) are of the same form. If in Eq. (4.73) the single unknown u be replaced by the unknown vector U and A is replaced by the Jacobian matrix $\left(\frac{\partial F_i}{\partial U_j}\right)$ then the Lax–Wendroff scheme for the system, Eq. (4.92) follows. Similarly the MacCormack scheme for the system may be obtained from (4.74) by replacing u by vector U.

4.8 SECOND-ORDER WAVE EQUATION

We consider next the Cauchy problem for the second-order wave equation

$$u_{tt} = c^2 u_{xx}, \quad c = \text{ real const.}, \tag{4.93}$$

with initial conditions

$$u(x, 0) = f(x), \quad \frac{\partial u(x, 0)}{\partial t} = g(x), \quad a \le x \le b, \tag{4.94}$$

$f(x),\ g(x)$ being continuous functions. Let the points $x = a$ and $x = b$ on the x-axis be denoted by A and B respectively. Let us subdivide the domain of integration into a network by straight lines $x_j = j\Delta x = jh$ and $t_n = n\Delta t,\ j = 0, 1, \ldots, N, \quad n = 1, 2, 3, \ldots$, as in Fig. 4.8.

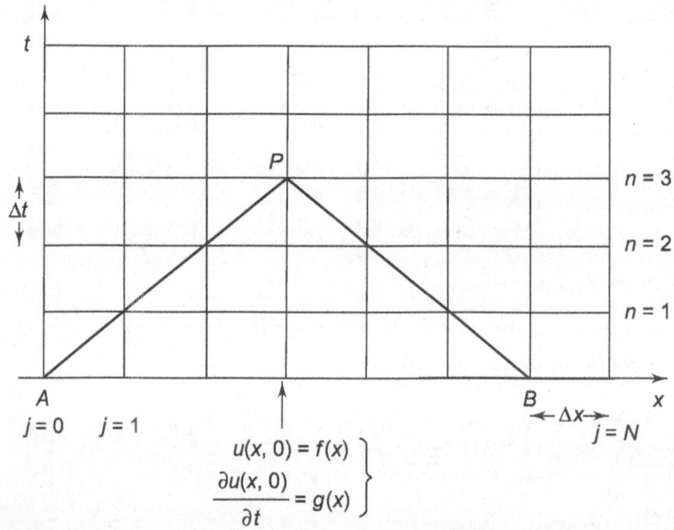

Figure 4.8 Cauchy problem for the second-order wave equation.

Let $u(x_j, t_n) = u_j^n$. Replacing both the time and space derivatives by central difference yields

$$\frac{u_j^{n+1} - 2u_j^n + u_j^{n-1}}{\Delta t^2} = c^2 \frac{u_{j+1}^n - 2u_j^n + u_{j-1}^n}{\Delta x^2}, \quad j = 0, 1, \ldots, N. \tag{4.95}$$

The initial conditions give

$$u_j^0 = f_j, \quad \text{and} \quad \frac{u_j^{n+1} - u_j^{n-1}}{2\Delta t} = g(jh) = g_j,$$

which yield

$$u_j^1 = u_j^{-1} + 2\Delta t\, g_j, \qquad j = 0, 1, 2, \ldots, N. \tag{4.96}$$

Equations (4.95) and (4.96) are the finite difference analogues of Eqs. (4.93) and (4.94), respectively.

Introducing the Courant number

$$\nu = \frac{c\Delta t}{\Delta x} = \frac{c\Delta t}{h},$$

in Eq. (4.95) we get on simplification

$$u_j^{n+1} = \nu^2 \left(u_{j+1}^n + u_{j-1}^n\right) + 2(1-\nu^2)u_j^n - u_j^{n-1}, j = 1, 2, \ldots, N. \tag{4.97}$$

This gives the solution at time level $n+1$ explcitly in terms of values of the previous time levels n and $n-1$. However, Eq. (4.96) contains u_j^{-1} which is outside the domain and has to be eliminated.

Observe that this scheme is really the *leap-frog scheme*. For parabolic-type equations, we have seen in Chapter 3, that the leap-frog scheme is unconditionally unstable. But for the wave equation it is stable for $\nu = \frac{c\Delta t}{\Delta x} < 1$, as we shall prove in the next section.

Putting $n = 0$ in Eq. (4.97), we get

$$\begin{aligned} u_j^1 &= \nu^2(u_{j+1}^0 + u_{j-1}^0) + 2(1-\nu^2)u_j^0 - u_j^{-1} \\ &= \nu^2(u_{j+1}^0 + u_{j-1}^0) + 2(1-\nu^2)u_j^0 - \{u_j^1 - 2\Delta t g_j\}. \end{aligned}$$

Using (4.96), this yields

$$u_j^1 = \frac{\nu^2}{2}(u_{j+1}^0 + u_{j-1}^0) + (1-\nu^2)u_j^0 + \Delta t g_j \tag{4.98}$$

$$= \frac{\nu^2}{2}(f_{j+1} + f_{j-1}) + (1-\nu^2)f_j + \Delta t g_j \tag{4.99}$$

However, note that f_{-1} and f_{N+1} are unknown as they lie outside the domain of integration. Hence, when initial values are prescribed between A and B ($j = 0, 1, \ldots, N$) in the first time level, the solution u_j^1 may be computed only for $j = 1, 2, \ldots, N-1$ according to (4.99). Then, putting $n = 1, 2, \ldots$ and continuing this process, we can compute the solution at all mesh points in the triangular region PAB (Fig. 4.9).

The lines PA and PB have slopes $\pm\dfrac{\Delta t}{\Delta x} = \pm\dfrac{\nu}{c}$. They are the numerical characteristics. Note that the analytical characteristics PA', PB', given by $x \mp ct =$ const., have slopes $\pm\frac{1}{c}$. If $\nu > 1$, then the analytical domain of dependence $A'B'$ includes the numerical domain of dependence AB. In this case, *the numerical solution, in general does not converge to the analytical solution.*

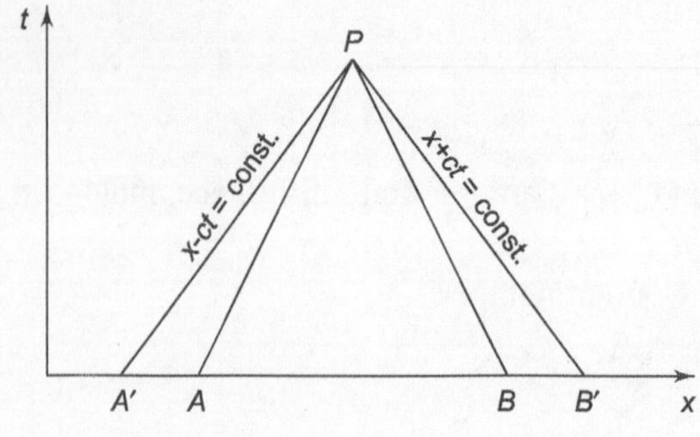

Figure 4.9 Analytical and numerical domain of dependence. Numerical characteristics PA, PB, analytical PA', PB'.

To see why, suppose that for a certain choice of initial values with $\nu > 1$, the numerical solution converges to the analytical solution. Now, let us change the initial values in the segment $A'A$ and $B'B$, such that continuity and differentiability are not disturbed. Then for the changed initial values, the analytical solution changes, while the numerical solution remains unchanged and the numerical solution cannot converge to the changed analytical solution.

Hence, for $\nu > 1$, in general, the numerical scheme does not converge.

4.8.1 Stability of the Leap-Frog Scheme for the Wave Equation

Let us take as a trial solution

$$u_j^n = \left(e^{\alpha\,\Delta t}\right)^n . e^{i\,k\,j\,\Delta x} \quad = Z^n \cdot e^{i\,\beta\,j}, \quad Z = e^{\alpha\,\Delta t}$$

and $k\Delta x = \beta$, k real, β real, α may be complex, $i = \sqrt{-1}$.

Putting it in the leap-frog scheme (4.97), we get on simplification,

$$e^{\alpha \Delta t} = \nu^2 \left(e^{i\beta} + e^{-i\,\beta}\right) + 2(1 - \nu^2) - e^{-\alpha\Delta t},$$

from which it follows that

$$Z^2 - \left(2 - 4\nu^2 \sin^2\frac{\beta}{2}\right) Z + 1 = 0.$$

Since the product of the roots of this quadratic equation equals unity, one of them must be greater than unity, leading to instability, unless both are equal to unity ($\alpha = 0$; degenerate case). Hence, for stability *complex roots must be admitted.*

Therefore, $(2 - 4\nu^2 \sin^2 \frac{\beta}{2})^2 < 4$.

Consequently, $-2 < 2 - 4\nu^2 \sin^2 \frac{\beta}{2} < 2$, which implies $4\nu^2 \sin^2 \frac{\beta}{2} < 4$, so that $\nu^2 < \dfrac{1}{\sin^2 \frac{\beta}{2}}$.

Hence, ν^2 must be less than the least value of $\frac{1}{\sin^2\frac{\beta}{2}}$ which is unity (attained when $\beta = \pi$), so that $\nu^2 < 1 \Longrightarrow \nu < 1$
Since $\nu > 0$, it requires for stability

$$\frac{c\Delta t}{\Delta x} < 1 \tag{4.100}$$

Here, condition (4.100) is the *CFL condition* (Courant–Friedrichs–Lewy condition) and the quantity $\frac{c\Delta t}{\Delta x}$ is the Courant number. So, *for stability, the Courant number must be less than unity.*

4.8.2 An Implicit Scheme for the Second-Order Wave Equation

The simplest implicit scheme for the second-order wave equation (4.93) is obtained by representing u_{tt} by central difference centred at (j, n) while u_{xx} is represented by the average of the central difference at the time-levels $n-1$ and $n+1$. We thus obtain

$$\frac{1}{\Delta t^2}\left[u_j^{n+1} - 2u_j^n + u_j^{n-1}\right] = \frac{c^2}{2\Delta x^2}\left[\left(u_{j+1}^{n+1} - 2u_j^{n+1} + u_{j-1}^{n+1}\right) + \left(u_{j+1}^{n-1} - 2u_j^{n-1} + u_{j-1}^{n-1}\right)\right]$$

On simplification, and writing $\nu = \frac{c\Delta t}{\Delta x}$, this yields

$$-\nu^2 u_{j+1}^{n+1} + 2(1+\nu^2)u_j^{n+1} - \nu^2 u_{j-1}^{n+1} = 4u_j^n + \nu^2 u_{j+1}^{n-1} - 2(1+\nu^2)u_j^{n-1} + \nu^2 u_{j-1}^{n-1} \tag{4.101}$$

For bounded space domains, this again yields a tridiagonal system which may be solved by a standard procedure such as the Thomas algorithm. It is next shown that the scheme (4.101) is unconditionally stable.

4.8.3 Stability of the Implicit Scheme

As before, let us take as a trial solution

$$u_j^n = e^{i\alpha n\Delta t} \cdot e^{i\beta j} = Z^n \cdot e^{i\beta j}, \quad Z = e^{\alpha\Delta t},$$

where the constant β is real and α may be complex, $i = \sqrt{-1}$, Z being the amplification factor. Substituting this in Eq. (4.101) we get

$$Z\left[-\nu^2 2\cos\beta + 2(1+\nu^2)\right] = 4 + [\nu^2 2\cos\beta - 2(1+\nu^2)]\frac{1}{Z}$$

which, when solved as a quadratic equation in Z yields, on simplification

$$Z = \left[1 \pm i2\nu\sin\frac{\beta}{2}\sqrt{1+\nu^2\sin^2\frac{\beta}{2}}\right] / \left(1 + 2\nu^2\sin^2\frac{\beta}{2}\right)$$

Thus the square of the modulus of the amplification factor is

$$|Z|^2 = \left[1 + 4\nu^2\sin^2\frac{\beta}{2}\left(1+\nu^2\sin^2\frac{\beta}{2}\right)\right] / \left(1 + 2\nu^2\sin^2\frac{\beta}{2}\right)^2 = 1.$$

Hence the implicit scheme is unconditionally stable.

4.9 METHOD OF CHARACTERISTICS FOR SECOND-ORDER HYPERBOLIC EQUATIONS

The method of characteristics is a graphical method for computing flow fields governed by *hyperbolic partial differential equations,* which may be quasi-linear. Extensive research has been made in this area in the pre-computer days, and even today these are the most dependable and efficient methods for computing solutions of hyperbolic-type partial differential equations, particularly if no jump discontinuities like shocks appear in the flow field. We outline here the method for the simple case of a second-order partial differential equation in two independent variables.

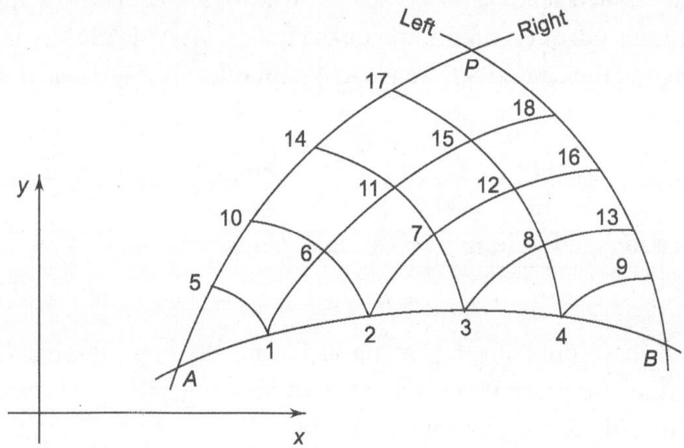

Figure 4.10 Computation of characteristic network. PA, PB are the right and left running characteristics through P.

We consider the second-order quasi-linear partial differential equation

$$a\,u_{xx} + 2b\,u_{xy} + c\,u_{yy} + e = 0 \tag{4.102}$$

where a, b, c and e are functions of x, y, u, u_x and u_y only in a certain domain Ω of the x, y-plane. Equation (4.102) is assumed to be of hyperbolic type. In Section 1.3.1 , we have seen that, through every point of the domain, there are two directions along which the integration of the partial differential equation reduces to the integration of an ordinary differential equation.

Thus, if Eq. (4.102) represents a hyperbolic-type partial differential equation, i.e. if $b^2 - ac > 0$ at a point of the domain of interest, a pair of real characteristic curves exist through the point whose slopes $\frac{dy}{dx}$ are given by

$$\frac{dy}{dx} = \frac{1}{a}\left[-b \pm \sqrt{b^2 - ac}\right] \tag{4.103}$$

Further, along the characteristic curves the partial derivatives $p = u_x, q = u_y$ of the unknown function $u = u(x, y)$ satisfy the so-called *compatibility relation*

$$a\frac{dp}{dx}\frac{dy}{dx} + c\frac{dq}{dx} + e\frac{dy}{dx} = 0 \tag{4.104}$$

giving a relation between the total differentials dp and dq. Note that, in view of Eq. (4.104), the slopes cannot be arbitrarily prescribed along a characteristic, but they must satisfy the compatibility relation (4.104).

A step by step graphical/numerical procedure, known as *method of characteristics* may now be developed to compute solution of Eq. (4.102) with appropriate initial and/or boundary conditions. The basic ideas of the method are as follows. Depending on the values of a, b, c at a point, and depending on the choice of the upper or lower sign, the characteristic curves, whose slopes are given by (4.103), would be left or right running when looked at along the positive y-direction. Accordingly, the characteristics are termed *left or right running*. For the sake of discussion, let us agree that

$$\text{for right-running}: \frac{dy}{dx} = \frac{1}{a}(-b + \sqrt{b^2 - ac}) = r(\text{say}) \tag{4.105}$$

$$\text{and for left-running}: \frac{dy}{dx} = \frac{1}{a}\left(-b - \sqrt{b^2 - ac}\right) = l(\text{say}) \tag{4.106}$$

Let us consider a non-characteristic curve AB along which the initial values of u, p and q are prescribed (Fig. 4.10) Then the solution can be computed in the entire triangular region PAB formed by the right-running characteristic PA through A and the left running characteristic PB through B.

For this, the segment AB is subdivided into a few small segments by the points 1, 2, 3,..... If the left and right running characteristics be drawn through these points we get the vertices, 5, 6, 7,....., and repeating this process the network in Fig. 4.10 is obtained. It is now easy to compute the state (i.e. the unknown u and its derivatives p, q etc.) at the points 5, 6, 7, ... For, consider the triangle with vertices 1, 2, 6 where the left-running characteristic through 2 and the right-running through 1 intersect at 6. If the points 1, 2,..... are taken sufficiently close then the arcs $1 \to 6$ and $2 \to 6$ may be approximated by straight segments.

Then, in the first approximation

$$\begin{aligned} y_6 - y_1 &= r_1(x_6 - x_1) \\ y_6 - y_2 &= l_2(x_6 - x_2) \end{aligned} \tag{4.107}$$

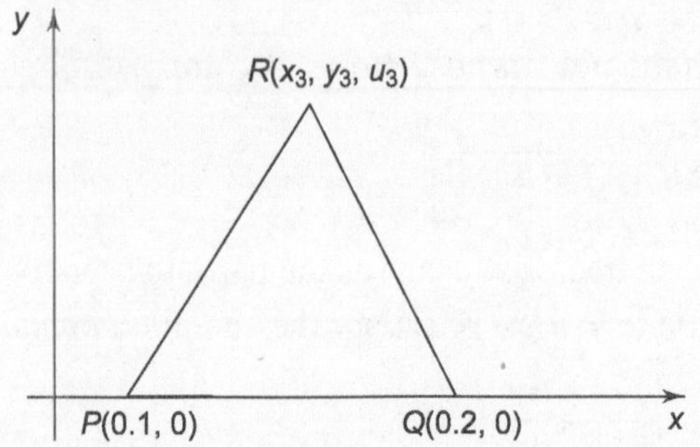

Figure 4.11 Computation of solution by the method of characteristics.

where r_1 is the value of the slope of the right-running characteristic r, given by Eq. (4.105) evaluated at the point 1. Similarly, l_2 denotes the slope of the left-running characteristic evaluated at the point 2. Noting that x_1, x_2, y_1, y_2, r_1 and l_2 are known quantities, the coordinates (x_6, y_6) of the point 6 may be computed, in the first approximation. The compatibility condition (4.104)

$$e\,\mathrm{d}x\frac{\mathrm{d}y}{\mathrm{d}x} + a\,\mathrm{d}p\frac{\mathrm{d}y}{\mathrm{d}x} + c\,\mathrm{d}q = 0 \tag{4.108}$$

may now be integrated approximately along $1 \to 6$ as

$$e_1(x_6 - x_1)\cdot r_1 + a_1(p_6 - p_1)\cdot r_1 + c_1(q_6 - q_1) = 0 \tag{4.109}$$

and along $2 \to 6$ as

$$e_2(x_6 - x_2)l_2 + a_2(p_6 - p_2)l_2 + c_2(q_6 - q_2) = 0 \tag{4.110}$$

Solving the linear system (4.109) and (4.110) the state (p_6, q_6) at the 6 may be obtained in the first approximation. In the first approximation the unknown u at the point 6 is then obtained from

$$u_6 = u_1 + \left(\frac{\partial u}{\partial x}\right)_{\text{Av.}} dx + \left(\frac{\partial u}{\partial y}\right)_{\text{Av.}} dy, \tag{4.111}$$

subscript Av. denoting the average value. Here, we approximate $\dfrac{\partial u}{\partial x}$ by the average value $\dfrac{p_1 + p_6}{2}$ and $\dfrac{\partial u}{\partial y}$ by the average value $\dfrac{q_1 + q_6}{2}$. Further, dx is replaced by $(x_6 - x_1)$ and dy by $(y_6 - y_1)$. So, we have

$$u_6 = u_1 + \frac{1}{2}(x_6 - x_1)(p_1 + p_6) + \frac{1}{2}(y_6 - y_1)(q_1 + q_6) \tag{4.112}$$

Thus the values of x_6, y_6, u_6, p_6 and q_6 are determined in the first approximation. We are now in a position to compute the slopes l_6 and r_6 corresponding to the point 6. The whole process is now repeated to obtain improved values of x_6, y_6, u_6, p_6 and q_6 in the second approximation with r_1 replaced by the average inclination $\frac{1}{2}(r_1 + r_6)$ and l_2 replaced by $\frac{1}{2}(l_2 + l_6)$, and the coefficients a, b, c, e being replaced by the corresponding average values $(a_1 + a_6)/2$, etc. Further iterations, if necessary, may now be carried out to obtain results correct to a prescribed tolerance, although in practice further iterations are not required when the length of the subdivisions are sufficiently small. When the values of x, y, u, u_x, u_y at the point 6 have been obtained in this way, the procedure may be continued to compute the solution at all the points of the triangular region PAB.

Prescribed boundary conditions are easily accommodated. If shock discontinuities appear, formed by intersection of characteristics of the same family, special treatment using the Rankine–Hugoniot shock conditions may be carried out. For details, the books Oswatitsch (1956), Niyogi (1977) may be consulted. These methods of characteristics were extensively used in the days before the digital computer. In modern terminology, they belong to what are known as *shock-fitting methods*.

It may be observed that the method of characteristics deliver the solution at the irregular shaped characteristic grid points. If we require the solution at rectangular grid points some kind of interpolation would be necessary. This is a little inconvenient. Secondly, inconvenience and inaccuracy arise, if the solution becomes discontinuous, for example, if shocks appear. In such cases of inviscid flow computation, particularly those involving strong shocks such as occur in high supersonic and hypersonic flows, the modern shock-capturing and TVD (total variation diminishing) methods are preferred. A rich literature exists on such methods. We mention only a few among them, such as Harten (1983), Hazra (1997), Hazra et al. (1998), Hazra (1999), Hazra et al. (1999), Jameson (1985), Lax(1972), Yee (1989). If no discontinuities are present, the various methods of characteristics are stable and robust and deliver results of high accuracy.

Example 4.4

Compute the solution of the partial differential equation $u_{xx} - u^2 u_{yy} = 0$ at the first characteristic grid point R with $y > 0$ formed by the characteristics through the points $P(0.1, 0)$ and $Q(0.2, 0)$, given the initial conditions

$$y = 0, \quad u = x, \quad u_y = 2x, \quad 0 \leq x \leq 1.$$

Let us call the points P, Q and R as the points 1, 2 and 3 (Fig. 4.11) Then it is given that

$$x_1{=}0.1,\ y_1{=}0,\ u_1{=}0.1,\ p_1{=}1,\ q_1{=}0.2 \text{ and } x_2{=}0.2,\ y_2{=}0,\ u_2{=}0.2,\ p_2{=}1,\ q_2{=}0.4$$

Then the slopes of the characteristics, according to Eq. (4.103) are $\dfrac{dy}{dx} = \pm u$. From the given initial values at P and Q, $u > 0$, so that, along PR : $\frac{\mathrm{d}y}{\mathrm{d}x} = u$ and along QR: $\frac{\mathrm{d}y}{\mathrm{d}x} = -u$. The compatibility condition (4.104) $\frac{\mathrm{d}y}{\mathrm{d}x}dp - u^2 dq = 0$, yields

$$\text{along}\quad PR : dp - udq = 0 \quad \text{and along QR} : dp + udq = 0$$

First approximation

$$\text{along}\quad PR : y_3 - y_1 = 0.1(x_3 - x_1) \quad \Rightarrow \quad y_3 = 0.1(x_3 - 0.1)$$

$$\text{along}\quad QR : y_3 - y_2 = -0.2(x_3 - x_2) \quad \Rightarrow \quad y_3 = -0.2(x_3 - 0.2)$$

$$\text{Solving, } x_3 = 0.1667, \quad y_3 = 0.0067$$

Compatibility condition

$$\text{along}\quad PR : p_3 - p_1 - u_1(q_3 - q_1) = 0 \Rightarrow p_3 - 1 - 0.1(q_3 - 0.2) = 0$$

$$\text{along}\quad QR : p_3 - p_2 + u_2(q_3 - q_2) = 0 \Rightarrow p_3 - 1 + 0.2(q_3 - 0.4) = 0$$

$$\text{solving,}\quad q_3 = 0.3333,\ p_3 = 1.0133,\ \mathrm{d}u = p\mathrm{d}x + q\mathrm{d}y$$

$$\text{along}\quad PR : u_3 - u_1 = \frac{p_3 + p_1}{2}(x_3 - x_1) + \frac{q_3 + q_1}{2}(y_3 - y_1)$$

so that $u_3 = 0.1689$.

Second approximation

$$y_3 - 0 = \frac{1}{2}(0.1+0.1689)(x_3 - 0.1),\ y_3 - 0 = -\frac{1}{2}(0.2 + 0.1689)(x_3 - 0.2),$$

which may be solved to obtain $x_3 = 0.1580$, $\quad y_3 = 0.0078$. Then, the compatibility conditions

$$p_3 - 1 - \frac{1}{2}(0.1689 + 0.1)(q_3 - 0.2) = 0 \quad \text{and}$$

$$p_3 - 1 + \frac{1}{2}(0.1689 - 0.2)(q_3 - 0.4) = 0, \quad \text{yield}$$

$$p_3 = 1.0156, \quad q_3 = 0.3157.$$

Now, $u_3 - 0.1 = \frac{1.0156 + 1}{2}(0.1580 - 0.1) + \frac{0.3157 + 0.2}{2}(0.0078 - 0)$, so that $u_3 = 0.1605$. Further iterations may be carried out as desired.

4.10 MODEL CONVECTION–DIFFUSION EQUATION

One dimensional heat-conduction equation (also known as the diffusion equation) studied in Chapter 3 is the model equation of parabolic type. In the previous sections of the present chapter we discussed the one dimensional convection equation, which is taken as the model equation of hyperbolic type. We now make a brief digression from our study of model equations of hyperbolic type in order to discuss the model convection–diffusion equation. This equation is another model equation of much importance in CFD (computational fluid dynamics) which is of the form

$$\frac{\partial \zeta}{\partial t} + u\frac{\partial \zeta}{\partial x} = \alpha \frac{\partial^2 \zeta}{\partial x^2}. \tag{4.113}$$

Here, u denotes the velocity of convection of some physical property ζ and the positive constant α denotes diffusion. Equation (4.113) is a second-order partial differential equation of parabolic type in the unknown ζ. If u be a known quantity or be independent of ζ then Eq. (4.113) is linear. On the other hand, if u be a function of ζ it is nonlinear. The term $\frac{\partial \zeta}{\partial t}$ is the *unsteady term*, $u\frac{\partial \zeta}{\partial x}$ is the *convection term* (also called *advection* in meteorology; incidentally, it may be noted that the terms convection and advection mean essentially the same thing) and $\alpha\frac{\partial^2 \zeta}{\partial x^2}$ denotes the diffusion term. We note that if the convection term be absent then Eq. (4.113) reduces to the model parabolic type heat-conduction equation (or diffusion equation)

$$\frac{\partial \zeta}{\partial t} = \alpha \frac{\partial^2 \zeta}{\partial x^2}, \tag{4.114}$$

discussed in detail in Chapter 3. On the other hand, if the diffusion term on the right hand side be absent, then Eq. (4.113) reduces to the model hyperbolic-type convection equation

$$\frac{\partial \zeta}{\partial t} + u\frac{\partial \zeta}{\partial x} = 0. \tag{4.115}$$

discussed in the previous sections.

Although the convection–diffusion equation (4.113) is of parabolic type, it has features distinct from that of the parabolic-type heat-conduction equation. Such features arise due to the presence of the convection term. For small values of α compared to the convection speed u, Eq. (4.113) behaves more like that of the convection equation (4.115), which is of hyperbolic type. Due to this, present discussions could not be included in Chapter 3, and we had to wait for the discussions on the convection equation.

We write the linear model convection–diffusion equation as

$$\frac{\partial u}{\partial t} + c\frac{\partial u}{\partial x} = \alpha \frac{\partial^2 u}{\partial x^2}. \tag{4.116}$$

where c denotes the constant velocity of convection that may be positive or negative.

In the steady case, the first term of Eq. (4.116) vanishes and it reduces to the ordinary differential equation

$$c\frac{\mathrm{d}u}{\mathrm{d}x} = \alpha \frac{\mathrm{d}^2 u}{\mathrm{d}x^2}, \tag{4.117}$$

representing balance between convection and diffusion. In many fluid flow problems for fluids with small viscosity, the effects of viscosity is found to be confined in a thin layer near the boundary, while the flow outside this region may be modelled like an inviscid fluid flow. Equation (4.117) is very useful in studying such features. We begin our discussion with the simplest model equation (4.117) in the following subsection.

4.10.1 Steady Convection–Diffusion Equation

Let us consider solution of the 1-D ordinary differential equation with constant coefficients

$$c\frac{du}{dx} - \alpha\frac{d^2u}{dx^2} = 0, \quad \alpha > 0,$$

together with boundary conditions

$$u(0) = 0, \text{ and } u(1) = 1, \tag{4.118}$$

in the domain $0 \leq x \leq 1$. The exact solution of this problem is easily found to be

$$u(x) = \frac{(e^{px} - 1)}{e^p - 1}, \quad p = \frac{c}{\alpha}. \tag{4.119}$$

The exact solution (4.119) for values of the parameter $p = 5, 8, 11, 14$ are shown in Fig. 4.12).

It may be seen that the solution is almost constant except in a very small region near the boundary $x = 1$ where steep gradient is present. We see that a *boundary layer* is formed near the boundary $x = 1$. In these cases, the convection speed c is positive, so that p is positive. If c be negative, that is, for $p < 0$, the boundary layer is formed near the end $x = 0$. The effect is pronounced when diffusion α is small compared to the speed of convection $|c|$, that is when p is large. For flow of fluids of very small viscosity such effects may be seen in a thin layer near the boundary (see Chapter 11 for detailed study).

Let us next compute the solution of Eq. (4.117) with boundary conditions Eq. (4.118) by finite difference. For this the domain $0 \leq x \leq 1$ is discretised as $x = j\Delta x$, with $j = 0, 1, \cdots, N+1$ with $(N+1)\Delta x = 1$. The boundaries $x = 0$ and $x = 1$ correspond respectively to the values $j = 0$ and $j = N + 1$.

Using central difference representation for the derivatives and using the notation $u(x) = u(j\Delta x) = u_j$, the problem is expressed as

$$c\frac{u_{j+1} - u_{j-1}}{2\Delta x} - \alpha\frac{u_{j+1} - 2u_j + u_{j-1}}{\Delta x^2} = 0,$$
$$j = 1, 2, \cdots, N \tag{4.120}$$

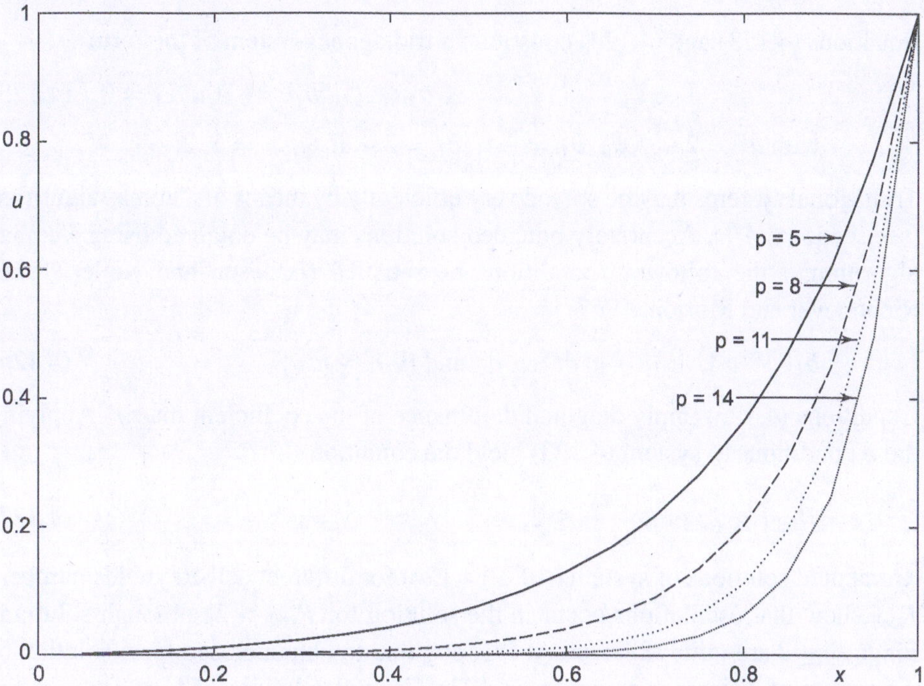

Figure 4.12 Exact solution of steady convection–diffusion equation for different values of $p = c/\alpha = 5, 8, 11, 14$.

and the boundary conditions are

$$u_0 = 0, \text{ and } u_{N+1} = 1. \tag{4.121}$$

We now introduce a very important dimensionless parameter R_{cell} called *cell-Reynolds number or* mesh Reynolds number (strictly, *Peclet number*) based on mesh-length Δx defined by

$$R_{\text{cell}} = \frac{c\Delta x}{\alpha} \tag{4.122}$$

Simplifying Eqs. (4.120) yields the linear system of algebraic equations

$$(2 + R_{\text{cell}})u_{j-1} - 4u_j + (2 - R_{\text{cell}})u_{j+1} = 0, \tag{4.123}$$

$$u_0 = 0, \quad u_{N+1} = 1. \tag{4.124}$$

Equations (4.123) and (4.124) constitute a tridiagonal system of the form

$$a_j u_{j-1} + b_j u_j + c_j u_{j+1} = 0, \quad (4.125)$$

$$\text{with } a_j = 2 + R_{\text{cell}},\ b_j = -4,\ c_j = 2 - R_{\text{cell}},\ j = 1, 2, \cdots, N$$

Tridiagonal systems may be solved very efficiently by means of Thomas algorithm (see section 3.4.1). Accurately bounded solutions may be obtained using Thomas algorithm if the following conditions be satisfied (Isaacson and Keller, 1966; Richtmeyer and Morton, 1967)

$$|b_1| > |c_1|, \ |b_j| \geq |a_j| + |c_j|, \text{ and } |b_N| > |a_N| \quad (4.126)$$

Conditions (4.126) imply diagonal dominance of the coefficient matrix. Applying these conditions to system (4.123) yield the condition

$$|R_{\text{cell}}| \leq 2, \text{ or } \frac{|c|\Delta x}{\alpha} \leq 2. \quad (4.127)$$

Computed solutions of system (4.123, 4.124) for different cell-Reynolds numbers R_{cell} show that oscillations occur in the solution for $R_{\text{cell}} > 2$, although solutions for $R_{\text{cell}} \leq 2$ are quite satisfactory. It is possible to obtain closed form solution of the system of difference equations (4.123). General solutions of (4.123) are of the form

$$u_j = A_1 q_1^j + A_2 q_2^j, \ A_1, A_2 \text{ arbitrary constants,}$$

where q_1, q_2 are the roots of the characteristic equation

$$(2 - R_{\text{cell}})q^2 - 4q + (2 + R_{\text{cell}}) = 0. \quad (4.128)$$

Solving Eq. (4.128) we get the roots

$$q_1 = 1, \text{ and } q_2 = \frac{2 + R_{\text{cell}}}{2 - R_{\text{cell}}} \quad (4.129)$$

Using boundary conditions (4.121) and simplifying yields the exact solution of the finite difference system (4.123)

$$u_j = \frac{q_2^j - 1}{q_2^{N+1} - 1}, \quad (4.130)$$

the quantity q_2 being given by Eq. (4.129). Note that for $R_{\text{cell}} > 2$, q_2 is negative; the solution indicates oscillatory behaviour.

Finite difference solution of system (4.124) obtained by Thomas algorithm has been compared with the exact solution (4.130), denoted by xuer in Fig. 4.13, for values of $R_{\text{cell}} = 1.5, 3.0$. Reasonable agreement may be seen for $R_{\text{cell}} = 1.5$, while those corresponding to $R_{\text{cell}} = 3$ give inaccurate results and clearly show nonphysical oscillations. Such oscillations cannot be tolerated in solutions of fluid

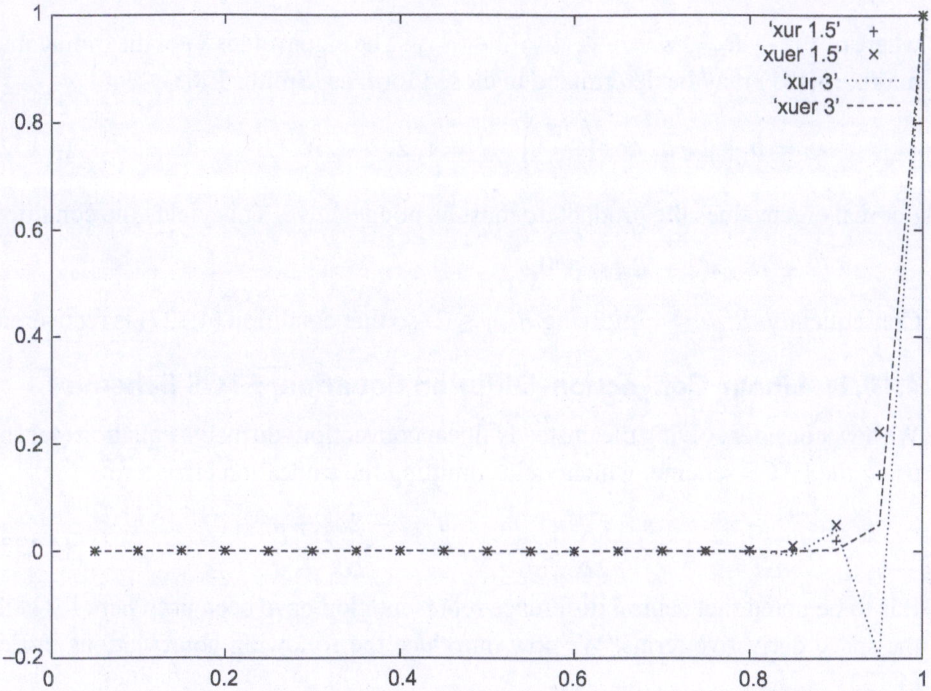

Figure 4.13 Finite difference (xur) and exact (xuer) solution of steady convection–diffusion equation for Cell Reynolds numbers 1.5 and 3.

flow problems. Experience shows that use of central difference in representation for the convective term generally leads to such oscillations for values of $R_{\text{cell}} > 2$, while for $R_{\text{cell}} \leq 2$ such oscillations disappear.

It is interesting to observe that for $R_{\text{cell}} \leq 2$ the eigenvalues of the coefficient matrix of system (4.123) are all real.

Example 4.5

Let us find the condition that all the eigenvalues of system (4.123) be real.

The coefficient matrix of system (4.123) is of the form

$$\begin{pmatrix} b & c & & & & & \\ a & b & c & & & O & \\ \cdots & \cdots & \cdots & \cdots & \cdots & \cdots & \cdots \\ \cdots & \cdots & \cdots & \cdots & \cdots & \cdots & \cdots \\ & O & & & a & b & c \\ & & & & & a & b \end{pmatrix} \tag{4.131}$$

where $a = 2 + R_{\text{cell}},\ b = -4,\ c = 2 - R_{\text{cell}}$. The eigenvalues λ_j of the tridiagonal matrix (4.131) may be determined in closed form as (Smith, 1965).

$$\lambda_j = b + 2\sqrt{ac}\cos\left(\frac{j\pi}{N}\right),\quad j = 1, 2, \cdots, N. \tag{4.132}$$

For real eigenvalues the product ac must be nonnegative. This yields the condition

$$(2 + R_{\text{cell}})(2 - R_{\text{cell}}) \geq 0,$$

Consequently, $R_{\text{cell}}^2 \leq 4$ implying $R_{\text{cell}} \leq 2$, so that condition (4.127) is recovered.

4.10.2 Linear Convection–Diffusion Equation: FTCS Scheme

We now consider solving the unsteady linear convection–diffusion equation (4.116) using the FTCS scheme, which yields omitting the truncation error terms

$$\frac{u_j^{n+1} - u_j^n}{\Delta t} + c\frac{u_{j+1}^n - u_{j-1}^n}{2\Delta x} = \alpha\frac{u_{j+1}^n - 2u_j^n + u_{j-1}^n}{\Delta x^2} \tag{4.133}$$

It is to be noted that central difference representation have been used here for both the space derivative terms. We now introduce the following abbreviations in Eq. (4.133)

$$R_{\text{cell}} = \frac{c\Delta x}{\alpha},\quad \nu = \frac{c\Delta t}{\Delta x}\quad r = \frac{\alpha\Delta t}{\Delta x^2},\ \text{and}\ s = \frac{1}{2}\nu, \tag{4.134}$$

where R_{cell} denotes the cell-Reynolds number, ν denotes the Courant number and r is the diffusion parameter introduced in Chapter 3. Simplifying Eq. (4.133) we get

$$u_j^{n+1} = (r - s)u_{j+1}^n + (1 - 2r)u_j^n + (r + s)u_{j-1}^n, \tag{4.135}$$

which is an explicit scheme. Without much difficulty, we may derive the modified equation for the FTCS scheme, (using Taylor expansion and eliminating the time derivative terms), which yields

$$u_t + cu_x = \left(\alpha - \frac{1}{2}c^2\Delta t\right)u_{xx} + \frac{c\Delta x^2}{3}\left(3r - \nu^2 - \frac{1}{2}\right)u_{xxx} + \cdots \tag{4.136}$$

We see that the scheme is consistent with the differential equation (4.116) and that its accuracy is $O(\Delta t, \Delta x^2)$.

In order to study stability by von Neumann method, we use trial solution

$$u_j^n = e^{\alpha n\Delta t}e^{i\beta j},\quad i = \sqrt{-1},\ \beta\ \text{real},\ \alpha\ \text{may be complex}, \tag{4.137}$$

in Eq. (4.135). It yields after simplification

$$e^{\alpha\Delta t} = (1 - 2r + 2r\cos\beta) - 2is\sin\beta. \tag{4.138}$$

Von Neumann stability condition $|e^{\alpha\Delta t}| \le 1$ yields

$$(1 - 2r + 2r\cos\beta)^2 + 4s^2\sin^2\beta \le 1. \tag{4.139}$$

Writing for the sake of abbreviation $z = \cos\beta$ in it and simplifying follows

$$(1 - z)[-2r(2 - 2r + 2rz) + 4s^2(1 + z)] \le 0. \tag{4.140}$$

Since $1 - z \ge 0$, we require $E(z) \le 0$, where

$$E(z) = -r(1 - r + rz) + s^2(1 + z). \tag{4.141}$$

However, $E(z)$ is a linear function of z, and $-1 \le z \le 1$. So it is sufficient to require

$$E(-1) \le 0, \text{ and } E(1) \le 0. \tag{4.142}$$

The former condition yields $-r + 2r^2 \le 0$ which imply

$$r \le \frac{1}{2} \text{ since } r > 0. \tag{4.143}$$

The latter condition delivers

$$\begin{aligned} 2s^2 - r \le 0, &\Rightarrow \nu^2 \le 2r, \\ \text{or, } &\frac{\nu}{r} \le \frac{2}{\nu} \end{aligned} \tag{4.144}$$

But, R_{cell} is the ratio of *convection to diffusion,* since,

$$R_{\text{cell}} = \frac{c\Delta x}{\alpha} = \frac{c\Delta t}{\Delta x}\frac{\Delta x^2}{\alpha\Delta t} = \frac{\nu}{r} \tag{4.145}$$

Consequently, combining conditions (4.142) and (4.144), we obtain

$$2\nu \le R_{\text{cell}} \le \frac{2}{\nu}. \tag{4.146}$$

We observed earlier that oscillations or wiggles may appear in the computed solution for $R_{\text{cell}} > 2$, so that for values of cell-Reynolds number in the range $2 < R_{\text{cell}} \le \frac{2}{\nu}$ the FTCS scheme, although stable may give rise to wiggles.

4.10.3 First-Order Upwind Scheme for Convection–Diffusion Equation

The wiggles encountered in FTCS solution, mentioned in the previous section, disappear if instead of central difference (which is symmetric) the first-order backward difference (which is one-sided) be used to represent the convective term in Eq. (4.116). This delivers for $c = \text{const.} > 0$,

$$\frac{u_j^{n+1} - u_j^n}{\Delta t} + c\frac{u_j^n - u_{j-1}^n}{\Delta x} = \alpha\frac{u_{j+1}^n - 2u_j^n + u_{j-1}^n}{\Delta x^2}. \tag{4.147}$$

Substituting $\nu = c\dfrac{\Delta t}{\Delta x}$, $r = \dfrac{\alpha\Delta t}{\Delta x^2}$ in (4.147) we get on simplification the first-order explicit upwind scheme

$$u_j^{n+1} = (r + \nu)u_{j-1}^n + (1 - \nu - 2r)u_j^n + ru_{j+1}^n. \tag{4.148}$$

von Neuman stability analysis shows that the scheme is stable for $\nu + 2r \leq 1$. Computed solutions show no wiggles or oscillations. Taylor's expansion shows that the scheme is consistent and is only first order accurate. The modified equation is found to be

$$u_t + cu_x = \left[\alpha(1 + \frac{1}{2}R_{\text{cell}}) - \frac{1}{2}c^2\Delta t\right]u_{xx} + \cdots, \tag{4.149}$$

so that, compared to the modified equation of FTCS scheme Eq. (4.136), the upwind scheme introduces an additional artificial viscosity amounting to $\frac{1}{2}\alpha R_{\text{cell}}u_{xx}$. For $R_{\text{cell}} > 2$, this term is greater than the natural viscosity present. Due to this, the scheme leads to inaccurate results and is not useful in practical fluid flow computations. Instead, higher order upwind schemes have been put forward (Fletcher, 1988a; Leonard, 1979) where the convective term has been discretised with four-point upwind representation

$$\frac{\partial u}{\partial x} = \begin{cases} \dfrac{u_{j+1} - u_{j-1}}{2\Delta x} + \dfrac{p(u_{j-2} - 3u_{j-1} + 3u_j - u_{j+1})}{3\Delta x}, & \text{for } c > 0, \\ \dfrac{u_{j+1} - u_{j-1}}{2\Delta x} + \dfrac{p(u_{j-1} - 3u_j + 3u_{j+1} - u_{j+2})}{3\Delta x}, & \text{for } c < 0, \end{cases} \tag{4.150}$$

Here p is a parameter. It may be established that for $p = \frac{1}{2}$, (4.150) approximates u_x to $O(\Delta x^3)$ while for any other value of p it is $O(\Delta x^2)$. (See problem 8, Exercise 4). Solutions computed with representation (4.150) lead to significant improvement in accuracy.

4.10.4 Burgers Equation

A model equation of much importance in fluid dynamics is the Burgers equation (Burgers, 1948; Hopf, 1950)

$$u_t + uu_x = \alpha u_{xx}, \ \alpha = \text{const.} > 0. \tag{4.151}$$

It is obtained from the linear convection–diffusion equation (4.116) by replacing the convection term cu_x by uu_x. It is a nonlinear parabolic-type equation and has the type of nonlinear features as that of 1-D Navier–Stokes equation. Burgers equation, (Fletcher, 1983; Whitham, 1974) is found very suitable in modelling essentially nonlinear features like turbulence and shock waves. Moreover, it is useful in problems where nonlinearity, convection and diffusion are of particular interest, like boundary-layer study (see Chapter 11).

For many combinations of boundary and initial conditions exact solutions of Burgers equation have been found. These are useful in evaluating the accuracy and performance of different numerical schemes. For example, the exact steady-state solution of Eq. (4.151) (that is the solution $u(x, t)$ as $t \to \infty$) for boundary conditions

$$u(0, t) = U, \text{ and } u(l, t) = 0, \tag{4.152}$$

is given by

$$u = Uu' \left\{ \frac{1 - \exp[u'R(x/l - 1)]}{1 + \exp[u'R(x/l - 1)]} \right\} \tag{4.153}$$

where $R = \dfrac{ul}{\alpha}$ and u' is a solution of

$$\frac{u' - 1}{u' + 1} = e^{-u'R}. \tag{4.154}$$

Finite difference FTCS scheme for numerical solution of Burgers equation (4.151) is

$$\frac{u_j^{n+1} - u_j^n}{\Delta t} + \frac{u_j^n(u_{j+1}^n - u_{j-1}^n)}{2\Delta x} = \alpha \frac{u_{j+1}^n - 2u_j^n + u_{j-1}^n}{\Delta x^2}, \tag{4.155}$$

which on simplification yields

$$u_j^{n+1} = -\frac{\Delta t}{2\Delta x} u_j^n(u_{j+1}^n - u_{j-1}^n) + (1 - 2r)u_j^n + r(u_{j-1}^n + u_{j+1}^n) \tag{4.156}$$

This is an explicit scheme. In order to carry out stability analysis, we note that von Neumann stability analysis is applicable to linear equations only. To gain insight into stability behaviour, the coefficients are assumed locally frozen and taken as constants. The idea is that if the scheme so treated be unstable, it is likely to be more so, under nonlinear conditions.

We consider next the conservative form of Burgers equation

$$\frac{\partial u}{\partial t} + \frac{\partial F}{\partial x} = \alpha \frac{\partial^2 u}{\partial x^2}, \quad F = \frac{u^2}{2}. \tag{4.157}$$

Schemes for hyperbolic-type convection equation written in conservation form as

$$u_t + F_x = 0, \tag{4.158}$$

have been discussed in the previous sections. These schemes may be generalised and applied to study (4.157). The FTCS scheme for (4.157) is

$$\frac{u_j^{n+1} - u_j^n}{\Delta t} + \frac{F_{j+1}^n - F_{j-1}^n}{2\Delta x} = \alpha \frac{u_{j+1}^n - 2u_j^n + u_{j-1}^n}{\Delta x^2}. \tag{4.159}$$

A more accurate treatment, suggested in the higher order upwind representation (4.150) is

$$\frac{\partial F}{\partial x} = \begin{cases} \dfrac{F_{j+1} - F_{j-1}}{2\Delta x} + \dfrac{p(F_{j-2} - 3F_{j-1} + 3F_j - F_{j+1})}{3\Delta x}, & \text{for } u > 0, \\ \dfrac{F_{j+1} - F_{j-1}}{2\Delta x} + \dfrac{p(F_{j-1} - 3F_j + 3F_{j+1} - F_{j+2})}{3\Delta x}, & \text{for } u < 0, \end{cases} \tag{4.160}$$

Note that for $p = \frac{1}{2}$, (4.160) is third order, and is of second order for any other value of p.

Among other second-order schemes, the two-step Lax–Wendroff and the MacCormack scheme have been extensively studied and may be applied to nonlinear equations like the Burgers equation. The MacCormack scheme has been applied successfully to Euler equations and Navier–Stokes equations. However, the solutions loose accuracy as the time grows.

Recently, *higher-order compact* schemes have been developed with weighted time discretisation, by Kalita et al. (2002). The schemes are second- or lower-order accurate in time depending on the weighted average parameter μ and fourth-order accurate in space. For $0 \leq \mu \leq 1$, the schemes are unconditionally stable and efficiently capture both transient and steady solutions of linear and nonlinear convection–diffusion equations with Dirichlet as well as Neumann boundary conditions. Using stream-function vorticity formulation, these schemes have been applied to study the flow in a thermally driven square cavity with adiabatic top and bottom walls and differentially heated vertical walls in Kalita et al. (2001).

4.11 SUMMARY

Using the first-order linear wave equation (also known as the convection or the advection equation) as model of hyperbolic-type partial differential equations, sevaral well-known finite difference schemes have been studied with respect to the properties: (a) consistency, (b) stability, (c) amplitude and phase error and (d) modified equation. It is found that

1. FTCS scheme is unconditionally unstable; of no use.
2. FTFS scheme is unconditionally unstable; of no use.

3. First order upwind scheme stable if and only if the Courant number $\nu = \dfrac{c\Delta t}{\Delta x}$ satisfies $0 \leq \nu \leq 1,\ c > 0$.
4. Leap-frog scheme.
5. Lax scheme not uniformly consistent; consistent for a fixed value of the Courant number.
6. Lax–Wendroff and variants, stable for $|\nu| \leq 1$.
7. MacCormack scheme stable for $|\nu| \leq 1$.
8. Warming–Beam upwind scheme and
9. Implicit schemes unconditionally stable.

Properties and methods of solution for first-order hyperbolic systems and hyperbolic system of conservation laws are discussed next. Solution of the Cauchy problem for second-order linear wave equation is discussed in the last section. It is found that the leap-frog scheme is stable for Courant number $\nu < 1$. Also, in general, the scheme does not converge for $\nu > 1$.

A brief digression is made next in order to investigate the basic properties and methods of solving the convection–diffusion equation. This equation, although of parabolic type show certain new features different from that of the parabolic-type heat-conduction equation. In particular, it is found that the cell-Reynolds number, R_{cell} plays an important role in determining properties of the solution. For $0 \leq R_{\text{cell}} \leq 2$, the solutions are well behaved while for $R_{\text{cell}} > 2$, solutions show nonphysical oscillations. Stability depends on two parameters, namely $r = \dfrac{\alpha \Delta t}{\Delta x^2}$ and the Courant number $\nu = \dfrac{c\Delta t}{\Delta x}$.

Programs – The program lxmc.f in FORTRAN (C-program in appendix) with test results is presented. It solves by Lax–Wendroff and MacCormack scheme, the one-dimensional, nonlinear Burger's equation.

4.12 KEY TERMS

Advection equation
Amplification factor
Amplitude error
Approximate factorisation method
Cauchy problem
Cell-Reynolds number
Characteristics
Characteristic cannonical form
FTCS scheme
Hyperbolic-type equation
Hyperbolic system of conservation laws
Implicit scheme
Inviscid Burger's equation
Jacobian matrix
Lax scheme
Lax–Wendroff scheme

Conservative equations	Leap-frog scheme
Conservation law form	MacCormack scheme
Consistency	Modified equation
Convection equation	Numerical flux function
Convection–diffusion equation	Numerical characteristics
Convergence	Pedet number
Courant number	Predictor-Corrector schemes
Cell Reynolds number	Upwind scheme
Diffusion equation	von Neuman stability analysis
First-order hyperbolic system	Wave equation.

4.13 EXERCISE 4

4.1 Compute the solution for the first two-time steps, of the first-order wave equation

$$u_t + cu_x = 0, \;\; c = \text{const.} > 0, \tag{4.161}$$

with initial condition

$$u(x,0) = \begin{cases} 0, & x < 0, \\ \frac{1}{2}x, & 0 \le x \le 1 \\ 1 - \frac{1}{2}x. & 1 \le x \le 2 \\ 0, & x > 2, \end{cases}$$

using the first-order upwind scheme with $\Delta x = 0.2, \;\; \nu = \frac{1}{2}$.

4.2 Compute the first two-time step solutions of problem 1, by the MacCormack's scheme, other conditions remaining unaltered.

4.3 Investigate by von Neumann method, stability of the fully implicit scheme (4.60) for the wave equation (4.161). Show that it is stable for all values of $\nu = \dfrac{c\Delta t}{\Delta x}$.

4.4 Show that the leap-frog scheme (both time and space derivatives represented by central difference) for (4.161) is second order accurate in both time and space. Show further, that this scheme is stable if and only if $\left|\dfrac{c\Delta t}{\Delta x}\right| \le 1$.

4.5 Consider computation of solution of the equation

$$u_t + cu_{xx} = f(x,t),$$

by the leap-frog scheme

$$\frac{u_j^{n+1} - u_j^n}{\Delta t} + c\frac{u_{j+1}^n - u_{j-1}^n}{2\Delta x} = f_j^n, \tag{4.162}$$

Study its stability.

Show that the changed scheme

$$\frac{1}{\Delta t}\left[u_j^{n+1} - \frac{1}{2}(u_j^{n+1} + u_j^{n-1})\right] + c\left[\frac{u_{j+1}^n - u_{j-1}^n}{2\Delta x}\right] = f_j^n \qquad (4.163)$$

is $O(\Delta t) + O(\Delta x^2)$ and is stable if and only if $|\frac{c\Delta t}{\Delta x}| \leq 1$.

4.6 Compute the solution of the initial boundary value problem

$$u_{tt} = c^2 u_{xx}, \quad c = \text{const.},\ 0 < x < 1,\ t > 0, \qquad (4.164)$$

with boundary conditions

$$u(0,t) = 0 = u(1,t)$$

and initial conditions

$$u(x,0) = 0.8 \sin \pi x,\ u_t(x,0) = 0,\ 0 \leq x \leq 1,$$

for the first three-time steps with $\Delta x = 0.2$ and $\nu = \dfrac{c\Delta t}{\Delta x} = 0.5$, $c = 1$, using the explicit scheme.

4.7 Compute the first three-step solution of the above problem using the implicit scheme, with $\nu = \dfrac{c\Delta t}{\Delta x} = 1$. Compare with the exact solution.

4.8 Establish that the modified equation for the FTCS scheme applied to the linear convection–diffusion equation (4.116) is

$$u_t + cu_x = \left(\alpha - \frac{1}{2}c^2\Delta t\right)u_{xx} + \frac{c\Delta x^2}{3}\left(3r - \nu^2 - \frac{1}{2}\right)u_{xxx} + \cdots.$$

4.9 Establish the modified equation

$$u_t + cu_x = \left[\alpha(1 + \frac{1}{2}R_{\text{cell}}) - \frac{1}{2}c^2\Delta t\right]u_{xx} + \cdots,$$

corresponding to the first-order upwind scheme (4.147) applied to the linear convection–diffusion equation (4.116).

4.10 Using von Neumann stability analysis show that the first-order upwind scheme (4.147) is stable for time-step

$$\Delta t \leq \frac{1}{\dfrac{c}{\Delta x} + \dfrac{2\alpha}{\Delta x^2}}.$$

4.11 Show that the modified equation for the leap-frog scheme, for the first-order convection equation is

$$u_t + cu_x = -\frac{c\Delta x^2(1-\nu^2)}{6}u_{xxx} - \frac{c\Delta x^4}{120}(9\nu^4 - 10\nu^2 + 1)u_{xxxxx} + \cdots.$$

4.12 Establish that the amplification factor G of the Lax–Wendroff scheme Eq. (4.43) is given by

$$|G|^2 = 1 - 4\nu^2(1-\nu^2)\sin^4\frac{\beta}{2}.$$

4.13 Establish that the amplification factor G of the Warming and Beam upwind scheme (4.57) may be expressed as

$$|G| = 1 - 4\nu(1-\nu)^2(2-\nu)\sin^4\frac{\beta}{2}.$$

4.14 Establish that the modified equation for the Crank–Nicholson type of implicit scheme (Section (4.4) is

$$u_t + cu_x = -\frac{c}{12}[c^2\Delta t^2 + 2\Delta x^2]u_{xxx}$$

$$-\frac{c}{240}[2\Delta x^4 + 10c^2\Delta t^2\Delta x^2 + 3c^3\Delta t^4]u_{xxxxx} + \cdots.$$

4.15 Assuming u to be sufficiently smooth, establish that the upwind representations of the convective term in (4.116)

$$\frac{\partial u}{\partial x} = \begin{cases} \dfrac{u_{j+1}-u_{j-1}}{2\Delta x} + \dfrac{p(u_{j-2}-3u_{j-1}+3u_j-u_{j+1})}{3\Delta x}, & \text{for } c > 0, \\ \dfrac{u_{j+1}-u_{j-1}}{2\Delta x} + \dfrac{p(u_{j-1}-3u_j+3u_{j+1}-u_{j+2})}{3\Delta x}, & \text{for } c < 0, \end{cases}$$

p being a parameter is $O(\Delta x^3)$ for $p = \frac{1}{2}$ and $O(\Delta x^2)$ for any other value of p.

4.16 Show that first-order upwind representation for the convection term in steady convection–diffusion equation (4.117) yields the upwind scheme

$$-(1 + R_{\text{cell}})u_{j-1} + (2 + R_{\text{cell}})u_j - u_{j+1} = 0. \tag{4.165}$$

Write a computer program to solve the system for $p = c/\alpha = 25$.

4.17 Compute finite difference solution of the steady convection–diffusion equation for $R_{\text{cell}} = 1, 2, 3$ and compare with the exact solution. Take $\Delta x =$ 0.1 and 0.05, corresponding to R_{cell} values 2.5 and 1.25, and boundary conditions (4.118) and compare with the exact solution (4.119).

5

Equations of Elliptic Type

Finite difference method for model hyperbolic-type equations were discussed in the previous chapter. In this chapter, finite difference method for model equations of elliptic type have been discussed. Steady-state problems or potential problems are governed by elliptic type equations. Laplace and Poisson equations, that occur abundantly in science and engineering are typical examples of elliptic type equation. A large sparse system of algebraic equations results through discretisation of elliptic equations. So methods for solving such systems of algebraic equations are important in the present discussion.

5.1 INTRODUCTION

As mentioned in Chapter 1, steady-state problems are governed by partial differential equations of elliptic type. In such problems the unknown quantities do not depend on the physical quantity time. So the governing equations are independent of time. For example, the continuity equation for an incompressible fluid is of elliptic type. The fluid flow problems governed by elliptic-type equations, in contrast to those governed by parabolic or hyperbolic type, do not show any directional bias. Consequently, it is suitable to use central-difference representation for the derivatives. The domains of interest are often closed, although they may be unbounded or infinite, while domains for parabolic or hyperbolic type equations are usually open in the time or time-like direction. The boundary value problems are either of the Dirichlet type, where the prescribed boundary values for the unknowns are continuous functions, or of Neumann type with derivative boundary conditions, or a linear combination of the two. The difficulties encountered in solving elliptic-type equations numerically arise from two requirements. First, we are faced with the need to solve a very large number of discretised equations, that is, a large algebraic system has to be solved efficiently. For nonlinear problems the discretised equations are nonlinear. Often the domain involves irregular shapes or complex geometries. Various acceleration techniques, like the approximate factorisation (AF) schemes or the multi-grid techniques are used. More and more powerful and/or parallel computers are used to tackle the first difficulty while body-fitted grid generation techniques (see Section 5.6 and Chapter 8) are used for overcoming the second difficulty.

We begin our study of elliptic-type equations with the simplest and most common equation viz. the Laplace equation in two space dimensions.

5.2 THE LAPLACE EQUATION IN TWO DIMENSION

Let us consider solution of the Laplace equation in 2-D (two dimensions)

$$u_{xx} + u_{yy} = 0 \tag{5.1}$$

in a simply connected domain Ω with a smooth boundary $\partial\Omega$, on which is prescribed a Dirichlet type boundary condition $u = f(x, y)$, f being a continuous function. As mentioned above, an irregular boundary shape is first transformed into a rectangle in the computational domain with edges parallel to the coordinate directions. Thus, for most of the common problems, the domain is a rectangle. A typical boundary value problem for 2-D Laplace equation was briefly discussed in Chapter 1, Section 1.1. Here, we consider the problem in detail, for a rectangular domain, assumed for the sake of simplicity, to be situated in the first quadrant of the x, y-plane.

For finite difference formulation, the rectangular domain is subdivided into a network by drawing straight lines $x = j\Delta x$ and $y = k\Delta y$, $j = 1, 2, \ldots, N$, $k = 1, 2, \ldots, M$, parallel to the coordinate axes, as shown in Fig. 5.1. The lines $j = 0$, $j = N + 1$, $k = 0$ and $k = M + 1$ are the boundaries of the domain. Using finite difference notation

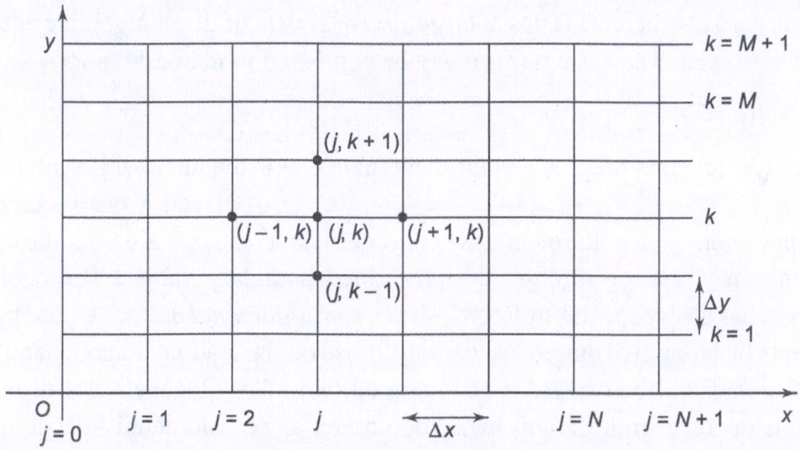

Figure 5.1 Finite difference discretisation.

$u(x, y) = u(j\Delta x, k\Delta y) = u_{j,k}$ the derivatives u_{xx} and u_{yy} may be represented by three-point central difference as

$$u_{xx}]_{j,k} = \frac{1}{\Delta x^2}(u_{j+1,k} - 2u_{j,k} + u_{j-1,k}) + O(\Delta x^2),$$

$$u_{yy}]_{j,k} = \frac{1}{\Delta y^2}(u_{j,k+1} - 2u_{j,k} + u_{j,k-1}) + O(\Delta y^2)$$

Neglecting the truncation error terms, and substituting these representations in Eq. (5.1) yields a system of algebraic equations, one equation for each of the mesh points (j, k)

$$\frac{1}{\Delta x^2}(u_{j+1,k} - 2u_{j,k} + u_{j-1,k}) + \frac{1}{\Delta y^2}(u_{j,k+1} - 2u_{j,k} + u_{j,k-1}) = 0, \quad (5.2)$$

$$j = 1, 2, \ldots, N, \quad k = 1, 2, \ldots, M.$$

The boundary conditions are appropriately discretised. Values of the unknowns at all the boundary grid points on the boundaries $j = 0$ and $j = N + 1$, for all k and on the boundaries $k = 0$ and $k = M + 1$ for all j, are prescribed for a Dirichlet problem. Our object is to solve Eqs. (5.2) efficiently for large values of N and M, that is, for large values of $n = NM$. We note that, the coefficient matrix of this system has a very large number of zero elements, each equation containing only five non zero coefficients. Thus, a *large sparse system* of linear algebraic equations has to be solved. The system (5.2) may be expressed in matrix notations as

$$Au = f \quad (5.3)$$

where $A = (a_{jk})$, is the $n \times n$ coefficient matrix, u is the unknown column vector $u = (u_{11}, \ldots, u_{1M}, u_{21}, \ldots, u_{2M}, \ldots, u_{N1}, \ldots, u_{NM})^{\mathrm{T}}$ and f is a n-component column vector. Most of the elements of the coefficient matrix A are zero; the nonzero elements in f arising through the prescribed boundary values. The coefficient matrix A has a very special structure—it is a *penta-diagonal matrix*, whose nonzero elements lie along five diagonals, namely, those on the leading diagonal and on the two diagonals on the two sides of it, and on two other diagonals at a distance N columns or rows from the left hand top corner, as seen in detail in Example 3.4 in Chapter 3. It may be mentioned that efficient direct elimination methods for solving such a system require of the order of MN^3 (assuming $N < M$) arithmetic operations. In order to decide on the efficiency of a particular method for solving the large sparse system (5.3) we calculate the number of multiplications and divisions required by a particular method to solve it and neglect the additions and subtraction. For moderate values of M and N, say $M = N = 100$, $MN^3 = 10^2 \cdot (10^2)^3 = 10^8$ which is quite a large number. We look for other efficient methods like iterative methods. Discussion on iterative methods for solving linear algebraic system may be found in standard literature, for example Golub and van Loan (1989), Isaacson and

Keller (1966), Niyogi (2003). For the sake of completeness, we make a digression here and discuss briefly the basic ideas of some of the important iterative methods.

5.3 ITERATIVE METHODS FOR SOLUTION OF LINEAR ALGEBRAIC SYSTEMS

We consider a system of n linear algebraic equations in n unknowns, expressed in matrix notation as

$$Au = f \tag{5.4}$$

where $A = (a_{ij})$ is the $n \times n$ coefficient matrix, and u and f are n component vectors, $u = (u_1, u_2, \ldots, u_n)^{\mathrm{T}}$ being the unknown vector and $f = (f_1, f_2, \ldots, f_n)^{\mathrm{T}}$ the known vector of inhomogeneous terms. In most problems of practical interest the matrix A is nonsingular, that is, $\det(A) \neq 0$, so that system (5.4) is uniquely solvable. We are particularly interested in the case where the number of unknowns n is large. The methods for solving such large systems are mainly divided into two categories, namely, *direct and iterative.* The elimination methods are the direct methods. They compute the exact solution of the system in a finite number of steps, although such solutions are generally, mixed with *round-off errors.* We do not discuss the direct methods here, but refer to Datta (1995), Golub and van Loan (1989), Isaacson and Keller (1966), Niyogi (2003) for a thorough discussion.

The iterative methods start from an initial guess $u^{(0)}$ of the solution which is improved step by step following a well-defined rule, called an *iterative scheme.* If the scheme converges, the computed solution $u^{(\nu)}$ approaches the exact solution of the system only as the number of iteration steps ν approaches infinity. However, solution to any desired degree of accuracy may be obtained in a finite number of steps. The question of convergence is very important for such schemes and practical considerations require the scheme to converge fast. We discuss here, briefly, some of the important features of the commonly used iterative methods.

Let the coefficient matrix A be broken up into two nonsingular matrices N and P of order $n \times n$, such that $A = N - P$. Then the given system (5.4) is rewritten as

$$Nu = Pu + f \tag{5.5}$$

We now define an iteration scheme which expresses the unknown vector $u^{(\nu)}$ at step ν in terms of that of the previous step according to the rule

$$Nu^{(\nu)} = P\,u^{(\nu-1)} + f, \quad u^{(0)} = \text{arbitrary}; \quad \nu = 1, 2, \cdots \tag{5.6}$$

By assumption, the matrix N is nonsingular, and so N^{-1} exists uniquely. Consequently, beginning from $u^{(0)}$ all the iterates $u^{(1)}, u^{(2)}, u^{(3)}, \ldots,$ may be uniquely determined according to the rule, Eq. (5.6). Since this equation is to be solved repeatedly at every iteration step, we immediately see that a practical requirement for efficiency is that the matrix N should be easily invertible.

Let the error vector $e^{(\nu)}$ be defined as

$$e^{(\nu)} = \text{true value} - \text{approximate value} = u - u^{(\nu)}.$$

Then, subtracting Eq. (5.6) from (5.5), we get

$$Ne^{(\nu)} = Pe^{(\nu-1)}. \tag{5.7}$$

Premultiplying both sides of Eq. (5.7) by N^{-1} yields

$$e^{(\nu)} = Me^{(\nu-1)}, \text{ where } M = N^{-1}P.$$

Since the matrix N is nonsingular, M is a unique matrix. Replacing here ν by $\nu - 1$, we get $e^{(\nu-1)} = Me^{(\nu-2)}$, so that substituting for $e^{(\nu-1)}$ in Eq. (5.7) yields

$$e^{(\nu)} = M(M\,e^{(\nu-2)}) = M^2\,e^{(\nu-2)}.$$

Repeating the process, we get

$$e^{(\nu)} = M^\nu\,e^{(0)}. \tag{5.8}$$

Consequently, *the necessary and sufficient condition* that $e^{(\nu)} \longrightarrow 0$(null vector) for arbitrary initial error vector $e^{(0)}$ is that

$$\lim_{\nu\to\infty} M^\nu = 0, \text{ (null matrix)}, \tag{5.9}$$

that is, all the elements of the matrix M^ν should approach zero.

A matrix satisfying the property (5.9) is said to be a *convergent matrix*. Convergence of a square matrix depends on the eigenvalues of the matrix. The following important result is stated in the form of a theorem. (Proof may be found in Niyogi, 2003).

Theorem: A necessary and sufficient condition that the matrix M be convergent is that all its eigenvalues have modulus less than unity.

If $\lambda_1, \lambda_2, \ldots, \lambda_n$ be the n eigenvalues of the matrix M, then the quantity $\hat{\lambda}$defined as

$$\hat{\lambda} = \max_i |\lambda_i|, \quad i = 1, 2, \ldots, n \tag{5.10}$$

is known as the *spectral radius of the matrix* M, often denoted by $\rho(M)$.

Rate of Convergence: For a convergent matrix M, the *rate of convergence* R of the iteration scheme Eq. (5.6) is defined as

$$R = -\log_{10}\hat{\lambda}, \tag{5.11}$$

where $\hat{\lambda}$is the *spectral radius* of M.

Some authors call R the asymptotic rate of convergence. However, we refer to it simply as the rate of convergence. The importace of the quantity R may be readily seen from the next theorem, stated without proof. (Proof may be found, for example, in Niyogi, (2003)).

Theorem: The least number of iteration steps needed to reduce an arbitrary initial error vector $e^{(0)}$ by a factor of $10^{-m}(m > 0)$ of itself, is inversely proportional to the rate of convergence R.

Note that the condition of convergence Eq. (5.10) as well as the rate of convergence R require the knowledge of the spectral radius $\hat{\lambda}$ of the matrix M. However, determination of the eigenvalues of a general $n \times n$ matrix is not easier than solving the system. In practice, we often have to be satisfied with simple sufficient conditions and simple bounds on the spectral radius $\hat{\lambda}$. Often the conditions are expressed in terms of various *matrix norms,* like the *row norm, column norm, Euclidean norm* etc. For example, it is known that a square matrix $M = (m_{ij})$, is convergent, if any of the following conditions on the norms of M hold (Isaacson and Keller, 1966):

$$\text{(a)}\quad ||M||_{\text{R}} = \text{row norm} = \max_{i=1}^{n} \sum_{j=1}^{n} \left|m_{ij}\right| < 1,$$

$$\text{(b)}\quad ||M||_{\text{C}} = \text{column norm} = \max_{j=1}^{n} \sum_{i=1}^{n} |m_{ij}| < 1, \tag{5.12}$$

$$\text{(c)}\quad ||M||_{\text{E}} = \text{Euclidean norm} = \left(\sum_{i=1}^{n}\sum_{j=1}^{n} m_{ij}^2\right)^{1/2} < 1$$

The conditions (5.12) are sufficient conditions. Since the condition $\hat{\lambda} < 1$ is necessary and sufficient, it follows that for a convergent matrix $\hat{\lambda}$ is smaller than any of the above matrix norms. In fact, in text books on linear algebra it is proved that among the various matrix norms, the spectral norm is the smallest. For a real symmetric matrix, $\hat{\lambda}$ is the spectral norm.

We now discuss two well known particular cases of the general iteration scheme (5.6).

5.3.1 The Jacobi and the Gauss–Seidel Schemes

Let us choose the matrix N in Eq. (5.6) as the diagonal matrix

$$N = (a_{ii}\,\delta_{ij}), \quad \delta_{ij} = \text{Krönecker delta} = \begin{cases} 1, & i = j, \\ 0, & i \neq j \end{cases}$$

Then the resulting scheme is known as the *Jacobi iteration scheme*. Componentwise, the scheme may be written as

$$u_i^{(\nu)} = \frac{1}{a_{ii}}\left[f_i - \sum_{\substack{j=1\\j\neq i}}^{n} a_{ij}u_j^{(\nu-1)}\right], \quad i = 1, 2, \cdots, n \tag{5.13}$$

$$u_i^{(0)} = \text{arbitrary.}$$

The matrix M corresponding to this case, denoted by M_{J} is

$$M_{\text{J}} = N^{-1}P = \left(\delta_{ij} - \frac{a_{ij}}{a_{ii}}\right). \tag{5.14}$$

It is often referred to as the *Jacobi iteration matrix*.

The sufficient conditions (a) and (b) of Eq. (5.12) then yield the conditions

$$\text{(a)} \quad ||M||_{\text{R}} = \max_{i=1}^{n} \sum_{\substack{j=1\\j\neq i}}^{n} \left|\frac{a_{ij}}{a_{ii}}\right| < 1,$$

$$\text{(b)} \quad ||M||_{\text{C}} = \max_{j=1}^{n} \sum_{\substack{i=1\\i\neq j}}^{n} \left|\frac{a_{ij}}{a_{ii}}\right| < 1 \tag{5.15}$$

The conditions (5.15) are quite easy to apply, and are known as the conditions for *diagonal dominance* of the matrix A. Thus the Jacobi iteration scheme for the linear algebraic system (5.4) is convergent if the coefficient matrix A is diagonally dominant. Note that the norms $\|M\|_{\text{R}}$ and $\|M\|_{\text{C}}$ in Eq. (5.15) provide a rough estimate of the rate of convergence of the Jacobi iteration scheme. Since $\hat{\lambda} \leq \|M\|_{\text{R}}$ and $\hat{\lambda} \leq \|M\|_{\text{C}}$ we obtain a lower bound on the rate of convergence R

$$R = \log_{10}\frac{1}{\hat{\lambda}} \geq \frac{1}{\min(\|M\|_{\text{R}}, \|M\|_{\text{C}})} \tag{5.16}$$

Noting from Eq. (5.13) that the Jacobi iteration scheme requires n^2 operations/step, it follows that when the Jacobi scheme converges, it takes about $\dfrac{mn^2}{R}$ number of operations to reduce the initial error by a factor of $10^{-m} (m > 0)$ of itself.

It may be observed from the Jacobi iteration scheme (5.13) that in the νth step of iteration while u_i is computed, the values $u_1, u_2, \ldots, u_{i-1}$ have already been computed and are available. But these values are not used by the Jacobi scheme. The values of the previous step $u_1^{(\nu-1)}, \ldots, u_{i-1}^{(\nu-1)}$ have been used instead. If on the other hand, the latest available values are used, the resulting scheme, known as the

Gauss–Seidel scheme follows, given by

$$u_i^{(\nu)} = \frac{1}{a_{ii}}\left[f_i - \sum_{j=1}^{i-1} a_{ij}\, u_j^{(\nu)} - \sum_{j=i+1}^{n} a_{ij}\, u_j^{(\nu-1)} \right], \nu = 1, 2, 3, \cdots \quad (5.17)$$

$$u_i^{(0)} = \text{arbitrary.}$$

The Gauss–Seidel scheme may be obtained from the general iteration scheme Eq. (5.6), by choosing the matrix N as the lower triangular matrix

$$N = \begin{pmatrix} a_{11} & & O \\ a_{21} & a_{22} & \\ \vdots & & \ddots \\ a_{n1} & a_{n2} & & a_{nn} \end{pmatrix}, \quad (5.18)$$

$$\text{with } A = N - P. \quad (5.19)$$

It may be established (Isaacson and Keller, 1966) that the conditions of diagonal dominance (5.15) are again sufficient conditions of convergence of the Gauss–Seidel scheme.

Although the Jacobi and the Gauss–Seidel schemes are of much theoretical importance, they turn out to be too slow for any practical use in CFD. Computations in CFD demand schemes with much faster rate of convergence. The SOR (successive over-relaxation) scheme (Varga, 1962) and its variants discovered in the late fifties are found to be faster than the Gauss–Seidel scheme, roughly by an order of magnitude (that is, about ten times faster). The SOR vector in the $(\nu + 1)$th step $u^{(\nu+1)}$ is a weighted average of the starting vector $u^{(\nu)}$ and the Gauss–Seidel vector ũ obtained with $u^{(\nu)}$ as starting value:

$$u^{(\nu+1)} = (1 - \omega)u^{(\nu)} + \omega\tilde{u} \quad (5.20)$$

where ω is a parameter, $0 < \omega < 2$, known as the *relaxation factor*. For values of $1 < \omega < 2$, the scheme is known as *over-relaxation*, while for $0 < \omega < 1$ it is under-relaxation, $\omega = 1$ delivering the Gauss–Seidel scheme.

The fastest convergence takes place for the optimum value ω_{opt} of the relaxation factor given by Isaacson and Keller (1966), Varga (1962).

$$\omega_{\text{opt}} = \frac{2}{1 + \sqrt{1 - (\hat{\lambda}_J)^2}} \quad (5.21)$$

where $\hat{\lambda}_J = \max\limits_i |\lambda_i|$, $\lambda_1, \lambda_2, \cdots, \lambda_n$ being the eigenvalues of the Jacobi iteration matrix defined in Eq. (5.14).

5.4 SOLUTION OF THE PENTADIAGONAL SYSTEM

After the brief digression, we now come back to the solution of the linear system (5.2) arising from the Dirichlet problem for the 2-D Laplace equation. For the sake of simplicity, let us take $\Delta x = \Delta y$. Then the system of Equations (5.2) may be rewritten as

$$u_{j,k} = \frac{1}{4}(u_{j+1,k} + u_{j-1,k} + u_{j,k+1} + u_{j,k-1}) - \frac{1}{4} f_{j,k} \qquad (5.22)$$
$$j = 1, 2, \cdots, N, \; k = 1, 2, \cdots, M.$$

The coefficient matrix A in this case is given by

$$A = \begin{pmatrix} -4 & 1 & & & 1 & O \\ 1 & -4 & 1 & & & 1 \\ & & & \ddots & & \\ & & 1 & -4 & & \\ 1 & & & & & \\ & 1 & & & & \\ O & & & 1 & & -4 \end{pmatrix},$$

$$N = -4I, \; P = -(A - N),$$

so that the Jacobi matrix

$$M_J = N^{-1}P = \frac{1}{4} \begin{pmatrix} 0 & 1 & & & & 1 & \\ 1 & 0 & 1 & & & & 1 \\ \vdots & \ddots & \ddots & \ddots & & & \ddots \\ 1 & \ddots & \ddots & \ddots & \ddots & & \end{pmatrix} \qquad (5.23)$$

It is possible to determine all the eigen values $\lambda_{m,n}$ of the Jacobi matrix (5.23) in closed form (Ferziger, 1981). They are

$$\lambda_{m,n} = \frac{1}{2}\left[\cos\frac{m\pi}{M+1} + \cos\frac{n\pi}{N+1}\right], \quad m = 1, 2, \cdots, M, n = 1, 2, \cdots, N \qquad (5.24)$$

Clearly, all the eigenvalues are numerically smaller than unity. Hence, the Jacobi iteration scheme (5.13) converges for the system (5.22). Corresponding to different values of n and m the largest value of $\hat{\lambda}$ is given by

$$\hat{\lambda} = \frac{1}{2}(\cos\frac{\pi}{M+1} + \cos\frac{\pi}{N+1})$$

$$\approx 1 - \frac{\pi^2}{4}\left[\frac{1}{(M+1)^2} + \frac{1}{(N+1)^2}\right]. \tag{5.25}$$

For large M and N, and for $M = N$, we have approximately

$$\hat{\lambda} \approx 1 - \frac{\pi^2}{2N^2} \approx e^{-\pi^2/2N^2}, \tag{5.26}$$

for which the rate of convergence

$$R = -\log_{10}\hat{\lambda} \approx \frac{\pi^2}{2N^2}\log_{10} e. \tag{5.27}$$

So, for an accuracy of, say 10^{-6}, the number ν of Jacobi iterations needed to reduce the error by a factor of 10^{-6} of itself, is

$$\nu \geq \frac{6}{\dfrac{\pi^2}{2N^2}\log_{10} e} \approx 3.2N^2 \tag{5.28}$$

Since each iteration takes four arithmetic operations (counting additions and subtractions as well), that is, a total of $4N^2$ operations per iteration, the number of arithmetic operations needed is $\approx 12.8\, N^4$, which is more than that needed by a direct elimination method. The Jacobi method is not a suitable choice and we look for better methods.

The Gauss–Seidel scheme (5.17) applied to the system (5.22) is

$$u_{j,k}^{(\nu+1)} = \frac{1}{4}\left[u_{j-1,k}^{(\nu+1)} + u_{j+1,k}^{(\nu)} + u_{j,k-1}^{(\nu+1)} + u_{j,k+1}^{(\nu)}\right] - \frac{1}{4} f_{j,k}, \tag{5.29}$$

$$j = 1, 2, \cdots, N;\ k = 1, 2, \cdots, M$$

$$u_{j,k}^{(0)} = \text{arbitrary}.$$

It can be shown that the eigenvalues of the Gauss–Seidel matrix are the squares of the eigenvalues of the Jacobi matrix M_J. Consequently, the Gauss–Seidel scheme converges to the desired accuracy in half the number of steps compared to that of the Jacobi scheme, with practically the same amount of labour per step. The advantage obtained through Gauss–Seidel scheme is due to the use of the latest available values. It may be noted that half the data in Eq. (5.29) is old while the other half is new and we require only half the number of iteration steps. It is natural to ask, whether any further improvement in the rate of convergence is possible by including more new data. The answer is "yes" and the schemes based on this idea are the *line relaxation schemes.* These are line versions of the Gauss–Seidel scheme, where the values of u at all the mesh points on a horizontal line or else on a vertical line are treated as unknown. For example, *the line Gauss–Seidel scheme along vertical lines* is given by

$$u_{j,k}^{(\nu+1)} = \frac{1}{4}\left[u_{j,k+1}^{\nu+1} + u_{j,k-1}^{\nu+1} + u_{j-1,k}^{\nu+1} + u_{j+1,k}^{\nu}\right] - \frac{1}{4}f_{j,k} \tag{5.30}$$

$$j = 1, 2, \cdots, N,\ k = 1, 2, \cdots, M$$

$$u_{j,k}^{(0)} = \text{arbitrary}$$

Note that while solving for the unknowns on the jth vertical line in the $(\nu + 1)$-th step, the previously computed value $u_{j+1,k}^{\nu}$ has to be used, since this is the latest available value. This results in a tridiagonal system which has to be solved, now. Note that the line Gauss–Seidel scheme, often called *line relaxation,* treats one line *implicitly.*

The rate of convergence of the line-relaxation scheme may be readily improved by introducing relaxation parameter ω, $0 < \omega < 2$, as follows:

$$\bar{u}_{j,k} = \frac{1}{4}\left[u_{j,k+1}^{\nu+1} + u_{j,k-1}^{\nu+1} + u_{j-1,k}^{\nu+1} + u_{j+1,k}^{\nu}\right] - \frac{1}{4}f_{j,k},$$

$$u_{j,k}^{\nu+1} = (1-\omega)u_{j,k}^{\nu} + \omega\bar{u}_{j,k},\ 1 < \omega < 2,\ j = 1, \ldots, N,\ k = 1, \cdots, M. \tag{5.31}$$

This is the well-known *successive line over-relaxation (SLOR)* scheme and is the fastest among the relaxation schemes. The value of the relaxation parameter ω depends on the nature of the problem and on the number of unknowns, and has to be found out by numerical experiments. For large number of unknowns a suitable value of ω would be a high value, possibly $\omega = 1.8$ or 1.9. In case of point successive over–relaxation (SOR) scheme given by

$$\bar{u}_{j,k} = \frac{1}{4}\left[u_{j,k+1}^{\nu} + u_{j,k-1}^{\nu+1} + u_{j-1,k}^{\nu+1} + u_{j+1,k}^{\nu}\right] - \frac{1}{4}f_{j,k},$$

$$u_{j,k}^{\nu+1} = (1-\omega)u_{j,k}^{\nu} + \omega\bar{u}_{j,k},\ j = 1, \cdots, N,\ k = 1, \cdots, M, \tag{5.32}$$

the optimum value of the relaxation parameter ω is given by

$$\omega_{\text{opt}} = \frac{2}{1 + \sqrt{1 - \hat{\lambda}^2}}, \tag{5.33}$$

where for the system of equations arising through discretisation of 2-D Laplace equation, $\hat{\lambda}$ is known in closed form as

$$\hat{\lambda}_j = \frac{1}{2}\left[\cos\frac{\pi}{M+1} + \cos\frac{\pi}{N+1}\right]. \tag{5.34}$$

Example 5.1

Solve the Dirichlet problem for 2-D Laplace equation $u_{xx} + u_{yy} = 0$, in the unit square in the first quadrant, with boundary values $u(x, 0) = 1,\ 0 < x < 1$, on the x-axis, $u(0, y) = 1, 0 \leq y < 1$, on the y-axis and 0 on the remaining sides. Compute by the Gauss-Seidel scheme, the solution correct to 2-D, with $\Delta x = \frac{1}{3}$.

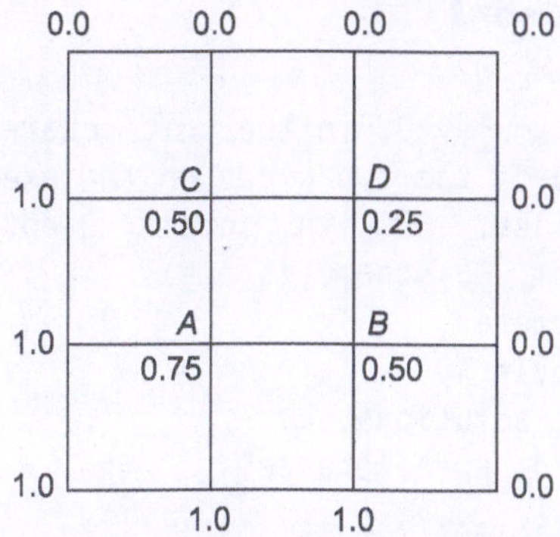

Figure 5.2 BVP 2-D Laplace equation with converged solution.

The boundary value problem and the final converged solution is shown in Fig. 5.2. As starting values, we take $u = 0.5$, at all the four-grid points A, B, C and D. Note that the iteration proceeds from left to right, from the lowest line upwards, that is along $A \to B$, then $C \to D$ in Fig. 5.2. Convergence correct upto two-decimal places is obtained in five-steps. The values at the respective grid points A, B, C, D are shown in Table 5.1.

Table 5.1 Solution of 2-D Laplace equation by G-S scheme

Step No.	A	B	C	D
0	0.5	0.5	0.5	0.5
1	0.750	0.562	0.562	0.281
2	0.781	0.516	0.516	0.258
3	0.758	0.504	0.504	0.252
4	0.752	0.501	0.501	0.250
5	0.750	0.500	0.500	0.250

As already observed, faster convergence may be achieved by using SOR, often by an order of magnitude. This is seen in the results of the following Program 5.1: sor.f. It computes solution of the Dirichlet problem for 2-D Laplace equation in the unit square in the first quadrant. The boundary values are: $u(x, y) = 0$ on the sides on the coordinate axes and $u(x, y) = x$ and $u(x, y) = y$ on the sides parallel to x and y axes respectively. This problem admits of an exact solution $u(x, y) = xy$ obtained by separation of variables. Solution correct to six-decimal places, have been computed with 128×128 grid points. It is found that Gauss–Seidel scheme requires 9634 steps, whereas SOR, with optimum relaxation parameter 1.9525 achieves this in 356 steps.

5.4.1 Program 5.1: sor.f

```
c......Program 5.1: sor.f
c       It solves 2-D Laplace eq. u_xx+u_yy=0, in the unit square
c......in the 1st quadrant. B.C: u=0 on the two sides on the axes,
c...... and u=x, u=y on remaining sides.  SOR itn. scheme  used.
c...... SOR relaxation parameter = om, GS-scheme is om=1.
c.......omt= optimum relaxation parameter.
c...... Known exact solution    u(x,y)= xy.
c us= starting, uf= final, ux= exact solutions....
c itmax=max iteration steps performed; ep=tolerance.
c..........................................................
parameter(ix=305, iy=305)
dimension uf(0:ix,0:iy), us(0:ix,0:iy), ux(0:ix,0:iy),x(0:ix)
dimension y(0:iy)
open(unit=8, file='sor.out')
c......Set Boundary values......................................
pi=3.1415926
itmax= 60000
ep=0.0000005
kmax=129
jmax=129
c om=1.5
write(*,*)'Input value of  relaxation parameter omega'
read(*,*) om
write(16,*) 'SOR relaxation parameter omega = ', om
alm=cos(pi/float(jmax))
omt=2.0/(1.0+sqrt(1.0-alm*alm))
jm=jmax-1
km=kmax-1
k2=kmax/2
dx=1.0/jmax
dy= dx
c........Setting boundary values.........
do 30 j=0,jmax
x(j)=j*dx
y(j)=j*dy
30 continue
do 21 j=0,jmax
us(j,0)=0.0
```

```
us(j,kmax)=x(j)
21 continue
do 22 k=0,kmax
us(0,k)=0.0
us(jmax,k)= y(k)
22 continue
c..... Evaluation of the exact solution................
do 26 k=1,km
do 27 j=1,jm
ux(j,k)=  j*k*dy*dx
27 continue
26 continue
c.......Testing dependence of convergence rate on omt.......
c-------------------------------------------------------
write(16,*)'Output of Program 5.1: sor.f'
write(16,*)'-------------------------------------'
write(16,*)' jmax= ',jmax,' kmax= ', kmax,' itmax= ',itmax
write(16,*) 'ep = ',ep,' omt= ',omt , '  om = ',om
write(16,*) '---------------------------------------'
c.......Starting guess  us(j,k)  values  .............
do 35 k=1,km
do 35 j=1,jm
us(j,k)=0.1
35 continue
c.......Main SOR iteration loop............................
111 continue
do 50 it=1,itmax
14 do 44 k=1,km
do 46 j=1,jm
uf(j,k)=us(j,k)+ om*(0.25*(uf(j-1,k)+us(j+1,k)+uf(j,k-1)+us(j,k+1))
1      -us(j,k))
46 continue
44 continue
c.......Convergence test  ..............................
47 do 45 k=1,km
do 45 j=1,jm
if(abs(uf(j,k)-us(j,k)).gt.ep)go to 65
45 continue
62 write(16,*)' SOR Conv. in Steps= ', it,' relax. parm.= ', om
write(16,*)' Converged solution  '
write(16,*) '  x(j)  ', '  y(k)  ', '  uf(j,k2)  ', '  ux(j,k2)  '
go to 71
65 do 48 k=1,km
do 48 j=1,jm
us(j,k)=uf(j,k)
```

```
48 continue
50 continue
write(16,*)No convergence in steps= ' , itmax
write(16,*) ' The final values are: '
71 do 61 j=1,jm,16
write(16,9) x(j),y(k2),  uf(j,k2),ux(j,k2)
9 format(1x,2f6.2,2f11.6)
61 continue
70 continue
close(16)
stop
 end
c----------------------------------------------------
```

```
 Output of Program 5.1: sor.f
 ---------------------------------------

  jmax=  129 kmax=  129 itmax=  50000
 ep =   4.99999999E-07 omt=   1.95245552  om =    1.95249999
 ---------------------------------------
  SOR Conv. in Steps=  356 relax. parm.=   1.95249999
  Converged solution
   x(j)  y(k)   u(j,k2)    uex(j,k2)
   0.01  0.50   0.003846   0.003846
   0.13  0.50   0.065379   0.065381
   0.26  0.50   0.126914   0.126915
   0.38  0.50   0.188449   0.188450
   0.50  0.50   0.249983   0.249985
   0.63  0.50   0.311517   0.311520
   0.75  0.50   0.373052   0.373055
   0.88  0.50   0.434587   0.434589
%.....................................
```

Table 5.2 Effect of relaxation parameter on the rate of convergence

ω	No. of steps
1.5	3901
1.6	3061
1.7	2276
1.8	1534
1.85	1170
1.9	799
1.95	365
1.953	356

It is interesting to see the effect of the choice of the relaxation parameter ω (denoted by 'om' in program 5.1) on the rate of convergence of SOR. Table 5.2 shows the number of iteration steps needed for convergence correct to six-decimal places for different values of the relaxation parameter ω. omt in the program denotes the optimum relaxation parameter.

5.5 APPROXIMATE FACTORISATION SCHEMES

As already seen, we obtain a system of algebraic equations through discretisation of partial differential equations by numerical methods like the finite difference method. Most frequently, these equations are solved iteratively by relaxation methods like SOR, SLOR or Gauss–Seidel, which suffer from the drawback that all of them become prohibitively slow, as the number of unknown increases. The approximate factorisation schemes and the multigrid schemes do not suffer from this drawback. Here, we explain the AF1 scheme, and apply it to a body-fitted grid generation problem, by way of illustration.

The approximate factorisation schemes are generalised forms of ADI schemes. The extension of ADI schemes for non-linear elliptic-type partial differential equations is the *AF1 scheme* and for steady transonic potential flow equation, which is non-linear and of mixed elliptic–hyperbolic type, *the scheme is AF2*, proposed by Ballhaus et al. (1977). Extension to non-conservative form of the potential equation is AF3 scheme (Baker, 1984). All the AF schemes (AF1, AF2, AF3) have remarkably fast rate of convergence, being very economic with respect to computational labour and have better stability behaviour. In order to understand the schemes better, it is helpful to analyse the behaviour of the line Gauss–Seidel scheme for the simplest elliptic type equation, namely, the Laplace equation in 2-D.

5.5.1 Analysis of Line Gauss–Seidel Scheme for the Laplace Equation

It is typical of relaxation methods like SOR, SLOR or the Gauss–Seidel scheme that they become increasingly slower as the number of unknowns, that is, mesh points increase, which happens if the grid is more and more refined. Here, we first analyse, following Baker (1984), why this is so and then suggest remedial measures. We present the ideas with reference to the Laplace equation which is quite easy to understand:

$$\frac{\partial^2 u}{\partial x^2} + \frac{\partial^2 u}{\partial y^2} = 0, \tag{5.35}$$

with Dirichlet boundary conditon.

Let $x = j\Delta x$ and $y = k\Delta y$. Using central difference representation for both the derivatives, we get

$$\frac{u_{j+1,k}-2u_{j,k}+u_{j-1,k}}{\Delta x^2}+\frac{u_{j,k+1}-2u_{j,k}+u_{j,k-1}}{\Delta y^2}+O(\Delta x^2)+O(\Delta y^2)=0 \quad (5.36)$$

To solve this system of equations iteratively by the line Gauss–Seidel scheme proceeding along vertical lines, left to right, the iteration scheme, omitting the truncation error terms, is:

$$\frac{u^{n}_{j+1,k}-2u^{n+1}_{j,k}+u^{n+1}_{j-1,k}}{\Delta x^2}+\frac{u^{n+1}_{j,k+1}-2u^{n+1}_{j,k}+u^{n+1}_{j,k-1}}{\Delta y^2}=0, \quad (5.37)$$

where $j = 1, 2, \ldots, M,\quad k = 1, 2, \ldots, N$, and the superscript n denotes the present iteration step and the superscript $(n + 1)$ the new iteration step. Let us define the error as

$$e^{(n)}_{j,k} = u^{(n)}_{j,k} - u_{j,k},$$

where $u_{j,k}$ denotes the exact solution of the difference equations (5.36). Then

$$\frac{1}{\Delta x^2}\left[e^{(n)}_{j+1,k}+u_{j+1,k}-2(e^{(n+1)}_{j,k}+u_{j,k})+e^{(n+1)}_{j-1,k}+u_{j-1,k}\right]$$
$$+\frac{1}{\Delta y^2}\left[e^{(n+1)}_{j,k+1}+u_{j,k+1}-2(e^{(n+1)}_{j,k}+u_{j,k})+e^{(n+1)}_{j,k-1}+u_{j,k-1}\right]=0.$$

Subtracting Eq. (5.36) from it and simplifying, we obtain

$$e^{(n)}_{j+1,k}=\left[2e^{(n+1)}_{j,k}-e^{(n+1)}_{j-1,k}\right]-\left(\frac{\Delta x}{\Delta y}\right)^2\left[e^{(n+1)}_{j,k+1}-2e^{(n+1)}_{j,k}+e^{(n+1)}_{j,k-1}\right] \quad (5.38)$$

We propose to study the growth or decay of the error spectrum, in a single time step, using von Neumann stability analysis. To do this let the error be expanded in a finite Fourier series, as

$$e^{(n)}_{j,k}=\sum_{p=1}^{N}\sum_{q=1}^{M}\rho^{(n)}(p,q)e^{ipj\Delta x}e^{iqk\Delta y} \quad (5.39)$$

where $i = \sqrt{-1}$, $N = \frac{2\pi}{\Delta x}$, $M = \frac{2\pi}{\Delta y}$, N and M being the number of grid points in the x and y directions respectively.
Here, we note that:

1. p and q are wave numbers,
2. $\dfrac{p}{2\pi}$ and $\dfrac{q}{2\pi}$ are frequencies of the waves,
3. values of p and q near 1 and 2π are low and those near $\frac{\pi}{h}$ are high.

Substituting one term of Eq. (5.39) in Eq. (5.38), we get

$$\rho^{(n)}e^{ip(j+1)\Delta x}e^{iqk\Delta y} = \rho^{(n+1)}\Big\{e^{iqk\Delta y}\left[2e^{ipj\Delta x} - e^{ip(j-1)\Delta x}\right] - \left(\frac{\Delta x}{\Delta y}\right)^2 e^{ipj\Delta x}\left[e^{iq(k+1)\Delta y} - 2e^{iqk\Delta y} + e^{iq(k-1)\Delta y}\right]\Big\}$$

which yields on simplification

$$\rho^{(n)}e^{i\xi} = \rho^{(n+1)}\left\{\left[2 - e^{-i\xi}\right] - \left(\frac{\Delta x}{\Delta y}\right)^2 [2\cos\eta - 2]\right\},$$

where $\xi = p\Delta x$, $\eta = q\Delta y$.

The amplification factor is given by $G(p,q) = \dfrac{\rho^{(n+1)}}{\rho^{(n)}}$. Therefore,

$$G(p,q) = \frac{e^{i\xi}}{(2 - e^{-i\xi}) + 2(1 - \cos\eta)}, \tag{5.40}$$

where $\xi = ph$, $\eta = qh$, $p, q = 1, 2, \ldots, \dfrac{\pi}{h}$, $\Delta x = \Delta y = h$,, $\dfrac{\Delta x}{\Delta y} = 1$.

The amplification factor measures the extent by which each frequency component of the error spectrum is increased or reduced after one iteration cycle. Taking absolute values

$$|G(p,q)| = \frac{|e^{i\xi}|}{|(2 - e^{i\xi}) + 2(1 - \cos\eta)|}$$

The von Neumann condition states that $|G(p,q)| \le 1$ is the necessary condition for stability and convergence. If the value of the amplification factor is less than but close to unity, then the iterative scheme may be stable but only converge at a very slow rate. It is therefore desirable to select a method for which the amplification factor is significantly less than unity throughout the range of frequencies in the error spectrum. In the above example, it can be seen that the expression for $|G(p,q)|$ is small for high frequencies, for example, if $\xi \sim \pi$ and $\eta \sim \pi$. Then,

$$G = \frac{e^{i\pi}}{2 - e^{-i\pi} + 2(1 - \cos\pi)} = \frac{-1}{2 - (-1) + 2(1 + 1)} = -\frac{1}{7}$$

Therefore, $|G| \sim \frac{1}{7}$. This shows that line *Gauss–Seidel method (and similarly the other relaxation methods) is capable of reducing the high frequency errors quite rapidly.* At the low frequency end of the error spectrum $\xi \sim h$ and $\eta \sim h$, we obtain on expansion, assuming h to be small,

$$\begin{aligned}
G &= \frac{e^{ih}}{2 - e^{-ih} + 2(1 - \cos h)} \\
&= \frac{\cos h + i\sin h}{4 - 3\cos h + i\sin h} \\
&\approx 1 - 2h^2 + \cdots
\end{aligned}$$

Therefore, $|G| \sim 1 - 2h^2$. As h becomes smaller and smaller, that is, as the grid is refined more and more, the amplification factor G approaches unity, which implies a slower and slower convergence rate. *Thus the low frequency errors take an enormously large number of iteration steps to get reduced.* Therefore, the Gauss–Seidel or any of the relaxation methods become extremely slow on finer grids. One looks for better methods. Among the current methods, the approximate factorisation and multigrid methods are the choice. We discuss first the AF1 scheme.

5.5.2 Time-Dependent Analogy

We consider again the line Gauss–Seidel scheme along vertical lines, sweeping from left to right. Central difference representation, for both the space derivatives of Laplace equation, neglecting the truncation error terms gives

$$\frac{u^n_{j+1,k} - 2u^{n+1}_{j,k} + u^{n+1}_{j-1,k}}{\Delta x^2} + \frac{u^{n+1}_{j,k+1} - 2u^{n+1}_{j,k} + u^{n+1}_{j,k-1}}{\Delta y^2} = 0. \tag{5.41}$$

Taking, for the sake of simplicity $\Delta x = \Delta y = h$, we get from Eq. (5.41),

$$\frac{1}{h^2}\left[u^n_{j+1,k} - 2u^{n+1}_{j,k} + u^{n+1}_{j-1,k}\right] + \frac{1}{h^2}\left[u^{n+1}_{j,k+1} - 2u^{n+1}_{j,k} + u^{n+1}_{j,k-1}\right] = 0$$

Let us define the residual $R_{j,k}$ at the n-th iteration step.

$$R_{j,k} = \frac{1}{h^2}\left[u^n_{j+1,k} + u^n_{j-1,k} + u^n_{j,k+1} + u^n_{j,k-1} - 4u^n_{j,k}\right].$$

It may be noted that while the unknowns on the j-th line are being computed, the values of u_j in the iteration cycle $n+1$ for $j = 1, 2, \ldots, j-1$ are available. So the iteration scheme is,

$$\frac{1}{h^2}\left[u^n_{j+1,k} + u^{n+1}_{j-1,k} + u^{n+1}_{j,k+1} + u^{n+1}_{j,k-1} - 4u^{n+1}_{j,k}\right] = 0,$$

which may be rewritten as

$$\frac{1}{h^2}\left[u^{n+1}_{j+1,k} - u^n_{j+1,k} - u^{n+1}_{j+1,k} - u^{n+1}_{j-1,k} - u^{n+1}_{j,k+1} - u^{n+1}_{j,k-1} + 4u^{n+1}_{j,k}\right] = 0. \tag{5.42}$$

Let $c_{j,k} = u^{n+1}_{j,k} - u^n_{j,k}$ = correction at mesh point (j, k) through one cycle, so that, $u^{n+1}_{j,k} = c_{j,k} + u^n_{j,k}$. Eliminating quantities with superscript $n+1$ from Eq. (5.42) yields

$$\frac{1}{h^2}\left[c_{j+1,k} - (c_{j+1,k} + u^n_{j+1,k}) - (c_{j-1,k} + u^n_{j-1,k}) - (c_{j,k+1} + u^n_{j,k+1})\right.$$
$$\left. -(c_{j,k-1} + u^n_{j,k-1}) + 4(c_{j,k} + u^n_{j,k})\right] = 0. \tag{5.43}$$

Then, in terms of the correction $c_{j,k}$ and residual $R_{j,k}$, Eq. (5.43) may be expressed as

$$\frac{1}{h^2}\left[c_{j,k}+(c_{j,k}-c_{j-1,k})-(c_{j,k+1}-2c_{j,k}+c_{j,k-1})\right]=R_{j,k}. \tag{5.44}$$

Equation (5.44) can be interpreted as a finite difference approximation to a time-dependent equation in which the final steady state solution represents the required result (Garabedian, 1956). Thus, if one iteration step corresponds to a step Δt in artificial time, we take $\Delta t = h^2$, then to the first order in h, Eq. (5.44) approximates the parabolic equation

$$u_t = u_{xx} + u_{yy}. \tag{5.45}$$

Note that by definition, the backward difference $\overline{\delta}_x c_{j,k} = c_{j,k} - c_{j-1,k}$, and the second central difference $\delta_{yy} c_{j,k} = c_{j,k+1} - 2c_{j,k} + c_{j,k-1}$, and $\dfrac{c_{j,k}}{h^2} = \dfrac{u^{n+1}_{j,k} - u^n_{j,k}}{h^2} = \left.\dfrac{\partial u}{\partial t}\right|_{j,k}$.

From this, it seems that the poor decay of low frequency error components is related to the small time steps that is taken for computation.

If we write $\Delta t = rh^2$, then it is possible to create a modified line Gauss–Seidel scheme in which the parameter r could be varied to change the effective time step. However, this explicit scheme is stable only for $r \le 1/2$. Consequently, it would not be possible to choose effectively large step size. An implicit scheme is expected to have better stability behaviour and remove this inconvenient stability restriction. The ADI scheme is the choice, which for solving Eq. (5.45) may be expressed as

$$\left(I-\frac{r}{2}\delta_{xx}\right)\left(I-\frac{r}{2}\delta_{yy}\right)u^{n+1}_{j,k}=\left(I+\frac{r}{2}\delta_{xx}\right)\left(I+\frac{r}{2}\delta_{yy}\right)u^{n}_{j,k},\ r=\frac{\Delta t}{h^2}. \tag{5.46}$$

The amplification factor $G(p, q)$ for this system is

$$G(p,q)=\frac{\{1-r(1-\cos\xi)\}\{1-r(1-\cos\eta)\}}{\{1+r(1-\cos\xi)\}\{1+r(1+\cos\eta)\}} \tag{5.47}$$

where $\xi = ph$, $\eta = qh$. It follows that $|G(p,q)| \le 1$, for all p, q and hence the method is stable for all positive r (unconditionally stable). Rearranging Eq. (5.46) and for the sake of convenience replacing r by α where $\alpha = \dfrac{2}{rh^2}$, leads to the difference scheme, expressed in terms of corrections $\Delta^n_{j,k}$ and residuals $Lu^n_{j,k}$

$$\left(\alpha-\frac{\delta_{xx}}{h^2}\right)\left(\alpha-\frac{\delta_{yy}}{h^2}\right)\Delta^n_{j,k}=2\alpha Lu^n_{j,k} \tag{5.48}$$

where

$$\Delta^n_{j,k} = \text{correction} = u^{n+1}_{j,k} - u^n_{j,k}, \quad Lu^n_{j,k} = \left(\frac{\delta_{xx}}{h^2} + \frac{\delta_{yy}}{h^2}\right) u^n_{j,k} \tag{5.49}$$

This is known as *the approximate factorisation scheme, AF1*. On examining the amplification factor Eq. (5.47), it can be seen that the choice $\alpha = \frac{2}{h^2}(1 - \cos ph)$ will reduce the p-th frequency component to zero. Thus, the finite sequence $\alpha_p = \frac{2}{h^2}(1 - \cos ph)$, $p = 1, 2, \ldots, \frac{\pi}{h}$ should remove all error components and hence produce a converged result. The α's are called *acceleration parameters.* Large values of α correspond to small time steps which are effective against high frequency error components. On the other hand, small values of α, which represents large time steps, reduce the low frequency end of the error spectrum.

In practice, it is found best to use a small sequence of acceleration parameters running between the minimum and maximum values given by Eq. (5.47). Minimum value $\alpha_{\text{L}} = 1$, occurs for $p = 1$. Maximum value $\alpha_{\text{H}} = \dfrac{4}{h^2}$, occurs for $p = \dfrac{\pi}{h}$. The following geometric sequence, which is repeated in cyclic fashion has proved very effective.

$$\alpha_k = \alpha_{\text{H}} \left(\frac{\alpha_{\text{L}}}{\alpha_{\text{H}}}\right) \frac{k-1}{M-1}, \tag{5.50}$$

where $k = 1, 2, \ldots, M$, $6 \le M \le 8$.

5.5.3 Program 5.2: afl.f

We explain the application of AF1 scheme to compute solution of the Dirichlet problem for Laplace equation in 2-D in the unit square in the first quadrant, discussed in the previous section.

```
c..... Program 5.2: af1.f.
c..... It solves 2-D Laplace equation in the unit square in the 1st
c......quadrant by approximate factorization scheme af1.
c...... Boundary conditions are: u(x,y)=0.0 on the two sides on
c......coordinate axes and u= x and u= y on the remaining two sides.
c.......The exact solution of this problem is  u(x,y)= x y.
c...... al(i) are the acceleration parameters of AF1 scheme.
c.......jmax, kmax are max no. of mesh points along x and y directions.
c.......ep=tolerance for convergence. It uses tridiagonal solver
c.......subroutine trid.f
c..........................................................
      parameter(ix=305, iy=305)
      dimension uf(0:ix,0:iy), us(0:ix,0:iy), ux(0:ix,0:iy),x(0:ix)
```

```
dimension y(0:iy),aa(ix),bb(ix),cc(ix),dd(ix), al(8)
dimension f(0:ix,0:iy), c(0:ix,0:iy)
open(unit=8, file='af1.out')
c......Set Boundary values...............................
itmax=500
ep=0.0000005
pi=3.1415927
kmax=129
jmax=129
aj=jmax
jm=jmax-1
km=kmax-1
k2=kmax/2
dx=1.0/aj
dy= dx
ddx=dx*dx
hs=1.0/(ddx)
ths=2.0*hs
h4=ddx/4.0
alh=4.*hs
all=ths*(1.0-cos(dx))
do 13 kk=1,6
al(kk)=alh*(all/alh)**(float(kk-1)/5.0)
13 continue
do 30 j=0,jmax
x(j)=j*dx
y(j)=j*dy
30 continue
c......Boundary values of temporary array f(j,k) set to zero.
do 21 j=0,jmax
uf(j,0)=0.0
us(j,0)=0.0
f(j,0)=0.0
us(j,kmax)= j*dx
uf(j,kmax)=j*dx
f(j,kmax)=0.0
21 continue
do 22 k=0,kmax
uf(0,k)=0.0
us(0,k)=0.0
f(0,k)=0.0
us(jmax,k)=k*dy
uf(jmax,k)=k*dy
f(jmax,k)=0.0
22 continue
```

```
c..... Evaluation of the exact solution denoted by ux.........
do 26 k=1,km
do 27 j=1,jm
ux(j,k)=  j*k*dy*dx
27 continue
26 continue
c...... Approximate-Factorization scheme AF1. ...........
write(8,*)'Output of af1.f: AF1 scheme '
write(8,*)'-----------------------------------'
write(8,*)' jmax= ',jmax,' kmax= ', kmax,' itmax= ',itmax
write(8,*) ' ep = ',ep, '  AF1  acceleration parameters are: '
write(8,*) 'al= ', (al(i),i=1,6)
write(8,*) '-------------------------------------'
c.......Starting values  .............................
do 35 k=1,km
do 35 j=1,jm
us(j,k)=0.0
35 continue
c.......AF1 iteration loop.......................
do 50 it=1,itmax
ik =mod(it,6)+1
alp=al(ik)
do 41 k=1,km
do 40 j=1,jm
alu =us(j+1,k)+us(j-1,k)+us(j,k+1)+us(j,k-1)-4.0*us(j,k)
cc(j)=-hs
aa(j)=-hs
bb(j)=alp+ths
dd(j)=ths*alp*alu
if(j. eq.jm) dd(j)=f(jmax,k)+dd(j)
40 continue
call trid(aa,bb,cc,dd,jm)
do 31 j=1,jm
f(j,k)=dd(j)
31 continue
41 continue
c......cross-sweep.............................
do 81 j=1,jm
do 82 k=1,km
cc(k)=-hs
aa(k)=-hs
bb(k)=alp+ths
dd(k)=f(j,k)
```

```
82 continue
call trid(aa,bb,cc,dd,km)
do 85 k=1,km
c(j,k)=dd(k)
uf(j,k)=us(j,k)+dd(k)
85 continue
81 continue
c ----  Convergence test  -------------------------
do 45 j=1,jm
do 45 k=1,km
if(abs(c(j,k)).ge.ep)go to 65
45 continue
go to 60
65 do 48 j=1,jm
do 47 k=1,km
us(j,k)=uf(j,k)
47 continue
48 continue
50 continue
write(8,*)'No convergence in steps= ' , itmax
go to 71
60 write(8,*)'Convergence achieved in steps= ',it
write(8,*)'Converged solution'
write(8,*) ' x(j)  ', ' y(k) ', ' uf(j,k) ', ' ux(j,k) '
71 do 61 j =1,jm,16
write(8,9) x(j),y(k2), uf(j,k2),ux(j,k2)
9 format(1x,2f6.2,2f11.6)
61 continue
70 continue
close(8)
stop
end

  subroutine trid(a,b,c,d,m)
parameter(ix=205)
dimension a(ix),b(ix),c(ix),d(ix),p(ix)
c  ... forward elimination.............
  p(1)=c(1)/b(1)
   d(1) = d(1)/b(1)
  do 1 j=2,m
 jm=j-1
factor = 1./(b(j)-a(jm)*p(jm))
p(j)=c(j)*factor
1        d(j) = (d(j) -a(jm)*d(jm))*factor
c       ....... back substitution sweep ..........
do 2 j= m-1,1,-1
```

```
jp=j+1
2          d(j) = d(j)- p(j)*d(jp)
return
end
```

%....................................

```
 Output of af1.f: AF1 scheme(6 acceleration parameters)
 ------------------------------------
  jmax=  129 kmax=  129 itmax=  500
  ep =   4.99999999E-07  AF1  acceleration parameters are:
 al=   66564.  7220.6377  783.270325  84.9665146  9.21687984  0.999815941
 ------------------------------------
 Convergence achieved in steps=  23
 Converged solution
  x(j)   y(k)  u(j,k)  uex(j,k)
   0.01  0.50   0.003846   0.003846
   0.13  0.50   0.065381   0.065381
   0.26  0.50   0.126916   0.126915
   0.38  0.50   0.188450   0.188450
   0.50  0.50   0.249985   0.249985
   0.63  0.50   0.311520   0.311520
   0.75  0.50   0.373054   0.373055
   0.88  0.50   0.434589   0.434589
```

We note that, although with eight-acceleration parameters it takes only 15 steps to converge correct to six-decimal places, while with six-parameters, 23 steps are needed, the latter solution comes out to be more accurate than the former. It may be noted that the same problem needs 9634 steps for convergence correct to six-decimal places, by the Gauss–Seidel scheme and 354 steps by the SOR scheme with optimum relaxation parameter, found to be $\omega_{\text{opt}} = 1.9525$.

5.6 GRID GENERATION EXAMPLE

As another application of the AF1 scheme, the problem of body-fitted grid generation by solving elliptic-type equations is considered, which delivers a pair of second order nonlinear elliptic type partial differential equations. Thompson et al. (1985) solved the resulting algebraic equations using relaxation method. Here, these equations are solved by AF1 scheme, as an illustration, only briefly. Theory and practice of grid generation is discussed in detail in Chapter 8.

Domains in the physical flow plane are in general irregular in shape. Use of Cartesian grids leads to difficulties in the satisfaction of boundary conditions, since the grid points may not lie on the boundary. Accuracy of a computed solution depends very strongly on how well the boundary conditions have been satisfied.

The basic problem of grid generation is a coordinate transformation by means of which an irregular shaped domain in the physical flow plane (x, y) changes over to a rectangle in the computational plane (ξ, η). Thus, a curvilinear body-fitted coordinate system may be obtained where the boundary grid data coincide with the physical boundary of the domain.

In a sense the problem of grid generation may be posed as a boundary-value problem:
Given $\xi = \xi_b(x, y)$, $\eta = \eta_b(x, y)$ on the boundary ∂R, generate $\xi = \xi(x, y)$ and $\eta = \eta(x, y)$ in the region R bounded by ∂R. The physical coordinates (x, y), which are typically Cartesian, are the independent variables and the generalised coordinates (ξ, η) are the dependent variables, Fig. 5.3(a).

In practice grid is generated with less computational effort by working in the computational plane. Thus, fixing the location of the points on the boundary $x = x_b(\xi, \eta)$ and $y = y_b(\xi, \eta)$, the generation of the grid in the interior is expressed as the following boundary-value problem :
Given $x = x_b(\xi, \eta)$, $y = y_b(\xi, \eta)$ on the boundary ∂R, generate $x = x(\xi, \eta)$ and $y = y(\xi, \eta)$, in the region R bounded by ∂R. (See Figs. 5.3(a) and 5.3(b)).

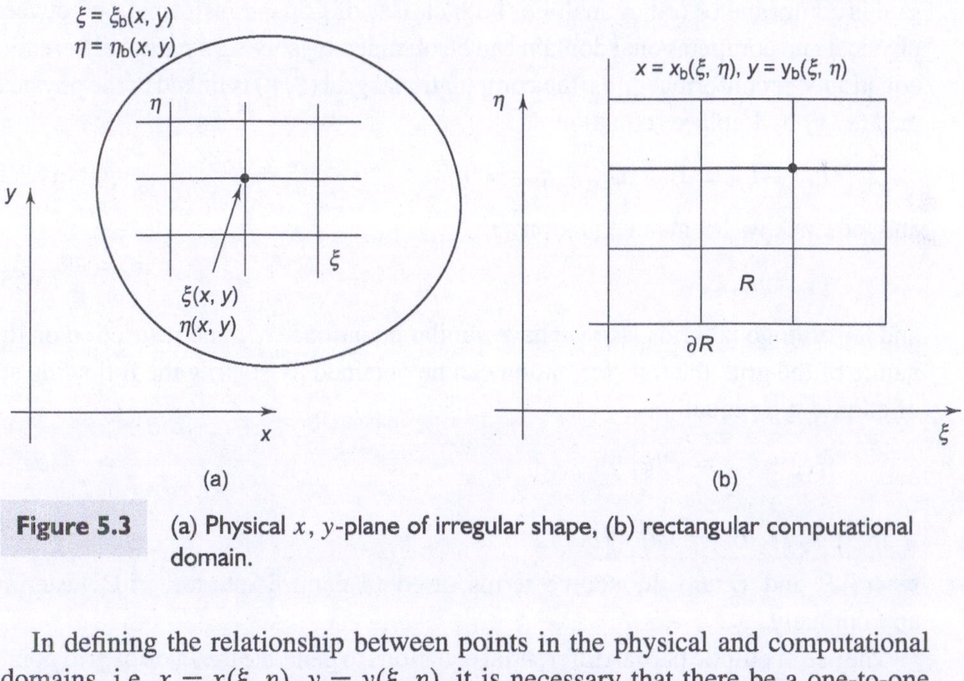

(a) (b)

Figure 5.3 (a) Physical x, y-plane of irregular shape, (b) rectangular computational domain.

In defining the relationship between points in the physical and computational domains, i.e. $x = x(\xi, \eta)$, $y = y(\xi, \eta)$, it is necessary that there be a one-to-one correspondence. It would be unacceptable for a single point in the physical domain to map into two points in the computatonal domain, and vice-versa. This is equivalent to the requirement that coordinate lines of the same family must not intersect and that coordinate lines of different families may intersect only once. Once the mapping $x = x(\xi, \eta)$, $y = y(\xi, \eta)$ has been established, the requirement of a one-to-one mapping can be determined by evaluating the determinant of the transformation Jacobian $|J|$, defined by

$$J = \begin{vmatrix} \dfrac{\partial \xi}{\partial x} & \dfrac{\partial \xi}{\partial y} \\ \dfrac{\partial \eta}{\partial x} & \dfrac{\partial \eta}{\partial y} \end{vmatrix} = \frac{\partial(\xi, \eta)}{\partial(x, y)}, \quad J^{-1} = \begin{vmatrix} \dfrac{\partial x}{\partial \xi} & \dfrac{\partial x}{\partial \eta} \\ \dfrac{\partial y}{\partial \xi} & \dfrac{\partial y}{\partial \eta} \end{vmatrix} = \frac{\partial(x, y)}{\partial(\xi, \eta)} \tag{5.51}$$

For the mapping to be one-to-one the Jacobian $|J|$ must be finite and nonzero. Depending on how the grid has been generated, $|J|$ can be evaluated at each grid point, analytically or numerically, in order to check for one-to-one mapping.

5.7 BODY-FITTED GRID GENERATION USING ELLIPTIC-TYPE EQUATIONS

In transforming the governing equations for fluid dynamics into generalised coordinates, we observe that the governing equations have a similar structure if the grid is conformal or orthogonal. For both classes of grids, transformation between physical and computational domain can be obtained by solving a partial differential equation. For conformal grids, the computational grid (ξ, η) is linked to the physical grid (x, y) by Laplaces equation

$$\xi_{xx} + \xi_{yy} = 0, \quad \eta_{xx} + \eta_{yy} = 0, \tag{5.52}$$

and the Cauchy–Riemann equations

$$\xi_x = \eta_y, \xi_y = -\eta_x, \tag{5.53}$$

and for orthogonal grids also we have similar equations. With no restriction on the nature of the grid, the transformation can be obtained by solving the following set of elliptic-type equations.

$$\xi_{xx} + \xi_{yy} = P(\xi, \eta), \tag{5.54}$$

$$\eta_{xx} + \eta_{yy} = Q(\xi, \eta), \tag{5.55}$$

where P and Q are the source terms, used to control interior grid clustering appropriately.

The use of elliptic partial differential equations to generate the interior grid points brings with it certain advantages. First, the grid will be smoothly varying even if the boundary of the domain has a slope discontinuity. Further, this guarantees a one-to-one mapping in view of the maximum principle for elliptic-type equations that the maximum and minimum of the unknowns occur on the boundary.

The actual solution of Eqs. (5.54) and (5.55) is carried out in the computational (ξ, η) domain. In this domain these equations transform to (see Section 7.2)

$$\alpha' \frac{\partial^2 x}{\partial \xi^2} - 2\beta' \frac{\partial^2 x}{\partial \xi \partial \eta} + \gamma' \frac{\partial^2 x}{\partial \eta^2} + \delta' \left(P \frac{\partial x}{\partial \xi} + Q \frac{\partial x}{\partial \eta} \right) = 0 \tag{5.56}$$

and,

$$\alpha' \frac{\partial^2 y}{\partial \xi^2} - 2\beta' \frac{\partial^2 y}{\partial \xi \partial \eta} + \gamma' \frac{\partial^2 y}{\partial \eta^2} + \delta' \left(P \frac{\partial y}{\partial \xi} + Q \frac{\partial y}{\partial \eta} \right) = 0 \tag{5.57}$$

where $\alpha' = x_\eta^2 + y_\eta^2$, $\beta' = x_\xi x_\eta + y_\xi y_\eta$, $\gamma' = x_\xi^2 + y_\xi^2$, $\delta' = (x_\xi y_\eta - x_\eta y_\xi)^2$.

It is desirable to perform all numerical computations in the uniform rectangular transformed plane R with equal mesh spacings $\Delta\xi = \Delta\eta = 1$ in the ξ and η directions, for ease of finite difference formulation. Further, this introduces least amount of numerical error through the grids.

The system of equations (5.56) and (5.57) is a quasi-linear elliptic system for the coordinate functions $x(\xi, \eta)$, $y(\xi, \eta)$ in the transformed plane. Solution of this linearly coupled system of equations gives the transformation which maps the distribution of coordinates in the physical space onto a uniform distribution of coordinates in the computational space (ξ, η).

Consider, for example, generation of an O-type body-fitted grid around a closed body ABC of prescribed shape in an unbounded fluid, as shown in Fig. 5.4(a). The chosen mapping into the (ξ, η)-plane, the so called *computational plane* is shown in Fig. 5.4(b). The infinite domain outside the body ABC is transformed into the rectangle $A'B'C'D'E'F'G'H'I'A'$. The boundary $DEFGHID$ is a circle of sufficiently large radius, where conditions at infinity are to be satisfied. On the contour $ABC(A'B'C')$, $\eta = \eta_1$ and $x = x_{ABC}(\xi)$, $y = y_{ABC}(\xi)$, for $\xi_1 \le \xi \le \xi_2$, where the functional relations, $x_{ABC}(\xi)$ and $y_{ABC}(\xi)$, are known and specify the distribution of grid points on ABC.

It is emphasised that boundary conditions must not be specified on $A'I'$ or $C'D'$ since the corresponding lines in the physical plane are interior lines (and coincide). Equation (5.56) can be discretised using central-difference formula to give

$$\alpha'(x_{j-1,k} - 2x_{j,k} + x_{j+1,k}) - 2\beta'(x_{j+1,k+1} - x_{j-1,k+1} - x_{j+1,k-1} + x_{j-1,k-1})$$

$$+\gamma'(x_{j,k-1} - 2x_{j,k} + x_{j,k+1}) + \delta' P(x_{j+1,k} - x_{j-1,k}) + \delta' Q(x_{j,k+1} - x_{j,k-1}) = 0, \tag{5.58}$$

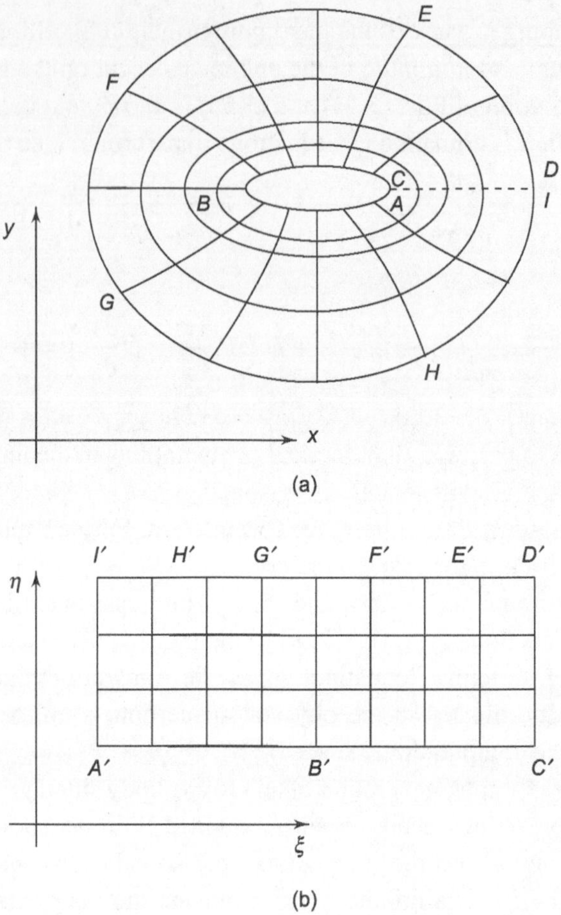

Figure 5.4 O-grid (a) multiply connected domain (b) computational plane.

where

$$\alpha' = 0.25\left[(x_{j,k+1} - x_{j,k-1})^2 + (y_{j,k+1} - y_{j,k-1})^2\right], \tag{5.59}$$

$$\beta' = 0.25\left[(x_{j+1,k} - x_{j-1,k})(x_{j,k+1} - x_{j,k-1}) + (y_{j+1,k} - y_{j-1,k})(y_{j,k+1} - y_{j,k-1})\right] \tag{5.60}$$

$$\gamma' = 0.25\left[(x_{j+1,k} - x_{j-1,k})^2 + (y_{j+1,k} - y_{j-1,k})^2\right] \tag{5.61}$$

$$\delta' = \frac{\left[(x_{j+1,k} - x_{j-1,k})(y_{j,k+1} - y_{j,k-1}) - (x_{j,k+1} - x_{j,k-1})(y_{j+1,k} - y_{j-1,k})\right]^2}{16}. \tag{5.62}$$

Equation (5.57) is discretised in a similar way. It has been assumed above that $\Delta\xi = \Delta\eta = 1$. This choice does not affect the grid in the physical domain. The appearance of the branch cut (AI/CD) in Fig. 5.4(a) introduce some increase in coding complexity for points on $A'I'(j = 1)$ or $C'D'(j = j\max)$. For example, $\dfrac{\partial^2 x}{\partial \xi^2}$ located on $A'I'$ is evaluated as

$$\frac{\partial^2 x}{\partial \xi^2} = x_{jmax-1,k} - 2x_{1,k} + x_{2,k} \tag{5.63}$$

In addition, the solution on $j = 1$ and $j = j\max$ is identical, so that iterative adjustment to the solution on $j = 1$ and $j = j\max$ should be made at the same time. Equations (5.58) and the corresponding equation for y are nonlinear simultaneous algebraic equations that can be solved iteratively. Thompson et al. (1977) apply point SOR for solving these equations. In the next subsection, these equations are solved by AF1 scheme.

5.7.1 Solution of the Algebraic Equations by AF1 Scheme

Holst (1979) and Chakrabartty et al. (1981), Mathur et al. (1994) used AF1 scheme to solve the discretised equations (5.58). In usual notations and assuming $P = Q = 0$, the transformed equations are

$$Ax_{\xi\xi} - 2Bx_{\xi\eta} + Cx_{\eta\eta} = 0, \tag{5.64}$$

$$Ay_{\xi\xi} - 2By_{\xi\eta} + Cy_{\eta\eta} = 0 \tag{5.65}$$

where $A = x_\eta^2 + y_\eta^2$, $B = x_\xi x_\eta + y_\xi y_\eta$, $C = x_\xi^2 + y_\xi^2$. The discretised equations are solved in two steps:

Step I:

$$(\alpha - A_{i,j}^n \delta_{\xi\xi}) f_{i,j}^n = \alpha w L x_{i,j}^n, \tag{5.66}$$

$$(\alpha - A_{i,j}^n \delta_{\xi\xi}) g_{i,j}^n = \alpha w L y_{i,j}^n. \tag{5.67}$$

Step II:

$$(\alpha - C_{i,j}^n \delta_{\eta\eta})(x_{i,j}^{n+1} - x_{i,j}^n) = f_{i,j}^n, \tag{5.68}$$

$$(\alpha - C_{i,j}^n \delta_{\eta\eta})(y_{i,j}^{n+1} - y_{i,j}^n) = g_{i,j}^n, \tag{5.69}$$

where $f_{i,j}$ and $g_{i,j}$ are intermediate results stored at each point in the finite difference mesh. In step I, f and g arrays are obtained by solving two tridiagonal matrix equations for each $\eta =$ constant line. The corrected values of x and y are then obtained in step II from f and g arrays respectively by solving two tridiagonal matrix equations for each $\xi =$ constant line. Here $\delta_{\xi\xi}$ denotes second order central difference in the direction of ξ

$$\delta_{\xi\xi}()_{i,j} = ()_{i+1,j} - 2()_{i,j} + ()_{i-1,j}, \tag{5.70}$$

$$\delta_{\xi\eta}()_{i,j} = \frac{1}{4}\left[()_{i+1,j+1} - ()_{i+1,j-1} - ()_{i-1,j+1} + ()_{i-1,j-1}\right], \text{ etc.} \tag{5.71}$$

$$L()_{i,j} = \{A_{i,j}\delta_{\xi\xi} - 2B_{i,j}\delta_{\xi\eta} + C_{i,j}\delta_{\eta\eta}\}()_{i,j}, \tag{5.72}$$

is the residual operator.

At the beginning of the first step at each iteration, boundary conditions for f and g can be taken as $f_\eta = g_\eta = 0$, since these two are intermediate functions and approach zero at convergence.

5.8 SOME OBSERVATIONS ON AF SCHEMES

Guidelines for the construction of AF schemes can be formulated by considering the following general form for a two-level iteration procedure,

$$NC^n + \omega L\phi^n = 0, \tag{5.73}$$

where C^n is the correction $(\phi^{n+1} - \phi^n)$, $L\phi^n$ is the residual, which is a measure of how well the finite difference equation is satisfied by the nth level solution (ϕ^n), and ω is a relaxation parameter. The iteration scheme in equation (5.73) can be regarded as an iteration in pseudo-time, where the superscript n indicates the time-step level of the solution. The operator N determines the type of iterative procedure and therefore, determines the rate at which the solution procedure converges. In the AF scheme N is chosen as a product of two or more factors indicated by

$$N = N_1 N_2 \ldots$$

The factors N_1 and N_2 are chosen so that

1. their product is an approximation of L,
2. only simple matrix operations are required, and
3. the overall scheme is stable.

For solving Eqs. (5.58) of grid generation, the operator N of Eq. (5.73) is chosen as

$$\alpha N()_{i,j} = -(\alpha - A^n_{i,j}\delta_{\xi\xi})(\alpha - C^n_{i,j}\delta_{\eta\eta})()_{i,j} \tag{5.74}$$

where α is a relaxation parameter. Multiplying out the two factors of Eq. (5.74) yields an approximation to the operator L (Eq. (5.72)) plus two error terms. The approximate L operator does not have the mixed derivative term $(\phi_{\xi\eta})$ contained in the exact L operator and has time-linearized coefficients (A and C). In spite of these approximations, unconditional linear stability exists for this scheme, proved by Mckee and Mitchell (1970). Because of the implicit conservation of this scheme, each point in the finite difference mesh influences every other point during each iteration. As a result, evolution of the solution proceeds at a much faster rate. A computer code *grid.f* has been developed following the above scheme. A typical O-grid for a NACA0012 airfoil is shown in Fig. 5.5 with 81 × 21 grid points. For six decimal place accuracy, AF scheme converges in 44 steps whereas SOR with optimum relaxation factor takes around four hundred steps for the same convergent solution.

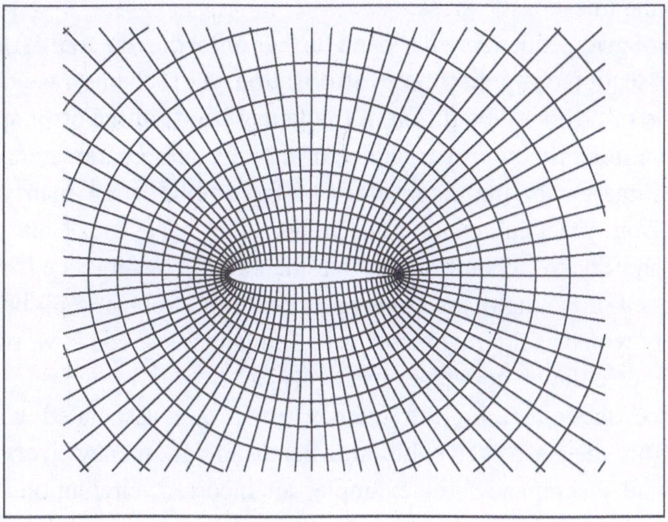

Figure 5.5 O-grid for a NACA0012 airfoil (81 × 21) using solution of Laplace equation.

5.9 MULTI-GRID METHOD

We explain here briefly, the basic concepts of the multi-grid method for solving linear partial differential equations. The method is also applicable to non-linear equations. For a thorough discussion of the topic reference may be made to Brandt (1977), Hackbusch (1985).

It is a common experience that the relaxation methods like the Gauss–Seidel or the SOR schemes for solving the discretised equations arising from elliptic-type partial differential equations, grow extremely slow as the number of mesh points increases. von Neumann stability analysis applied to Laplace equation show that the rate of convergence is related to the mesh spacings. The high frequency components of the error spectrum decay relatively fast while the low frequency components take an enormously large number of steps to get reduced (see Section 5.5.1). As a remedy, Fedorenko (1964) considered *grid sequencing,* in which fine grid solution of an elliptic-type boundary-value problem, was obtained by first solving it by relaxation on a coarse grid till convergence to desired accuracy and then proceeding with the converged values as the starting values on a finer grid. Appropriate interpolations are made to determine the starting values at the unknown mesh points. By repeating the process, one can proceed to compute the solution on the finest grid. It was found that such a procedure is significantly faster than computation by relaxation of the solution on the finest of the grids.

The idea of grid-sequencing has been further refined in the multigrid method. In the previous section on approximate factorisation (AF) schemes, we have seen that the reduction of errors in the different frequency bands of the error spectrum may be achieved efficiently by proper choice of the relaxation parameters α. However, a precise estimation of the eigenvalues of the amplification matrix is not very convenient from practical standpoint. On the other hand, von Neumann stability analysis of an iterative scheme shows that the rate of convergence is related to the grid spacings. For example, the convergence rate of a relaxation scheme becomes progressively worse as the number of grid points increases. It is worth noting that a coarse grid calculation converges relatively faster.

Experience show that high frequency errors are associated with localised deviations from the converged solution, whereas low frequency errors reflect some kind of global discrepancy, for example, an incorrect circulation. The idea of multigrid acceleration is to reduce the error for different frequency bands at different grid levels, from fine to coarse.

The multi-grid method or multi-level adaptive technique (MLAT) as named by Brandt 1977) is a numerical strategy of solving continuous problems by cycling between coarser and finer levels of discretisation. For general partial differential equations with approprate boundary conditions, this technique provides a very fast solver of the discrete equations. A sequence of uniform grids (or "levels"), with geometrically decreasing mesh sizes, is employed. All the grid levels are used in computing the solution comprising the steps:

1. relaxation sweeps over each of the grid levels;
2. coarse-grid to fine-grid interpolations of corrections and
3. fine-grid to coarse-grid transfer of residuals.

In this technique, it is required to construct difference approximations on uniform grids only. We also assume the boundary to coincide with the grid lines. Furthermore, the questions of accuracy and stability are effectively separated in multi-grid algorithms. Stable approximations are needed in the relaxation phase, while accuracy is determined by the approximation used in the residual transfers, which itself need not be stable. The stability of difference operators is intimately connected to their fast multi-grid solution. Specifically, the error can be efficiently smoothed by relaxation, particularly for elliptic-type equations. However, the method has been found to be equally successful for other types of equations as well. Depending on the problem, the ADI schemes may also be used as a *smoother* instead of the relaxation schemes.

The multigrid method consists of *fine grid smoothing, followed by coarse grid correction, used repetitively to reach the coarsest grid beginning from the finest grid.* The iterations then return back from the coarsest grid to the finest grid. At these stages transfer of corrections take place. This completes one cycle. The cycles are repeated till convergence to desired accuracy is achieved on the finest of the grids.

We discuss the *full approximate scheme (FAS)* only. A Dirichlet problem for Laplace equation, discussed earlier, has been presented as illustration, at the end of the section. Let us consider, a sequence of l-$grid$ levels with mesh spacings $h_1, h_2, \cdots, h_l$, where $h_k = \frac{1}{2}h_{k+1}$, $k = 1, 2, \cdots, l-1$. Here, h_l is the coarsest grid spacing.

Let the boundary-value problem to be solved be discretised on the finest grid having mesh spacing h as

$$L^h U^h = G^h, \tag{5.75}$$

L being some suitable finite difference operator.

Let U^h be the exact solution vector of the discretised difference equations (including discretised boundary conditions). Let u^h denote an approximation to U^h, obtained by performing a few iterations of (5.75). This is called a *smoother* because it reduces the high frequency components of the error spectrum. Then,

$$L^h u^h = G^h. \tag{5.76}$$

Let

$$U^h = u^h + v^h, \tag{5.77}$$

where v^h is the *correction vector.* Then,

$$L^h(u^h + v^h) = G^h,$$

so that

$$L^h v^h = G^h - L^h u^h. \tag{5.78}$$

Let $H = 2h$ be the next coarser grid spacing. The correction v^h is evaluated on the coarser grid with spacing H. This is achieved by approximating (5.78) on the coarser grid H. This is called a *restriction operation,* denoted by I_h^H. Clearly, the simplest kind of restriction would be simply to drop every alternate mesh point.

The approximate version of (5.78) on the coarser grid is

$$L^h v^h = I_h^H (G^h - L^h u^h). \tag{5.79}$$

It is important to note that the success of the method depends much on the removal of the high frequency error contents by the fine grid iterations (5.76). It is expected that the right hand side of (5.79) mainly contains low frequency error components. We may use the *injection mapping* defined as

$$I_h^H : F_j^H = F_{2j}^h, \ j = 0, 1, \cdots, N. \tag{5.80}$$

Requisite number of iterations may be carried out with (5.79), in order to obtain v^H.

From grid level H we go to the next coarser grid of spacing $2H$, in the same way as in the previous step, and this is repeated till we reach the coarsest grid.

After this, we return back, one after the other, to the finest grids. From one grid to the next finer grid we require *interpolation operation* I_H^h, with $H = 2h$. Often, a simple linear interpolation may be used for I_H^h. Now, suppose, u_{old}^h be the old values on grid with mesh spacing h. Then, the new values are obtained as

$$u_{\text{new}}^h = u_{\text{old}}^h + I_H^h v^h. \tag{5.81}$$

This is repeated till we reach the finest grid level, where we test for convergence. This completes one cycle. If convergence to desired accuracy be not acieved the cycles are to be repeated iteratively.

The labour involved is measured in terms of *work-units.* A work unit can be defined as the computational effort required to carry out one fine grid computation. If for the AF1 scheme a set of six acceleration parameters α_{p} be used, then one AF1 iteration with six parameter values constitutes six work units. From a rough estimate it may be seen that one multigrid cycle involves about $1\frac{2}{3}$ work units, assuming one iteration is performed on the finest grid and two iterations on all the coarser grids $\left(1 + 2\left[\frac{1}{4} + \frac{1}{16} + \frac{1}{64} + \cdots\right] \leq 1\frac{2}{3}\right)$. Since coarser grid calculations are less laborious, it is estimated that multigrid iterations are roughly three times faster than the corresponding AF1 iterations.

A rich literature now exists on application of multi-grid method to computational study of various non-linear problems of fluid dynamics, initiated among others, by Brandt (1977) and Hackbusch (1985). Applications may be found, for example, in Ghia et al. (1982), Jameson (1979, 1983, 1985b, 1985c), Jameson and Yoon (1985), Jameson (1993), Martinelli and Jameson (1988), South and Brandt (1977). For a thorough discussion of the topic reference may be made to Hackbusch (1985).

5.9.1 Program 5.3: mgc.f

As illustration of multigrid method, we present here a Program mgc.f. It solves the Dirichlet problem for 2-D Laplace equation in the unit square in the first quadrant by multigrid method:

$$u_{xx} + u_{yy} = 0, \;\; u = 0 \text{ on } x = 0 \text{ and } y = 0; \tag{5.82}$$

$$u = 4x, \;\; \text{on } y = 1 \text{ and } u = 4y \text{ on } x = 1. \tag{5.83}$$

A similar problem has been numerically solved earlier by Gauss–Seidel, SOR and AF schemes, the corresponding programs being sor.f and af1.f. Representing the derivatives by three-point central difference and taking, both the mesh lengths in x and y directions as h, omitting truncation error terms, the discretised equations are

$$\frac{u_{j+1,k} - 2u_{j,k} - u_{j-1,k}}{h^2} + \frac{u_{j,k+1} - 2u_{j,k} + u_{j,k-1}}{h^2} = 0, \tag{5.84}$$
$$j = 1, \cdots, j\text{max} - 1; \; k = 1, \cdots, k\text{max} - 1.$$

Point Gauss–Seidel iteration for the above system is defined as

$$\frac{u^n_{j+1,k} - 2u^{n+1}_{j,k} + u^{n+1}_{j-1,k}}{h^2} + \frac{u^n_{j,k+1} - 2u^{n+1}_{j,k} + u^{n+1}_{j,k-1}}{h^2} = 0, \tag{5.85}$$
$$j = 1, \cdots, j\text{max} - 1; \; k = 1, \cdots, k\text{max} - 1.$$

If an approximate solution $u^n_{j,k}$ is substituted in Eq. (5.85), in general, it will not satisfy the equation. We define *the residual $R_{j,k}$ as the quantities required to be added to the left hand side in order that the equations may be identically satisfied.* Thus on simplification,

$$R_{j,k} = -\frac{1}{h^2}\left[u^n_{j+1,k} + u^n_{j-1,k} + u^n_{j,k+1} + u^n_{j,k-1} - 4u^n_{j,k}\right] \tag{5.86}$$
$$j = 1, \cdots, j\text{max} - 1; \; k = 1, \cdots, k\text{max} - 1.$$

Let us define *corrections* $v_{j,k}$ at any point (j, k) as

$$u^{n+1}_{j,k} = u^n_{j,k} + v_{j,k}. \tag{5.87}$$

Substituting these in Eqs. (5.85), the equations for determining the corrections in terms of the residuals are

$$v_{j,k} = \frac{1}{4}\left(v_{j,k-1} + v_{j-1,k} - h^2 R_{j,k}\right) \tag{5.88}$$

In the following program the prolongation operations are simple average of neighbouring mesh points and restriction operations simply drop the intermediate points.

```
c.....Program 5.3: mgc.f prepared by Subhadip Niyogi.
c......It solves by multigrid method Dirichlet problem for
c..... 2-D Laplace equation in the unit-square in the 1st quadrant,
c......with boundary conditions: u=0 on the axes and u=4x and u=4y
c..... on the remaining sides. The known  exact solution is u=4xy.
c..... n=no. of fine grid mesh points in either direction,m=grid levels.
c......Corrections=v and residuals=r, point Gauss-Seidel smoother used.
c-------------------------------------------------------------
parameter(ix=257)
dimension us(0:ix,0:ix),v(0:ix,0:ix), r(0:ix,0:ix),x(0:ix)
dimension y(0:ix), uex(0:ix,0:ix)
open (unit=18,file='mgc.out')
write(*,*) 'Input no. of grid points, levels,tolnc. '
c read(*,*) n,m,eps
c n =16
n=128
c m=3
m=5
eps=0.0000005
write(18,*) ' Output of Program mgc.f '
write(18,*) '--------------------------------------'
write(18,*) 'grid pts= ',n,' levels= ',m, ' eps= ', eps
write(18,*) '---------------------------------------------'
c.......Parameters and boundary conditions............
h=1.0/float(n)
k2=n/2
do 10 i=0,n
x(i)=i*h
y(i)=i*h
10 continue
c........Starting  values of u, denoted by us.........
do 12 i=0,n
do 12 j=0,n
us(i,j)=0.0
12 continue
do 56 i=0,n
do 56 j=0,n
v(i,j)=0.0
56 continue
do 15 i=0,n
us(i,0)=0.0
us(i,n)=4.0*x(i)
us(0,i)=0.0
us(n,i)=4.0*y(i)
15 continue
```

```
c.......Computation of exact solution, denoted by uex.....
do 55 i=1,n
do 55 j=1,n
uex(i,j)= 4.0*x(i)*y(j)
55 continue
c..... Start iteration....................
do 31 i=1,n-1
do 31 j=1,n-1
us(i,j)= 0.25*(us(i+1,j)+us(i-1,j)+us(i,j+1)+us(i,j-1))
31 continue
do 11 ic=1,30
call residual(us,r,n)
call solve(us,r,v,n)
it=it+1
do 17 k=1,m
n=n/2
call restriction(us,r,v,n)
call solve(us,r,v,n)
it=it+1
17 continue
c.......Lowest grid-level reached. Initiate prolongation....
do 21 kk=1,m
call prolongation(us,r,v,n)
n=n*2
call solve(us,r,v,n)
it=it+1
21 continue
rms=0.0
do 25 i=1,n-1
do 25 j=1,n-1
rms= rms+v(i,j)*v(i,j)
25 continue
rmss=sqrt(rms/n)
do 125 i=1,n-1
do 125 j=1,n-1
if(abs(v(i,j)).gt.eps)go to 11
if(rmss.gt.eps) go to 11
125 continue
write(18,*)  ' kk= ', kk,'  rmss= ', rmss
write(18,*)' Convergence reached in cycles= ',ic, 'it= ', it
write(18,*) ' Converged solution '
go to 113
11 continue
write(18,*) ' No convergence in cycles = ',ic,' it= ', it
go to 113
7 format(1x,i3,2f12.6)
```

```
113 do 111 j=1,n-1,8
write(18,7) j,  us(j,k2), uex(j,k2)
111 continue
112 continue
close(18)
stop
end
c..............................................
subroutine residual(us,r,n)
parameter(ix=257)
dimension us(0:ix,0:ix),r(0:ix,0:ix)
h=1.0/float(n)
do 25 i=1,n-1
do 25 k=1,n-1
r(i,k)=-(us(i-1,k)+us(i+1,k)+us(i,k-1)+us(i,k+1)-4.0*us(i,k))/(h*h)
25 continue
do 26 i=0,n
r(0,i)=0.0
r(i,0)=0.0
r(n,i)=0.0
r(i,n)=0.0
26 continue
return
end
c............................................
subroutine restriction(us,r,v,n)
c--------------------------------------------
parameter (ix=257)
dimension us(0:ix,0:ix), r(0:ix,0:ix),v(0:ix,0:ix),tmp1(0:ix,0:ix)
dimension tmp2(0:ix,0:ix),tmp3(0:ix,0:ix)
do 20 i=0,n
do 30 j=0,n
i2=2*i
j2=2*j
tmp1(i,j)=us(i2,j2)
tmp2(i,j)=v(i2,j2)
tmp3(i,j)=r(i2,j2)
us(i,j)=tmp1(i,j)
v(i,j)=tmp2(i2,j2)
r(i,j)=tmp3(i2,j2)
30 continue
20 continue
return
end
```

```
c.........................................
subroutine solve(us,r,v,n)
c-----------------------------------------
parameter (ix=257)
dimension us(0:ix,0:ix), r(0:ix,0:ix),v(0:ix,0:ix)
h=1.0/float(n)
call residual(us,r,n)
do 40 i=0,n
v(0,i)=0.0
v(i,0)=0.0
v(n,i)=0.0
v(i,n)=0.0
40 continue
c do 35 kk=1,2
call residual(us,r,n)
do 41 j=1,n-1
do 41 k=1,n-1
v(j,k)=(v(j-1,k)+v(j,k-1)-h*h*r(j,k))/4.0
41 continue
do 42 i=1,n-1
do 42 j=1,n-1
us(i,j)=us(i,j)+v(i,j)
42 continue
c35 continue
call residual(us,r,n)
return
end
c.....................................................
subroutine prolongation(us,r,v, n)
parameter(ix=257)
dimension us(0:ix,0:ix), v(0:ix,0:ix),r(0:ix,0:ix)
dimension temp1(0:ix,0:ix),temp2(0:ix,0:ix),temp3(0:ix,0:ix)
do 25 i=0,n
do 25 j=0,n
temp1(i,j)=us(i,j)
temp2(i,j)=v(j,k)
temp3(i,j) = r(i,j)
25 continue
do 26 i=0,n
do 27 j=0,n
i2=i*2
j2=j*2
us(i2,j2)=temp1(i,j)
v(i2,j2)=temp2(i,j)
r(i2,j2)=temp3(i,j)
```

```
27 continue
26 continue
do 35 i=0,n
do 36 j=0,n
i2=i*2
j2=j*2
us(i2,j2+1)=( temp1(i,j)+temp1(i,j+1))/2.0
us(i2+1,j2)= (temp1(i,j)+temp1(i+1,j))/2.0
us(i2+1,j2+1)=(temp1(i+1,j+1)+temp1(i,j)+temp1(i+1,j)+temp1(i,j+1))/4.
v(i2,j2+1)=( temp2(i,j)+temp2(i,j+1))/2.0
v(i2+1,j2)= (temp2(i,j)+temp2(i+1,j))/2.0
v(i2+1,j2+1)=(temp2(i+1,j+1)+temp2(i,j)+temp2(i+1,j)+temp2(i,j+1))/4.
r(i2,j2+1)=( temp3(i,j)+temp3(i,j+1))/2.0
r(i2+1,j2)= (temp3(i,j)+temp3(i+1,j))/2.0
r(i2+1,j2+1)=(temp3(i+1,j+1)+temp3(i,j)+temp3(i+1,j)+temp3(i,j+1))/4.
36 continue
35 continue
do 40 i=1,n
n2 = 2*n
i2p = i*2+1
i2m=i*2-1
us(n2,i2m)=4*i2m /float(n2)
us(i2m,n2)=4*i2m/float(n2)
v(n2,i2m)=0.0
v(i2m,n2)=0.0
r(n2,i2m)=0.0
r(i2m,n2)=0.0
40 continue
return
end
c---------------------------------------------
  Output of Program mgc.f
 --------------------------------------
 grid pts=  128 X 128, levels=  5 eps=   4.99999999E-07
 --------------------------------------------
  kk=  6   rmss=   2.8221487E-07
  Convergence reached in cycles=  8
  Converged solution
  j     us(j,k2)     uex(j,k2)
   1      0.003904     0.003906
  17      0.066372     0.066406
  33      0.128850     0.128906
  49      0.191341     0.191406
  65      0.253843     0.253906
  81      0.316354     0.316406
```

```
 97    0.378871    0.378906
113    0.441389    0.441406

 Output of Program mgc.f
--------------------------------------
grid pts=  128 X 128, levels=  6 eps=   4.99999999E-07
-------------------------------------------
 kk=  7    rmss=    0.
 Convergence reached in cycles=  1
 Converged solution
 j     us(j,k2)     uex(j,k2)
  1    0.003906    0.003906
 17    0.066406    0.066406
 33    0.128906    0.128906
 49    0.191406    0.191406
 65    0.253906    0.253906
 81    0.316406    0.316406
 97    0.378906    0.378906
113    0.441406    0.441406
```

We note that it takes eight cycles to converge correct to six decimal places when the number of prescribed grid levels is five while for six grid levels it requires only one cycle.

5.10 SUMMARY

1. Laplace equation in 2D has been studied as model elliptic-type equation. Three-point central difference discretisation may be used; number of equations to be solved is generally very large. A pentadiagonal linear system has to be solved. Iterative methods like SOR or line SOR recommended.
2. Grid generation techniques reduce irregular boundaries to a rectangle in the computational domain, so that boundary conditions may be correctly imposed. For the sake of simplicity, grid generation has been discussed here very briefly. The reader may consult Chapter 8 and cited references.
3. For larger number of equations, relaxation methods become very slow to converge. Convergence can be satisfactorily accelerated by employing methods like approximate factorisation or multigrid technique, discussed very briefly.

The following programs in Fortran (programs in C in Appendix) with test results presented.

1. sor.f: solves using Gauss–Seidel and SOR a Dirichlet problem for Laplace equation in the unit square in the first quadrant.
2. af1.f: solves Dirichlet problem for 2-D Laplace equation in the unit square in the first quadrant by AF1 (approximate factorisation) scheme.
3. mgc.f: solves a Dirichlet problem for 2-D Laplace Equation by multigrid technique in terms of corrections and residuals.

5.11 KEY TERMS

Amplification factor
Approximate factorisation method
Conservative equations
Consistency
Convergence
Dirichlet problem
Implicit scheme
Jacobian matrix
Laplace equation
Multigrid technique
Poisson equation
Successive overrelaxation
Successive line overrelaxation
von Neuman stability analysis.

5.12 EXERCISE 5

5.1 Compute the solution of the Dirichlet problem for Laplaces equation

$$u_{xx} + u_{yy} = 0,$$

in the unit square in the first quadrant with boundary condition

$$u(x, y) = e^{0.1\pi x} \sin(0.1\pi y)$$

on the boundary of the square. Take $\Delta x = \Delta y = \frac{1}{4}$ and use SOR scheme with optimum relaxation factor. Compute correct to 2-D.

5.2 Using three-point central difference representation for the second derivatives, compute the solution of the boundary-value problem for the Poisson equation

$$u_{xx} + u_{yy} = -4, \quad |x| \leq 1, \ |y| \leq 1, \tag{5.89}$$

with Dirichlet boundary condition $u(x, y) = 0$ on the boundary of the square. Take $\Delta x = \Delta y = 0.5$, and compute by Gauss–Seidel iteration scheme solution correct to two-decimals.

5.3 Using three-point central difference representation for the second derivatives, compute the solution of the boundary-value problem for the Poisson equation

$$u_{xx} + 4u_{yy} = -2, \quad |x| \leq 1, \ |y| \leq 1, \tag{5.90}$$

with Dirichlet boundary condition $u(x, y) = 0$ on the boundary of the square $|x|=1$, $|y|=1$. Take $\Delta x = \Delta y = 0.5$, and compute by Gauss–Seidel iteration scheme solution correct to two-decimals.

5.4 Compute the solution of the boundary-value problem for the Poisson equation

$$u_{xx} + u_{yy} = 2x^2 - 2, \quad |x| \le 1,\ |y| \le 1, \tag{5.91}$$

with Dirichlet boundary condition $u(x, y) = 0$ on the boundary of the square. Take $\Delta x = \Delta y = 0.5$, compute the first three steps by Gauss–Seidel iteration scheme and continue till convergence, correct to 2-D, by SOR.

5.5 Compute the solution of the boundary-value problem for the Poisson equation

$$u_{xx} + u_{yy} = -4(x^2 + y^2), \quad 0 < x < 1,\ 0 < y < 1, \tag{5.92}$$

with Dirichlet boundary condition $u(x, y) = 0$ on the boundary of the square. Take $\Delta x = \Delta y = 0.2$, and compute the solution correct to 2-D.

5.6 Compute the solution of the boundary-value problem for the Poisson equation

$$u_{xx} + u_{yy} = -10xy, \quad 0 < x < 1,\ 0 < y < 1, \tag{5.93}$$

with Dirichlet boundary condition $u(x, y) = 0$ on the boundary of the unit square in the first quadrant. Take $\Delta x = \Delta y = 0.2$, and compute the solution correct to 2-D.

5.7 Use multigrid method to solve the BVP

$$\frac{\partial^2 u}{\partial x^2} = 0,\ u(0) = 0, u(1) = 1,$$

The exact solution of this problem is $u = x$.

Hint: Write the subroutines gs, restric, prolong and start. Here, gs is the Gauss–Seidel solver, restric is restriction operator, prolong is the prolongation operator. The start subroutine starts with a rough initial guess of the solution which is then improved. The solution method is formulated in terms of residual and corrections. The iteration stops when the root-mean-square correction is smaller than the prescribed tolerance. Results show that for 128×128 points, Gauss–Seidel scheme requires 7685 iteration steps, while multigrid needs only three cycles to compute the solution correct to 6-D.

5.8 Using three-point central difference representation for the second derivatives, compute the solution of the boundary-value problem for the Poisson equation

$$u_{xx} + 4u_{yy} = -1, \quad |x| \le 1,\ |y| \le 1, \tag{5.94}$$

with Dirichlet boundary condition $u(x, y) = x - y$, on the boundary of the square $|x|=1$, $|y|=1$. Take $\Delta x = \Delta y = 0.5$, compute the first three steps by Gauss–Seidel iteration scheme and continue till convergence, correct to 2-D, by SOR.

5.9 Compute the solution of the boundary-value problem for the Poisson equation

$$u_{xx} + u_{yy} = x^2 - 1, \quad |x| \le 1, \ |y| \le 1, \tag{5.95}$$

with Dirichlet boundary condition $u(x, y) = x + y$, on the boundary of the square. Take $\Delta x = \Delta y = 0.5$, compute the first three steps by Gauss–Seidel iteration scheme and continue till convergence, correct to 2-D, by SOR.

5.10 Compute the solution of the boundary-value problem for the Poisson equation

$$u_{xx} + u_{yy} = -4(x^2 + y^2), \quad 0 < x < 1, \ 0 < y < 1, \tag{5.96}$$

with Dirichlet boundary condition $u(x, y) = xy$, on the boundary of the square. Take $\Delta x = \Delta y = 0.2$, and compute the solution correct to 2-D.

6

Equations of Mixed Elliptic–Hyperbolic Type

Model linear equations of parabolic, hyperbolic and elliptic type have been studied respectively in Chapters 3, 4 and 5. The present chapter is devoted to model equations of mixed elliptic–hyperbolic type. *Tricomi equation* is the model second-order linear partial differential equation of mixed elliptic–hyperbolic type. Formulation of well-posed boundary-value problems for the Tricomi equation has been discussed. A closely related equation of much practical importance is the *transonic small perturbation (TSP) equation*, which is an approximate equation, governing the steady inviscid transonic flow past a thin profile. It is elliptic in the subsonic part of the flow field, hyperbolic in the supersonic part and parabolic on the sonic line. This second-order non-linear equation exhibits all the typical features of such a mixed-type equation. For this reason, this equation is used as a model for study of mixed type equations. For high-speed inviscid compressible flow, the steady Euler equations are a system of mixed elliptic–hyperbolic type. In order to understand the nature of such flow fields, it is illuminating to study quantitatively the simpler TSP equation. This chapter undertakes such an investigation, found useful in Chapter 10, on study of inviscid compressible flow fields based on the potential and Euler models.

6.1 INTRODUCTION

Numerical solution of linear equations of parabolic, hyperbolic and elliptic types were discussed in the previous chapters. The present chapter is devoted to the study of partial differential equations of mixed elliptic–hyperbolic type. The canonical form of linear second-order equations is the *Tricomi equation*. Tricomi (1923) established existence and uniqueness of well-posed problems for the equation, now known after him, viz. the Tricomi equation. Such equations are of much interest in the study of transonic flow problems, discussed in Chapter 10. The transonic small perturbation equation, governing the steady flow of an inviscid fluid past a thin body, is one of the simplest examples of a quasi-linear second-order partial differential equation of mixed elliptic–hyperbolic type.

Over, rather a long period of time, during the Second World War and the subsequent quarter of a century, transonic flow fields baffled scientists and engineers by eluding quantitative study. Even in the simplest case of steady high subsonic flow past a thin profile at small incidence, where a small supersonic region is formed, embedded in an otherwise subsonic flow, the governing small perturbation equations are nonlinear (strictly speaking quasi-linear) and of mixed elliptic-hyperbolic type (Figs. 6.1 and 6.2). The line of demarcation between the subsonic and supersonic regions, i.e. the sonic line, is not known *a priori* and must be found out as a part of the solution.

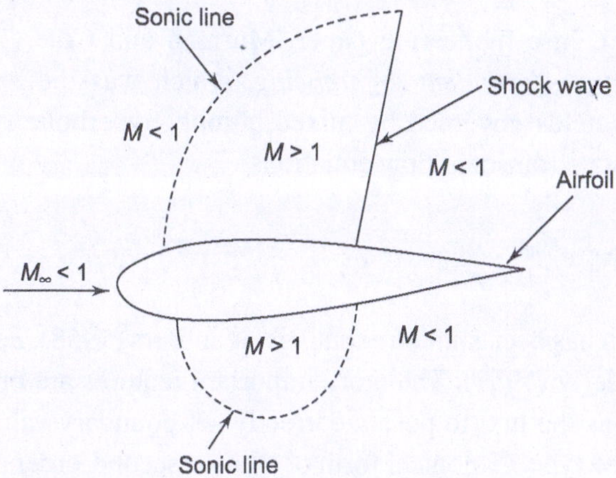

Figure 6.1 Transonic flow past an airfoil terminating in a shock.

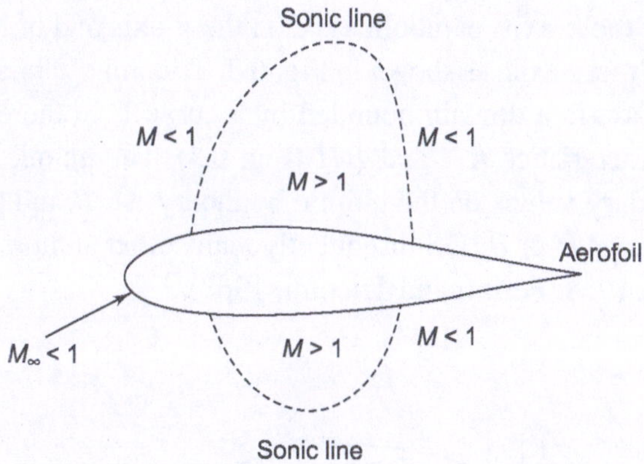

Figure 6.2 Shock-free transonic flow past an airfoil.

Experiments carried out during fifties and early sixties, showed the embedded supersonic region to terminate in a shock wave (Fig. 6.1). On the other hand, exact solutions like Ringleb's solution (Niyogi, 1977; Oswatitsch, 1956) showed the existence of shock-free solution, both acceleration and deceleration through the sonic speed were continuous (for example, see Fig. 6.2). The first successful computational methods of adequate accuracy for a steady small perturbation flow are the finite difference results of Magnus and Yoshihara (1970) and Murman and Cole (1971). In a pioneering paper, Murman and Cole (1971), introduced the concept of *type-dependent differencing,* which was the key to success for computing flow fields governed by mixed elliptic–hyperbolic type of equations, like the steady-state transonic flow equations.

6.2 TRICOMI EQUATION

Tricomi equation has been studied exhaustively in Bers (1958), Ferrari and Tricomi (1968), and Guderly (1957). The more important features are briefly summarised here. Tricomi was the first to pose a correctly set boundary value problem for an equation of mixed type. Canonical form of a linear second-order partial differential equation of mixed elliptic–hyperbolic type is the *Tricomi equation* (Ferrari and Tricomi, 1968; Guderly, 1957; Prasad and Ravindran, 1985)

$$yu_{xx} + u_{yy} = 0. \tag{6.1}$$

In a domain having the x-axis inside it, it is of elliptic type for $y > 0$, that is on the upper part of the x-axis, parabolic type on the x-axis and of hyperbolic type in the lower part of the x-axis, as shown in Fig. 6.3. Tricomi established that Eq. (6.1) has unique solution in a domain bounded by a curve Γ in the elliptic part $y > 0$ and the two characteristics AH and BH (Fig. 6.3). The unique solution assumes prescribed boundary values on the elliptic boundary ACB and on any one of the two characteristics AH or BH. Subsequently many other authors generalised these problems, (Bers, 1958; Ferrari and Tricomi, 1968).

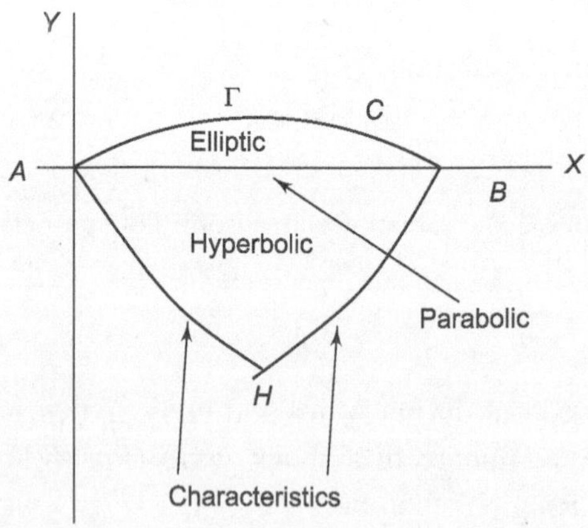

Figure 6.3 A typical domain for the Tricomi equation.

Tricomi showed that under certain appropriate smoothness assumptions, any linear second-order partial differential equation

$$au_{xx} + 2bu_{xy} + cu_{yy} + \cdots = 0, \tag{6.2}$$

for which the quantity $b^2 - ac$ changes sign along a curve, can be reduced to a certain canonical form

$$yu_{xx} + u_{yy} + \mathrm{d}_1 u_x + \mathrm{d}_2 u_y + \mathrm{d}_3 u = 0. \tag{6.3}$$

Tricomi equation (6.1) is a special form of the *Chaplygin type equation*

$$K(y)u_{xx} + u_{yy} = 0, \tag{6.4}$$

where $K(y)$ is a monotone function such that

$$\text{for } y \neq 0,\ yK(y) > 0.$$

Such equations occur frequently in gasdynamics (Niyogi, 1977; Oswatitsch, 1956, 1977.)

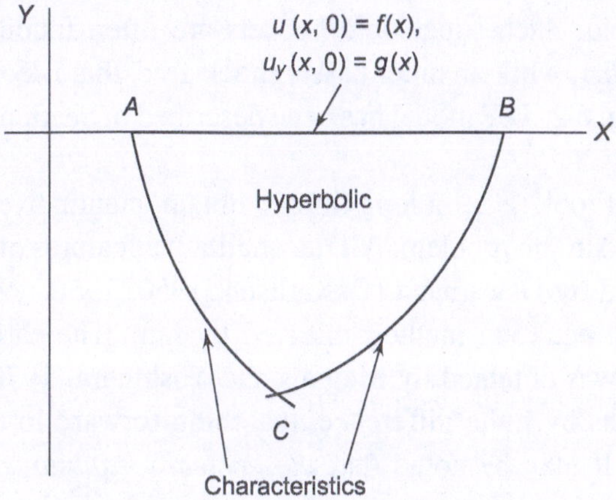

Figure 6.4 Cauchy problem for Chaplygin-type equation.

It is known that Dirichlet problem and other typical elliptic boundary value problems are correctly posed for Eq. (6.2), even if the domain contains a segment of the parabolic line $y = 0$. Bers (1958) established that the Cauchy problem for Eq. (6.4) with data on a segment of the parabolic line

$$u(x, 0) = f(x), \quad u_y(x, 0) = g(x), \tag{6.5}$$

is correctly posed. The solution exists in the characteristic triangle ABC (Fig. 6.4) and satisfies the inequality

$$|u| \leq \max|f| + |y|\max|g|. \tag{6.6}$$

It is interesting to note that the inequality (6.6) does not involve the function $K(y)$.

6.3 TRANSONIC COMPUTATIONS BASED ON TSP MODEL

Our computational study of transonic flow begins with the relatively simple problem of steady inviscid irrotational transonic flow field computation past a thin symmetric profile at zero incidence. The transonic small perturbation (TSP) model is considered, according to which the body is assumed to be thin, with continuously turning tangents inclined at small angles everywhere with the free-stream direction. Clearly, the small-perturbation assumption is violated at the forward and rear stagnation points. The free-stream subsonic Mach number is assumed to be sufficiently high, such that a local supersonic pocket is formed in an otherwise subsonic flow field. Such supersonic pockets are often found to terminate in a shock, as in Fig. 6.1, while in other cases, shock-free, that is continuous flow may be found, as in Fig. 6.2. TSP model has been described in Sections 7.10.3 and 7.10.5 of Chapter 7.

We note that, it took rather a long time to obtain quantitatively correct solution of this apparently simple problem. All the qualitative features of the solution were, however, obtained by Oswatitsch (Oswatitsch, 1950, 1956, 1977; Niyogi, 1982) using the integral equation method initiated by him. The earliest quantitatively correct solution, was obtained by Magnus and Yoshihara (1970), who solved the unsteady equations by finite difference, marching forward in time until a steady state is reached. It may be noted that *the unsteady equations are of hyperbolic type. In contrast, the steady-state equations are of mixed elliptic–hyperbolic type.* Murman and Cole (1971), introduced the *concept of type-dependent differencing* and successfully solved by finite difference steady-state equations, thereby reducing the computational time (CPU time) by an order, compared to that of Magnus and Yoshihara. The basic ideas of the Murman–Cole scheme is presented here, for the computation of steady inviscid irrotational transonic flow past a thin symmetric profile at zero incidence, based on TSP assumptions. Through this simple problem one can learn many key concepts necessary for successful computation of more complex problems.

A body-fixed rectangular Cartesian coordinate axes is chosen, with origin at the tip of the profile and x-axis along the chord, as shown in Fig. 6.5. Let the upper part of the profile shape be given by

$$y = \begin{cases} \tau F(x), & 0 \le x \le 1, \\ 0, & \text{otherwise}, \end{cases} \tag{6.7}$$

where τ denotes the *thickness-ratio,* that is the ratio of the maximum thickness to chord-length of the profile. The high subsonic free-stream flow, with Mach number $M_\infty < 1$, is along the profile chord. We are particularly interested in *super-critical flow* for which there is at least one point on the surface of the profile such that the local flow speed is supersonic there.

The governing equations, as given in Section 7.10.3 of Chapter 7, are Eqs. (7.119) and the tangency boundary condition (7.120) (with $\alpha = 0$), repeated here for ready reference:

$$\frac{\partial}{\partial x}\left[K\phi_x - \frac{\gamma + 1}{2}\phi_x^2\right] + \frac{\partial}{\partial \bar{y}}\phi_{\bar{y}} = 0, \tag{6.8}$$

$$\phi_{\bar{y}}(x, 0+) = F'(x), \quad 0 \le x \le 1. \tag{6.9}$$

Here, the velocity components u, v are related to the perturbation potential ϕ by

$$\phi_x = \frac{u - u_\infty}{\tau^{\frac{2}{3}} u_\infty}, \quad \phi_{\bar{y}} = \frac{v - v_\infty}{\tau u_\infty}, \tag{6.10}$$

Note that, for flow at zero incidence, v_∞ vanishes. Further, at large distances from the profile, the perturbation velocity components must vanish,

$$\phi_x,\ \phi_{\bar{y}} \to 0, \ \text{for} \sqrt{x^2 + \bar{y}^2} \to \infty. \tag{6.11}$$

Because of symmetry, it is enough to consider the solution in the semi-infinite domain $\bar{y} \ge 0$. A symmetry condition $\phi_{\bar{y}} = 0$ has to be applied on $\bar{y} = 0$, $0 > x > 1$.

6.3.1 Finite Difference Discretisation

Here, the domain is a semi-infinite one, and some kind of a transformation is required to map it into a finite domain. However, in practice, for TSP computations, most authors, replace the domain approximately by a large finite rectangle, taken as the computational domain (Fig. 6.5) with one side of the x-axis. The computational domain is subdivided into a network by straight lines $x = i\Delta x$, $\bar{y} = j\Delta\bar{y}$, the mesh spacings Δx, $\Delta\bar{y}$ being taken to be uniform, $i, j = 1, 2, \ldots$. (In practice, however, non-uniform mesh spacings are preferred, with a view to computational economy, although it is easier to understand the basic ideas with uniform mesh spacings.) The perturbation velocity components ϕ_x, $\phi_{\bar{y}}$ are computed by central difference according to

$$\phi_x|_{i+\frac{1}{2},j} = \frac{1}{\Delta x}\left[\phi_{i+1,j} - \phi_{i,j}\right], \quad \phi_{\bar{y}}|_{i,j+\frac{1}{2}} = \frac{1}{\Delta \bar{y}}\left[\phi_{i,j+1} - \phi_{i,j}\right], \tag{6.12}$$

where the suffix $i+\frac{1}{2},\ j$, denote the value at the point mid-way between the mesh points $(i+1,\ j)$ and $(i,\ j)$ and similarly for the other fractional suffixes.

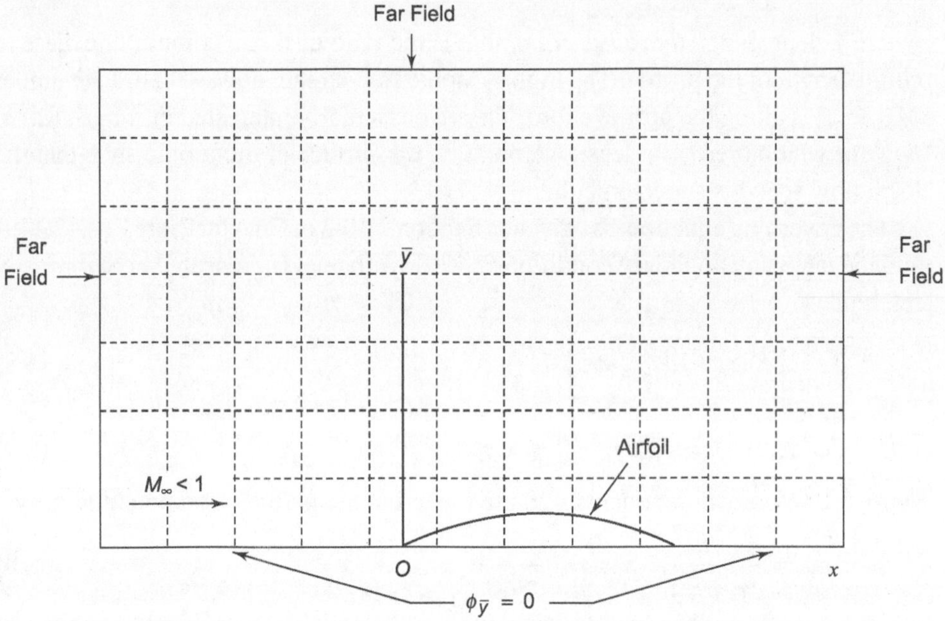

Figure 6.5 Computational domain and boundary values for the TSP model.

At a subsonic point, where the governing equation (6.8) is of elliptic type, it may be discretised by central difference as

$$\frac{1}{\Delta x}\left[\left\{K\phi_x - \frac{\gamma+1}{2}\phi_x^2\right\}_{i+\frac{1}{2},j} - \left\{K\phi_x - \frac{\gamma+1}{2}\phi_x^2\right\}_{i-\frac{1}{2},j}\right]$$
$$+\frac{1}{\Delta\bar{y}}\left[(\phi_{\bar{y}})_{i,j+\frac{1}{2}} - (\phi_{\bar{y}})_{i,j-\frac{1}{2}}\right] = 0. \tag{6.13}$$

Substituting from Eq. (6.12) in Eq. (6.13) and factorising, yields

$$\left[K-(\gamma+1)\frac{\phi_{i+1,j}-\phi_{i-1,j}}{2\Delta x}\right]\left[\frac{\phi_{i+1,j}-2\phi_{i,j}+\phi_{i-1,j}}{\Delta x^2}\right]+\frac{\phi_{i,j+1}-2\phi_{i,j}+\phi_{i,j-1}}{\Delta\bar{y}^2}=0. \tag{6.14}$$

It may be noted that the left hand side of Eq. (6.14) is a second order accurate difference operator. Let us define the quantity $(V_e)_{i,j}$ by

$$(V_e)_{i,j} = K - (\gamma+1)\frac{\phi_{i+1,j}-\phi_{i-1,j}}{2\Delta x}. \tag{6.15}$$

Then, it may be verified by von Neumann stability analysis, that at a subsonic point, where $(V_e)_{i,j} > 0$, the difference operator in Eq. (6.14) is stable. On the other hand, at a supersonic point a backward or upwind operator may be defined at the point (i, j) by

$$\frac{1}{\Delta x}\left[\left\{K\phi_x - \frac{\gamma+1}{2}\phi_x^2\right\}_{i-\frac{1}{2},j} - \left\{K\phi_x - \frac{\gamma+1}{2}\phi_x^2\right\}_{i-\frac{3}{2},j}\right] + \frac{1}{\Delta\bar{y}}\left[(\phi_{\bar{y}})_{i,j+\frac{1}{2}} - (\phi_{\bar{y}})_{i,j-\frac{1}{2}}\right] = 0, \qquad (6.16)$$

which may be rewritten as

$$\left[K-(\gamma+1)\frac{\phi_{i,j}-\phi_{i-2,j}}{2\Delta x}\right]\left[\frac{\phi_{i,j}-2\phi_{i-1,j}+\phi_{i-2,j}}{\Delta x^2}\right]+\frac{\phi_{i,j+1}-2\phi_{i,j}+\phi_{i,j-1}}{\Delta\bar{y}^2}=0. \qquad (6.17)$$

As a result of using the backward (i.e. upwind) difference, the difference operator in Eq. (6.17) is *only first order accurate and linearly stable at a hyperbolic point* for which the quantity $(V_{\rm h})_{i,j}$ defined by

$$(V_{\rm h})_{i,j} = K - (\gamma+1)\frac{\phi_{i,j} - \phi_{i-2,j}}{2\Delta x}, \qquad (6.18)$$

is negative.

When we proceed to solve the resulting discretised algebraic system of equations by iteration, two other cases may arise, depending on the values of the quantities $(V_{\rm e})_{i,j}$ and $(V_{\rm h})_{i,j}$. During the iteration, at each mesh point (i, j) the quantity $(V_{\rm e})_{i,j}$ is computed. If $(V_{\rm e})_{i,j} > 0$, the flow at the point is subsonic (Eq. (6.8) of elliptic type) and we use the elliptic operator Eq. (6.14). On the other hand, if $(V_{\rm e})_{i,j} < 0$ and $(V_{\rm h})_{i,j} < 0$, the flow at the point is supersonic (Eq. (6.8) of hyperbolic type) and the hyperbolic operator Eq. (6.17) is used. When the flow accelerates through the sonic velocity, from subsonic to a supersonic flow, a point may be reached where $(V_{\rm e})_{i,j} < 0$ as also $(V_{\rm h})_{i,j} > 0$. (Note that $(V_{\rm h})_{i,j} = (V_{\rm e})_{i-1,j}$.) At such a point, called the *parabolic point*, neither Eq. (6.14) nor Eq. (6.17) is stable. Murman and Cole (1971), removed the difficulty by introducing *the concept of a parabolic point operator* for which $(V_{\rm e})_{i,j} = 0$. This yields

$$\frac{\phi_{i,j+1} - 2\phi_{i,j} + \phi_{i,j-1}}{\Delta\bar{y}^2} = 0. \qquad (6.19)$$

On the other hand, when the flow decelerates through the sonic point, a point may be reached where

$$(V_{\rm e})_{i,j} > 0, \text{ and } (V_{\rm h})_{i,j} < 0. \qquad (6.20)$$

Both the elliptic and the hyperbolic operators are locally linearly stable in this case. At such a point, *called a shock-point*, which corresponds to a shock discontinuity, Murman and Cole (1971), used the elliptic operator Eq. (6.14). Treatment of boundary conditions is discussed next.

6.3.2 Implementation of Boundary Conditions

The tangency boundary condition Eq. (6.9) on the boundary $\overline{y} = 0$ may be implemented using the approximate representation

$$\phi_{\overline{y}\overline{y}}\Big|_{i,0} = \frac{\phi_{\overline{y}}|_{i,\frac{1}{2}} - \phi_{\overline{y}}|_{i,0}}{\Delta\overline{y}/2} = \frac{2}{\Delta\overline{y}}\left[\frac{\phi_{i,1} - \phi_{i,0}}{\Delta\overline{y}} - \phi_{\overline{y}}|_{i,0}\right]. \tag{6.21}$$

In this equation, the value of $\phi_{\overline{y}}|_{i,0}$ is taken from the tangency boundary condition Eq. (6.9), the body boundary being placed at $j = 0$. Further, on $\overline{y} = 0$ outside the body,

$$\phi_{\overline{y}}|_{i,0} = 0. \tag{6.22}$$

For the implementation of the far-field boundary condition, the computational boundary may be extended to infinity by some coordinate transformation. Alternatively, most authors working on the TSP model use asymptotic expansion, for large values of the distance from the airfoil. The asymptotic expansion is carried out in the integral representation for the perturbation velocity potential Klunker (1971), Murman and Cole (1971), Niyogi (1982) Oswatitsch (1977). Thus, the far-field boundary condition reduces, on simplification to

$$\phi(x, \overline{y}) = \frac{1}{2\pi\sqrt{K}}\frac{Dx}{x^2 + K\overline{y}^2} + \cdots, \tag{6.23}$$

where D denotes the doublet strength, given by

$$D = 2\tau\int_{-1}^{1} F(\xi)\mathrm{d}\xi + \frac{\gamma+1}{2}\int_{-\infty}^{\infty}\int_{-\infty}^{\infty} u^2(\xi, \eta)\,\mathrm{d}\xi\mathrm{d}\eta, \tag{6.24}$$

the profile $\tau F(x)$ being situated between $-1 \leq x \leq 1$, τ denoting its thickness-ratio, (Section 7.10.5).

The resulting integral equation is discretised and solved, along with the other discretised equations.

The outer boundary of the rectangle in Fig. 6.5, often chosen at a distance of about 3 or 4 times the chord, gives reasonable accuracy. Otherwise, in the transonic speed range, the perturbations are found to die out, with reasonable accuracy, at a distance of about 6 or 7 times the chord.

6.3.3 Iterative Solution of the Discretised Equations

Murman and Cole (1971) solved the discretised equations by successive line over-relaxation method, proceeding in a vertical line, sweeping from left to right. The scheme is a generalisation of the classical successive over-relaxation scheme, adapted to transonic discretised equations. The details are presented in the next subsection, in connection with the conservative scheme. The linearised subsonic solution (Prandtl solution) for the same problem, may be taken as the starting value at each mesh point of the computational domain. In the nth step of iteration, the value $(V_e)_{i,j}$ and $(V_h)_{i,j}$ are computed from Eqs. (6.15) and (6.18). Depending on its value the appropriate equation from Eqs. (6.14), (6.17) and (6.19), and that corresponding to Eq. (6.20) is chosen. The iterations are continued till a prescribed accuracy is achieved.

Numerical solutions computed with the above type-dependent scheme, show good agreement with available exact solutions for continuous subcritical and super-critical shock-free flows. Results with shock also agree well with experimental results. However, the jump in pressure across a shock at the profile surface is found to be less than the theoretical value for a normal shock. This drawback was subsequently removed by Murman (1974), through the introduction of shock-point operator, thereby making the scheme conservative. The original Murman and Cole scheme was found to be non-conservative, giving rise to spurious source terms through the numerical truncation error. In view of its importance, this concept has been discussed here in some detail, in the next subsection.

6.3.4 Artificial Viscosity and Conservative Schemes

If a numerical scheme is not conservative, spurious source terms may be generated through the numerical truncation error, thereby leading to erroneous results. Due to this reason, for dependable results the numerical schemes should be conservative.

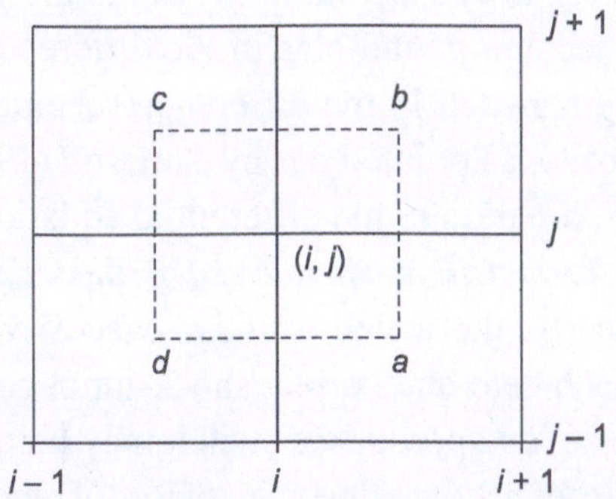

Figure 6.6 Flux computation for a conservative scheme.

Basic ideas of conservative schemes have been discussed in Chapter 4, for hyperbolic conservation laws. As explained in Jameson (1978), Lax and Wendroff (1960), the finite difference scheme for a conservation law of the form

$$\frac{\partial g}{\partial x} + \frac{\partial h}{\partial y} = 0, \tag{6.25}$$

should be such that the flux is conserved in each elementary cell of the computational domain. Let us consider the flux balance in the elementary cell abcd in Fig. 6.6, the points a, b, c, d being the centroids of the four neighboring cells of the point (i, j). This yields

$$\frac{G_{i+\frac{1}{2},j} - G_{i-\frac{1}{2},j}}{\Delta x} + \frac{H_{i,j+\frac{1}{2}} - H_{i,j-\frac{1}{2}}}{\Delta y} = 0, \tag{6.26}$$

where G and H are numerical flux functions, which should converge respectively to g and h in the limit as the mesh widths tend to zero. Further, for this purpose, it is sufficient if the functions G and H be chosen such that $G = g + O(\Delta x)$ and $H = h + O(\Delta x)$. Consequently, in constructing an approximate difference scheme, artificial viscosity of the form

$$\frac{\partial R}{\partial x} + \frac{\partial S}{\partial y}, \tag{6.27}$$

may be added to the scheme, provided R and S are of the order Δx. Then, instead of Eq. (6.25), we are considering an approximation to it, viz.

$$\frac{\partial}{\partial x}(g + R) + \frac{\partial}{\partial y}(h + S) = 0, \tag{6.28}$$

which reduces to the original conservation law as the mesh widths approach zero. Thus, in order to produce an upwind bias in the difference scheme at supersonic points, instead of using a switch in the difference scheme, we can add explicitly requisite artificial viscosity, as pointed out by Jameson (1975, 1978). If we have a central difference approximation to the differential equation in conservation form, then *the conservation form will be preserved if the added viscosity is also in conservation form.* Clearly, the added artificial viscosity would tend to zero as the mesh widths approach zero and correct shock-jump conditions also, would be satisfied. For this purpose, an appropriate switch may be used such that the added artificial viscosity vanishes in the subsonic region of the flow field. In the light of the above discussions, the following points may be noted, while constructing a suitable finite difference scheme for transonic flow study:

1. The partial differential equation be written in conservation form, i.e. in divergence form.
2. A central difference approximation to the partial differential equation be constructed.

3. In order to produce the requisite directional bias in the hyperbolic region, artificial viscosity may be added.
4. The added artificial viscosity should be in conservation form.

Note that the directional bias at a point of the flow field, may be found from the direction of the characteristics at that point. For the TSP equations, they are equally inclined to the x-axis, so that at all supersonic points we have to use upwind differencing only in the x-direction.
In the light of the above guidelines, let us relook into the Murman and Cole scheme, introduced through Eqs. (6.14), (6.17), (6.19) and (6.20). We use the following abbreviations

$$r_{i,j} = A_{i,j}\frac{\phi_{i+1,j} - 2\phi_{i,j} + \phi_{i-1,j}}{\Delta x^2}, \tag{6.29}$$

$$A_{i,j} = K - (\gamma + 1)\frac{\phi_{i+1,j} - \phi_{i-1,j}}{2\Delta x}, \tag{6.30}$$

and

$$s_{i,j} = \frac{\phi_{i,j+1} - 2\phi_{i,j} + \phi_{i,j-1}}{\Delta \bar{y}^2}. \tag{6.31}$$

Note that, the function $A_{i,j}$ in Eq. (6.30) is the same as the elliptic operator $(V_e)_{i,j}$, defined in Eq. (6.15). Further, we define a switching function ν with the value unity at a supersonic point and zero at a subsonic point,

$$\nu_{i,j} = \begin{cases} 0, & \text{if } A_{i,j} > 0, \\ 1, & \text{if } A_{i,j} < 0 \end{cases} \tag{6.32}$$

The original scheme of Murman and Cole, as discussed in the previous subsection, may be rewritten as

$$r_{i,j} + s_{i,j} - \nu_{i,j}(r_{i,j} - r_{i-1,j}) = 0. \tag{6.33}$$

Let us now consider the quantity R,

$$R = \Delta x\frac{\partial}{\partial x}\left(K\phi_x - \frac{\gamma + 1}{2}\phi_x^2\right) \sim \Delta x A\phi_{xx}, \tag{6.34}$$

where A is the nonlinear coefficient $A = K - (\gamma + 1)\phi_x$. Then, assuming A to be locally constant the added terms are an approximation to

$$-\nu\frac{\partial R}{\partial x} = -\nu\Delta x A\phi_{xxx}. \tag{6.35}$$

This term may be considered as an artificial viscosity of order Δx, which is added at all points in the supersonic region. Since the coefficient $-A$ of $\phi_{xxx} = u_{xx}$ is positive in the supersonic region, we observe that the artificial viscosity term is similar to the viscous terms in the Navier–Stokes equation. Moreover, since the artificial viscosity term is not in conservation form, the computed solution cannot converge to the correct weak solution. However, the remedy to the problem is simply to rewrite the artificial viscosity term in conservation form, as $\frac{\partial}{\partial x}(\nu R)$. This yields the conservative scheme of Murman (1974),

$$r_{i,j} + s_{i,j} - \nu_{i,j} r_{i,j} + \nu_{i-1,j} r_{i-1,j} = 0. \tag{6.36}$$

It may be noted that, *at a subsonic point,* $\nu_{i-1,j} = 0,\ \nu_{i,j} = 0$, *while at a supersonic point* $\nu_{i-1,j} = 1,\ \nu_{i,j} = 1$. *Further, noting that at a parabolic point,* $\nu_{i-1,j} = 0$ *and* $\nu_{i,j} = 1$, *and at a shock point,* $\nu_{i-1,j} = 1$ *and* $\nu_{i,j} = 0$ (the flow velocity after a normal shock being subsonic), the respective equations at the parabolic and shock points are

$$s_{i,j} = 0, \tag{6.37}$$

and

$$r_{i,j} + s_{i,j} + r_{i-1,j} = 0. \tag{6.38}$$

For supercritical flow, the system of algebraic Eqs. (6.36)–(6.38) is nonlinear. It is solved by a generalisation, of the classical relaxation method, *called SLOR (successive line over-relaxation).* The method is expressed in terms of *corrections* $C_{i,j}$ *and residuals* $R_{i,j}$. After the nth cycle of iteration is complete, the values of $\phi^n_{i,j}$ is known at all mesh points of the computational domain, the residuals $R_{i,j}$ are evaluated by substituting $\phi^n_{i,j}$ in place of $\phi_{i,j}$ on the left hand side of the relevant equation (6.36) to yield

$$R_{i,j} = \text{residual} = s^n_{i,j} + (1 - \nu_{i,j}) r^n_{i,j} + \nu_{i-1,j} r^n_{i-1,j}, \tag{6.39}$$

where

$$s^n_{i,j} \equiv \frac{1}{\Delta \bar{y}^2} \left[\phi^n_{i,j+1} - 2\phi^n_{i,j} + \phi^n_{i,j-1}\right]$$

and similar meaning for $r^n_{i,j}$ and $r^n_{i-1,j}$. The corrections $C_{i,j}$ are defined as

$$\phi^{n+1}_{i,j} = \phi^n_{i,j} + C_{i,j}, \tag{6.40}$$

Let $\tilde{\phi}_{i,j}$ denote provisional values determined by line Gauss–Seidel scheme proceeding along vertical lines, sweeping from left to right, defined by

$$\frac{1}{\Delta \bar{y}^2}\left[\phi^{n+1}_{i,j+1} - 2\phi^{n+1}_{i,j} + \phi^{n+1}_{i,j-1}\right] + (1 - \nu_{i,j})\left[K - (\gamma + 1)\frac{\phi^n_{i+1,j} - \phi^n_{i-1,j}}{2\Delta x}\right]$$
$$\frac{\phi^n_{i+1,j} - 2\tilde{\phi}_{i,j} + \phi^{n+1}_{i-1,j}}{\Delta x^2} + \nu_{i-1,j} A^n_{i-1,j} \frac{\phi^{n+1}_{i,j} - 2\phi^{n+1}_{i-1,j} + \phi^{n+1}_{i-2,j}}{\Delta x^2} = 0. \tag{6.41}$$

An over-relaxation parameter ω is introduced, which relates the new values $\phi^n_{i,j}$ to the old values by

$$\phi^{n+1}_{i,j} = \phi^n_{i,j} + \omega(\tilde{\phi}_{i,j} - \phi^n_{i,j}), \quad \text{or, } \tilde{\phi}_{i,j} = \phi^n_{i,j} + \frac{1}{\omega} C_{i,j}. \tag{6.42}$$

For a subsonic flow, $1 \le \omega \le 2$. Note that, at a supersonic point $\nu_{i,j}$ being equal to unity, no provisional values are used and Eq. (6.41) is a marching scheme, typical of hyperbolic-type equations. Eliminating $\tilde{\phi}_{i,j}$ from Eq. (6.41) with Eq. (6.42) we obtain the SLOR scheme for the unknown corrections

$$\frac{1}{\Delta \bar{y}^2}\left[C_{i,j+1} - 2C_{i,j} + C_{i,j-1}\right] + (1 - \nu_{i,j})A^n_{i,j}\frac{-\frac{2}{\omega}C_{i,j} + C_{i-1,j}}{\Delta x^2} +$$

$$\nu_{i-1,j}A^n_{i-1,j}\frac{C_{i,j} - 2C_{i-1,j} + C_{i-2,j}}{\Delta x^2} + R_{i,j} = 0. \tag{6.43}$$

The iteration scheme (6.42) behaves in a subsonic region, as a classical relaxation scheme for systems obtained through discretisation of elliptic-type equations, while in a supersonic region, it behaves as a scheme marching from left to right, as is typical of systems obtained through discretisation of hyperbolic-type equations. Some caution is needed in choosing the iteration scheme for solving the algebraic equations obtained through discretisation of partial differential equations of mixed elliptic–hyperbolic type, because otherwise a slope discontinuity may occur in the computed solution at the parabolic point.

6.3.5 Computational Results

A FORTRAN program $tsc.f$ based on the conservative scheme of Murman (1974) using the formulation of Jameson (1978) as discussed above, is presented here. It computes transonic flow past a thin symmetric profile at zero incidence. As a typical example, transonic flow past a 6% thick circular-arc airfoil at a free-stream Mach number of $M_\infty = 0.85$, is computed and shown in Fig. 6.7. The airfoil shape is taken as

$$y = \begin{cases} \tau(1 - x^2), & \text{for } |x| < 1, \\ 0, & \text{otherwise,} \end{cases} \tag{6.44}$$

τ denoting the thickness-ratio of the airfoil. The requisite input values are taken as, $M = 150$, $N = 195$, $EPS = 0.000005$, LFBndry $= -3.01$, RTBndry $=$ 3.02, UpperBndry $= 3.10$, $AMINF = 0.85$, with uniform grid spacings in both the directions, the LFbndry, RTBndry and UpperBndry respectively denoting the left, right and the upper far-field boundaries. The iteration scheme converges correct to six-decimal places in 2117 steps with the relaxation parameter $\omega = 1.85$. The non-conservative solution requires 2177 steps for the same accuracy. The program tsc.f compututes the solution according to the formulation in Section

7.10.5, as explained above. The final solution is subsequently expressed in terms of the reduced u-velocity distribution, denoted by uu, according to Oswatitsch reduction (7.10.4). These reduced perturbation velocities are $O(1)$ and the sonic line corresponds to $uu = 1$, the surface pressure coefficient being simply $-2uu$. For easier comprehension, Fig. 6.7 shows the reduced velocity distribution uu. From the figure, it may be seen that the non-conservative solution delivers a weaker shock (shock strength is measured by the jump in the physical quantities like pressure, velocity, density or entropy). The shock is situated upstream compared to the conservative shock. Further, we see clearly, an expansion immediately after the shock in the conservative solution which is absent from the non-conservative one. Such an expansion after the shock, attached to any curved wall was predicted by Oswatitsch and Zierep (1960).

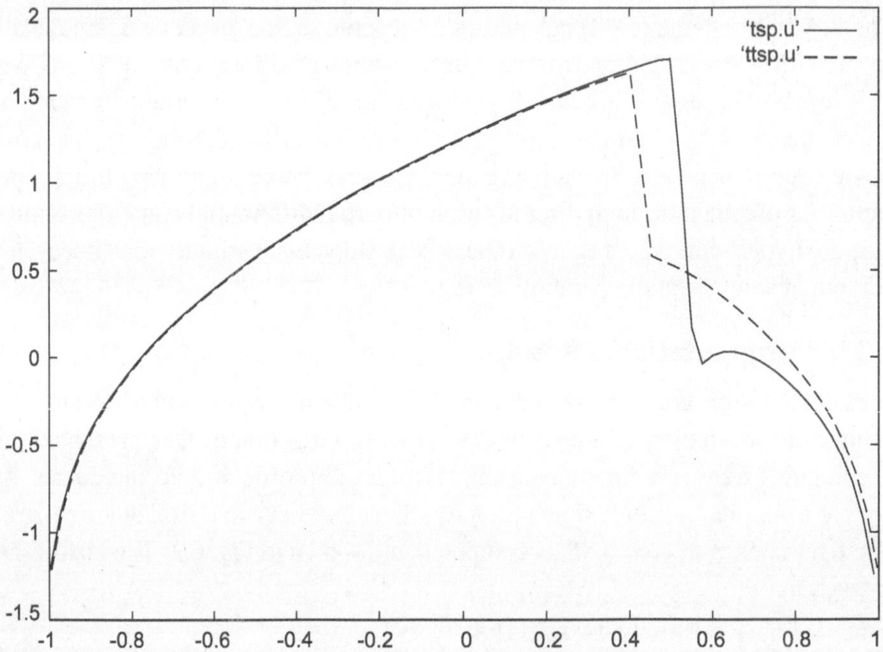

Figure 6.7 Reduced velocity distribution uu on profile axis, for a symmetric 6% circular-arc airfoil at zero incidence, for $M_\infty = 0.85$ using conservative and non-conservative TSP scheme.

6.3.6 Program 6.1 tsc.f

The following program tsc.f computes conservative transonic solution for flow past a thin circular-arc airfoil using TSP model. Results shown in Fig. 6.7.

```
c..............................................................
c......Program tsc.f.It computes transonic flow past a thin circular-arc
C profile at zero incidence under TSP model by  Murman-Cole, Murman
C Conservative Scheme, using formulation  of Jameson(1978).
C.......System of Equations solved by slor.  .......................
C.......Farfield boundary conditions calculated in subroutine FARF.
C ..........................................................
PARAMETER(IX=300, IY=200)
DIMENSION PHI(IX,IY),SAN(IX,IY),AA(IX),BB(IX),CC(IX)
DIMENSION DD(IX),A(IX,IY),XX(IX),U(IX,IY),UU(IX,IY)
DIMENSION AN(IX,IY),YY(IY)
COMMON U,PHI,XX,YY,AH,AK,BK,GP,DB,CONST,M,N,L
OPEN(UNIT=14, FILE='tsc.out')
OPEN(UNIT=15,FILE='tsc.u')
OPEN(UNIT=16,FILE='tsc.uu')
C.......M,N : No. of subdivisions in y and x-directions.
C.......AH,AK= mesh spacings along x,y. BK= trans.similarity parameter.
C.......EPS conv. tolerance of SLOR scheme. AL1,AL2,U1
C.......left, right and upper far-field boundaries respectively.
c...... AMINF, the free-stream Mach number.
c...... STAR= reduction factor of Oswatitsch reduction,
c...... T=reduced thickness-ratio. OM=relax.param. of SLOR iteration.
c..............................................................
WRITE(*,4)
4 FORMAT(1X,'input values of: M,N,EPS,AL1,AL2,U1,AMINF')
READ(*,*)M,N,EPS,AL1,AL2,U1,AMINF
WRITE(14,7)M,N,EPS,AL1,AL2,U1,AMINF
7 FORMAT(1X,'M= ',I3,2X,'N = ',I3,2X,'EPS= ',F10.7,
     1 2X, 'LF BNDRY=',F10.7/'RT BNDRY= ',F7.3,
     2 2X,'UPPER BNDRY= ',F7.3,2X,'AMINF=',F8.4/)
GAMMA=1.4
GP= GAMMA+1.0
PI=4.0*ATAN(1.0)
TAU=0.06
DB=8.0*TAU/3.0
OM=1.85
BETA=1.0-AMINF**2
BK=BETA/(TAU**(2.0/3.0))
AMSTR=SQRT(AMINF**2/(1.0-0.4/GP*BETA))
STAR=(1.0/AMSTR-1.0)
T=TAU/(SQRT(BETA)*STAR)
AH=(AL2-AL1)/FLOAT(N)
AHH=AH*AH
AH2=AH*2.0
AK=U1/FLOAT(M)
AKK=AK*AK
```

```
CONST=1.0/(PI*BK**0.5)
WRITE(14,6)BETA,BK,AMSTR,STAR,TAU,T,AH,AK,CONST,OM
6 FORMAT(1X,'BETA= ',F8.5,2X,'BK=',F8.5,2x,'AMSTAR=',
      1 F8.5,2X/'STAR= ',F8.5,2X,'TAU = ',F6.3,2X,'T=',F6.3,2X,
      2 'AH=',F7.3,2x,'AK=',F7.3,2X, 'CONST= ',F7.4,2X,'OM=',F6.3/)
c.............................................................
C.......computation of starting values for a circular arc airfoil
C....... y=TAU*(1-x**2), |x|<=1, =0, otherwise. Linearized solution
C........is taken as the starting solution.
c...... The far-field boundaries are: I=1, I=N+1, J=M+1.
c.......The far-f  values are modified through iteration.........
C....... J=1 denotes body boundary.
c......  PHI is  perturbation potential, new values  are SAN.
c.............................................................
 DO 102 J=1,M+1
YY(J)=(J-1)*AK
102 CONTINUE
DO 103 I=1,N+1
XX(I)=AL1+(I-1)*AH
103 CONTINUE
c WRITE(14,*) (XX(I),I=1,N+1)
c WRITE(14,*) (YY(J),J=1,M+1)
DO 11 I=1,N+1
X=XX(I)
X1=1.-X
X2=1.+X
PHI(I,1)=CONST*(2.0*X-X1*X2/2.0*ALOG(X1**2/X2**2))
SAN(I,1) =PHI(I,1)
11 CONTINUE
BK1=SQRT(BK)
DO 101 I=1,N+1
DO 101 J=2,M+1
A1=BK1*YY(J)
B1=A1*A1
X1=1.0-XX(I)
X2=1.0+XX(I)
PHI(I,J)=CONST*(2.0*XX(I)-2.0*XX(I)*A1*(ATAN(X1/A1)+ATAN(X2/A1))
1 -(X1*X2+B1)/2.*ALOG((X1**2+B1)/(X2**2+B1)))
SAN(I,J)=PHI(I,J)
101 continue
C.......Velocity computation subroutine SUBU.Also needed for far-f.
C........SUBU called after each 10-iteration steps.
L=0
CALL SUBU
718 CONTINUE
C.....   L  denotes iteration step................................
```

```
300 IK=0
K=0
C.......Farfield boundary values are now calculated,which involves
C.......double integral evaluation. ............................
CALL FARF
200 IK=IK+1
K=K+1
L=L+1
C......  Main Iteration Loop ..............................
C.......Parabolic,supersonic and subsonic and shock points tested.
C.......Velocities at points on vertical lines before and after
C....... the profile are not tested, but taken as subsonic.
50 DO 32 I=2,N
LB=1
IF(ABS(XX(I)).GT.1.0) LB=0
DO 65 J=1,M
JM=J-1
V1=BK-GP*(PHI(I,J)-PHI(I-2,J))/AH2
V2=BK-GP*(PHI(I+1,J)-PHI(I-1,J))/AH2
SS=(PHI(I,J+1)-2.0*PHI(I,J)+PHI(I,J-1))/AKK
RR=V2*(PHI(I+1,J)-2.0*PHI(I,J)+PHI(I-1,J))/AHH
CM=SAN(I-1,J)-PHI(I-1,J)
IF(I.EQ.2)GO TO 51
CMM=SAN(I-2,J)-PHI(I-2,J)
BR=V1*(PHI(I,J)-2.0*PHI(I-1,J)+PHI(I-2,J))/AHH
GO TO 52
51 CMM=SAN(I-1,J)-PHI(I-1,J)
BR=V1*(PHI(I,J)-2.0*PHI(I-1,J))/AHH
52 XA=ABS(XX(I))
IF(XA.GE.1.0.OR.YY(J).GT.1.0) GO TO 440
C.......velocity testing  ..............................
IF(V1.GT.0.AND.V2.GT.0) GO TO 440
IF(V1.GT.0.AND.V2.LT.0) GO TO 450
IF(V1.LT.0.AND.V2.GT.0) GO TO 460
C......  HYPERBOLIC POINT  ..........................
470 NU1=1
NU2=0
GO TO 480
C......  SUBSONIC POINT...........................
440 NU1=0
 NU2=1
 GO TO 480
C......... SONIC POINT ..............................
450 NU1=0
```

```
NU2=0
GO TO 480
C-------------  SHOCK POINT ---------------------------
460 NU1=1
NU2=1
C.......case testing over.............................
480 CM1=(2.0*NU1*V1-NU2*V2)*CM/AHH
CMM1=V1*CMM*NU1/AHH
RSR=SS+RR*NU2+BR*NU1
IF(J.EQ.1)GO TO 444
AA(JM)=1.0/AKK
BB(J)=-2.0*(1.0/AKK+V2/(OM*AHH)*NU2)+NU1*V1/AHH
CC(J)=1.0/AKK
DD(J)= -RSR+CM1-CMM1
C......COEFFICIENTS FOR J.NE.1  are as above............
GO TO 65
C.........coefficients for j=1 : Body Boundary ...........
444 BB(J)=-2.0/AKK-2.0*NU2*V2/(OM*AHH)+V1*NU1/AHH
CC(J)=2.0/AKK
BD=2.0*(PHI(I,2)-PHI(I,1))/AKK+4.0*LB*XX(I)/AK
BDR=NU2*RR+BD+NU1*BR
CMM2=CM1-CMM1
DD(J)=-BDR+CMM2
65 CONTINUE
CALL STRID(AA,BB,CC,DD,M)
DO 67 J=1,M
SAN(I,J)=PHI(I,J)+DD(J)
67 CONTINUE
32 CONTINUE
c..........................................
C......CONVERGENCE TESTING. One cycle complete.
c..........................................
IF(IK.GT.10) GO TO 750
DO 12 I=2,N
DO 12 J=1,M
W=ABS(SAN(I,J)-PHI(I,J))
IF (W.GT.EPS) GO TO 700
12 CONTINUE
C.......Convergence achieved........................
C.......Print output...............................
CALL SUBU
GO TO 900
700 CONTINUE
 C.......Next cycle begins ..........................
L=L+1
```

```
DO 13 I=1,N+1
DO 13 J=1,M+1
PHI(I,J)=SAN(I,J)
13 CONTINUE
c CALL SUBU
750 IF (K.GE.5) GO TO 500
GO TO 200
500 CALL SUBU
IK=0
IF (L.GT.10000) GO TO 901
GO TO 300
900 WRITE(14,5)L
5 FORMAT(1X,'The Scheme Converges after iteration steps= ',I5)
WRITE(14,2)
2 FORMAT(1X,'XX ',10X,'  U  ',10X,'  OSW.Red. UU '/)
905 DO 555 I=2,N
IF(ABS(XX(I)).GT.1.0) GO TO 555
C.......................................................
UU(I,1)=U(I,1)*TAU**(2.0/3.0)/STAR
c write(14,2)
write(14,3) xx(i),u(i,1),uu(i,1)
3 FORMAT(1X, F6.3,5X,3E14.4)
write(15,8) xx(i),u(i,1)
write(16,8)xx(i),uu(i,1)
8 format(1x, f8.4,f14.4)
555 CONTINUE
c.....................................................
GO TO 1000
901 WRITE(14,55) L
55 FORMAT(1X,' Scheme does N O T converge in iteration step= ',I5)
GO TO 905
1000 STOP
END
C.........................................................
C.......Velocity Calculation SUBU ......................
C.....................................................
SUBROUTINE SUBU
PARAMETER(IX=300,IY=200)
DIMENSION PHI(IX,IY),U(IX,IY),XX(IX),YY(IY)
COMMON U,PHI,XX,YY,AH,AK,BK,GP,DB,CONST,M,N,L
DO 20 I=2,N
DO 20 J=1,M+1
U(I,J)=(PHI(I+1,J)-PHI(I-1,J))/(2.0*AH)
20 CONTINUE
```

```
DO 25 J=1,M+1
U(1,J)=(PHI(2,J)-PHI(1,J))/AH
I=N+1
U(I,J)=(PHI(I,J)-PHI(I-1,J))/AH
25 CONTINUE
RETURN
END
C..................................................
C.......Calculation of Far-field:  Farf.................
C..................................................
SUBROUTINE FARF
PARAMETER(IX=300,IY=200)
DIMENSION PHI(IX,IY),U(IX,IY),XX(IX),YY(IY)
C COMMON /C1/ TAU,GP,AL1
COMMON U,PHI,XX,YY,AH,AK,BK,GP,DB,CONST,M,N,L
PI=3.1415926
D=DB+GP/2.0*ANTG(X)
DO 10 I=1,N+1
J=M+1
10 CONTINUE
DO 20 J=2,M
I=1
PHI(I,J)=0.5*CONST*D*XX(I)/(XX(I)**2+BK*YY(I)**2)
I=N+1
PHI(I,J)=0.5*CONST*D*XX(I)/(XX(I)**2+BK*YY(I)**2)
20 CONTINUE
RETURN
END
C...............................................................
C.......Function subroutine to calculate Double-Integral: ANTG(A)
c...... used in far-field boundary condition evaluation...........
C...............................................................
FUNCTION ANTG(A)
PARAMETER(IX=300,IY=200)
DIMENSION PHI(IX,IY),U(IX,IY),XX(IX),YY(IY)
C COMMON /A1/ U
C COMMON /B1/ AH,AK,BK,M
COMMON U,PHI,XX,YY,AH,AK,BK,GP,DB,CONST,M,N,L
J=1
TSUM=SUM(J)
J=M+1
TSUM=TSUM+SUM(J)
K=2
DO 20 J=2,M
K=6-K
```

```
TSUM=TSUM+K*SUM(J)
20 CONTINUE
ANTG=TSUM*AH*AK/9.0
RETURN
END
C..............................................
FUNCTION SUM(LL)
PARAMETER(IX=300,IY=200)
DIMENSION PHI(IX,IY),U(IX,IY),XX(IX),YY(IY)
C COMMON /A1/ U
COMMON U,PHI,XX,YY,AH,AK,BK,GP,DB,CONST,M,N,L
SU=U(1,LL)**2+U(N+1,LL)**2
K=2
DO 70 I=2,N
K=6-K
SU=SU+K*U(I,LL)**2
70 CONTINUE
SUM=SU
RETURN
END
C.....................................................
C.......Print Subroutine   .........................
C...................................................
SUBROUTINE SUBPRT
PARAMETER(IX=300,IY=200)
DIMENSION PHI(IX,IY),U(IX,IY),UE(IX,IY),XX(IX),YY(IY)
COMMON U,PHI,XX,YY,AH,AK,BK,GP,DB,CONST,M,N,L
WRITE (14,7) ITEST
7 FORMAT(1X,'values of U(I,1) after ITERATION STEP= ',I4/
     1 5X,'X',12X,'U(I,1)',12X,'UE')
DO 234 I=1,N+1
X=AL1+(I-1)*AH
IF (ABS(X).GT.1.0) GO TO 234
UE(I,1)=2.0*U(I,2)-U(I,3)
WRITE(14,11) X,U(I,1),UE(I,1),U(I,M+1)
11 FORMAT(1X,F8.4,3X,2F12.4)
234 CONTINUE
RETURN
END
c.......................................................
C.........Subroutine for solving tridiagonal system: strid  ...
c.........................................................
SUBROUTINE STRID(A,B,C,D,M)
PARAMETER(IY=200)
DIMENSION A(IY),B(IY),C(IY),D(IY),P(IY)
C........B=leading diagonal,A= left off-diagonal,C=right off-diagonal,
```

```
C.......D= RHS elements, solution array and used as dummy...........
C.......M=number of equations. D(I), I=1,M, solution vector.........
C...........................................................
C.......Forward elimination.................................
P(1)=C(1)/B(1)
D(1)=D(1)/B(1)
DO 1 J=2,M-1
JM=J-1
FACTOR=1.0/(B(J)-A(JM)*P(JM))
P(J)=C(J)* FACTOR
D(J)=(D(J)-A(JM)*D(JM))*FACTOR
1 CONTINUE
D(M)=(D(M)-A(J-1)*D(J-1))/(B(M)-P(M-1)*A(M-1))
C.......Back substitution Sweep.............................
DO 2 J=M-1,1,-1
JP=J+1
2 D(J)=D(J)-P(J)*D(JP)
RETURN
END
```

6.4 SUMMARY

Properties of the Tricomi equation, which is the canonical equation of mixed elliptic–hyperbolic type and formulation of associated well-posed boundary-value problems have been discussed. A closely related equation of much practical interest is the TSP (transonic small perturbation) equation. It is an approximate equation, governing steady inviscid transonic flow past a thin profile. Type-dependent differencing is necessary for computing successfully, solution of a well-posed problem for a mixed-type equation. The basic ideas of *type-dependent differencing* have been explained in this chapter, by means of TSP equation as an example. A Fortran program tsc.f has been presented. It computes solution of steady transonic flow past a thin symmetric airfoil at zero incidence.

6.5 KEY TERMS

Cauchy problem
Chaplygin equation
Corrections
Compressible flow
Conservative scheme
Elliptic point
Far-field boundary
Mixed elliptic–hyperbolic
Parabolic point
Residuals
Shock point
SLOR
Thin aerofoil
Tricomi equation
TSP equation
Cole's form
Oswatitsch form
Type-dependent differencing.

6.6 EXERCISE 6

6.1 Compute the surface pressure distribution for transonic flow past the thin symmetric airfoil at zero incidence,

$$h(x) = 2\tau(x - x^2), 0 \leq x \leq 1,$$

at $M_\infty = 0.85$, $\tau = 0.08$, using the TSP model. Take the left-boundary (far-field) at $x = -1.5$, right-boundary at $x = 2.5$ and the upper boundary at $y = 2$, taking $\phi_x = 0$ on the left and right boundaries and $\phi = 0$ on the upper boundary. Modify the program tsc.f in order to compute the solution of the problem.

6.2 Compute the surface pressure distribution for transonic flow past the thin symmetric airfoil at zero incidence,

$$h(x) = \tau(x^2 - 1)^2, -1 \leq x \leq 1,$$

at $M_\infty = 0.8$, $\tau = 0.08$, using the TSP model. Take the left-boundary (far-field) at $x = -2.5$, right-boundary at $x = 2.5$ and the upper boundary at $y = 2$, taking $\phi_x = 0$ on the left and right boundaries and $\phi = 0$ on the upper boundaries. Write a computer program or else modify the program tsc.f in order to compute the solution of the problem.

6.3 What is the thickness-ratio of the symmetric profile

$$h(x) = \begin{cases} A(x - x^3), & 0 < x < 1, \\ 0, & \text{otherwise ?} \end{cases}$$

Where is the maximum thickness situated? Use the computer program prepared for problem 6.1 to compute the surface pressure distribution based on TSP model, for $M_\infty = 0.80$, $\tau = 0.06$, $\alpha = 0$, and compare with the linearised solution.

6.4 Compute the surface pressure distribution for the thin symmetric profile shape at zero incidence:

$$h(x) = \begin{cases} A(x^2 - x^3), & 0 < x < 1, \\ 0, & \text{otherwise}, \end{cases}$$

Taking $M_\infty = 0.8$, $\tau = 0.08$, compare with the corresponding linearised solution at the surface (shifted to body-axis in this approximation it is given by)

$$u(x, 0) = \frac{1}{\pi\beta} \int_0^1 \frac{v(\xi, 0+)}{x - \xi} \mathrm{d}\xi, \tag{6.45}$$

$v(x, 0+)$ being determined by the tangency boundary condition at the profile, shifted to the body-axis in this approximation.

6.5 Compute the surface pressure distribution for the symmetric $NACA0012$ airfoil at zero incidence,

$$h(x) = \frac{T}{0.20}\left(0.29690\sqrt{x} - 0.12600x - 0.35160x^2 + 0.28430x^3 - 0.10150x^4\right)$$

at $M_\infty = 0.8$, $\tau = 0.12$, using the TSP model. Use the computer program already developed for problems 6.2 and 6.3.

6.6 Compute solution for transonic flow past a thin unsymmetric airfoil at small incidence α at high subsonic free-stream $M_\infty < 1$. Use the TSP model in Section 7.10.5 of Chapter 7. Write a computer program for this and present typical test results.

Part II
Computational Fluid Dynamics

Computational fluid dynamics (abbreviated CFD) complements experimental and theoretical fluid dynamics by providing an alternative, cost-effective means of simulating real flows. With rapid developement in computer technology and advancement in numerical analysis and algorithms over the past few decades, today CFD finds its application in a large variety of fields including aircraft and missile design, car design, ship design, studies of blood flow, oceanography, meteorology and astrophysics.

However, its role in aircraft design is crucial because of the narrow margins within which aircrafts can operate efficiently. Moreover, the tremendous increase in the cost-effectiveness of CFD (compared to expensive wind-tunnel testing) has given it a key role in aircraft design. Thus one of the main use of CFD in aeronautical science is to provide reliable aerodynamic predictions that will enable the aerodynamic designer to produce better airplanes, thereby avoiding increasing experimental costs. It is also possible to compute complex flow situations numerically that are not readily accessible to experimental measurements. This can provide new insights into the underlying physical process.

In order to compute high speed compressible flow, different mathematical models of varying complexity have been studied in the various chapters of the present work. The most accurate model for representation of fluid flow problems is governed by the unsteady Navier–Stokes equations. However, at high Reynolds numbers the flow becomes complicated with the onset of turbulence. The disparity of scales in a turbulent flow is so large that even with the new generation of super computers and with significant improvement of computational techniques, direct numerical simulation (DNS) of Navier–Stokes equations is not feasible. Thus, the problems of turbulent flow are treated by the solution of Reynolds averaged Navier–Stokes (RANS) equations with a suitable turbulence model for closure. The unsteady Navier–Stokes equations may be solved directly for laminar flow problems only. These are discussed in Chapter 12 for incompressible flow and in Chapter 13 for compressible flow. Viscous flow study based on boundary layer model has been presented in Chapter 11. In fluid flow problems with attached flows, the viscous effects are mainly confined to the boundary layer. Consequently, they have a relatively small effect on the global flow pattern and inviscid flow predictions can serve a useful role. Therefore, keeping in mind the large computational expense for solving the more complicated Navier–Stokes equations, we have confined ourselves to inviscid flow in the remaining chapters.

The most accurate model for inviscid compressible flow is represented by the Euler equations in which the conservation of mass, momentum and energy are guaranteed. Most part of Chapter 10 has been devoted to the solution of the Euler equations and the full potential equation. Also, transonic small perturbation model has been discussed there. The simplest model, viz. inviscid incompressible fluid flow has been presented in Chapter 9. A recapitulation of the basic equations and common boundary conditions of fluid dynamics may be found in Chapter 7 while Chapter 8, deals with the problems of grid generation.

Before we go for numerical computation, two important steps of CFD should be introduced. The first is the *pre-processing* where we start the processing of the input boundary data to get the grid and see whether the numerically generated surface grid approximates the given geometry and the flow boundary well. The grid should have sufficient number of points where important flow features like boundary layer, shock, wake etc. are resolved and also examine the grid qualitatively with regard to grid stretching, skewness, etc. This activity needs care and skill since inaccuracy introduced at this stage may lead to wrong results.

The last but most important part of CFD activity is the *post-processing* of the results obtained by numerical solution of the governing equations for a given problem. The CFD activity as a whole is the numerical simulation of the flow, analysis and interpretation of the results. The *flow-solver* solves a system of equations (for example, Euler, Navier–Stokes) governing the flow with proper boundary conditions where the free-stream Mach number, angle of attack and the free-stream Reynolds number are specified. The solution file contains highly detailed information of the flow field like density, velocity components, temperature etc. Then it is the graphic routines which turn these voluminous data into graphs and pictures so that the results can be seen and interpreted. These will enhance the ability of engineers and scientists to visualize and understand complex flow fields.

Visualisation methods may be broadly classified according to whether they treat vector or scalar properties. Streamlines, skin-friction lines, velocity vectors, etc. belong to the first category while iso-surfaces, iso-lines and graphs to compare a particular parameter distribution with either experimental or some other available exact/numerical solution (*validation*) belong to the second category. Now-a-days many commercial softwares are available for this purpose.

7

The Basic Equations of Fluid Dynamics

7.1 INTRODUCTION

The first part of the book consisting of the first six chapters is intended to make the reader familiar with mathematical foundation needed for the computational methods discussed. The first chapter briefly recapitulates the more important properties of partial differential equations together with boundary and/or initial conditions needed in our subsequent study. Chapter 2 introduces the reader to the computational methods studied in detail in this book, namely the finite difference and the finite volume methods. Chapters 3 to 6 give the foundation of finite difference study of partial differential equations of parabolic, hyperbolic, elliptic and mixed elliptic–hyperbolic types, respectively. The boundary and initial value problems change with the type of the partial differential equation. Consequently, the computational methods are also different for different types of the governing partial differential equations. The basic concepts needed in our subsequent study have been explained in these chapters.

A fundamental assumption in fluid dynamics is that the fluid medium is considered to be a *continuum.* This means that the *mean-free-path* of the molecules of the fluid is much smaller than some characteristic length scale of the flow. This assumption is true for most of the cases of practical interest. Nowadays, it is possible to directly simulate viscous turbulent flows with large Reynolds numbers using very accurate numerical techniques and resolve very small turbulent scales. The development of modern computers with large memory and speed has made it possible to solve the unsteady Navier–Stokes equations by the *spectral methods.* These flow simulations have made it possible to understand better the physics of turbulence and transition. Research is going on to develop new numerical algorithms with higher accuracy and faster rate of convergence. These developments have kept pace with the rapid advances in computer technology.

As already mentioned, the main intention of this book is to deal with the basic finite difference and finite volume techniques to solve numerically the fluid dynamic equations together with appropriate boundary and initial conditions. For the sake of completeness a recapitulation of the basic equations and boundary conditions, for the description of continuum motion are presented here, following the Eulerian approach. The detailed derivation of the fluid dynamic equations can be had from any standard text books on the subject (Batchelor, 1967; Landau and Lifshitz, 1989; Oswatitsch, 1956; Schlichting and Gersten, 2000).

7.2 BASIC CONSERVATION PRINCIPLES

The motion of a continuous medium is governed by the principles of classical mechanics and thermodynamics. The governing equations are mathematical statements of the conservation of mass flux, momentum flux and energy flux, usually called the *Navier–Stokes equations* (text books often name only the unsteady viscous momentum equations as the Navier–Stokes equations). Conservation law forms of these equations are generally considered for numerical computations because of their ability to treat the flow discontinuities accurately and automatically. It may be noted that the integral forms of the conservation laws are naturally conservative. As pointed out in Chapter 1, a partial differential equation is said to be of the *conservative form* if it can be expressed as

$$\frac{\partial}{\partial t}u(x) + \vec{\nabla}.\vec{f}(u) = 0. \tag{7.1}$$

This means that the quantity $u(x)$ is to be conserved and $\vec{f}(u)$ which is some vector function of u, is the flux of that quantity across the boundary surface. To compute flows with discontinuities such as *shock waves or contact discontinuities,* the property of conservative differential equations possessing discontinuous *weak solutions* is exploited. The non-conservative forms of equations do not admit weak solutions and hence cannot give the correct solution with shocks. For the set of equations to be complete for compressible fluid flows, the density and temperature variations along with the thermodynamic equation of state are to be considered. Under the assumption of a continuous medium these are valid fundamental equations for compressible viscous flows.

7.3 UNSTEADY NAVIER–STOKES EQUATIONS IN INTEGRAL FORM

The conservation laws discussed in the previous section can be expressed in integral, as well as in differential forms. The differential form of the equations is suitable for its numerical solution using *finite difference* type of methods. To compute the flow past complex geometries nowadays the finite volume method has become popular. This method discretises the equations in integral form and hence is naturally conservative.

Now, considering a control volume Ω, assuming the internal forces and heat sources to be zero, the conservation laws (mass, momentum and energy) can be written in the integral form as

$$\frac{\partial}{\partial t}\int_{\Omega} \vec{W}\,\mathrm{d}\Omega + \int_{\partial\Omega} \overline{\overline{F}}.\vec{n}\,\mathrm{d}s = 0, \tag{7.2}$$

where the solution vector is

$$\vec{W} = (\rho, \rho u, \rho v, \rho w, \rho E)', \tag{7.3}$$

the superscript $'$ denoting the transpose of a vector. Here u, v, w are the Cartesian components of the velocity vector, ρ the density, E the total energy per unit mass, and Ω denotes the arbitrary control volume with boundary $\partial\Omega$ and $\vec{n}$ is the unit outward normal vector to the elementary surface $\mathrm{d}S$. The total flux tensor $\overline{\overline{F}}$ can be rewritten as the sum of the inviscid flux $\overline{\overline{F}}_e$ and viscous flux $\overline{\overline{F}}_v$

$$\overline{\overline{F}}_e = \begin{pmatrix} \rho\vec{V} \\ \rho u\vec{V} + p\,i_x \\ \rho v\vec{V} + p\,i_y \\ \rho w\vec{V} + p\,i_z \\ \rho H\vec{V} \end{pmatrix} \tag{7.4}$$

$$\overline{\overline{F}}_v = \begin{pmatrix} 0 \\ \overline{\overline{\tau}}i_x \\ \overline{\overline{\tau}}i_y \\ \overline{\overline{\tau}}i_z \\ \overline{\overline{\tau}}.\vec{V} - \vec{q} \end{pmatrix} \tag{7.5}$$

where the velocity vector $\vec{V}$ and the stress tensor $\overline{\overline{\tau}}$ are given by

$$\vec{V} = u\,i_x + v\,i_y + w\,i_z, \quad \overline{\overline{\tau}} = \begin{pmatrix} \sigma_x & \tau_{xy} & \tau_{xz} \\ \tau_{yx} & \sigma_y & \tau_{yz} \\ \tau_{zx} & \tau_{zy} & \sigma_z \end{pmatrix} \tag{7.6}$$

The components of the stress tensor $\overline{\overline{\tau}}$, can be expressed for a *Newtonian fluid* as

$$\begin{aligned} \tau_{xx} = \sigma_x &= -\lambda\left(\frac{\partial u}{\partial x} + \frac{\partial v}{\partial y} + \frac{\partial w}{\partial z}\right) - 2\mu\frac{\partial u}{\partial x}, \\ \tau_{yy} = \sigma_y &= -\lambda\left(\frac{\partial u}{\partial x} + \frac{\partial v}{\partial y} + \frac{\partial w}{\partial z}\right) - 2\mu\frac{\partial u}{\partial y}, \\ \tau_{zz} = \sigma_z &= -\lambda\left(\frac{\partial u}{\partial x} + \frac{\partial v}{\partial y} + \frac{\partial w}{\partial z}\right) - 2\mu\frac{\partial u}{\partial z}, \end{aligned} \tag{7.7}$$

$$\tau_{xy} = \tau_{yx} = -\mu\left(\frac{\partial u}{\partial y} + \frac{\partial v}{\partial x}\right), \tag{7.8}$$

$$\tau_{yz} = \tau_{zy} = -\mu\left(\frac{\partial u}{\partial z} + \frac{\partial w}{\partial y}\right), \tag{7.9}$$

$$\tau_{zx} = \tau_{xz} = -\mu\left(\frac{\partial u}{\partial z} + \frac{\partial w}{\partial x}\right). \tag{7.10}$$

The heat-flux vector $\vec{q}$, the total energy E and the specific enthalpy H are given by

$$\vec{q} = -k\nabla T = -k\left(\frac{\partial T}{\partial x}i_x + \frac{\partial T}{\partial y}i_y + \frac{\partial T}{\partial z}i_z\right), \tag{7.11}$$

$$E = e + \frac{1}{2}\left(u^2 + v^2 + w^2\right), \quad H = E + p/\rho. \tag{7.12}$$

Here i_x, i_y and i_z are the unit vectors of the Cartesian co-ordinate system (x, y, z), e is the specific internal energy per unit mass. The quantities λ and μ are the first and second coefficients of viscosity, respectively.

7.4 NAVIER–STOKES EQUATIONS IN DIFFERENTIAL FORM

Assuming the properties of the medium to be continuous functions of space and time, the conservation equations in integral form, Eq. (7.2), can be transformed into an equivalent set of partial differential equations by means of the divergence theorem of vector calculus,

$$\text{Equation of continuity} \quad \frac{\partial \rho}{\partial t} + \nabla\cdot(\rho\vec{V}) = 0, \tag{7.13}$$

$$\text{Equation of motion} \quad \frac{\partial}{\partial t}(\rho\vec{V}) + \nabla\cdot(\rho\vec{V}\otimes\vec{V} - \overline{\overline{\sigma}}) = \vec{f}, \tag{7.14}$$

$$\text{Equation of energy} \quad \frac{\partial}{\partial t}(\rho E) + \nabla\cdot(\rho E\vec{V} - \overline{\overline{\sigma}}\vec{V} + \vec{q}) = \vec{f}\cdot\vec{V}. \tag{7.15}$$

The operator $\otimes$ used in Eq. (7.14) implies a Cartesian product (or dyadic product) of two vectors giving a tensor, e.g., if $\overline{\overline{A}} = \vec{a}\otimes\vec{b}$, then the elements of the tensor $\overline{\overline{A}}$ are $A_{ij} = a_i b_j$. In these equations t is the time, ρ the density, $\vec{V}$ the velocity of a material particle in the frame of reference, E the specific total energy

$$E = e + \frac{1}{2}V^2 \tag{7.16}$$

where e is the specific internal energy per unit mass, $\overline{\overline{\sigma}}$ is the stress tensor, $\vec{q}$ the heat flux vector and $\vec{f}$ the body force per unit volume. The energy equation is valid under the assumption that there is no source or sink of energy in the control volume. The above set of equations (7.13)–(7.15) can be written in matrix form as

$$\frac{\partial \vec{W}}{\partial t} + \nabla.(\overline{\overline{F}}_e + \overline{\overline{F}}_v) = 0, \tag{7.17}$$

where $\overline{W}$, $\overline{\overline{F}}_e$ and $\overline{\overline{F}}_v$ are defined by Eqs. (7.3)–(7.5).

The basic dependent variables, ρ, $\vec{V}$ and E (or e) may be normalized with respect to their free-stream values. Constitutive relationships for the stress tensor $\overline{\overline{\sigma}}$ and for the heat-flux vector $\vec{q}$ must be added to these equations to obtain a closed system. We are concerned here with the case of *Newtonian fluids*, i.e. fluids for which the stress tensor is a linear function of the velocity gradients. From this definition, excluding the existence of distributed force couples, results Newton's law, also called the Navier–Stokes law for tensor $\overline{\overline{\sigma}}$, i.e.

$$\overline{\overline{\sigma}} = -p\overline{\overline{I}} + \overline{\overline{\tau}} \tag{7.18}$$

where p is the pressure, $\overline{\overline{I}}$ the unit tensor and $\overline{\overline{\tau}}$ is the viscous stress tensor given by

$$\overline{\overline{\tau}} = \lambda \text{ div } \vec{V}\overline{\overline{I}} + 2\mu \text{ def } \vec{V} \tag{7.19}$$

Here,

$$\text{def } \vec{V} = 1/2[\text{grad}\vec{V} + (\text{grad}\vec{V})']$$

is the rate of deformation tensor, λ and μ are the two coefficients of viscosity and the superscript $'$ denotes the transpose of a tensor. Newtonian fluids, particularly air and water, also follow Fourier's law of heat conduction for the heat–flux vector

$$\vec{q} = -k \text{ grad } T \tag{7.20}$$

where T is the absolute temperature and k is the coefficient of thermal conductivity. The state variables ρ, e, T and p are connected by thermodynamic relationships (assuming local thermodynamic equilibrium). For a simple gas the thermodynamic state can be defined by only two variables, say, ρ and e. The pressure p and the temperature T can be obtained as functions of these two variables. These functions are known as the equations of state. An important case is a perfect gas with constant specific heats c_p and c_v. For such a gas, the equations of state are

$$p = (\gamma - 1)\rho e, \quad \gamma = c_\text{p}/c_\text{v}, \quad e = c_\text{v}T. \tag{7.21}$$

The viscosity and thermal conductivity coefficients also depend on the local thermodynamic state. However, in most conditions they depend only on the temperature, i.e., $\lambda = \lambda(T)$, $\mu = \mu(T)$ and $k = k(T)$. Since the dissipation function can not be negative by the second law of thermodynamics, it can be shown (Hollanders and Viviand, 1980; Landau and Lifshitz, 1989; Oswatitsch, 1956), that this leads to the conditions

$$3\lambda + 2\mu \geq 0$$

and in the absence of internal relaxation phenomenon which would involve departure from local thermodynamic equilibrium, the Stokes relationship

$$3\lambda + 2\mu = 0, \tag{7.22}$$

is generally accepted as a valid approximation.
Further, Sutherland's formula for viscosity is often used to relate μ and T and is given by

$$\mu = C_1 \frac{T^{3/2}}{T + C_2},$$

where C_1, C_2 are constants for a given gas. Once μ is known, the Prandtl number $P_r = \dfrac{\mu c_p}{k}$ is often used to determine the coefficient of thermal conductivity k since the ratio $\dfrac{c_p}{P_r}$ is approximately constant for most gases.

7.4.1 Compressible Two-Dimensional Equations in Vector Form

In the case of two-dimensional plane flow, the compressible Navier–Stokes equations in Cartesian coordinates without body forces or external heat addition can be written in a compact vector form as

$$\frac{\partial \vec{W}}{\partial t} + \frac{\partial \vec{F}_1}{\partial x} + \frac{\partial \vec{F}_2}{\partial y} = \frac{\partial \vec{G}_1}{\partial x} + \frac{\partial \vec{G}_2}{\partial y} \tag{7.23}$$

where $\vec{W}$ is the vector of dependent variables and $\vec{F}_1$ and $\vec{F}_2$ are the convective flux vectors given by

$$\vec{W} = \begin{bmatrix} \rho \\ \rho u \\ \rho v \\ \rho E \end{bmatrix}, \vec{F}_1 = \begin{bmatrix} \rho u \\ \rho u^2 + p \\ \rho u v \\ (\rho E + p)u \end{bmatrix} \quad \text{and} \quad \vec{F}_2 = \begin{bmatrix} \rho v \\ \rho u v \\ \rho v^2 + p \\ (\rho E + p)v \end{bmatrix}, \tag{7.24}$$

u, v being the Cartesian components of the velocity vector $\vec{V}$. Here $\vec{G}_1$ and $\vec{G}_2$ represent the flux vectors for the viscous terms and are written as

$$\vec{G}_1 = \begin{bmatrix} 0 \\ \tau_{xx} \\ \tau_{xy} \\ u\tau_{xx} + v\tau_{xy} + k\frac{\partial T}{\partial x} \end{bmatrix}, \quad \vec{G}_2 = \begin{bmatrix} 0 \\ \tau_{xy} \\ \tau_{yy} \\ u\tau_{xy} + v\tau_{yy} + k\frac{\partial T}{\partial y} \end{bmatrix}, \tag{7.25}$$

where the viscous stresses are given by

$$\tau_{xx} = \frac{2}{3}\mu\left(2\frac{\partial u}{\partial x} - \frac{\partial v}{\partial y}\right),$$
$$\tau_{yy} = \frac{2}{3}\mu\left(2\frac{\partial v}{\partial y} - \frac{\partial u}{\partial x}\right), \tag{7.26}$$
$$\tau_{xy} = \mu\left(\frac{\partial u}{\partial y} + \frac{\partial v}{\partial x}\right) = \tau_{yx}.$$

Although the Navier–Stokes equations are the basic equations governing the motion of a compressible fluid, depending on the intended application useful simplification might be achieved with a range of mathematical models of varying complexity.

7.4.2 Incompressible Navier–Stokes Equations in Cartesian Coordinates

In rectangular Cartesian coordinates, the differential form of 3-D incompressible Navier–Stokes (N–S) equations are

$$\text{continuity,} \quad \frac{\partial u}{\partial x} + \frac{\partial v}{\partial y} + \frac{\partial w}{\partial z} = 0, \tag{7.27}$$

$$x\text{-momentum,} \quad \rho\frac{\mathrm{d}u}{\mathrm{d}t} = -\frac{\partial p}{\partial x} + \left(\frac{\partial \tau_{xx}}{\partial x} + \frac{\partial \tau_{xy}}{\partial y} + \frac{\partial \tau_{xz}}{\partial z}\right) + f_x, \tag{7.28}$$

$$y\text{-momentum,} \quad \rho\frac{\mathrm{d}v}{\mathrm{d}t} = -\frac{\partial p}{\partial y} + \left(\frac{\partial \tau_{yx}}{\partial x} + \frac{\partial \tau_{yy}}{\partial y} + \frac{\partial \tau_{yz}}{\partial z}\right) + f_y, \tag{7.29}$$

$$z\text{-momentum,} \quad \rho\frac{\mathrm{d}w}{\mathrm{d}t} = -\frac{\partial p}{\partial z} + \left(\frac{\partial \tau_{zx}}{\partial x} + \frac{\partial \tau_{zy}}{\partial y} + \frac{\partial \tau_{zz}}{\partial z}\right) + f_z, \tag{7.30}$$

the operator $\frac{\mathrm{d}}{\mathrm{d}t}$ denoting differentiation following the fluid, defined as

$$\frac{\mathrm{d}}{\mathrm{d}t} \equiv \frac{\partial}{\partial t} + u\frac{\partial}{\partial x} + v\frac{\partial}{\partial y} + w\frac{\partial}{\partial z}, \tag{7.31}$$

(f_x, f_y, f_z), being the components of body forces per unit volume.
The viscous stress tensor for incompressible flow reduces to (Schlichting and Gersten, 2000):

$$\tau_{xx} = 2\mu\frac{\partial u}{\partial x}, \quad \tau_{yy} = 2\mu\frac{\partial v}{\partial y}, \quad \tau_{zz} = 2\mu\frac{\partial w}{\partial z},$$

$$\tau_{xy} = \tau_{yx} = \mu\left(\frac{\partial u}{\partial y} + \frac{\partial v}{\partial x}\right), \quad \tau_{xz} = \tau_{zx} = \mu\left(\frac{\partial u}{\partial z} + \frac{\partial w}{\partial x}\right). \tag{7.32}$$

Substituting Eqs. (7.32) in Eqs. (7.28)–(7.30), yields on simplification the Navier–Stokes equations for incompressible flow in rectangular Cartesian coordinates

$$\rho\frac{du}{dt}=f_x-\frac{\partial p}{\partial x}+\frac{\partial}{\partial x}\left(2\mu\frac{\partial u}{\partial x}\right)+\frac{\partial}{\partial y}\left[\mu\left(\frac{\partial u}{\partial y}+\frac{\partial v}{\partial x}\right)+\frac{\partial}{\partial z}\left[\mu\left(\frac{\partial u}{\partial z}+\frac{\partial w}{\partial x}\right)\right]\right.\tag{7.33}$$

$$\rho\frac{dv}{dt}=f_y-\frac{\partial p}{\partial y}+\frac{\partial}{\partial x}\left(\mu\left(\frac{\partial u}{\partial y}+\frac{\partial v}{\partial x}\right)\right)+\frac{\partial}{\partial y}\left(2\mu\frac{\partial v}{\partial y}\right)+\frac{\partial}{\partial z}\left[\mu\left(\frac{\partial v}{\partial z}+\frac{\partial w}{\partial y}\right)\right]\tag{7.34}$$

$$\rho\frac{dw}{dt}=f_z-\frac{\partial p}{\partial z}+\frac{\partial}{\partial x}\left[\mu\left(\frac{\partial w}{\partial x}+\frac{\partial u}{\partial z}\right)\right]+\frac{\partial}{\partial y}\left[\mu\left(\frac{\partial v}{\partial z}+\frac{\partial w}{\partial y}\right)\right]+\frac{\partial w}{\partial z}\left(2\mu\frac{\partial w}{\partial z}\right)\tag{7.35}$$

Note that, using continuity equation (7.27), the incompressible N–S equations in conservative form follow from Eqs. (7.33)–(7.35) as

$$\rho\left[\frac{\partial u}{\partial t}+\frac{\partial u^2}{\partial x}+\frac{\partial uv}{\partial y}+\frac{\partial uw}{\partial z}\right]=-\frac{\partial p}{\partial x}+\Delta u,\tag{7.36}$$

$$\rho\left[\frac{\partial v}{\partial t}+\frac{\partial vu}{\partial x}+\frac{\partial v^2}{\partial y}+\frac{\partial vw}{\partial z}\right]=-\frac{\partial p}{\partial y}+\Delta v,\tag{7.37}$$

$$\rho\left[\frac{\partial w}{\partial t}+\frac{\partial wu}{\partial x}+\frac{\partial wv}{\partial y}+\frac{\partial w^2}{\partial z}\right]=-\frac{\partial p}{\partial y}+\Delta w,\tag{7.38}$$

where Δ denotes the *Laplacian*

$$\Delta\equiv\frac{\partial^2}{\partial x^2}+\frac{\partial^2}{\partial y^2}+\frac{\partial^2}{\partial z^2}.\tag{7.39}$$

7.4.3 Dimensionless Form of the Basic Equations

For considerations of *dynamic similarity,* it is very useful to obtain dimensionless form of the basic equations. For this purpose, dimensionless quantities, denoted by an asterisk on the corresponding symbols, may be defined as

$$u^*=\frac{u}{U_0},\ v^*=\frac{v}{U_0},\ w^*=\frac{w}{U_0},\ x^*=\frac{x}{L},\ y^*=\frac{y}{L},\ z^*=\frac{z}{L}$$

$$t^*=\frac{t}{L/U_0},\ \mu^*=\frac{\mu}{\mu_0},\ \rho^*=\frac{\rho}{\rho_0},\ p^*=\frac{p}{\rho_0U_0^2},\ T^*=\frac{T}{T_0},\ e^*=\frac{e}{U_0^2}\tag{7.40}$$

$$E^*=\frac{E}{U_0^2},\ k^*=\frac{k}{k_0},\ \vec{q}^*=\frac{\vec{q}}{\rho_0U_0^3}$$

$$\bar{\bar{\sigma}}^*=\frac{\bar{\bar{\sigma}}}{\rho_0U_0^2},\ \bar{\bar{\tau}}^*=\frac{\bar{\bar{\tau}}}{(\mu_0U_0/L)}.\tag{7.41}$$

The quantities with subscript '0' are some suitable reference quantity, for example, the free-stream velocity may be denoted by U_0, L is some characteristic length used to define the Reynolds number

$$Re_0 = \frac{\rho_0 U_0 L}{\mu_0}. \tag{7.42}$$

Introducing these in the basic equations, we see after a little calculation and simplification that Eqs. (7.2)–(7.19) maintain the same form as the original dimensional equations; that is there is no change in the equations except that the original symbols in the equations are replaced by the corresponding dimensionless symbols with asterisk superscript. The constitutive relationship (7.18) becomes in dimensionless form (denoting the dimensionless quantities by the same symbols)

$$\bar{\bar{\sigma}} = -p\bar{\bar{I}} + \frac{1}{Re}\bar{\bar{\tau}}, \tag{7.43}$$

while (7.19) remains unchanged. For a perfect gas Eq. (7.20) changes over to

$$\vec{q} = \frac{\gamma}{RePr} k\nabla e, \tag{7.44}$$

where the *Prandtl number* Pr is defined as $Pr = \dfrac{\mu_0 C_p}{k_0}$.

It appears convenient to use the same symbols for the dimensionless quantities and drop the asterisk symbol. From now on, we use dimensionless quantities and omit, for the sake of convenience, the asterisks on the symbols.

7.4.4 Incompressible Two-Dimensional Equations: Dimensionless Form

In the subsequent chapters, we frequently need the dimensionless form of the basic equations for incompressile flow, which we present here for 2-D flow. The basic equations may be expressed in dimensionless form as

$$\text{Continuity,}\quad \frac{\partial u}{\partial x} + \frac{\partial v}{\partial y} = 0, \tag{7.45}$$

$$x\text{-momentum,}\quad \frac{\partial u}{\partial t} + u\frac{\partial u}{\partial x} + v\frac{\partial u}{\partial y} = -\frac{\partial p}{\partial x} + \frac{1}{Re}\left(\frac{\partial^2 u}{\partial x^2} + \frac{\partial^2 u}{\partial y^2}\right) \tag{7.46}$$

$$y\text{-momentum,}\quad \frac{\partial v}{\partial t} + u\frac{\partial v}{\partial x} + v\frac{\partial v}{\partial y} = -\frac{\partial p}{\partial y} + \frac{1}{Re}\left(\frac{\partial^2 v}{\partial x^2} + \frac{\partial^2 v}{\partial y^2}\right). \tag{7.47}$$

Here, u, v denote the dimensionless velocity components along x and y-axial directions, p and ρ the dimensionless pressure and density and Re the Reynolds number.

7.4.5 Observations on the Basic Equations

It may be observed in the non-conservative form of the Navier–Stokes equations that, in each equation, the highest order derivatives of one of the basic variables ρ, V and e with respect to the space variables are present. These equations are nonlinear and strongly coupled. However, it may also be noted that to each equation one basic variable is naturally associated, the time as also the material derivatives of which are given explicitly by that equation. For example, ρ for the continuity equation, $\vec{V}$ for the vectorial momentum equation and e for the energy equation. Furthermore, the space derivatives of the highest order, i.e. second order, are solely derivatives of the basic variables associated with the equation. Thus it is possible to gain some insight into the mathematical nature of these equations looking at each of them separately as an equation for the determination of the associated basic variable, assuming the other basic variables in that equation to be known quantities.

The continuity equation, considered as an equation for unknown ρ, is a first-order equation whose characteristic curves are the trajectories of fluid particles. This equation is of hyperbolic character for steady as well as for unsteady flows (Courant and Hilbert, 1953; Prasad and Ravindran, 1985).

The momentum equation is in fact a system of two or three scalar equations depending on the dimensionality of the problem. The corresponding second-order differential operator of these equations are of elliptic type (except for $\mu = 0$ or for $\lambda + 2\mu = 0$ which is not the case for a viscous fluid). Thus the momentum equation for the unknown $\vec{V}$ is of elliptic type for steady flows and of parabolic type with respect to time for unsteady flows. The energy equation, for the unknown e, has the same mathematical character as the momentum equation: i.e. elliptic for steady flows and parabolic for unsteady flows.

Now, granting that the system of Navier–Stokes equations retains the mathematical properties of each equation considered separately, we can say that this system is of hybrid elliptic–hyperbolic type for steady flows and of hybrid parabolic–hyperbolic type for unsteady flows, the hyperbolic character being due, in both the cases, to the continuity equation.

The parabolic or elliptic character of the equations is due to the dissipative terms. In flow regions where the dissipative effects are very small, the solutions of the Navier–Stokes equations exhibit all the properties of solutions of the Euler equations, and we can say that, in such regions, the Navier–Stokes equations lose their parabolic or elliptic character and reduce to the Euler equations. But this happens only so as far as the local properties of the solutions are concerned. The reason why the global properties of the solutions of the Navier–Stokes equations do not approach those of the Euler equations as the viscosity becomes very small is that the choice of the boundary conditions is always governed by the mathematical nature of the equations. The mathematical nature of the equation itself depends only on the highest order derivatives with respect to the space co-ordinates and in the unsteady case, with respect to time, without regard to the order of magnitude of these derivatives. In other words, the solution of the Navier–Stokes equations always depends on the dissipative terms in a global way even if, for large Reynolds numbers, these solutions include large quasi-inviscid flow regions in which they are also solutions of the Euler equations. For the dissipative terms to play a negligible part everywhere in a solution of the Navier–Stokes equations at high Reynolds number, it would be necessary that the boundary conditions be compatible with a quasi-inviscid flow, which means that they should be determined from a previous inviscid flow calculation. These special conditions do occur in unsteady, time accurate flow computations.

7.5 BOUNDARY CONDITIONS FOR NAVIER–STOKES EQUATIONS

A significant problem in computational fluid dynamics is to choose the conditions at the boundaries of a computational domain. In practice, one encounters the following type of boundaries:

1. natural boundaries and
2. artificial boundaries.

For an external flow past a solid body, *the solid wall, the far field* and for an internal flow, the *inlet and outlet boundaries* along with the solid wall, are the examples of natural boundaries. Artificial boundaries are those introduced due to different types of grid topologies used and by artificial cuts introduced either to divide the computational domain into different blocks or to make a multiply connected domain a simply connected one. So, in general we have to deal with the following types of boundaries,

1. solid boundary; the flow velocity relative to the solid boundary is zero (no slip conditions),
2. far field boundary; where the disturbance due to the solid boundary is vanishingly small, except probably the circulation produced at the wall,
3. cut boundary; where the unknown quantities at one side of the cut are related to those at the other side of it,
4. symmetry boundary; which represents a line or a plane of symmetry in the flow field used in order to reduce the computational effort,
5. degenerate boundary; which appears in some grids where the grid lines or planes become degenerate,

6. inlet and outlet boundaries; where the flow conditions are to be prescribed from physical considerations.

The number and the type of boundary conditions to be imposed depend on the mathematical nature of the governing equations. Assuming that the velocity field is known, the continuity equation may be used to obtain the material derivative of the density. Therefore the density of a fluid particle located at a boundary point M at time t can be calculated from this equation only if it was known just before the particle arrived at M, i.e. if the particle was inside the domain or on its boundary immediately before arrival at M. In other words, the density at a boundary point M must be calculated from the continuity equation (hence no condition can be imposed) if the fluid *leaves* the computation domain at M or if its velocity is tangent to the boundary; on the contrary the density at M must be obtained from a boundary condition (i.e the value must be prescribed) if the fluid *enters* the computational domain.

The momentum equation, considered as an equation for $\vec{V}$, and the energy equation, considered as an equation for e, are of elliptic type with respect to the space variables for steady flow and of parabolic type with respect to time for unsteady flow; therefore one (vectorial) boundary condition on $\vec{V}$ and one condition on e must be imposed at all points of the boundary. Because of the coupling of the equations together, this statement makes no pretense to mathematical rigour, and it should be considered only as a reasonable conjecture which provides a guideline for the boundary conditions. For the *Navier–Stokes equations in two dimensions* this means that we have to prescribe four boundary conditions at an inflow boundary and three at an outflow boundary. In the case where the boundary is an impermeable wall and when rarefaction effects are not present, the usual no-slip condition is

$$\vec{V} = \vec{V}_{\rm w} \text{ and } T = T_{\rm w} \text{ or } \frac{\partial T}{\partial n} = 0. \tag{7.48}$$

i.e. zero relative velocity and either temperature or heat transfer rate specified. Here $\vec{V}_{\rm w}$ and $T_{\rm w}$ are velocity and temperature of the wall and $\vec{n}$ is the unit vector normal to the wall. In the case of a perfect gas with constant specific heats, the above conditions on temperature are equivalent to that on the internal energy $e = c_{\rm v}T$.

For the conditions at the far field, it is necessary to differentiate between inflow and outflow behavior of the fluid particle at the boundary points which can be done by considering the sign of the normal velocity. The unsteady Euler equations (inviscid flow approximation to the Navier–Stokes equations, discussed later) are hyperbolic and it is straightforward to construct characteristics and then to require that as many boundary conditions must be specified as there are characteristics entering the domain. For three dimensional flow with two thermodynamic variables

(ρ, u, v, w, p, T), this implies that five boundary conditions must be specified at inflow if the flow is supersonic. Gustafsson and Sundstrom (1978) have shown that the number of boundary conditions shown in Table 7.1 are necessary and sufficient to produce a well posed problem.

Table 7.1 Number of far-field boundary conditions to be specified for Euler and Navier–Stokes equations

	Inflow		Outflow	
System of equations	Supersonic	Subsonic	Supersonic	Subsonic
Euler	5	4	0	1
Navier–Stokes	5	5	4	4

It must be noted from the above table that unlike the Euler equations the Navier–Stokes equations do not distinguish between supersonic and subsonic flow with respect to the number of conditions. It is a fundamental question whether, at high Reynolds numbers ($Re \to \infty$,) the Navier–Stokes solutions converge to the Euler solutions and the answer will depend on the geometry and data. In particular, near the solid wall, there is a boundary layer which becomes thinner as Re increases. The limit process is complicated by the boundary conditions, since for any finite Reynolds number all velocity components are zero, while in the limit only the normal component is zero.

At open boundaries with smooth flow the situation is different. There the second derivatives are bounded and hence the viscosity and heat conduction terms are negligible. In that case we can expect the limit solution to be the solution of the Euler equations. The compatibility conditions associated with the characteristics of the equivalent Euler equations provide suitable Dirichlet boundary conditions. So, the boundary conditions for the Navier–Stokes equations can be constructed starting with those of the Euler equations. The additional conditions required by the compressible Navier–Stokes equations are then imposed as Neumann boundary conditions.

Many flows at high Reynolds number, far away from an impermeable rigid body behave as though they were governed by the Euler equations rather than the Navier–Stokes equations. This suggests choosing boundary conditions accordingly and this often works in practice. When this leads to less than the required number of conditions then one may fail to obtain a unique solution. On the other hand, more than the required number of conditions usually produce a solution with a severe unphysical boundary layer adjacent to the boundary in question. Ideally the far field boundary conditions should be chosen so as to make the boundary appear transparent to the solution *(non-reflecting boundary conditions)*, i.e. the same solution ought to be obtained if the boundary locations were moved nearer to or farther away from the body.

7.6 REYNOLDS AVERAGED NAVIER–STOKES EQUATIONS

The unsteady Navier–Stokes equations apply equally to turbulent and laminar flows. However, computation of turbulent flows, with the above equations, is a difficult task because the time and space scales of turbulent motion are too small. As long as the scales of interest are many times the mean free paths of the fluid molecules, the Navier–Stokes equations described in the previous sections are the valid fundamental equations for viscous flow, whereas for turbulent motion, one often takes resort to time-averaging of the rapidly fluctuating components. This yields the *Reynolds averaged* equations of motion which are derived by decomposing the dependent variables in conservation equations into a time-averaged quantity ($\overline{g}$) and a fluctuating component ($g' = g - \overline{g}$) and then time-averaging the entire equation. A time-averaged quantity $\overline{g}$ is defined as

$$\overline{g} = \frac{1}{\Delta t}\int_{t_0}^{t_0+\Delta t} g\,\mathrm{dt}. \tag{7.49}$$

Here Δt should be large compared to the period of the random fluctuations associated with turbulence, but small with respect to the time constant for any slow variations in the flow field associated with ordinary unsteady flows. However, for treatment of compressible flows, the mass-weighted averaging suggested by Favre (1965) is more convenient. In this approach, the mass-averaged variables are defined as

$$\tilde{g} = \frac{\overline{\rho g}}{\overline{\rho}}$$

and the dependent variables are split into average and fluctuating parts as $g = \tilde{g} + g''$. However, it may be noted that only the velocity components and thermal variables are mass-averaged. Fluid properties such as density and pressure are treated as before. The detailed procedures involved in mass-averaging are described by Cebeci and Smith (1974). Application of this averaging process produces the following Reynolds equations

$$\frac{\partial\overline{\rho}}{\partial t} + \frac{\partial}{\partial x_j}\left(\overline{\rho}\,\tilde{u}_j\right) = 0 \tag{7.50}$$

$$\frac{\partial}{\partial t}(\bar{\rho}\,\tilde{u}_i)+\frac{\partial}{\partial x_j}(\bar{\rho}\,\tilde{u}_i\tilde{u}_j)=-\frac{\partial\bar{p}}{\partial x_i}+\frac{\partial}{\partial x_j}\left(\bar{\tau}_{ij}-\overline{\rho u_i'' u_j''}\right) \tag{7.51}$$

$$\frac{\partial}{\partial t}(\bar{\rho}\,\tilde{H})+\frac{\partial}{\partial x_j}(\bar{\rho}\,\tilde{u}_j\tilde{H})=\frac{\partial\bar{p}}{\partial t}+\frac{\partial}{\partial x_j}\tilde{u}_i\bar{\tau}_{ij}+\overline{u_i''\tau_{ij}}-\overline{\rho u_j'' H''}+k\frac{\partial\bar{T}}{\partial x_j}\Bigg) \tag{7.52}$$

where $\tilde{H}$ denotes mass averaged mean value of the total enthalpy per unit mass given by

$$\tilde{H}=\tilde{E}+\frac{\overline{p}}{\bar{\rho}}, \tag{7.53}$$

u_i represent the Cartesian velocity components and $\tau_{i,j}$ are the components of the viscous stress tensor $\bar{\bar{\tau}}$.

As a consequence of such Reynolds decomposition and averaging, extra terms are introduced in the equations which physically represent the turbulent transport of momentum and energy due to velocity and pressure fluctuations. However, it may be noted that if some small fluctuating terms are neglected, then Eqs. (7.51)–(7.52) are reduced to the same form as the instantaneous Navier–Stokes equations, except that the stress tensor is augmented by additional stresses called *Reynolds stresses* and the heat flux vector is augmented by the additional turbulent heat flux. Now, due to the presence of the Reynolds stresses and the turbulent heat flux quantities, the Eqs. (7.51)–(7.52) are not closed and turbulence modelling is required to form a closed and solvable system. Such closure is possible only through empirical information on turbulence as functions of mean flow parameters. The empirical formulation may be purely through algebraic relations connecting Reynolds stresses/fluxes to mean flow parameters or their gradients; or it may also be in the form of differential equations for transport of turbulence parameters which determine the Reynolds stresses/fluxes. In the present work, an algebraic turbulence model and other simple models have been discussed in Chapters 11–13.

7.7 BOUNDARY-LAYER, THIN-LAYER AND ASSOCIATED APPROXIMATIONS

The complete set of equations of motion for viscous compressible flows (Navier–Stokes equations) have been known for a long time. However, owing to the great mathematical difficulties in solving these equations (except for a small number of particular cases), rigorous theoretical treatment has not been possible. Even now, with the advances in computer technology and numerical algorithms, it is very difficult to solve the full Navier–Stokes equations due to limitations of computing resources. Now, many important viscous fluids, such as air, water and the like, have very small viscosity. By making an order of magnitude analysis of various terms of the Navier–Stokes equations and several simple experiments, Prandtl (1904) showed that the flow about a solid body can be divided into two regions: a very thin layer in the neighbourhood of the body (the *Boundary layer*) where viscous forces play an important part, and the rest of the fluid where the viscous forces may be neglected. Under this approximation equations are derived from the Navier–Stokes equations by retaining the lowest order terms in an expansion in inverse powers of the *Reynolds number*, Re. The Reynolds number, an important parameter in viscous fluid mechanics is the ratio of *inertial force to viscous force* and is defined as

$$Re = \frac{\rho U L}{\mu} = \frac{UL}{\nu}, \tag{7.54}$$

where μ is the coefficient of viscosity and $\nu = \mu/\rho$ is called the *coefficient of kinematic viscosity*. This formal procedure ultimately leads to

1. neglect of diffusion process parallel to the body surface, and
2. replacement of the momentum equation normal to the surface with assumption of zero normal pressure gradient throughout the boundary layer.

For a two-dimensional compressible flow over a flat plate the following governing equations can be derived (Oswatitsch, 1956),

$$\frac{\partial \rho}{\partial t} + \frac{\partial}{\partial x}(\rho u) + \frac{\partial}{\partial y}(\rho v) = 0, \tag{7.55}$$

$$\frac{\partial u}{\partial t} + u\frac{\partial u}{\partial x} + v\frac{\partial u}{\partial y} = -\frac{1}{\rho}\frac{\partial p}{\partial x} + \frac{1}{\rho}\frac{\partial}{\partial y}(\mu\frac{\partial u}{\partial y}) \tag{7.56}$$

$$\frac{\partial p}{\partial y} = 0 \tag{7.57}$$

$$\rho\left(\frac{\partial c_p T}{\partial t} + u\frac{\partial c_p T}{\partial x} + v\frac{\partial c_p T}{\partial y}\right) = \frac{\partial p}{\partial t} + u\frac{\partial p}{\partial x} + \frac{\partial}{\partial y}\left(k\frac{\partial T}{\partial y}\right) + \mu\left(\frac{\partial u}{\partial y}\right)^2 \tag{7.58}$$

$$p = \rho R T \tag{7.59}$$

These equations are still nonlinear, but simpler in the sense that the number of viscous terms have been reduced and the pressure is no longer an unknown and its variation across the boundary layer can be neglected. For a curved plate only, Eq. (7.57) should be modified to

$$-Ku^2 = -\frac{1}{\rho}\frac{\partial p}{\partial y}, \tag{7.60}$$

provided that the surface curvature K and its variation along the curvilinear coordinate x, are of order unity, and y is the perpendicular distance. It may be observed from Eq. (7.60) that the variation of pressure across the boundary layer is of the order of boundary layer thickness, δ and can be neglected, along with v-component of the fluid just outside the boundary layer. As a result we get (Pai, 1959)

$$-\frac{1}{\rho_e}\frac{\partial p}{\partial x} = \frac{\partial u_e}{\partial t} + u_e\frac{\partial u_e}{\partial x} \tag{7.61}$$

where the subscript e refers to its value at the edge of the boundary layer. The boundary conditions at the outer edge of the boundary layer (Pai, 1959) are,

$$y = y_e : u = u_e, \quad \frac{\partial u}{\partial y} = \frac{\partial^2 u}{\partial y^2} = \cdots = 0, \tag{7.62}$$

and at the surface of the wall

$$y = 0 : u = v = 0, \quad \text{and} \quad \frac{\partial p}{\partial x} = \frac{\partial}{\partial y}\left(\mu\frac{\partial u}{\partial y}\right)_w \tag{7.63}$$

can be obtained from Eq. (7.56). Differentiating Eq. (7.63) with respect to y and employing Eq. (7.57) we get

$$\left[\frac{\partial}{\partial y}\left(\frac{\partial}{\partial y}\mu\frac{\partial u}{\partial y}\right)\right]_w = \left[\frac{\partial^2}{\partial y^2}\left\{\frac{\partial}{\partial y}\left(\mu\frac{\partial u}{\partial y}\right)\right\}\right]_w = \cdots = 0. \tag{7.64}$$

The quantity $\partial u/\partial y$ determines the shearing stress in the boundary layer and its value at the wall $(\partial u/\partial y)_w$ is what we are interested in determining. Due to the presence of adverse pressure gradients this quantity may be zero or negative as x increases from the leading edge. When this happens the flow separates from the solid boundary and the boundary layer approximations become invalid. Hence, we can compute the boundary layer flow *only upto the separation point* by this method.

On the other hand the *thin-layer approximation* neglects the diffusion process parallel to a body surface but retains all the momentum equations and makes no assumption about the pressure. One advantage of retaining the normal momentum equation becomes evident in the application to high Reynolds number, separated turbulent flows. Its use removes the troublesome singularities at the separation points and permits the straight forward computation of separated and reverse flow regions. In this process streamwise like differences appearing for the first and second derivative terms in viscous flux quantities can be neglected. Much computational efforts can be saved by this approximation if one follows the classical space discretization technique (Chakrabartty, 1989, 1990; Radespiel, 1989; Swanson and Turkel, 1985). An improved space discretisation technique (Chakrabartty, 1990a) can be developed with appropriate choice of control volumes for computing the first derivatives, such that computational efforts can be kept to the same order, irrespective of whether one wants to compute the flow with or without thin layer approximations.

Another type of approximation often used for computing flows with high Reynolds number and particularly for supersonic and hypersonic flows is called *parabolisation of the Navier–Stokes equations.* Along with the streamwise diffusion terms one neglects the unsteady terms also in the Navier–Stokes equations. For a body fitted curvilinear coordinate system (ξ, η, ζ) this leads to a system of equations which are parabolic in the ξ direction with some additional restrictions

1. there should be no axial separation in the flow region, and
2. the pressure gradient in the boundary layer be treated properly.

The treatment of the pressure gradient term is an important factor in this approach, in order to avoid the ill–posedness of the initial boundary value problem (Rakich et al. 1984).

7.8 EULER EQUATIONS FOR INVISCID FLOWS

For a non-viscous and non-heat-conducting flow, the viscous flux tensor, $\overline{\overline{F}}_v$ vanishes identically and $\mu = 0$ and $k = 0$. We obtain the integral form of the Euler equations by substituting these values in Eqs. (7.5) and (7.11) . Thus the integral form of the Euler equations is given by

$$\frac{\partial}{\partial t}\int_{\Omega} \vec{W}\,\mathrm{d}\Omega + \int_{\partial\Omega} \overline{\overline{F}}.\vec{n}\,\mathrm{d}s = 0. \tag{7.65}$$

The total flux tensor $\overline{\overline{F}}$ now consists of only the inviscid flux $\overline{\overline{F}}_e$, the viscous flux $\overline{\overline{F}}_v$ vanishing identically.

$$\overline{\overline{F}} = \begin{pmatrix} \rho\vec{V} \\ \rho u\vec{V} + p\vec{i}_x \\ \rho v\vec{V} + p\vec{i}_y \\ \rho w\vec{V} + p\vec{i}_z \\ \rho H\vec{V} \end{pmatrix} \tag{7.66}$$

and

$$E = e + \frac{1}{2}\left(u^2 + v^2 + w^2\right), \qquad H = E + p/\rho.$$

Componentwise, the Euler equations (7.65) are

$$\int_{\Omega} \frac{\partial}{\partial t}\rho \mathrm{d}\Omega + \int_{\partial\Omega} \rho(\vec{v}.\vec{n})\mathrm{d}s = 0,$$

$$\int_{\Omega} \frac{\partial}{\partial t}(\rho u)\mathrm{d}\Omega + \int_{\partial\Omega} \left[\rho u(\vec{v}.\vec{n}) + p(\vec{i}_x.\vec{n})\right] \mathrm{d}s = 0,$$

$$\int_{\Omega} \frac{\partial}{\partial t}(\rho v)\mathrm{d}\Omega + \int_{\partial\Omega} \left[\rho v(\vec{v}.\vec{n}) + p(\vec{i}_y.\vec{n})\right] \mathrm{d}s = 0,$$

$$\int_{\Omega} \frac{\partial}{\partial t}(\rho E)\mathrm{d}\Omega + \int_{\partial\Omega} \rho H(\vec{v}.\vec{n})\mathrm{d}s = 0. \tag{7.67}$$

The differential form of Euler equations follows from Eqs. (7.14) and (7.15), with the above values of $\mu = 0$ and $k = 0$. For a two-dimensional flow in conservative form, the equations may be rewritten in a compact form as

$$\frac{\partial W}{\partial t} + \frac{\partial F}{\partial x} + \frac{\partial G}{\partial y} = 0 \tag{7.68}$$

where the vectors W, F and G are given by

$$W = \begin{pmatrix} \rho \\ \rho u \\ \rho v \\ \rho E \end{pmatrix}, \quad F = \begin{pmatrix} \rho u \\ \rho u^2 + p \\ \rho uv \\ (\rho E + p)u \end{pmatrix}, \quad G = \begin{pmatrix} \rho v \\ \rho uv \\ \rho v^2 + p \\ (\rho E + p)v \end{pmatrix}. \tag{7.69}$$

These equations are to be supplemented by the appropriate boundary conditions.

7.8.1 Certain Observations on Euler and Navier–Stokes Equations

In solving a differential equation, a serious difficulty occurs when the highest space derivative terms of this equation are multiplied by a small parameter, such as with the Navier–Stokes and related equations for high Reynolds number flow. We are considering a physical theory, viz. that of inviscid flow governed by the Euler equations, as an approximation to another physical theory, that of viscous flow governed by the Navier–Stokes equations. The solutions of the Euler equations admit discontinuities of different kinds, e.g. discontinuity in velocity from one streamline to an adjacent streamline, although the pressure is continuous. Such discontinuities, known as *contact discontinuities* are consistent with the theory of characteristics and with the integral conservation laws whose differential forms are the Euler equations. Another form of discontinuity, known as a *shock wave* may exist in which the normal velocity, pressure, density, entropy and absolute temperature jump across a surface, called shock surface (Courant and Hilbert, 1953;

Niyogi, 1977; Oswatitsch, 1956). Since such discontinuities may not occur in the Navier–Stokes equations, we expect the perturbation problem to be singular (van Dyke, 1964). This implies that the Euler limit of a Navier–Stokes solution may not be uniformly valid. The solution of the Euler equations for a given set of boundary conditions is not unique. Often, it is the inherent numerical dissipation in the solution scheme that makes the solution unique. We shall call a solution of the Euler equation, which is the Euler limit of the Navier–Stokes solution for the same conditions, a *relevant* Euler solution.

We emphasise again the fundamental difference, in principle, between the cases of small viscosity and large viscosity. The physical theory expressed by the Euler equations *assumes that the viscosity is zero*. An asymptotic expansion for small μ must tend to the Euler solution in the limit $\mu \to 0$. There is no physical theory for compressible flow in which the viscosity is infinite. The *Stokes flow* is valid for incompressible flow. The governing equations for large values of the coefficient of viscosity μ are still the Navier–Stokes equations.

7.9 BOUNDARY CONDITIONS FOR EULER EQUATIONS

Proper specification of boundary conditions is essential for numerical solution of the gasdynamic equations. Their improper implementation can result in inaccurate solutions and can also lead to instabilities. One of the major problems in the computation of solutions to the Euler equations is the numerical treatment of boundary conditions along the boundaries in the physical domain. Incorrect conditions can substantially degrade the accuracy of the computed solution and slow down the convergence to steady state.

On the solid boundary, the physical situation requires that the normal component of velocity should vanish so as to satisfy the principle of conservation of mass. Conditions should also be imposed on the artificial far-field boundaries that are introduced in external flows in order to restrict the computational domain to a finite region. However, it is difficult to specify the far-field boundary conditions in a way that will facilitate computation and accelerate convergence to steady state in the minimum number of time steps. One should also distinguish between inflow and outflow boundary conditions. The sign of the normal velocity helps to determine the type of the far-field boundary.

The unsteady Euler equations are of hyperbolic type for all Mach numbers. Characteristics play a decisive role for studying such problems of hyperbolic type since, physically, information is carried along the characteristics. According to the theory of Kreiss (1970), the number of conditions to be imposed at a far-field boundary point should be equal to the number of incoming characteristics and the remaining conditions should be determined from the interior solution by extrapolation. The unsteady two-dimensional Euler equations have eigenvalues given by q_n, q_n, $(q_n - c)$ and $(q_n + c)$, where q_n represents the normal velocity and c is the local speed of sound. According to the signs of the eigenvalues, there are three incoming characteristics and one outgoing characteristic for a subsonic inflow $(-c < q_n < 0)$, while there is one incoming and three outgoing characteristics for a subsonic outflow $(0 < q_n < c)$. Therefore, three conditions in subsonic inflow and one condition in subsonic outflow are specified by the free stream values. The choice, as to which of the physical quantities are to be defined by their free stream values, is still open. In a similar manner, for supersonic inflow it may be observed that all characteristics point into the domain so that boundary conditions are required on all the four variables. On the other hand, at a supersonic outflow, all the four characteristics are outgoing so that no boundary condition may be prescribed.

However, for fluid flow problems exterior to a body, there are no unique far-field boundary conditions known in advance. On physical grounds, it is desirable to construct far-field conditions such that no outgoing wave incident on the boundary is reflected back into the flow field or at least, care should be taken so that the amplitude of waves reflected from these artificial boundaries is minimum. In the literature, there exists a number of theories of locally absorbing the outgoing waves at the artificial boundaries, thus yielding stable far-field boundary conditions. The theoretical basis for such non-reflecting boundary conditions stems from a paper by Engquist and Majda (1977) which discusses both ideal non-reflecting boundary conditions and a method for constructing approximate forms for the general class of wave equations (linear hyperbolic systems in several space dimensions). These conditions not only guarantee stable approximations but also minimise the non-physical artificial reflections which occur at these boundaries.

Based on the theory of Riemann invariants, non-reflecting boundary conditions for one-dimensional nonlinear hyperbolic systems were developed by Hedstrom (1979). A procedure based on the Riemann invariants for one-dimensional flow normal to the boundary was used by Jameson (1985) for the solution of time-dependent Euler equations in two space dimensions. Different procedures for treating artificial boundaries to yield well-posed problems and stable difference approximations were analyzed by several authors, for example, Oliger and Sundström (1978), Bayliss and Turkel (1980), Blaschak and Kriegsmann (1988), Ferm (1988), Keller and Givoli (1989), Giles (1990), and so on.

7.9.1 Far-field Boundary Conditions for Euler Equations

In aerodynamics the viscous effects are often confined to thin boundary layers so that the viscous fluxes in Eqs. (7.2) and (7.17) can be neglected and the governing equations can be simplified to the *Euler equations* in integral or in differential forms. The Euler equations have two important features:

1. conservation and
2. signal propagation.

Strong conservation laws for Euler equations in integral form, hide the hyperbolicity or the wave nature. For the Euler equations, the treatment of far-field boundaries is based on Riemann invariants for one dimensional flow normal to the boundary. The Riemann invariants for one dimensional flow are given in Garabedian (1964), Niyogi (1977); a short review of the derivation is given here. By eliminating pressure from the momentum equation, the Euler equations for one dimensional inviscid isentropic flow can be written as

$$\begin{aligned} &\rho_t + u\rho_x + \rho u_x = 0 \\ &u_t + uu_x + \frac{c^2}{\rho}\rho_x = 0 \end{aligned} \tag{7.70}$$

In matrix notation it can be written as

$$\frac{\partial \vec{W}}{\partial t} + A\frac{\partial \vec{W}}{\partial x} = 0, \tag{7.71}$$

where

$$\vec{W} = \begin{pmatrix} \rho \\ u \end{pmatrix}, \quad A = \begin{pmatrix} u & \rho \\ \dfrac{c^2}{\rho} & u \end{pmatrix} \tag{7.72}$$

The matrix A is called the *Jacobian flux matrix*, having real eigenvalues

$$\nu_1 = u - c \quad \text{and} \quad \nu_2 = u + c. \tag{7.73}$$

So, the set of Eqs. (7.70) or (7.71) is of hyperbolic type having two families of characteristics,

$$\frac{dx}{dt} = u - c \quad \text{and} \quad \frac{dx}{dt} = u + c. \tag{7.74}$$

Introducing characteristic variables (ξ, η) in two directions, we obtain the following two relations

$$\begin{aligned} \frac{\partial}{\partial \xi} &= \frac{\partial}{\partial t} + (u + c)\frac{\partial}{\partial x} \\ \frac{\partial}{\partial \eta} &= \frac{\partial}{\partial t} + (u - c)\frac{\partial}{\partial x}. \end{aligned} \tag{7.75}$$

Multiplying the first equation of (7.70) by $\pm$ (c/ρ) and adding it to the second it can be rewritten as

$$\left[\frac{\partial}{\partial t}+(u \pm c)\frac{\partial}{\partial x}\right]u \pm \frac{c}{\rho}\left[\frac{\partial}{\partial t}+(u \pm c)\frac{\partial}{\partial x}\right]\rho = 0. \tag{7.76}$$

Equations (7.74), (7.75) and (7.76) take the form

$$x_\xi = (u+c)t_\xi,\, x_\eta = (u-c)t_\eta,\, u_\xi + \frac{c}{\rho}\rho_\xi = 0,\, u_\eta - \frac{c}{\rho}\rho_\eta = 0. \tag{7.77}$$

The last two equations of the above *canonical system* can be replaced (Garabedian, 1964) by

$$u_\xi + w_\xi = 0, \quad u_\eta + w_\eta = 0$$

where $w(\rho) = \int \frac{c}{\rho} d\rho = \frac{2c}{\gamma - 1}$, which on integration yields

$$u + w = R^+(\eta)$$

$$u - w = R^-(\xi).$$

R^- and R^+ given by

$$R^- = u - \frac{2c}{\gamma-1}, \quad \text{and} \quad R^+ = u + \frac{2c}{\gamma-1} \tag{7.78}$$

are functions of ξ and η respectively and are known as *Riemann invariants*. For one dimensional flow normal to the boundary the Riemann invariants of incoming and outgoing characteristics are defined respectively as

$$R^- = q_n - \frac{2c}{\gamma-1} \quad \text{and} \quad R^+ = q_n + \frac{2c}{\gamma-1}, \tag{7.79}$$

where q_n is the normal velocity and c is the speed of sound. The invariants R^- and R^+ can be calculated using free–stream conditions and by extrapolation from the interior of the field respectively. Thus

$$R^-_\infty = q_{n\infty} - \frac{2c_\infty}{\gamma-1} \tag{7.80}$$

$$R^+_e = q_{ne} + \frac{2c_e}{\gamma-1}. \tag{7.81}$$

Boundary values for the normal velocity component and speed of sound are constructed by adding and subtracting these invariants. At an inflow boundary, the velocity components tangential to the boundary, and the entropy are prescribed from their free stream values, whereas at an outflow boundary they are extrapolated from the interior.

The far field conditions discussed above *assume zero circulation*. Therefore, for the computation of lifting flows the far field boundary should be placed sufficiently far away from the aerofoil so that the flow field remains undisturbed. In order to reduce the extent of the far field boundary the far field effect of a single vortex in compressible medium centered at the aerofoil can be added to free stream flow. Using the circulation, Γ, due to a vortex at the aerofoil (obtained from the Kutta–Joukowsky theorem) in the expressions for the velocity components at the far field one gets the modified free stream velocities referred above in Eq. (7.80). The modified free-stream pressure and density can be obtained by using the steady state Bernoulli equation and the equation of state. Subsequently, free stream entropy and velocity of sound can be calculated.

7.10 THE FULL POTENTIAL EQUATION

The Euler equations (7.68), obtained using the assumption of inviscid flow, may be further simplified by *assuming the flow to be irrotational*. Such flows are known as *potential flows*. (In the literature this is commonly called the full-potential equation, with a view to distinguish it from the small-perturbation equations). The inaccuracy introduced through the assumption of irrotationality (which is justified only for weak normal shocks) in potential flow is often small and has been accepted in order to save computer storage and time that would otherwise be required to obtain solution of the more accurate, but computationally more difficult, Euler and Navier–Stokes equations.

A flow field is said to be *irrotational* if the vorticity vector curl$\vec{V}$ vanishes everywhere in the flow field. For an irrotational flow, there exists a potential function ϕ such that

$$\vec{V} = \text{grad}\,\phi. \tag{7.82}$$

If we now assume that the specific entropy s maintains the same constant value in the entire upstream region upto the appearance of the first shock and call it *isentropic* flow, the equation of state takes the simpler form

$$p/\rho^{\gamma} = \text{const.} \tag{7.83}$$

where γ is the ratio of specific heats of the gas, assumed to be a constant quantity. For isentropic flow the speed of sound c is defined by

$$c^2 = \gamma\frac{p}{\rho}. \tag{7.84}$$

Eliminating the pressure and density from the Euler equations, we obtain the basic gas dynamic equations for isentropic flow

$$\frac{dc^2}{dt} = -(\gamma - 1)c^2 \text{ div } \vec{V}$$
$$\frac{d\vec{V}}{dt} = -\frac{1}{\gamma - 1} \text{ grad } c^2, \tag{7.85}$$

which is a system of quasilinear equations of hyperbolic type for the unknowns c^2 and $\vec{V}$. The above system delivers the well known *Helmholtz equation* for the vorticity vector $\vec{\omega}$ $(= \frac{1}{2} \text{ curl } \vec{V})$,

$$\frac{d\vec{\omega}}{dt} = (\vec{\omega} \cdot \nabla)\vec{V} - \vec{\omega}(\nabla\vec{V}) \tag{7.86}$$

valid for isentropic flow. The *circulation* Γ round any closed curve C consisting of fluid particles is defined as

$$\Gamma = \oint_C (u\ dx + v\ dy + w\ dz), \tag{7.87}$$

where x, y, z are rectangular Cartesian coordinate directions and $\vec{V} = (u, v, w)$. By Stokes's theorem, the vorticity vector is related to the circulation

$$\Gamma = \iint_S \text{curl } \vec{V} \cdot d\vec{S}, \tag{7.88}$$

where S denotes a surface having C as boundary curve and the vector $d\vec{S}$ denotes a directed surface element of S. For isentropic flows, it follows from the Euler equations, the well known *Kelvin's theorem* that the circulation around any fluid curve does not change with time, as it moves with the fluid. This implies that if the circulation vanishes at any time, it will remain so for all time as the curve moves. In aerodynamic applications, most frequently we come across the case where the free–stream condition is a uniform state; so, it follows that *an isentropic flow with uniform free-stream condition must be irrotational.*

For steady isentropic flow, Eqs. (7.85) become

$$\vec{V} \cdot \nabla c^2 + (\gamma - 1)c^2 \nabla \cdot \vec{V} = 0 \tag{7.89}$$

and

$$(\vec{V}\nabla)\vec{V} + \frac{1}{\gamma - 1}\nabla c^2 = 0 \tag{7.90}$$

Expanding the vector $(\vec{V}\cdot\nabla)\vec{V}$ and forming scalar product with $\vec{V}$, it follows from Eq. (7.90) that

$$\vec{V}\nabla(\frac{1}{2}V^2) = -\frac{1}{\gamma-1}\vec{V}\nabla c^2. \tag{7.91}$$

Integrating (7.91) along a streamline we get *Bernoulli's equation*

$$V^2 + \frac{2}{\gamma-1}c^2 = \text{const.} = V_\infty^2 + \frac{2}{\gamma-1}c_\infty^2, \tag{7.92}$$

suffix ∞ denoting free-stream conditions. From the above equation we can determine the speed of sound, c as

$$c^2 = \frac{\gamma-1}{2}(V_\infty^2 - V^2) + c_\infty^2 \tag{7.93}$$

which serves to calculate the fluid pressure p and the density ρ. Equations (7.90) and (7.92) deliver the *basic gasdynamic equation for steady isentropic flow*

$$c^2\nabla\cdot\vec{V} = \vec{V}\cdot\nabla\left(\frac{1}{2}V^2\right). \tag{7.94}$$

A second basic equation for such flow can be obtained from Eq. (7.90) by eliminating the speed of sound by (7.92)

$$(\vec{V}\cdot\nabla)\vec{V} = \nabla(\frac{1}{2}V^2),$$

which on simplification may be put to the form

$$\vec{V}\times\text{curl}\,\vec{V} = 0, \tag{7.95}$$

so that the irrotationality condition is regained, apart from the exceptional case where the flow velocity $\vec{V}$ is parallel to the vorticity vector curl$\vec{V}$.

Instead of the isentropic assumption one can assume an *isoenergetic flow*, where the stagnation enthalpy is constant over the whole flow field. *The stagnation enthalpy does not change across shocks.* Then, one gets a second basic equation (Niyogi, 1982; Oswatitsch, 1977),

$$\vec{V}\times\text{curl}\,\vec{V} = -T\,\text{grad}\,s, \tag{7.96}$$

instead of the irrotationality condition. This is known as the *Crocco's vortex theorem*, where T is the absolute temperature and s is the entropy. Crocco's vortex theorem (7.96), implies that *a flow field with variable entropy is necessarily rotational.*

If, however, the shock strength is constant, then the discontinuity of entropy across it is also constant, so that the flow after it is isentropic, i.e. grad $s = 0$. It then follows from Crocco's vortex theorem Eq. (7.96) that either the flow is irrotational or the vectors $\vec{q}$ and curl $\vec{q}$ are parallel everywhere. The second alternative is however impossible, since at the shock wave $\vec{q}$ has a nonzero normal component, but the normal component of curl $\vec{q}$ is zero, since it is given by the tangential derivatives of the tangential velocity components which are continuous across a shock.

On the other hand, *if a curved shock is present, the flow field after the shock can no more be irrotational.* However, for weak shocks, the entropy change across the shock is of third order in terms of the shock strength and it has been established by Guderly (1957), Cole and Messiter (1957), Cole (1975), (Oswatitsch, 1956) that, neglecting higher order quantities, the irrotationality condition is valid in the transonic range even in the flow field after a curved shock and the entire flow field may be treated as isentropic. This approximation is used in the *TSP model,* and will be discussed later in Section 7.10.3.

For irrotational flow, where curl $\vec{V} = 0$, there exists a potential function ϕ, such that $\vec{V}$ = grad ϕ. Then the gasdynamic equation leads to

$$c^2\nabla^2\phi = \vec{\nabla}\phi \cdot \vec{V}(\frac{1}{2}V^2). \tag{7.97}$$

Equation (7.97) is a quasilinear equation having characteristic polynomial

$$(c^2)^2(c^2 - V^2) = \left(c_\infty^2 + \frac{\gamma - 1}{2}\left[V_\infty^2 - (\nabla\phi)^2\right]\right)^2 \left(c_\infty^2 + \frac{\gamma - 1}{2}V_\infty^2 - \frac{\gamma + 1}{2}(\nabla\phi)^2\right). \tag{7.98}$$

Thus, Eq. (7.97) is of elliptic type for $V < c$, parabolic for $V = c$ and hyperbolic for $V > c$. In terms of rectangular Cartesian coordinates x, y, z, Eq. (7.97) becomes

$$\left(c^2 - \phi_x^2\right)\phi_{xx} + \left(c^2 - \phi_y^2\right)\phi_{yy} + \left(c^2 - \phi_z^2\right)\phi_{zz} - 2\phi_x\phi_y\phi_{xy} - 2\phi_y\phi_z\phi_{yz} - 2\phi_z\phi_x\phi_{zx} = 0. \tag{7.99}$$

The full potential equation (7.99) is a *quasilinear and mixed type partial differential equation.* It is in *quasi-linear, i.e. non-conservative form.* The conservative form is discussed in the next subsection. One advantage of this equation over the Euler equations is that it involves only one unknown, i.e. ϕ, so that its solution is expected to save much computational effort. Many numerical techniques used currently are the outcome of efficient and accurate numerical solution of this equation, particularly in the case of transonic flow with shocks, developed during the seventies. The numerical solution procedures for this equation has reached a mature stage and a robust and reliable solution may be obtained for transonic flow past airfoils, wings and even for complete aircrafts (Chakrabarty, 1990, 1992).

7.10.1 Potential Equation in Conservative Form

For steady, inviscid, two-dimensional irrotational flow, the gasdynamic equation (7.99) reduces to the quasilinear equation (Jameson, 1978; Niyogi, 1977; Oswatitsch, 1956)

$$\left(c^2 - u^2\right)\phi_{xx} - 2uv\phi_{xy} + \left(c^2 - v^2\right)\phi_{yy} = 0, \tag{7.100}$$

In terms of the the stagnation sound speed c_0, the local sound speed c can be determined from the energy relation

$$c^2 = c_0^2 - \frac{\gamma - 1}{2}v^2. \tag{7.101}$$

It may be noted that the *energy relation is valid across a shock.* Consequently, the stagnation sound speed does not change across a shock. Further, the density ρ and the pressure p can be determined by the relations

$$\rho^{\gamma-1} = M_\infty^2 c^2 \tag{7.102}$$

and

$$p = \frac{\rho^\gamma}{\gamma M_\infty^2}. \tag{7.103}$$

For a continuous flow, i.e. when no discontinuities appear, Eq. (7.100) implies the conservation of both mass and momentum, as may be verified by multiplying it by ρ/c^2 and simplifying it when it reduces to the equation for conservation of mass :

$$\frac{\partial}{\partial x}(\rho u) + \frac{\partial}{\partial y}(\rho v) = 0. \tag{7.104}$$

Again, on multiplying Eq. (7.100)) by $\rho u/c^2$ or by $\rho v/c^2$, it can be reduced to the equations for conservation of the x or y components of momentum respectively.

Continuous transonic flows are known to exist only in special cases. In general, shock waves appear. Supersonic flows, with embedded subsonic regions, are also accompanied by shock waves. Thus, one must admit weak solutions with suitable discontinuities. Now since, according to Crocco's vortex theorem Eq. (7.96) irrotational flow is isentropic, it is consistent to replace shock waves by discontinuities across which entropy is conserved. However, in an isentropic flow in which the energy is conserved, it is not possible to conserve both mass and momentum across a discontinuity. Most potential flow models prefer mass conservation, and this yields relatively good approximation to shock waves of moderate strength. The corresponding momentum deficiency then yields an approximation to the wave drag (Steger and Baldwin, 1972). Thus, we look for a solution ϕ, which is such that:

1. ϕ is continuous everywhere in the flow field;
2. the velocity components are piecewise continuous, and satisfy the conservation law Eq. (7.104) at points where the flow is continuous;
3. the jump condition

$$[\rho v] - \frac{dy}{dx}[\rho u] = 0, \tag{7.105}$$

is satisfied across a discontinuity, where $[\rho v]$ denotes the jump and $\frac{dy}{dx}$ is the slope of the discontinuity.

In other words, ϕ should be a weak solution (see Chapter 1, Section 9; and also Lax (1954, 1972)), of the conservation law Eq. (7.104) satisfying the condition

$$\iint \rho \left(\phi_x w_x + \phi_y w_y\right) \, \mathrm{d}x\mathrm{d}y = 0 \tag{7.106}$$

for any smooth test function w, which vanishes in the far-field.

Under the assumption of small perturbation, the potential equation may be simplified further, as discussed in a later section on the TSP model.

7.10.2 Boundary Conditions for the Full Potential Equation

The full potential equation must be solved subject to the following boundary conditions :

1. On a rigid surface, the flow tangency condition, i.e., the physical condition of zero normal flow through the surface, is enforced.
2. Theoretically, the flow approaches a uniform free-stream at infinity. However, numerical computation of flow exterior to a body requires a bounded domain. Hence, artificial far-field boundaries have to be introduced to limit the area of computation. Boundary conditions have to be specified at these artificial boundaries, so that a well-posed problem for the partial differential equation is constituted. Care should also be taken that the conditions at these artificial boundaries affect the solution in a manner such that it closely approximates the free space solution that exists in the absence of these boundaries. For potential flow computation, the velocity potential and density at the far field boundary are kept fixed at the initial free-stream values. For bodies that produce lift, the potential at the outer boundary is updated by the value of that due to a vortex of prescribed circulation or lift.

7.10.3 Transonic Small Perturbation Model

This model assumes that:

1. the body produces small perturbation of the flow field and
2. the local flow speed is near the local speed of sound, everywhere in the flow field.

Then the steady potential flow model, discussed above, may be considerably simplified. Several authors carried out such simplifications, the earliest one being that due to Oswatitsch (1956, 1977), Niyogi (1982). We present briefly the ideas involved.

A steady inviscid transonic flow past a thin profile is considered. Then, $\left|\frac{v}{c}\right| \ll 1$ and Eq. (7.100) reduces to

$$(1 - \frac{u^2}{c^2})\frac{\partial u}{\partial x} + \frac{\partial v}{\partial y} = 0. \tag{7.107}$$

Let us introduce in it the perturbation quantities,

$$u' = \frac{u - u_\infty}{u_\infty}, \quad v' = \frac{v - v_\infty}{u_\infty}, \tag{7.108}$$

where u', v' are small quantities such that second and higher powers of u', v' are negligible. Then, using Bernoulli's equation (7.93) we get, neglecting second and higher order terms and simplifying, from Eq. (7.107)

$$\left[1 - M_\infty^2 - \bar{K}u'\right]\frac{\partial u'}{\partial x} + \frac{\partial v'}{\partial y} = 0, \tag{7.109}$$

where $\bar{K}$ is a function of the Mach number, given by

$$\bar{K} = M_\infty^2\left[2 + (\gamma - 1)M_\infty^2\right]. \tag{7.110}$$

Approximate representations of $\bar{K}$ in use, due respectively to Spreiter and Oswatitsch are

$$\bar{K} = (\gamma + 1)M_\infty^2, \text{ and } \bar{K} = \frac{(1 - M_\infty^2)}{(\frac{1}{M_\infty^*} - 1)}, \quad M_\infty^* = \frac{u_\infty}{c^*}, \tag{7.111}$$

where c^* denotes the critical speed of sound. The state corresponding to which the flow speed is equal to the sound speed is called the *critical state*. It is to be noted that, in the transonic range $(1 - M_\infty^2) \sim u'$, which is a small quantity and both the terms in the square bracket in Eq. (7.109) are to be retained, being of the same order.

7.10.4 Oswatitsch Reduction

Oswatitsch (1950, 1977) introduced the following reduced variables in his TSP model :

$$M_\infty < 1, \quad X = x, \quad Y = y\sqrt{1 - M_\infty^2},$$

$$\Phi = \frac{\overline{K}}{1 - M_\infty^2}\left[\phi - u_\infty - v_\infty y\right],$$

$$\Phi_X = U = \frac{\overline{K}}{1 - M_\infty^2}\frac{u - u_\infty}{u_\infty}, \quad \Phi_Y = V = \frac{\overline{K}}{(1 - M_\infty^2)^{\frac{3}{2}}}\frac{v - v_\infty}{u_\infty}, \tag{7.112}$$

where the reduced quantities are denoted by the capital letters and the corresponding lower case letters denote their true values. Using Eqs. (7.112) and (7.109) changes over to the following nonconservative and conservative forms, respectively

$$(1 - U)\frac{\partial U}{\partial X} + \frac{\partial V}{\partial Y} = 0, \quad \text{or,} \quad \frac{\partial}{\partial X}(U - \frac{1}{2}U^2) + \frac{\partial V}{\partial Y} = 0. \tag{7.113}$$

In terms of the reduced perturbation potential Φ, it may be expressed in conservation form as

$$\frac{\partial}{\partial X}(\Phi_X - \frac{1}{2}\Phi_X^2) + \frac{\partial}{\partial Y}\Phi_Y = 0. \tag{7.114}$$

It may be noted that the small perturbation Eqs. (7.113) or (7.114) are quasi–linear and of mixed elliptic–hyperbolic type.

The tangency boundary condition at the surface of the profile, may be shifted to the body axis $Y = 0$, (Niyogi, 1982; Oswatitsch, 1977).

7.10.5 Cole's and Other Forms of the TSP Equation

Various alternative forms of the TSP equation, other than Eq. (7.109) or Eq. (7.113), may be found in the literature. Their accuracy and applicability have been discussed in (Vander Vooren et al., 1976). The form derived by Cole (1975) and used later by many authors, results from an asymptotic expansion and involves proceeding to the double limits, viz. the free-stream Mach number $M_\infty \to 1$ and *the thickness-ratio (ratio of the maximum thickness to chord-length)* $\tau \to 0$ in the Euler equations, (Ashley and Landahl, 1965; Chakrabartty and Subramanian, 1980; Murman and Cole, 1971). It is found to be a singular perturbation problem. Let us define a parameter K, as

$$\lim_{\substack{M_\infty \to 1, \\ \tau \to 0}} \frac{1 - M_\infty^2}{\tau^{\frac{2}{3}}} = K = \text{ a finite quantity.} \tag{7.115}$$

The quantity K is a *transonic similarity parameter*. A coordinate transformation $\overline{y} = \tau^{1/3}y$ is used to take care of the large lateral extent in transonic flow. It may be noted that the definitions of K and $\overline{y}$ are not unique, and may be multiplied by functions like $f(M_\infty^2) = O(1)$, with $f(1) = 1$. Further, noting that, $(1 - M_\infty^2) \sim O(\tau^{\frac{2}{3}})$, the following series expansions for the velocity components, pressure and density:

$$
\begin{aligned}
\frac{u}{u_\infty} &= 1 + \tau^{\frac{2}{3}}\overline{u} + \tau^{\frac{4}{3}}u_2 + \cdots, \\
\frac{v}{u_\infty} &= \tau\overline{v} + \tau^{\frac{5}{3}}v_2 + \cdots, \\
\frac{p}{p_\infty} &= 1 + \tau^{\frac{2}{3}}\overline{p} + \tau^{\frac{4}{3}}p_2 + \cdots, \\
\frac{\rho}{\rho_\infty} &= 1 + \tau^{\frac{2}{3}}\sigma + \tau^{\frac{4}{3}}\sigma_2 + \cdots,
\end{aligned} \tag{7.116}
$$

are substituted in the Euler equations. On simplification, the first order terms yield *the TSP equations in Cole's formulation,* as

$$
\begin{aligned}
\frac{\partial}{\partial x}\left[K\overline{u} - \frac{\gamma+1}{2}\overline{u}^2\right] + \frac{\partial \overline{v}}{\partial \overline{y}} &= 0, \\
\frac{\partial \overline{u}}{\partial \overline{y}} - \frac{\partial \overline{v}}{\partial x} &= 0.
\end{aligned} \tag{7.117}
$$

The flow is irrotational (in fact upto second order), and introducing the perturbation potential ϕ by

$$\phi_x = \overline{u}, \quad \phi_{\overline{y}} = \overline{v}, \tag{7.118}$$

yields the *TSP equation in conservation form*

$$\frac{\partial}{\partial x}\left[K\phi_x - \frac{\gamma+1}{2}\phi_x^2\right] + \frac{\partial}{\partial \overline{y}}\phi_{\overline{y}} = 0. \tag{7.119}$$

The flow is subsonic at a point for $\phi_x < \frac{K}{\gamma+1}$ where Eq. (7.119) is of elliptic type, while it is supersonic, the equation being of hyperbolic type, at a point for $\phi_x > \frac{K}{\gamma+1}$. The shock jump relations are contained in the TSP equations in the sense that weak solutions to Eqs. (7.117) yield a consistent approximation to the Rankine–Hugoniot shock jump relations (Bailey, 1975; Cole, 1975).

For flow past a thin profile , *the tangency boundary condition on the surface of the profile may be shifted, in this approximation, to the profile axis* $\overline{y} = 0$. We get

$$\phi_{\overline{y}}(x, 0+) = F'_+(x) - \frac{\alpha}{\tau}, \quad \phi_{\overline{y}}(x, 0-) = F'_-(x) - \frac{\alpha}{\tau}, \tag{7.120}$$

where $y = \tau F_\pm(x)$, $\quad 0 \le x \le 1$, is the profile shape and α is the angle of incidence of the profile. For lifting airfoils, the Kutta condition is satisfied by requiring that the pressure ϕ_x be continuous across the line $\overline{y} = 0, \quad x \ge 1$, and the flow angle $\phi_{\overline{y}}$

be continuous across $\overline{y} = 0, \quad x \geq 1$. The perturbation potential is made single–valued by introducing a cut along $\overline{y} = 0$ across which the ϕ jumps by an amount Γ, where the circulation Γ is defined in terms of line integral over the airfoil,

$$\Gamma = -\oint d\phi, \tag{7.121}$$

For the exterior problem, a far-field boundary condition, viz. vanishing of perturbation velocities at infinity, is needed, together with the condition on the perturbation potential

$$\phi = \frac{\Gamma}{2\pi}\theta, \tag{7.122}$$

where θ denotes the angle between the position vector and the positive direction of x-axis.

Computation with this model has been presented in Chapter 6.

Making approximations at the functional level and using the concept of weak solution, Chakrabartty and Subramanian (1980) formulated the small perturbation equation and the boundary conditions for moderate aspect ratio finite wings for the unsteady case. For three dimensional steady case this equation can be written as

$$(1 - M_\infty^2)\phi_{xx} - \lambda\phi_x\phi_{xx} - 2M_\infty^2\phi_y\phi_{xy} - 2M_\infty^2\phi_z\phi_{xz} + (1 - M_\infty^2\phi_x)\phi_{yy} + (1 - M_\infty^2\phi_x)\phi_{zz} = 0. \tag{7.123}$$

where ϕ in Eq. (7.123) is the perturbation velocity potential defined as

$$\vec{V} = (u, v, w), \quad u = U_\infty + \phi_x, \quad v = \phi_y \quad \text{and} \quad w = \phi_z \tag{7.124}$$

and the parameter λ can be defined in many ways according to the perturbation procedure and one such is (Chakrabarty and Subramanian, 1980)

$$\lambda = \left[(\gamma + 1)M_\infty^2 + 3(1 - M_\infty^2)\right] M_\infty^2. \tag{7.125}$$

Equation (7.123) is also a quasilinear equation of mixed type. One advantage of using this equation is that the boundary condition on the solid surface may be linearized by satisfying it on the planform instead of the actual surface of the thin body.

7.11 INVISCID INCOMPRESSIBLE IRROTATIONAL FLOW

In this model, the density is assumed to be a constant quantity and viscosity is assumed to be zero. The continuity equation then delivers

$$\text{div } \vec{V} = 0.$$

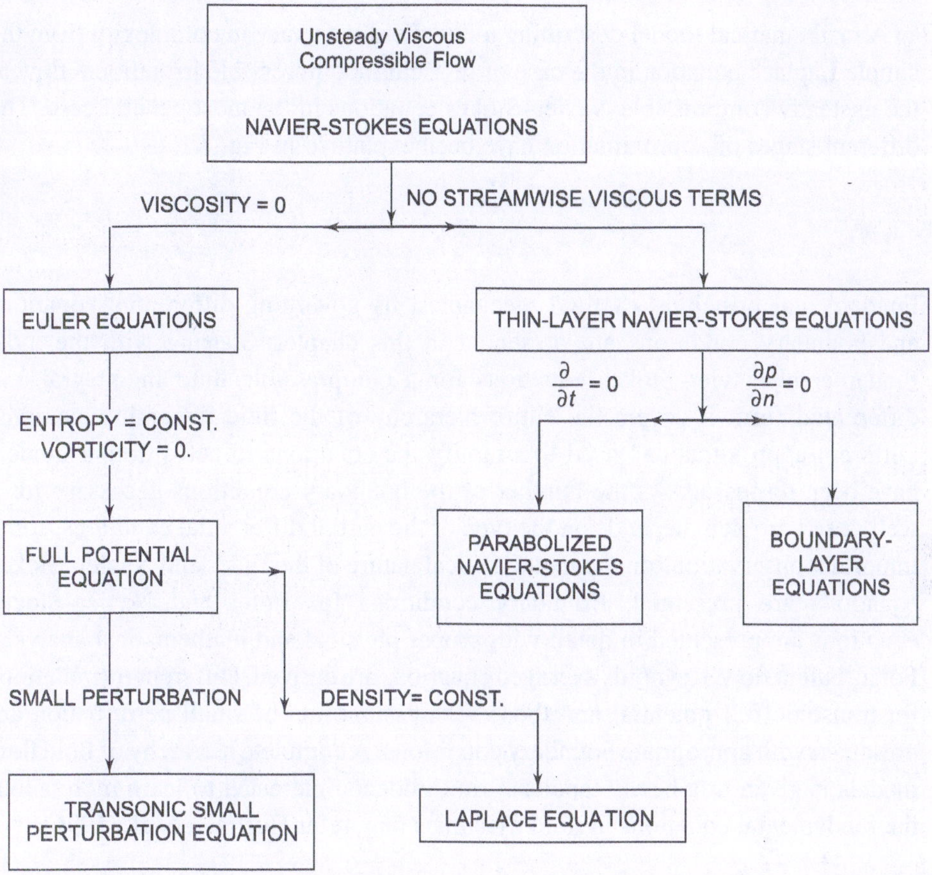

Figure 7.1 Hierarchy of fluid flow models.

In view of the irrotationality assumption, a velocity potential ϕ exists such that $\vec{V} = \text{grad}\,\phi$. Consequently the continuity equation reduces to Laplace equation

$$\nabla^2\phi = 0. \tag{7.126}$$

At a solid boundary, the condition of no normal flow relative to the boundary is imposed. For the exterior problem of flow past a body, a regularity condition, like the vanishing of perturbation velocities far away from the body is required to be satisfied. The pressure may be determined from Bernoulli's equation, as discussed in detail in Chapter 9, where relevant boundary conditions are also discussed.

A mathematical model describing a flow field may vary in complexity from the simple Laplace equation in the case of inviscid incompressible irrotational flow to the unsteady compressible Navier–Stokes equations in the most general case. The different stages of approximation have been explained in Fig. 7.1.

7.12 SUMMARY

Fundamental principles of fluid mechanics, its governing differential equations and boundary conditions are presented in this chapter. Starting with the most fundamental Navier–Stokes equations for a compressible fluid in integral and differential form it covers the entire hierarchy of the fluid dynamic equations. Different approximations used to simplify the equations to get different models have been discussed. As the number of the boundary conditions necessary for a well posed problem depends on the type of the partial differential equations, some important observations on the mathematical nature of the Euler and Navier–Stokes equations are presented. Boundary conditions for Euler and Navier–Stokes equations are presented in detail with proper physical and mathematical analysis. For turbulent flow, Reynolds averaged equations are derived. Different formulations for transonic full potential and those with assumption of small perturbation are presented with appropriate boundary conditions. A complete hierarchy of fluid flow models is given in a handy tabular form. Students interested to learn more about the fundamental equations in fluid dynamics may refer Hoffman et al. (1996).

7.13 KEY TERMS

Bernaolli's equation
Boundary layer
Canonical system
Circulation
Conservative form of
Partial differential equations
Contact discontinuity
Crocco's vortex theorem
Dynamic similarity
Euler equations
Gasdynamic equations

Navier–Stokes equations
Newtonian fluids
Non-reflecting boundary conditions
Parabolized Navier–Stokes equations
Potential flow
Prandtl number,
Reynolds Averaged Navier–Stokes (RANS) equations
Reynolds number
Reynolds stresses
Riemann invariants

Helmholtz equation
Inertial force
Irrotational flow
Isentropic flow
Isoenergetic flow
Jacobian flux matrix
Kelvin's theorem
Kinematic viscosity
Laminar flow
Mean-free-path
Separated flow
Shock waves
Stagnation enthalpy
Stokes flow
Thin-layer Navier–Stokes equations
Transonic similarity parameter
Transonic small perturbation equation
Turbulent flow
Viscous force
Weak solutions.

8

Grid Generation

Hierarchy of governing differential equations of fluid mechanics and boundary conditions have been discussed in the previous chapter. The choice of the governing differential equations depends on the complexity of the problem and the requirement of the user. Once the governing equations to be solved to suit the problem of interest are fixed, the next step involved is called the *pre-processing*. In the first part of this step, usually called the *geometrical data processing*, a comprehensive geometry in a global coordinate system is obtained from the component-wise data or a drawing sheet supplied by the user. Next is the discretization of the flow domain called the grid generation. As an example of solving elliptic type differential equations a simple grid generation problem has been introduced in chapter-5. Different grid generation methods will be discussed here in detail.

8.1 INTRODUCTION

Exact satisfaction of the wall boundary condition is of crucial importance in solving fluid flow problems using numerical methods. This motivates one to transform the governing equations of fluid dynamics from physical to a curvilinear coordinate system in the computational domain. For a two dimensional problem of flow past a single body, the physical (say, Cartesian) (x, y) coordinate system is mapped into a rectangle in the computational (ξ, η) coordinate system such that the body forms one of the coordinate lines (Fig. 8.1). This is known as generation of body-fitted coordinate system. Here, the body surface ABC, the outer boundary DEF and the cut lines FA and CD in the physical plane are mapped into the coordinate lines $\eta = \eta_{\min}$ (ABC), $\eta = \eta_{\max}$ (DEF), $\xi = \xi_{\min}$ (FA) and $\xi = \xi_{\max}$ (CD) lines respectively in the computational plane. Accuracy of the numerical methods also improves if the fluid dynamics equations are solved in the computational plane using uniform mesh.

Generation of grids is an integral part of any type of numerical solution of either integral or differential equations and is as old as the origin of the subject numerical analysis or numerical methods. Even for evaluating an integral numerically, we have to divide the interval into a finite number of sub-intervals which may be considered as a kind of grid generation. The terms *grid* and *mesh* have identical meaning and are used here interchangeably. Let us consider a problem to be solved numerically over a finite domain Ω. Then a set of uniquely identifiable infinitesimal domains $\omega_{i,j,k}$ for $i = 1, \cdots, I$, $j = 1, \cdots, J$ and $k = 1, \cdots, K$ will form a grid in three dimensional space if their union $\cup\omega_{ijk} = \Omega$ and intersection $\cap\omega_{ijk} = 0$ for all i, j, k. These infinitesimal domains are called *intervals*, *cells* or *volumes* for one, two and three dimensional cases respectively. The topology of the grids can be looked at in two different ways, (i) *structured* and (ii) *unstructured*. In case of structured grids, it is possible to define a curvilinear co-ordinate system that spans over the entire domain and hence the connectivity of the individual sub-domains is implicit. Whereas for unstructured grid, the connectivity of the individual sub-domains must be explicitly specified. In this case one can not span the entire domain using the coordinate directions but the infinitesimal domains can be identified individually such that $\cup\omega_i = \Omega$ and $\omega_i \cap \omega_j = 0$ for all $i \neq j$. Unstructured grids provide maximum flexibility in controlling the discretization of the domain for complex geometry and is widely used in finite element and finite volume methods. A sample unstructured grid with triangular elements over an aerofoil is shown in Fig. 8.2. To write a computer code to generate the unstructured grid for complex geometry with proper book-keeping for the connectivity of the grid cells is difficult and numerical algorithms of the flow-solver codes which use unstructured grids require more computational power as compared to their structured counterparts.

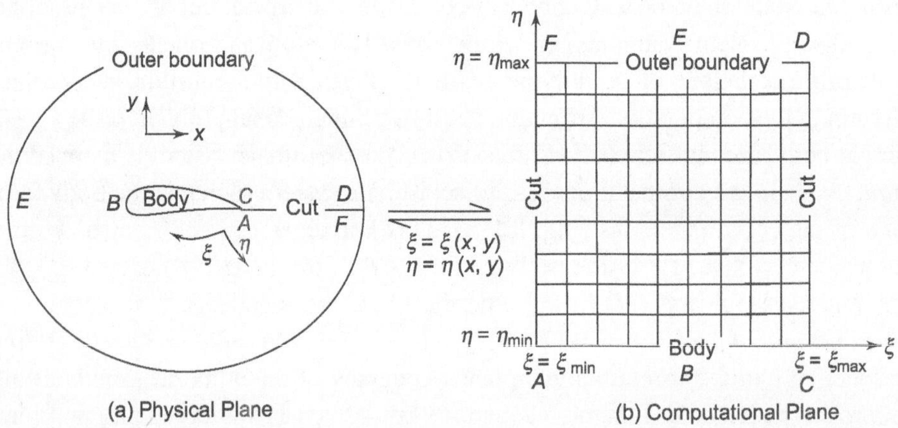

Figure 8.1 Transformation from physical plane to computational plane.

For realistic geometries, consisting of a single body or multiple bodies, the generation of a suitable computational grid is a major task, and it is generally not possible to generate a structured single-block grid. For two dimensional problems the flow field in the physical space is a multiply connected domain which can be mapped uniquely into a simply connected computational domain by introducing a cut. Whereas for a three dimensional flow field surrounding a closed body is simply connected and its mapping to a single computational domain is singular. It is similar to the classical topological *globe to map* problem. Hence a common approach is the use of *structured multi-block* grids, where the physical space is divided into a number of smaller blocks and a structured grid is generated in each block. However, numbering of the blocks can be unstructured. So, in our notation Ω can be divided into a number of smaller blocks Ω_l such that $\Omega = \cup\Omega_l$ and $\Omega_l \cap \Omega_m = 0$ for all $l \neq m$ if there is no overlapping. Each $\Omega_l = \cup\omega_{ijk}$ and $\cap\omega_{ijk} = 0$ for all i, j, k. This approach is very popular to generate extremely complex grids for complex geometries like multi-element aerofoils, a complete aircraft etc. (Chakrabartty et al., 2003a, 2003b; Mathur et al., 2003; Weatherill, 1990)

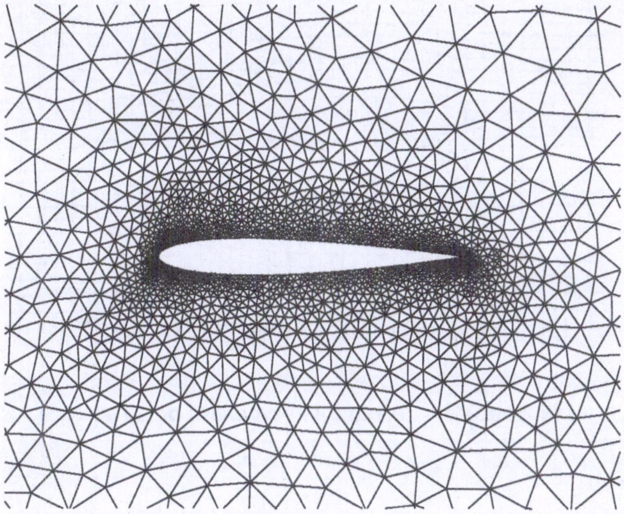

Figure 8.2 A typical unstructured grid on an aerofoil (courtesy: Dr. J. S. Mathur).

In two dimensions, the structured grids can be of H-type, O-type or C-type. These three types of grids with quadrilateral elements are shown in Figs. 8.3, 8.4 and 8.5, respectively. In three dimensions, grids may be a combination of these three types as viewed on three different coordinate planes. Usually, the type is designated as viewed in $z = constant$ and $x = constant$ planes only. Sample O-H and C-H type grids over an ONERA-M6 wing along with the coordinate directions are shown in Figs. 8.6 and 8.7 respectively. We can see that on $z = constant$ planes they look like O-type and C-type respectively but on $x = constant$ planes both look like H-type.

The construction of a suitable grid becomes an important part of the solution procedure. The quality of the grid used in the computations directly influences the solution obtained. An overall structured grid can be of a *single structured grid, a patched structured subgrids (multi-block)* or *an overlapping structured subgrids* (*chimera type*). For complex geometries the domain is divided into subdomains where one can generate structured grids and join either by patching, where there are common boundaries or by overlapping, where the intersection of two subdomains also forms another subdomain.

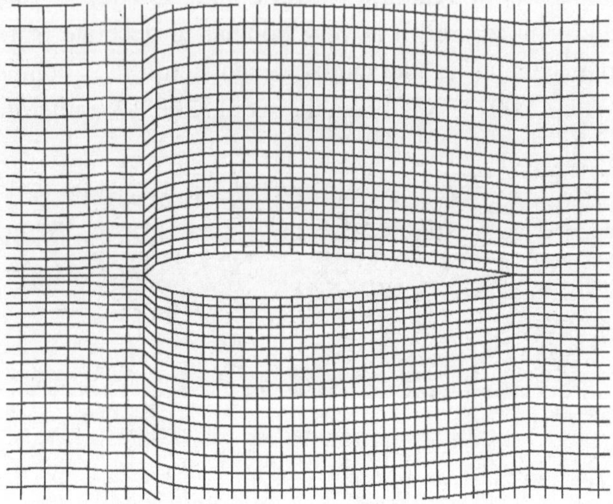

Figure 8.3 A typical H-type grid on an aerofoil.

A vast amount of literature is available to deal with the numerical generation of grids. Among them the most popular methods are (i) *Differential equation methods*

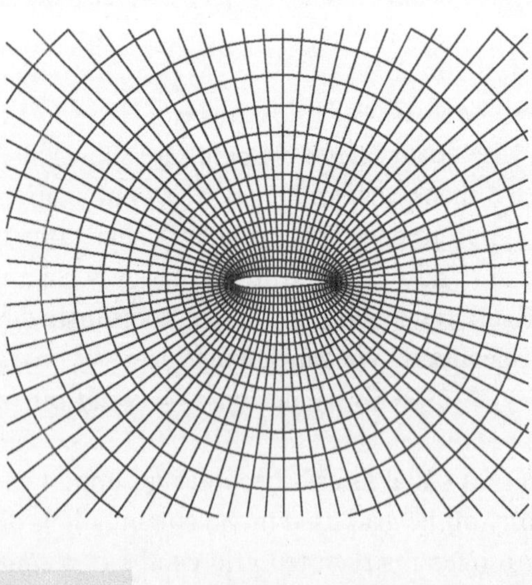

Figure 8.4 A typical O-type grid on an aerofoil.

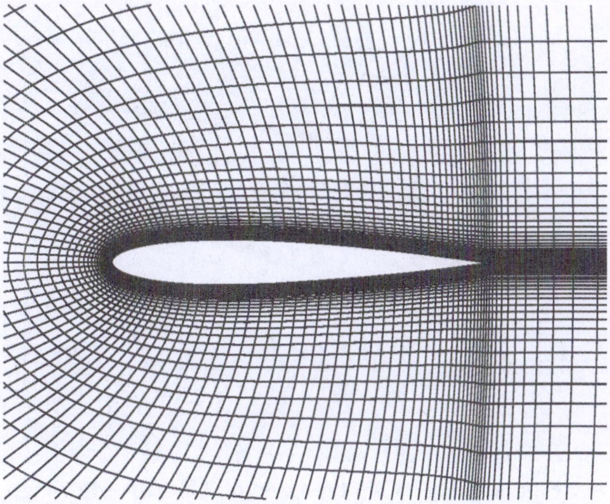

Figure 8.5 A typical C-type grid on an aerofoil.

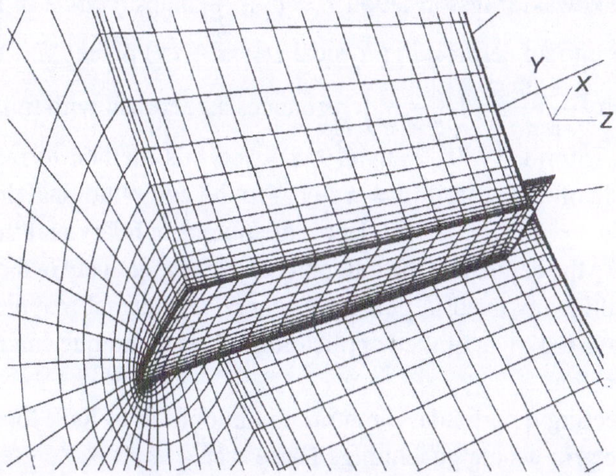

Figure 8.6 O-H type grid on ONERA-M6 wing.

(by solving elliptic partial differential equations) (ii) *Algebraic methods* and (iii) *Transfinite interpolation methods*. Apart from these three, another method using analytical conformal mapping functions based on the theory of complex

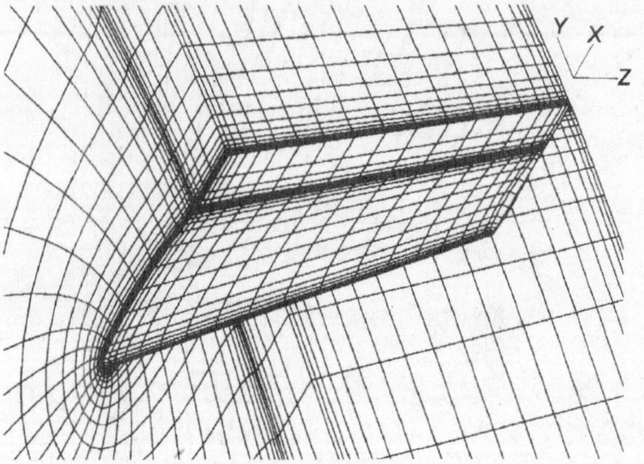

Figure 8.7 C-H type grid on ONERA-M6 wing.

variables has also been used to get a body-fitted grid for simple geometries. For examples: Joukowski transformation $z = \zeta + \frac{c^2}{\zeta}$ maps a circle of radius c in $\zeta(= \xi + i\eta)$ plane into an aerofoil in physical $z(= x + iy)$ plane, Von Karman-Trefftz transformation $\frac{z - nc}{z + nc} = \left(\frac{\zeta - c}{\zeta + c}\right)^n$ produces an aerofoil with trailing edge angle $\pi(2 - n)$ etc. (Smith, 1980). Sells (1968) used this method to map the physical domain outside of an aerofoil ($z = x + iy$) to the computational domain inside of a unit circle ($\sigma = re^{i\theta}$) (Fig. 8.8). Here, the outer boundary and the aerofoil have been mapped to the centre and the unit outer circle respectively in the computational plane. Difficulties involved in getting a general conformal mapping for complex geometries have led to its limited applications. This technique can not be extended to three dimensions.

Before selecting a particular method, it is important to look for some desirable features in the grid we are generating. These are:

1. *Orthogonality*—this is desirable at least near the boundaries.
2. *Control of spacing*—this is a must in order to resolve the flow in a better way, particularly near the solid wall boundary, viscous boundary layer, wake, shock wave etc., where there are large gradients of flow parameters.
3. *Skewness*—this should be minimized in order to get better accuracy of the numerical schemes to be employed for solving fluid dynamic equations and in order to match/overlap various regions of the flow domains.

Apart from these, very high or very low *aspect-ratio* (ratio of the base to height of a cell) of the grid cells and highly stretched grids should be avoided.

Before we go for different grid generation methods in detail let us discuss the techniques to transform the system from physical (x, y) to body-fitted curvilinear (ξ, η) coordinates in the following section.

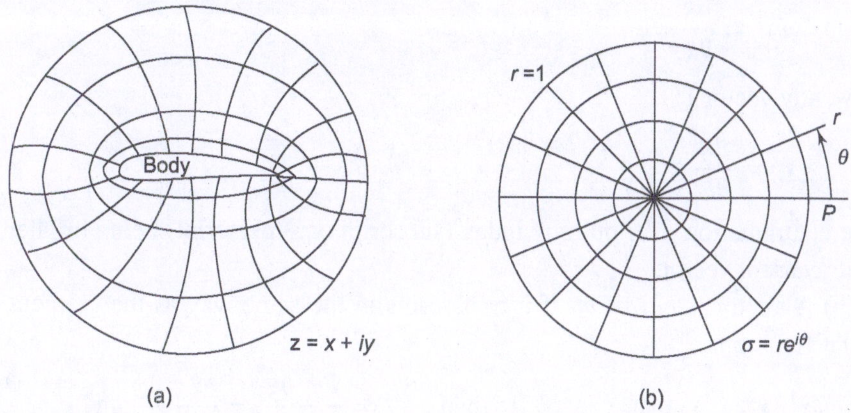

Figure 8.8 (a) Grid in the physical z-plane and (b) Uniform polar grid in computational ζ-plane.

8.2 CO-ORDINATE TRANSFORMATION

In finite difference methods usually the governing equations of fluid dynamics (say, in two dimensions) are being solved in a body-fitted coordinate system (ξ, η). So, the original equations written in Cartesian coordinate system (x, y) are to be transformed into the curvilinear system (ξ, η). The concept of the coordinate transformation is a very useful tool both for the generation of grids and for solving the governing equations in the transformed coordinate system. Two types of coordinate transformations commonly used in computational fluid dynamics are discussed below.

First type attempts to transform a governing differential equations in divergence form from one coordinate system to another. Let us define the new coordinates (ξ^i) in three dimension

$$\xi^i = \xi^i(x^j) \quad \text{for} \quad i, j = 1, 2, 3, \tag{8.1}$$

as functions of x^j. A vector is called a *contravariant* or a *covariant* according to its transformation behavior: If $\underline{v}^i$ are the components of a vector in the coordinate system described by x^i and v^i the components in the system ξ^i, then

(a) A vector is a *contravariant* vector if its components transform in the same way as the co-ordinate differentials: For $x^i = x^i(\xi^j)$,

$$dx^i = \frac{\partial x^i}{\partial \xi^j} d\xi^j.$$

Now, any vector $\underline{v}^i$

$$\underline{v}^i = \frac{\partial x^i}{\partial \xi^j} v^j,$$

where summation over dummy index (subscript or superscript) being implied is a contravariant vector.

(b) A vector is *covariant* if it transforms in the same way as the gradient of a scalar function:

Let $\underline{\phi}(x^1, x^2, x^3) = \phi(\xi^1, \xi^2, \xi^3)$, then $\dfrac{\partial \underline{\phi}}{\partial x^i} = \dfrac{\partial \phi}{\partial \xi^j}\dfrac{\partial \xi^j}{\partial x^i} = \dfrac{\partial \xi^j}{\partial x^i}\dfrac{\partial \phi}{\partial \xi^j}$ and $\underline{v}_i = \dfrac{\partial \xi^j}{\partial x^i} v_j$ is a covariant vector.

Let us consider the two dimensional transonic full potential equation as an example for transformation to the curvilinear coordinate system. This equation can be written as

$$\frac{\partial}{\partial x}(\rho\phi_x) + \frac{\partial}{\partial y}(\rho\phi_y) = 0, \tag{8.2}$$

where ρ is the density and ϕ is the velocity potential. The contravariant velocity vectors U and V can be defined as

$$U = \xi_x\phi_x + \xi_y\phi_y \quad \text{and} \quad V = \eta_x\phi_x + \eta_y\phi_y. \tag{8.3}$$

So, by chain rule

$$\begin{aligned} U &= \xi_x(\phi_\xi\xi_x + \phi_\eta\eta_x) + \xi_y(\phi_\xi\xi_y + \phi_\eta\eta_y) \\ &= \phi_\xi\left(\xi_x^2 + \xi_y^2\right) + \phi_\eta\left(\xi_x\eta_x + \xi_y\eta_y\right) \end{aligned} \tag{8.4}$$

Similarly,

$$V = \phi_\eta\left(\eta_x^2 + \eta_y^2\right) + \phi_\xi\left(\xi_x\eta_x + \xi_y\eta_y\right) \tag{8.5}$$

Solving for ϕ_x and ϕ_y from two equations in Eq. (8.3) we get

$$\frac{\phi_x}{-\xi_y V + \eta_y U} = \frac{\phi_y}{-\eta_x U + \xi_x V} = \frac{1}{J}, \tag{8.6}$$

where J is called the *Jacobian of transformation* and defined as

$$J = \xi_x\eta_y - \xi_y\eta_x. \tag{8.7}$$

Now substituting Eq. (8.6) into Eq. (8.2) and using the chain rules

$$\frac{\partial}{\partial x} = \xi_x \frac{\partial}{\partial \xi} + \eta_x \frac{\partial}{\partial \eta} \quad \text{and} \quad \frac{\partial}{\partial y} = \xi_y \frac{\partial}{\partial \xi} + \eta_y \frac{\partial}{\partial \eta} \tag{8.8}$$

yields after simplification

$$\frac{\partial}{\partial \xi}\left(\frac{\rho U}{J}\right) \cdot \{\xi_x \eta_y - \xi_y \eta_x\} + \frac{\partial}{\partial \eta}\left(\frac{\rho V}{J}\right) \cdot \{\xi_x \eta_y - \xi_y \eta_x\} = 0$$

$$\text{or} \quad \frac{\partial}{\partial \xi}\left(\frac{\rho U}{J}\right) + \frac{\partial}{\partial \eta}\left(\frac{\rho V}{J}\right) = 0. \tag{8.9}$$

The contravariant velocity vectors U and V can be expressed as

$$U = A_1 \phi_\xi + A_2 \phi_\eta \quad \text{and} \quad V = A_2 . \phi_\xi + A_3 . \phi_\eta \tag{8.10}$$

where the metric terms are

$$A_1 = \xi_x^2 + \xi_y^2, \quad A_2 = \xi_x \eta_x + \xi_y \eta_y \quad \text{and} \quad A_3 = \eta_x^2 + \eta_y^2. \tag{8.11}$$

It is to be observed that the divergence form of equation (8.2) remains unchanged in (8.9).

Second type of transformation appears in the process of generating a computational grid, where one has to find a mapping from the physical coordinate system (x, y) to a computational coordinate system (ξ, η) such that the boundaries of the physical space lies on the boundaries of the computational space. Usually it needs to interchange the dependent and independent variables of the governing partial differential equations to get the unknown coordinates (x, y) in terms of the known independent variables (ξ, η). For example, Laplace equations produce a smooth grid and the equations

$$\xi_{xx} + \xi_{yy} = 0, \quad \text{and} \quad \eta_{xx} + \eta_{yy} = 0 \tag{8.12}$$

are to be transformed so that x and y become dependent variables and then solved for x and y to get a two dimensional grid. To get the transformed equations let

$$x = x(\xi, \eta) \quad \text{and} \quad y = y(\xi, \eta). \tag{8.13}$$

Differentiating both the relations of Eq. (8.13) with respect to x and y we get

$$x_\xi \xi_x + x_\eta \eta_x - 1 = 0 \tag{8.14}$$

$$x_\xi \xi_y + x_\eta \eta_y - 0 = 0 \tag{8.15}$$

$$y_\xi \xi_x + y_\eta \eta_x - 0 = 0 \tag{8.16}$$

$$y_\xi \xi_y + y_\eta \eta_y - 1 = 0 \tag{8.17}$$

From (8.14) and (8.16) we can solve for ξ_x and η_x and from (8.15) and (8.17) for ξ_y and η_y in terms of $x_\xi,\ x_\eta,\ y_\xi,\ y_\eta$ and the Jacobian of transformation J to get the following relations :

$$y_\eta = J\xi_x,\quad x_\eta = -J\xi_y,\quad y_\xi = -J\eta_x,\quad x_\xi = J\eta_y \text{ and } J = x_\xi y_\eta - y_\xi x_\eta. \tag{8.18}$$

Now the Jacobian yields on differentiation

$$\begin{aligned} J_x &= J_\xi.\xi_x + J_\eta.\eta_x \\ &= \xi_x.\left\{x_{\xi\xi}y_\eta + x_\xi y_{\xi\eta} - y_{\xi\xi}x_\eta - y_\xi x_{\xi\eta}\right\} \\ &+\eta_x.\left\{x_{\xi\eta}y_\eta + x_\xi y_{\eta\eta} - y_{\xi\eta}x_\eta - y_\xi x_{\eta\eta}\right\} \end{aligned}$$

Substituting the relations for ξ_x and η_x from (8.18) in the above equation, we get

$$\begin{aligned} J_x &= \frac{1}{J}\left\{x_{\xi\xi}y_\eta^2 + x_\xi y_\eta y_{\xi\eta} - y_{\xi\xi}x_\eta y_\eta - y_\xi y_\eta x_{\xi\eta}\right\} \\ &-\frac{1}{J}\left\{x_{\xi\eta}y_\xi y_\eta + x_\xi y_\xi y_{\eta\eta} - x_\eta y_\xi y_{\xi\eta} - y_\xi^2 x_{\eta\eta}\right\} \\ &= \frac{1}{J}\left\{A_1 + A_2 y_{\xi\eta} - A_3\right\} \end{aligned} \tag{8.19}$$

where

$$A_1 = x_{\xi\xi}y_\eta^2 - 2x_{\xi\eta}y_\xi y_\eta + x_{\eta\eta}y_\xi^2,$$

$$A_2 = x_\xi y_\eta + x_\eta y_\xi,\qquad A_3 = x_\eta y_\eta y_{\xi\xi} + x_\xi y_\xi y_{\eta\eta}.$$

Similarly,

$$J_y = \frac{1}{J}\left\{B_1 + A_2 x_{\xi\eta} - B_3\right\}, \tag{8.20}$$

where

$$B_1 = y_{\eta\eta}x_\xi^2 - 2y_{\xi\eta}x_\xi x_\eta + y_{\xi\xi}x_\eta^2 \quad\text{and}\quad B_3 = x_\eta y_\eta x_{\xi\xi} + x_\xi y_\xi x_{\eta\eta}.$$

Now

$$\begin{aligned} \frac{\partial}{\partial x}(\xi_x) &= \frac{\partial}{\partial x}\left(\frac{1}{J}y_\eta\right) = y_\eta\frac{\partial}{\partial x}\frac{1}{J} + \frac{1}{J}\frac{\partial}{\partial x}(y_\eta)7 = y_\eta\frac{-1}{J^2}J_x + \frac{1}{J}\left(y_{\eta\xi}\xi_x + y_{\eta\eta}\eta_x\right) \\ &= -\frac{1}{J^2}J_x y_\eta + \frac{1}{J^2}\left(y_{\xi\xi}y_\eta - y_{\eta\eta}y_\xi\right) \\ &= -\frac{1}{J^3}y_\eta\left\{x_{\xi\xi}y_\eta^2 - 2x_{\xi\eta}y_\xi y_\eta + x_{\eta\eta}y_\xi^2\right. \\ &\left.- y_{\xi\xi}x_\eta y_\eta + y_{\xi\eta}\left(x_\xi y_\eta + x_\eta y_\xi\right) - y_{\eta\eta}x_\xi y_\xi\right\} \\ &+\frac{1}{J^2}\left(y_{\eta\xi}y_\eta - y_{\eta\eta}y_\xi\right) \end{aligned} \tag{8.21}$$

and

$$
\begin{aligned}
\frac{\partial}{\partial y}(\xi_y) &= -\frac{\partial}{\partial y}\left(\frac{1}{J}x_\eta\right) = -x_\eta \frac{\partial}{\partial y}\left(\frac{1}{J}\right) - \frac{1}{J}\frac{\partial}{\partial y}(x_\eta) \\
&= -x_\eta \frac{-1}{J^2} J_y - \frac{1}{J}\left(x_{\xi\eta}\xi_y + x_{\eta\eta}\eta_y\right) \\
&= \frac{1}{J^2} x_\eta J_y - \frac{1}{J^2}\left(-x_{\xi\eta}x_\eta + x_{\eta\eta}x_\xi\right) \\
&= \frac{1}{J^3} x_\eta \left[-x_{\xi\xi}x_\eta y_\eta + x_{\xi\eta}\left(y_\xi x_\eta + x_\xi y_\eta\right)\right. \\
&\quad \left. - x_{\eta\eta}x_\xi y_\eta + y_{\eta\eta}x_\xi^2 - 2y_{\xi\eta}x_\xi x_\eta + y_{\xi\xi}x_\eta^2\right] \\
&\quad - \frac{1}{J^2}\left(x_{\eta\eta}x_\xi - x_{\xi\eta}x_\eta\right).
\end{aligned} \tag{8.22}
$$

So, from Eqs. (8.21) and (8.22)

$$
\begin{aligned}
\xi_{xx} + \xi_{yy} &= \frac{1}{J^3}\left[-y_\eta\left(A_1 + A_2 y_{\xi\eta} - A_3\right) + x_\eta(B_1 + A_2 x_{\xi\eta} - B_3)\right] \\
&\quad + \frac{1}{J^2}\left(y_{\xi\eta}y_\eta - y_{\eta\eta}y_\xi - x_{\eta\eta}x_\xi + x_{\xi\eta}x_\eta\right) = 0.
\end{aligned}
$$

After a little algebraic simplification, we get

$$
\begin{aligned}
\xi_{xx} + \xi_{yy} = y_\eta &\left[A x_{\xi\xi} - 2B x_{\xi\eta} + C x_{\eta\eta}\right] \\
-x_\eta &\left[A y_{\xi\xi} - 2B y_{\xi\eta} + C y_{\eta\eta}\right] = 0,
\end{aligned} \tag{8.23}
$$

where,

$$
A = x_\eta^2 + y_\eta^2, \quad B = x_\xi x_\eta + y_\xi y_\eta \quad \text{and} \quad C = x_\xi^2 + y_\xi^2. \tag{8.24}
$$

Proceeding in a similar way we obtain

$$
\begin{aligned}
\eta_{xx} + \eta_{yy} = -y_\xi &\left[A x_{\xi\xi} - 2B x_{\xi\eta} + C x_{\eta\eta}\right] \\
+x_\xi &\left[A y_{\xi\xi} - 2B y_{\xi\eta} + C y_{\eta\eta}\right] = 0.
\end{aligned} \tag{8.25}
$$

Equations (8.23) and (8.25) represent two homogeneous equations for the unknowns

$$
A x_{\xi\xi} - 2B x_{\xi\eta} + C x_{\eta\eta} \quad \text{and} \quad A y_{\xi\xi} - 2B y_{\xi\eta} + C y_{\eta\eta},
$$

whose coefficient determinant $y_\eta x_\xi - x_\eta y_\xi = J \neq 0$. Consequently, it follows that

$$
\left.
\begin{aligned}
A x_{\xi\xi} - 2B x_{\xi\eta} + C x_{\eta\eta} = 0, \\
\text{and} \qquad\qquad \\
A y_{\xi\xi} - 2B y_{\xi\eta} + C y_{\eta\eta} = 0.
\end{aligned}
\right\} \tag{8.26}
$$

One can proceed in a similar way to get the corresponding transformed equations for the three dimensional case.

8.3 DIFFERENTIAL EQUATION METHODS

Many techniques have been developed to generate computational grids required for the numerical solutions of the governing equations of fluid dynamics. The method we describe in this section (Thompson et al. 1977) is quite popular. This involves the solution of a set of elliptic type partial differential equations to generate the numerical grid. Grid smoothness is one of the features which is guaranteed in this method. We know that two dimensional, inviscid, steady, incompressible and irrotational flows can be described as a solution of Laplace equations

$$\phi_{xx} + \phi_{yy} = 0 \quad \text{or} \quad \psi_{xx} + \psi_{yy} = 0, \tag{8.27}$$

where ϕ and ψ are the velocity potential and stream function respectively. Solutions of these equations (8.27) with appropriate boundary conditions represent the potential and stream-function respectivly. If $\phi - const.$ and $\psi - const.$ lines are plotted together they will resemble mutually perpendicular grid lines in the flow field. This motivates to solve two Laplace's equations in two directions to get a smooth two dimensional grid. The control of the grid spacing can be achieved by adding two source terms. Here a particular method developed by Thomas and Middlecoff (1980) for the source terms along with an efficient ADI scheme (Mathur and Chakrabartty, 1994) to solve the equations will be discussed.

For two dimensional grid generation, the following system of elliptic type partial differential equations

$$\begin{aligned} \xi_{xx} + \xi_{yy} &= P(\xi, \eta) \\ \eta_{xx} + \eta_{yy} &= Q(\xi, \eta) \end{aligned} \tag{8.28}$$

is solved in the transformed plane, such that we can apply boundary conditions to x and y which have known boundary values in the physical space. The functions P and Q are suitably prescribed functions to control the grid spacing. To make x and y dependent variables, these equations are transformed to the following system as explained in the previous section

$$\begin{aligned} Ax_{\xi\xi} - 2Bx_{\xi\eta} + Cx_{\eta\eta} &= -J^2(Px_\xi + Qx_\eta) \\ Ay_{\xi\xi} - 2By_{\xi\eta} + Cy_{\eta\eta} &= -J^2(Py_\xi + Qy_\eta) \end{aligned} \tag{8.29}$$

where

$$A = x_\eta^2 + y_\eta^2, \qquad B = x_\xi x_\eta + y_\xi y_\eta,$$

$$C = x_\xi^2 + y_\xi^2, \quad \text{and} \quad J = x_\xi y_\eta - x_\eta y_\xi. \tag{8.30}$$

Here J is the Jacobian of transformation, and P and Q are the grid control functions. The solution of these equations with appropriate boundary conditions produces a smooth grid. For $P = Q = 0$, these correspond to Laplace equations, solutions of which also produce smooth grid (Holst, 1979), suitable for inviscid flows over a simple aerofoil. For complex geometries and for solving Navier-Stokes equations we need to prescribe P and Q in order to get suitable stretching of the grids. The boundary conditions required to solve these equations are data on the aerofoil and outer boundary with proper distribution as per requirements, i.e. (x, y) values on $\eta = \eta_{\min}$ and $\eta = \eta_{\max}$ (Fig. 8.1) lines. On the cut, periodicity has been assumed for $\xi = \xi_{\min}$ and $\xi = \xi_{\max}$ lines. In general, a successive over relaxation (SOR) method is used to solve these equations, which is computationally expensive particularly for three dimensional grids. We discuss here an efficient alternating direction implicit (ADI) scheme with approximate factorization (AF) (Mathur and Chakrabartty, 1994).

Equation (8.29) can be written in the form

$$A(r_{\xi\xi} + \phi r_{\xi}) - 2Br_{\xi\eta} + C(r_{\eta\eta} + \psi r_{\eta}) = 0 \tag{8.31}$$

where

$$r = (x, y)^{\mathrm{T}}, \quad \phi = PJ^2/A, \quad \text{and} \quad \psi = QJ^2/C \tag{8.32}$$

A general iteration scheme for the solution of Eq. (8.31) is of the form

$$N\Delta r^n = \omega L(r^n) \tag{8.33}$$

where, $\Delta r^n = r^{n+1} - r^n$, ω is a relaxation parameter (≈ 1.8). Here r^n is the value of r at the nth iteration and $L(r^n)$ denotes the residual, L being the difference operator acting on the unknown r. The operator N determines the type of iteration procedure. For our problem, L is of the form

$$L = A(\delta_{\xi\xi} + \phi\delta_{\xi}) - 2B\delta_{\xi\eta} + C(\delta_{\eta\eta} + \psi\delta_{\eta}) \tag{8.34}$$

where $\delta_{\xi\xi}$, $\delta_{\eta\eta}$, $\delta_{\xi\eta}$, δ_{ξ} and δ_{η} are usual central difference operators. We introduce an acceleration parameter α and rewrite (8.33) as

$$-\alpha N\Delta r^n = \alpha\omega L(r^n) \tag{8.35}$$

and choose N to be of the form

$$-\alpha N = \left[\alpha - A(\delta_{\xi\xi} + \phi\delta_{\xi})\right]\left[\alpha - C(\delta_{\eta\eta} + \psi\delta_{\eta})\right] \tag{8.36}$$

so that (8.35) can now be rewritten as

$$\left[\alpha - A(\delta_{\xi\xi} + \phi\delta_{\xi})\right]\left[\alpha - C(\delta_{\eta\eta} + \psi\delta_{\eta})\right]\Delta x^n = \alpha\omega L(x^n) \tag{8.37}$$

and

$$\left[\alpha - A(\delta_{\xi\xi} + \phi\delta_{\xi})\right]\left[\alpha - C(\delta_{\eta\eta} + \psi\delta_{\eta})\right]\Delta y^n = \alpha\omega L(y^n). \tag{8.38}$$

The scheme now involves a two step solution procedure as follows :

$$Step - I \quad \begin{bmatrix} \alpha - A^n_{ij}(\delta_{\xi\xi} + \phi\delta_\xi) \end{bmatrix} f^n_{ij} = \alpha\omega L(x^n_{ij}) \\ \begin{bmatrix} \alpha - A^n_{ij}(\delta_{\xi\xi} + \phi\delta_\xi) \end{bmatrix} g^n_{ij} = \alpha\omega L(y^n_{ij}) \tag{8.39}$$

$$Step - II \quad \begin{bmatrix} \alpha - C^n_{ij}(\delta_{\eta\eta} + \psi\delta_\eta) \end{bmatrix} \Delta x^n_{ij} = f^n_{ij} \\ \begin{bmatrix} \alpha - C^n_{ij}(\delta_{\eta\eta} + \psi\delta_\eta) \end{bmatrix} \Delta y^n_{ij} = g^n_{ij}, \tag{8.40}$$

where f^n and g^n are intermediate results stored at each point of the mesh. In step-I, the f and g arrays are obtained by solving two tridiagonal matrix equations for each η = const. line. The corrected values of x and y are then obtained in the second step from the f and g arrays respectively by solving again two tridiagonal matrix equations for each ξ = const. line. During each iteration at the beginning of the first step, boundary conditions for f and g must be specified. These are intermediate functions with little physical meaing, so the specification of their value is difficult. Since f and g approach zero after convergence, these boundary conditions can not affect solution accuracy. However a poor choice can slow down the process or even cause instability. For acceptable accuracy, $f_\eta = g_\eta = 0$ can be used.

The acceleration parameter α introduced in the scheme can be considered as $1/\Delta t$, in a pseudo-unsteady formulation of the iterative scheme. As suggested in (Holst, 1979) a sequence of α's, α_k where k is the level of the frequency bands where the errors can be effectively reduced can be used here. The highest and lowest values of α, α_h and α_l are the estimates of the highest and lowest eigen values of the amplification matrix. Low and high frequency errors can be effectively reduced by the low and high values of α. A geometric sequence

$$\alpha_k = \alpha_k(\alpha_l/\alpha_h)^{\frac{k-1}{M-1}}, \quad k = 1, 2, \ldots, M \tag{8.41}$$

where M is the number of levels of frequency bands to cover the entire eigenvalue spectrum. Usually M is set equal to eight. This scheme is analogous to multigrid scheme, where the errors in the different frequency bands are reduced at different grid levels starting from the fine grid to coarse grid.

The control functions P and Q can be chosen as desired. A detailed account of the various possibilities of choosing P and Q are given by Thompson et al. (1977). However, the formulation of Thomas and Middlecoff (1980), where the control functions are derived from the boundary data and reflect the boundary spacing into the field has become popular because of its simplicity. These control functions are

$$\phi = -(x_\xi x_{\xi\xi} + y_\xi y_{\xi\xi})/(x^2_\xi + y^2_\xi) \text{ along } \eta = \text{const. boundaries,} \tag{8.42}$$

and

$$\psi = -(x_\eta x_{\eta\eta} + y_\eta y_{\eta\eta})/(x_\eta^2 + y_\eta^2) \text{ along } \xi = \text{const. boundaries.} \quad (8.43)$$

The control functions are evaluated at the boundaries and then interpolated to get the values in the interior. The use of these control functions enables the clustering of grid lines by an appropriate distribution of points on the boundaries.

8.4 ALGEBRAIC METHODS

To generate a C-type grid for an arbitrary two dimensional body, an algebraic method developed by Rizzi (1981) and later modified by Jain (1983) will be discussed here. This is very simple and easy to use to get a C-type grid having adequate controls to suit the requirements of the problem. Let us define a physical space in Cartesian coordinate system (x, y), as shown in Fig. 8.9a. Let (ξ, η) denote

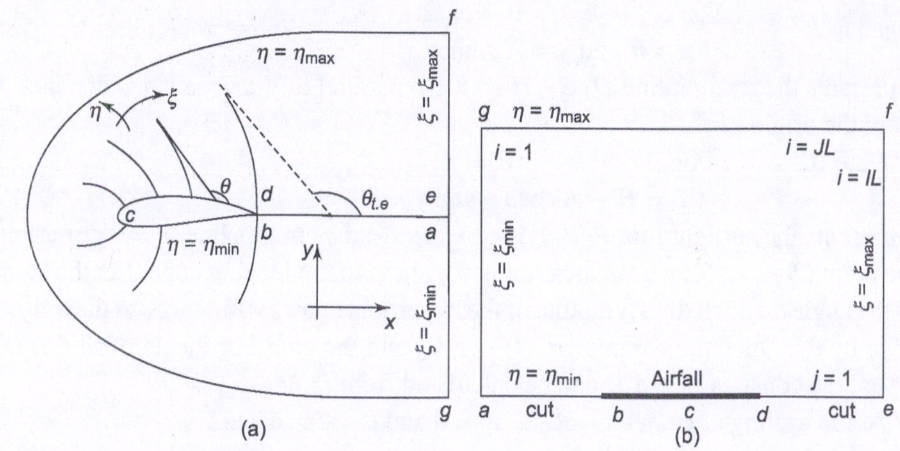

Figure 8.9 Coordinate system for C-type grid generation (a) physical plane and (b) computational plane.

a body-fitted curvilinear coordinate system, as shown in Fig. 8.9b. The curves ξ = const. emanate from the aerofoil and the wake line (cut-line behind the aerofoil). The curve $\eta = \eta_{\min}$ represents the aerofoil contour and the cut line, whereas $\eta = \eta_{\max}$ represents outer boundary. The family of coordinate curves (ξ = const.) are represented by a family of hyperbolas. These hyperbolas are characterized by their asymptotes. Let θ be the angle between the x-axis and the asymptote of the hyperbola, then ξ = const. curves are identical with θ = const. curves. Thus, x and y are obtained as

$$\begin{aligned} x &= B + A\cosh\eta\cos\theta \\ y &= A\sinh\eta\sin\theta \end{aligned} \tag{8.44}$$

where A and B are the constants for the family of hyperbolas. Now for numerical computation (θ, η) is the actual coordinate system.

The representation of the coordinate curves (ξ = const.) as a family of hyperbolas has a number of advantages, e.g. (i) The hyperbolas are almost straight lines and radial in the far field; the aerofoil then looks like a point source as observed from the far field boundary. (ii) The degenerate case of the hyperbola is a straight line and hence downstream far field boundary, which is to be a straight line for C-type mesh can be obtained as a member of the same family of hyperbolas. (iii) Patching of two grid regions obtained from two families of hyperbolas having a degenerate hyperbola as a common interface is very simple. Two degenerate cases of the hyperbolas can be obtained as follows :

Case I, For $\theta = \pi/2$,

$$x = \mathrm{B} \text{ and } y = \mathrm{A}\sinh\eta,$$

represents the straight line Q_1Q_2, (Fig. 8.10) parallel to y-axis and at a distance B from the origin.

Case II, For $\theta = \pi$,

$$x = B - A\cosh\eta \text{ and } y = 0$$

represents the straight line P_1P_2 lying on the x-axis. On this line, $\eta = 0$ represents the point $(B - A, 0)$, whose location is very near to the leading edge. Let the value of the angle θ which the asymptote of the hyperbola makes with x-axis at the trailing edge by θ_{te}. Once the parameters $B - A$, θ_{te} and the chord of the aerofoil CL are given, the constats A and B can be calculated from (8.44) as :

At the trailing edge, let $x = CL,\ \eta = 0$ and $\theta = \theta_{\text{te}}$. Then

$$CL = B + A\cos\theta_{\text{te}}$$

or

$$A = \frac{CL - (B - A)}{1 + \cos\theta_{te}}, \qquad B = (B - A) + A.$$

Now the origins of the chosen hyperbolas $(\xi_i, \eta_{\text{min}})$ lie on the cut line and the aerofoil surface. One can control the spacing of the hyperbolas as desired on the aerofoil by parameterizing the aerofoil by its arc length, s and then prescribing a desired distribution of elemental arc length Δs at any particular portion of the aerofoil. This is done by using a geometrical progression series for the elemental arc lengths giving expanding or condensing mesh density on any part of the arc length of the aerofoil by prescribing the number of elements and the length of the first element.

The angle θ of the asymptote of a particular hyperbola can be obtained, once the coordinate (x, y) of the point is known by the expression

$$\left(\frac{x-B}{A\cos\theta}\right)^2 - \left(\frac{y}{A\sin\theta}\right)^2 = 1. \tag{8.45}$$

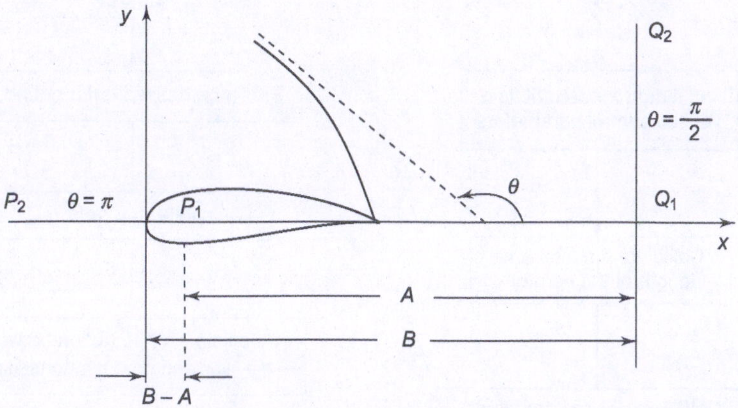

Figure 8.10 Degenerate cases of hyperbolas.

This has been obtained by eliminating η from (8.44). In general we need closer point distribution near leading edge, near trailing edge, and in the region where sharp gradient/discontinuity of the flow parameters are present.

To obtain $\eta = \text{const.}$ curves, it is necessary to prescribe the first mesh interval (arc length) in the η-direction from the aerofoil surface and the cut ($\eta = \eta_{\min}$ curve). Now on each hyperbola, an expanding mesh interval in the η-direction can be obtained using a geometrical progression series by prescribing the first interval on the surface of the aerofoil and the arc length of the curve from the aerofoil surface to the far field boundary. This distribution was obtained only up to the trailing edge of the aerofoil and the same can be followed downstream up to far field by translating these along the cut line. The cut emanating from the trailing edge is necessary to make the computational domain simply-connected and can be obtained by fitting an exponential curve with prescribed angle θ_{cut}, the curve makes at the trailing edge with the x-axis and the y_{cut}, the y-coordinate of the point at which it meets the downstream infinity. For better resolution of the wake it is necessary that this curve should simulate the wake centre line behind the aerofoil. After getting the network of the two families of curves the coordinates (x, y) of the nodal points may be obtained from (8.44). Various steps involved in this process are summarized in the flowchart (Fig. 8.11). Detailed explanation of each step has been given after that.

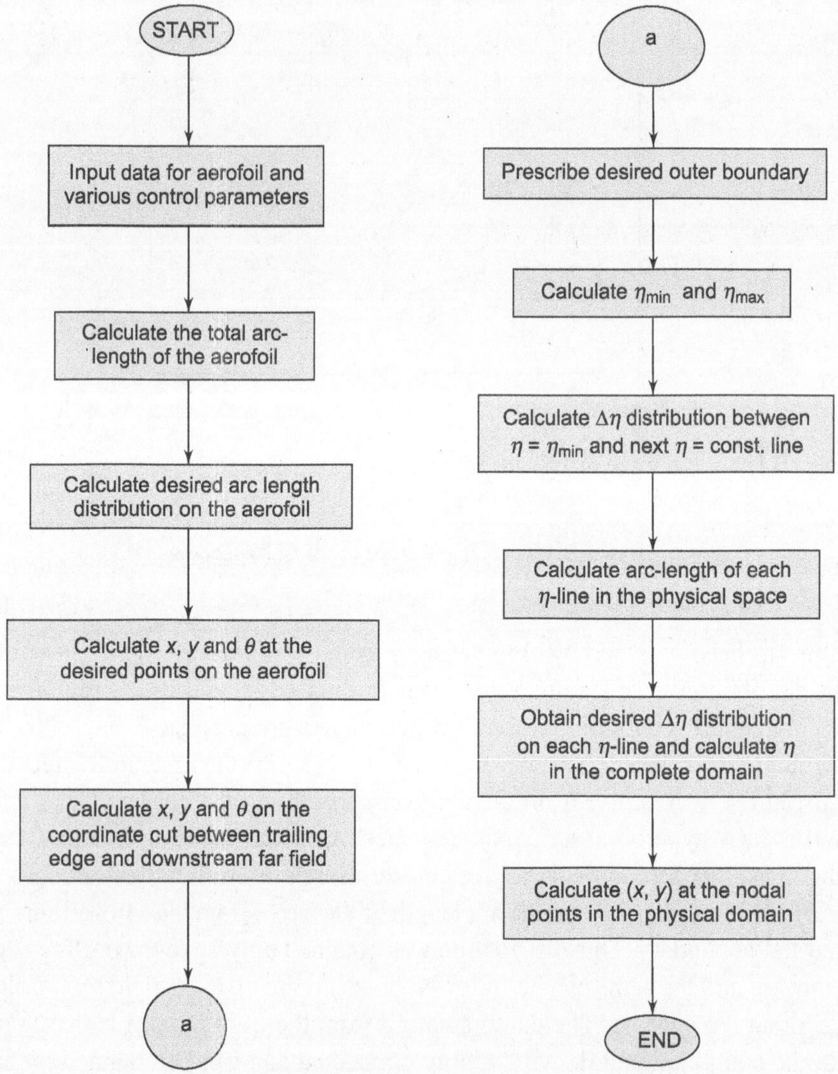

Figure 8.11 Flow chart for algebraic grid generation.

8.4.1 Calculation of the Arc Length

Let the coordinates of the data points $\{(x_i, y_i), i = 1, \ldots, I_{\text{te}})\}$ on the aerofoil be given starting from the lower surface trailing edge to the upper surface trailing edge (in a clock-wise sense). The arc length can be calculated from the same starting point for each of elements by simply joining each point to its neighbouring point by straight segments and then calculating the straight-line distance between the two points. This we call the zeroth order approximation for the elemental arc length. Hence,

$$\begin{aligned} S_i^{(0)} &= S_{i-1}^{(0)} + \Delta S_i^{(0)} \\ &= S_{i-1}^{(0)} + \sqrt{(\Delta x_i)^2 + (\Delta y_i)^2} \end{aligned}$$

Therefore,

$$S_i^{(0)} = S_{i-1}^{(0)} + \sqrt{(x_i - x_{i-1})^2 + (y_i - y_{i-1})^2} \quad \text{for} \quad i = 2, \ldots, I_{\text{te}} \tag{8.46}$$

where the superscript (0) indicates the zeroth approximation with $S_1^{(0)} = 0$. The accuracy can be further improved by taking the curvature effect of the aerofoil into account. This is done by fitting two cubic splines (Ahlberg et al. 1967), discussed in Section 8.4.8; one between the arc length $S^{(0)}$ and x-coordinates and the other between $S^{(0)}$ and y-coordinates. The spline routine gives the three derivatives $dx/ds,\ d^2x/ds^2$ and d^3x/ds^3 for $S - x$ spline and $dy/ds,\ d^2y/ds^2$ and d^3y/ds^3, for $S - y$ spline at each (x_i, y_i). Now the arc between i and $i - 1$ is divided into nine equal sub-elements with indices say, $l = 1, 10$. For each l, these derivatives are obtained by using a Taylor series. Now the elemental arc length on each of the subintervals is evaluated as

$$(\Delta s_l) = \left[\frac{\Delta s_i}{9}\right]_{\text{old}} \sqrt{\left(\frac{dx}{ds}\right)_{\text{av}}^2 + \left(\frac{dy}{ds}\right)_{\text{av}}^2}, \qquad l = 1, 2, \ldots, 9$$

where the subscript 'av' represents the average value. Thus,

$$(\Delta s_l)_{\text{new}} = \left(\frac{\Delta s_i}{9}\right)_{\text{old}} \frac{1}{2} \sqrt{\left\{\left(\frac{dx}{ds}\right)_l + \left(\frac{dx}{ds}\right)_{l-1}\right\}^2 + \left\{\left(\frac{dy}{ds}\right)_l + \left(\frac{dy}{ds}\right)_{l-1}\right\}^2}.$$

Therefore, the first approximation to the arc length at any point i is given by

$$S_i^{(1)} = S_{i-1}^{(1)} + \sum_{l=1}^{9} (\Delta s_l)_{\text{new}} \tag{8.47}$$

This process may be continued to whatever order of approximation one desires.

8.4.2 Desired Arc Length Distribution

A desirable point distribution on the aerofoil can be achieved in the following way. Let, $SL1$, $SL2$, $SU1$, $SU2$, be the given arc lengths, from lower surface trailing edge to the point where trailing end stretching ends, the extent of arc length on the lower surface from where the leading edge stretching begins, the arc length extent on the upper surface where the leading edge stretching ends and the arc length extent on the upper surface from where the trailing edge stretching begins respectively. Let $NL1$, $NL2$, $NU1$, and $NU2$ be the number of points to be distributed over $SL1$, $SL2$, $SU1$ and $SU2$ respectively. The arc length spacing at the trailing and leading edges are also prescribed as $\Delta\xi_{\text{te}}$ and $\Delta\xi_{\text{le}}$ respectively. So, both the lower and upper surface of the aerofoil has been divided into three sectors each, where for the first and third sectors the arc length, number of points and the starting elementary arc length are given. Let us first describe the procedure for a desirable distribution of points in the first sector, i.e. over $SL1$. A similar procedure follows for the other sectors.

Starting from the trailing edge an expanding mesh, i.e., $\Delta S_1 < \Delta S_2 < \ldots < \Delta S_{NL1-1}$ is desirable. Thus,

$$SL1 = \Delta S_1 + \Delta S_2 + \ldots + \Delta S_{NL1-1}, \quad \text{with} \quad \Delta S_1 = \Delta\xi_{\text{te}}.$$

A geometrical progression series for SL1 is assumed as,

$$SL1 = a + ar + ar^2 + \cdots + ar^{(NL1-1)} \tag{8.48}$$

$$\text{with} \quad a = \Delta\xi_{\text{te}} \quad \text{and} \quad r = \frac{\Delta S_2}{\Delta S_1} = \frac{\Delta S_3}{\Delta S_2} = \cdots \quad \text{and so on.}$$

Now, once the sum $SL1$, the first term a and the number of terms $NL1$ are given for the geometrical progression series (8.48), the common ratio or the stretching factor r can be solved from the following sum formula using Newton-Raphson method.

$$\text{sum} = a\left(\frac{r^n - 1}{r - 1}\right) \quad \text{for} \quad r > 1 (\text{expanding})$$

or

$$\text{sum} = a\left(\frac{1 - r^n}{1 - r}\right), \quad \text{for} \quad r < 1 (\text{condensing}).$$

This procedure can be used for the starting and ending sectors on the lower and upper surfaces of the aerofoil and also on the cut line. On the cut line, a smooth variation of θ from $\theta = \theta_{\text{cut}}$ at the trailing edge to $\theta = \pi/2$ at the far downstream can also be achieved in a similar way. For the middle sector, where starting and ending $\Delta\xi$'s are to be matched, a cubic polinomial of the form

$$a_0 + a_1 s + a_2 s^2 + a_3 s^3$$

where, a_0, a_1, etc. are determined by matching various derivatives at the last point of starting and the first point of ending sectors. With this, the arc lengths of the desired distribution of points on the aerofoil and along the cut line can be obtained. Now, the x and y coordinates of these points can be computed using two cubic splines, one for $S_i \sim x_i$ and another for $S_i \sim y_i$ where x_i, y_i and S_i correspond to the given aerofoil coordinates.

Another way of distributing points on a given interval when the required spacings at the two ends are given, can be achieved by using a combination of a third- and a fourth-degree polynomial. Let there be N points to be distributed in the interval $0 \le x \le x_{\max}$, such that $x(1) = 0, \quad x(N) = x_{\max}$ and let $x(2) - x(1) = x_a$ and $x(N) - x(N-1) = x_b$ are given. We choose a third degree polynomial f, as

$$f(i) = a\left(\frac{i^3}{3} - (2+N)\frac{i^2}{2} + 2Ni\right) + c, \quad i = 2, 3, \cdots, N$$

where the constants a and c are determined from the boundary conditions

$$f(2) = x_a, \quad f(N) = x_b.$$

We also choose a fourth degree polynomial h, as

$$h(i) = (i-2)(i-N)(i^2 + bi + c), \quad i = 2, \cdots, N$$

where the constants b and c can be determined from the boundary conditions $h'(2) = h'(N) = 0$. Note that $f'(2) = f'(N) = h(2) = h(N) = 0$.

The points are now given by

$$x(1) = 0,\ x(2) = x_a,\ x(i) = x(i-1) + f(i) + h(i)F,$$
$$i = 3, \cdots, N-1, \text{ and } x(N) = x_{\max}$$

where

$$F = \frac{x_{\max} - \sum_{i=2}^{N} f(i)}{\sum_{i=2}^{N} h(i)}.$$

The geometrical progression, cubic polynomial and a combination of third and fourth degree polynomials discussed above are very effective tools to distribute the points on a curve with desired stretching.

8.4.3 Calculation of the Angle θ on the Aerofoil and Cut

The angle θ, which the asymptotes of the hyperbolas make with the x-axis can be calculated by using (8.45), once the coordinates (x, y) is known. At this stage we know the aerofoil coordinates, (x_i, y_i) and the arc lengths S_i. We also know the desired arc length distribution and say, (x_1, y_1) be the desired coordinate of a point where the arc length is S_1. We want to know the angle θ at this point. Then from (8.45)

$$\left(\frac{x_1 - B}{A\cos\theta}\right)^2 - \left(\frac{y_1}{A\sin\theta}\right)^2 = 1.$$

After simplification we get the quadratic in terms of $(\cos^2\theta)$ as

$$A^2\cos^4\theta - \cos^2\theta\left[(x_1 - B)^2 + y_1^2 + A^2\right] + (x_1 - B)^2 = 0.$$

Solution of this equation gives $\cos^2\theta$ and θ is given by

$$\theta = \cos^{-1}\left[-\left\{\frac{A^2+(x_1-B)^2+y_1^2-\sqrt{|(A-x_1+B)^2+y_1^2||(A+x_1-B)^2+y_1^2|}}{2A^2}\right\}^{\frac{1}{2}}\right] \tag{8.49}$$

If the quantity under {} is greater than one, then

$$\theta = \cos^{-1}\left[-\frac{A^2+(x_1-B)^2+y_1^2+\sqrt{|(A-x_1+B)^2+y_1^2||(A+x_1+B)^2+y_1^2|}}{2A^2}\right]^{\frac{1}{2}} \tag{8.50}$$

Equation (8.49) works well for all $\theta \geq \frac{\pi}{2}$, which is the case generally for a C-type grid. For getting O-type grid as a special case using this method θ becomes $< \frac{\pi}{2}$. In that case, the first −ve sign after [should be changed to +ve sign. This method can be used to get O-type grid also by changing the input data properly, but at the trailing edge it is difficult to get proper clustering of the ξ = const. lines around the wake. Anand Kumar and Dhanalakshmi (1985) overcame this difficulty by using the transformation

$$x = B + \left(\frac{A}{2}\right)\left(\eta + \frac{1}{\eta}\right)\cos\xi$$
$$y = \left(\frac{A}{2}\right)\left(\eta - \frac{1}{\eta}\right)\sin\xi \tag{8.51}$$

instead of (8.44). This transforms the slit $y = 0$, $|x - B| < A$ in the physical (x, y) plane to a unit circle in the computational polar (ξ, η) plane. The aerofoil will therefore be transformed to an approximate circle. They also used the transformation

$$x = B + \frac{1}{2}(\xi^2 - \eta^2)$$
$$y = \xi\eta \tag{8.52}$$

to get a C-type grid, where the $\xi = const.$ lines are parabolas in the physical (x, y) plane. The other parameters corresponding to the transformations (8.51) and (8.52) can be obtained following similar steps as discussed here for (8.44).

The x- and y-coordinates of all the points lying on the cut line are obtained as follows

$$(x_i)_{\text{cut}} = x_{\text{te}} + \sum \Delta\xi_i \tag{8.53}$$

$$(y_i)_{\text{cut}} = YCUT + (y_{\text{te}} - YCUT)e^{-\frac{\sum \Delta\xi_i}{YCUT}\tan(\theta_{\text{cut}})}. \tag{8.54}$$

Where $YCUT$ is the y-coordinate of the point of intersection of the cut line with downstream far-field boundary and θ_{cut} is the angle that this cut line makes with the positive x-axis at the trailing edge. Along the cut line the $\Delta\xi_i$ are determined using a geometric progression with $\Delta\xi_1 = \Delta\xi_{te}$ and then the x-coordinates are determined using (8.53). The y-coordinates of these points are determined using the exponential distribution (8.54). It should be noted in equation (8.54) that at large distance downstream $\sum \Delta\xi_i$ is large and the y-coordinate of the cut line becomes $y = YCUT$ exponentially. The hyperbolas are defined at each point on the cut line by varying θ from $\theta = \theta_{\text{te}}$ at the trailing edge of the aerofoil to $\theta = \frac{\pi}{2}$ at the far field downstream using similar geometrical progression series as described above. Up to now, it has been described how to get the curve representing the cut line and aerofoil surface with desirable distribution of points.

8.4.4 Calculation of η_{min} and η_{max}

The line $\eta = \eta_{\text{min}}$ represents the coordinates (x, y) of the inner boundary consisting of the aerofoil and the cut. The outer boundary will be represented by the line $\eta = \eta_{\text{max}}$. Once we know the values of x, y and θ on the aerofoil and the cut, we can obtain η_{min} using (8.44) as follows:

We get

$$\eta_{\text{min}} = \sinh^{-1}\left(\frac{y}{A\sin\theta}\right) = \ln\left[\left(\frac{y}{A\sin\theta}\right) + \sqrt{\left(\frac{y}{A\sin\theta}\right)^2 + 1}\right] \tag{8.55}$$

and

$$\eta_{\text{min}} = \ln\left[\left(\frac{x-B}{A\cos\theta}\right) \pm \sqrt{\left(\frac{x-B}{A\cos\theta}\right)^2 - 1}\right] \tag{8.56}$$

from the second and the first expression of the equation (8.44) respectively. Use of the above two expressions to obtain η_{min} depends on the quadrant in which θ lies. We can impose the condition that if $|\tan\theta| \leq 1$ use (8.56) otherwise use (8.55). In (8.56), the sign is decided by the sign of $(y\ A\sin\theta)$. The values of η_{max}

corresponding to a particular value of θ_i of the hyperbola at the ith point can be calculated from the point of intersection of the given outer boundary and the hyperbola. Let us consider first the far field boundary given by an ellipse

$$\left(\frac{x-B}{XFAR}\right)^2 + \left(\frac{y}{YFAR}\right)^2 = 1, \tag{8.57}$$

with origin at $x = B$, $y = 0$, where $XFAR = B + |XUP|$, and the parameters XUP, $XDOWN$ and $YFAR$ are given to represent the far field bounday defining the x-coordinates of the upstream, downstream and the maximum y-coordinate of the outer ellipse respectively. The coordinates (x, y) of a far field point lying on the hyperbola are obtained from (8.44) as

$$x = B + A\cosh\eta_{\max}\cos\theta_i \tag{8.58}$$

$$y = A\sinh\eta_{\max}\sin\theta_i. \tag{8.59}$$

Substituting (8.58) and (8.59) into (8.57), $\eta_{\max}$ can be found out from

$$\left(\frac{A\cosh\eta_{\max}\cos\theta_i}{XFAR}\right)^2 + \left(\frac{A\sinh\eta_{\max}\sin\theta_i}{YFAR}\right)^2 = 1. \tag{8.60}$$

Now expressing hyperbolic functions in terms of exponential functions and using the approximation $\cosh\eta \simeq \sinh\eta \simeq \dfrac{e^\eta}{2}$ for large values of $\eta(=\eta_{\max})$, (8.60) reduced to

$$e^{\eta_{\max}} = \frac{2}{A}\frac{1}{\sqrt{\dfrac{\cos^2\theta_i}{(XFAR)^2} + \dfrac{\sin^2\theta_i}{(YFAR)^2}}} = \frac{2}{A}.RFAR \tag{8.61}$$

Hence,

$$\eta_{\max} = \ln\left(\frac{2}{A}RFAR\right), \quad \text{where } RFAR = \frac{1}{\sqrt{\left(\dfrac{\cos\theta_i}{XFAR}\right)^2 + \left(\dfrac{\sin\theta_i}{YFAR}\right)^2}}. \tag{8.62}$$

Beyond the trailing edge same $\eta_{\max}$ will be continued upto the downstream boundary at $x = XDOWN$. It is to be noted here that the approximation $\cosh\eta \simeq \sinh\eta \simeq \frac{e^\eta}{2}$ used to get the expression for $\eta_{\max}$ works very well for external aerodynamic problems where the far-field boundary is actually situated at a large distance from the aerofoil. The outer boundary can be given arbitrarily as required and the corresponding points of intersection with the ξ = const. lines are to be calculated to get the expression for $\eta_{\max}$. For the cases where $\eta_{\max}$ is small like simulation of flow through wind tunnel where the upper and lower outer boundaries are close to each other, or for the flow through cascades of aerofoils where the aerofoils are placed close to each other (see Fig. 13.22), this approximation will not be valid. For these cases, the expressions (8.58) and (8.59) are to be substituted in the given equation of outer boundary and the resulting equation for $\eta_{\max}$ is to be directly solved using methods like method of bi-section (Niyogi, 2003). Between the two values of $\eta_{\min}$ and $\eta_{\max}$ representing inner and outer boundaries, the intermediate ξ-lines (C-curves) would be evaluated next.

8.4.5 $\Delta\eta$-Distribution on the Aerofoil and the Cut

With the two input parameters $\Delta\eta_{\text{le}}$ and $\Delta\eta_{\text{te}}$, at the leading and trailing edges respectively, other values of $\Delta\eta_i$ on the remaining aerofoil are obtained by fitting a quadratic

$$\Delta\eta_i = \Delta\eta_{\text{le}} + (\Delta\eta_{\text{te}} - \Delta\eta_{\text{le}})\left(\frac{i - IN}{N - 1}\right)^2, \tag{8.63}$$

where IN is the index for the leading edge and N is the number of mesh points on the lower (or upper) surface of the aerofoil. At $i = IN$, $\Delta\eta_i = \Delta\eta_{le}$ and at trailing edge, $i = IN - N + 1$ (lower surface) or $i = IN + N - 1$ (upper surface), where $\Delta\eta_i = \Delta\eta_{\text{te}}$. Distribution of $\Delta\eta$ on the coordinate cut is assumed to be constant and equal to its value at the trailing edge.

8.4.6 Mesh Spacing in η-Direction

This is done in two steps. First, the arc length of each hyperbola between $\eta_{\min}$ and $\eta_{\max}$ is calculated in the physical space and then an expanding mesh interval on each hyperbola is achieved by fitting a geometrical progression series. The elementary length Δs of the curve can be expressed in terms of Δx and Δy which can be obtained from (8.44) for any two consecutive values of η. Thus

$$\Delta s = \sqrt{(\Delta x)^2 + (\Delta y)^2} \tag{8.64}$$

$$= \left[A^2 \cos^2\theta_i\,(\cosh\eta_2 - \cosh\eta_1)^2\right. \tag{8.65}$$

$$\left. + A^2 \sin^2\theta_i\,(\sinh\eta_2 - \sinh\eta_1)^2\right]^{\frac{1}{2}} \tag{8.66}$$

After simplification, the length of each η-curve can be written as

$$L_i = A \int_{\eta_{\min}}^{\eta_{\max}} \sqrt{\sinh^2\eta + \sin^2\theta_i}\; d\eta, \tag{8.67}$$

where, $\eta_2 = \eta_1 + \Delta\eta$, and the $\lim_{\Delta\eta\to 0} \sinh(\Delta\eta) = d\eta$ have been used and η_1 has been replaced by η. Interval between $\eta_{\min}$ and $\eta_{\max}$ is divided into 50 subintervals and expression (8.67) is evaluated using Simpson's rule. Knowing L_i of any ith hyperbola and the initial interval $\Delta\eta_1$ a geometrical progression series can be used to get proper $\Delta\eta_i$'s along the curve. With (8.67) and the geometrical progression series we can get the jth elementary length ΔL_{ij} of the ith hyperbola. Now to get the η distribution along the curve, ΔL_{ij} can be written as

$$\Delta L_{ij} = \int_{\eta_1}^{\eta_2} \sqrt{\sinh^2\eta + \sin^2\theta_1}\, d\eta \tag{8.68}$$

We carry out this integration begining with $\eta_1 = \eta_{\min}$ and proceeding for subsequent η's. The value of η_2 can be found in an iterative way as follows. For simplicity, let us consider an integral

$$I = \int_a^b f(x)\mathrm{d}x$$

where the integral I, the lower limit a and the function $f(x)$ are given. The unknown is the upper limit b. Let us assume a small unifom Δx such that for some n, $a + (n-1)\Delta x \leq b \leq a + n\Delta x$. Then

$$S = \Delta x\,[f(a) + f(a+\Delta x) + \cdots + f(a+n\Delta x)] \geq I,$$

and $b = a + n\Delta x - \frac{S-I}{f(a+n\Delta x)}$. This procedure is repeated for all j-values from $j = 2$ to $j = JL - 1$ on each hyperbola. Thus we know the η distribution at the nodal point (i, j).

8.4.7 Calculation of x and y at Nodal Points

Once we know θ_i of the hyperbolas and η_{ij} on these, the x- and y-coordinates at the nodal point (i, j) are given by

$$x_{ij} = B + A\cosh\eta_{ij}\cos\theta_i \tag{8.69}$$

$$y_{ij} = A\sinh\eta_{ij}\sin\theta_i. \tag{8.70}$$

Thus we obtain a complete procedure to generate a body-fitted C-type grid. The control of the mesh spacing on both ξ and η directions can be achieved efficiently by choosing proper values of $\Delta\xi_{\text{le}}$, $\Delta\xi_{\text{te}}$, $\Delta\eta_{\text{le}}$, $\Delta\eta_{\text{te}}$, SL1, SL2, SU1, SU2, NL1, NL2, NU1 and NU2.

8.4.8 Cubic Spline

Let $f_i = f(x_i)$, $i = 0, 1, \cdots, n$, be the $n+1$ data points given on the partition $a = x_0 < x_1 < x_2 < \cdots < x_n = b$ in the interval $a \leq x \leq b$ along with the first derivatives $f'(a)$ and $f'(b)$ at the end points $x = a$ and $x = b$ respectively, then there exists one and only one cubic spline $g(x)$ corresponding to this partition satisfying the following two requirements:

$$g(x_0) = f(x_0),\ g(x_1) = f(x_1),\ \cdots,\ g(x_n) = f(x_n) \tag{8.71}$$

and

$$g'(x_0) = f'(a) = k_0 \text{ (say)}, \; g'(x_n) = f'(b) = k_n \text{ (say)}, \tag{8.72}$$

where the superscript $'$ refers to the derivative of the function with respect to the argument. On every subinterval I_j given by $x_j \le x \le x_{j+1}$, the spline $g(x)$ must agree with a cubic polynomial $p_j(x)$ such that

$$p_j(x_j) = f(x_j), \quad p_j(x_{j+1}) = f(x_{j+1}) \tag{8.73}$$

and

$$p'_j(x_j) = k_j, \quad p'_j(x_{j+1}) = k_{j+1}, \tag{8.74}$$

where k_0 and k_n are given and $k_1, \cdots, k_{n-1}$ are unknown to be determined. There are four conditions in (8.73) and (8.74) for $p_j(x)$. By direct calculation one can verify that the unique cubic polynomial $p_j(x)$ satisfying (8.73) and (8.74) is

$$\begin{aligned} p_j(x) = {} & f(x_j)c_j^2(x-x_{j+1})^2\left[1+2c_j(x-x_j)\right]+f(x_{j+1})c_j^2(x-x_j)^2\left[1-2c_j(x-x_{j+1})\right] \\ & +k_jc_j^2(x-x_j)(x-x_{j+1})^2+k_{j+1}c_j^2(x-x_j)^2(x-x_{j+1}), \end{aligned} \tag{8.75}$$

where $c_j = 1/(x_{j+1} - x_j)$. Differentiating (8.75) twice we obtain

$$p_j''(x_j) = -6c_j^2 f(x_j) + 6c_j^2 f(x_{j+1}) - 4c_jk_j - 2c_jk_{j+1} \tag{8.76}$$

and

$$p_j''(x_{j+1}) = 6c_j^2 f(x_j) - 6c_j^2 f(x_{j+1}) + 2c_jk_j + 4c_jk_{j+1} \tag{8.77}$$

at $x = x_j$ and $x = x_{j+1}$ respectively. By definition, $g(x)$ has continuous second derivatives. This gives the condition

$$p''_{j-1}(x_j) = p''_j(x_j), \quad j = 1, \cdots, n-1.$$

If we use (8.77) with j replaced by $j-1$, and equate with (8.76), we get $n-1$ equations

$$\begin{aligned} c_{j-1}k_{j-1} + 2(c_{j-1}+c_j)k_j + c_jk_{j+1} = {} & 3\left[c_{j-1}^2\left\{f(x_j) - f(x_{j-1})\right\}\right. \\ & \left. +c_j^2\left\{f(x_{j+1}) - f(x_j)\right\}\right], \text{ for } \quad j = 1, \cdots, n-1. \end{aligned} \tag{8.78}$$

This gives a tridiagonal matrix to be solved for $k_1, \cdots, k_{n-1}$ with given k_0 and k_n as the boundary values. So, from (8.75) we can get the desired polynomial and also the derivatives.

8.5 TRANSFINITE INTERPOLATION METHODS

Transfinite interpolation belongs to the class of algebraic grid generation methods. What we have seen in the previous section of algebraic grid generation is that it provides a direct functional description of the transformation between the computational and physical planes. Different interpolation methods were used skillfully to get the desired distribution of points on a curve and finally to get the grid. The term *transfinite* usually means non-denumerable. So, it describes a class of interpolation schemes using some simple functions (interpolants) matching over a transfinite number of points on the given curve. This involves simple interpolation of data from the boundary data. In two dimensional problems data are known on four sides of the domain. Interpolation is to be done between the opposite pairs of boundaries. Let the boundaries be defined by the lines $\xi = 0,\ \xi = 1,\ \eta = 0$ and $\eta = 1$. Then along any $\xi =$ constant curve, the simple interpolation function (Eriksson, 1982; Weatherill, 1990) is

$$R_a(\xi, \eta) = (1 - \eta)R(\xi, 0) + \eta R(\xi, 1) \tag{8.79}$$

and that along any $\eta =$ constant curve is

$$R_b(\xi, \eta) = (1 - \xi)R(0, \eta) + \xi R(1, \eta). \tag{8.80}$$

The functions $\xi,\ (1 - \xi),\ \eta$, and $(1 - \eta)$ are called blending functions. $R(\xi, \eta)$ is the required function (interpolant) of ξ and η and can be expressed as the sum,

$$R(\xi, \eta) = R_a(\xi, \eta) + R_b(\xi, \eta), \tag{8.81}$$

with $R(\xi, 0),\ R(\xi, 1),\ R(0, \eta)$ and $R(1, \eta)$ are the given values of R on the four boundaries. This however does not recover the boundary values. For example, for $\xi = 0$, (8.81) gives

$$R(0, \eta) = (1 - \eta)R(0, 0) + \eta R(0, 1) + R(0, \eta) \tag{8.82}$$

and similar results for other boundaries. To overcome this, a third expression must be included and that can be written as

$$R_c(\xi, \eta) = (1-\xi)(1-\eta)R(0, 0) + \xi(1-\eta)R(1, 0) + (1-\xi)\eta R(0, 1) + \xi\eta R(1, 1).$$

The total interpolant is now

$$R = R_a + R_b - R_c. \tag{8.83}$$

These basic concepts can be extended to three dimensions. For more details about the implementation of this method one should see Eriksson (1982).

8.6 UNSTRUCTURED GRID GENERATION

A grid basically consists of a set of points connected in some manner. In a structured grid, this connectivity need not be specified explicitly, since the points are stored in a matrix such that the neighbouring points in the matrix are connected to each other. However, this connectivity is to be defined explicitly as a *connectivity matrix* for an unstructure grid. A structured grid can always be represented as an unstructured grid.

Let us consider a 3×3 structured grid stored in two arrays $x(i, j)$ and $y(i, j)$ with $i, j = 1, 2, 3$ containing x and y values of all the nine points respectively. To represent the same grid as an unstructured one, the points are numbered from 1 to 9, as shown in Fig. 8.12. The numbering may be arbitrary. The x and y values of each node are stored in two arrays $x(i)$ and $y(i)$, with $i = 1, \cdots, 9$. As we see in Fig. 8.12, there are nine points (nodes), twelve edges and four cells. Each edge is formed by connecting two nodes. Considering the grid as a set of edges the connectivity matrix, shown in Table 8.1 would define the forming nodes of each edge. Alternatively, we can consider the grid as a set of cells formed by joining four nodes. So, the connectivity matrix would define the forming nodes of each cell as shown in Table 8.2.

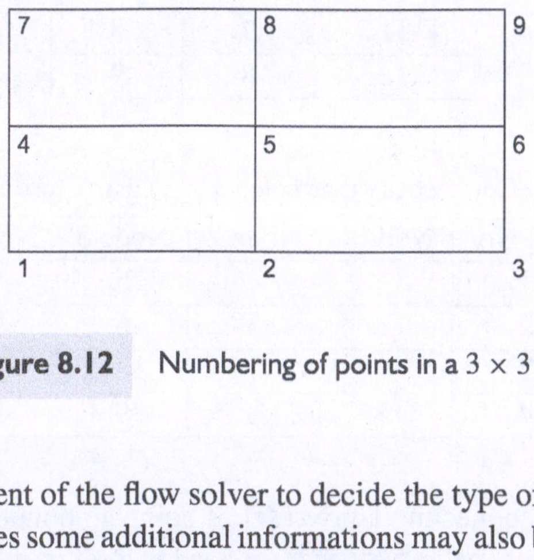

Figure 8.12 Numbering of points in a 3×3 grid.

It is the requirement of the flow solver to decide the type of connectivity matrix to be used. Sometimes some additional informations may also be required and those are to be supplied through the connectivity matrix.

Methods of *advancing front* and *Delaunay triangulation* (Weatherill, 1990) are very popular ones for unstructured grid generation. The first method is based upon the simultaneous point generation and connection. Given a geometry, say a two dimensional aerofoil, and a measure of the local spacing within the domain, the method extends or advances the boundary connectivity into the field. It first generates one layer of triangles close to the surface, joins the vertices of the triangles to form the next surface. Use that surface to generate the second layer and continue the process until the far field boundary is reached. The second method usually starts from a set of points (generated by any structured grid generation method) and join the points to get a triangular grid. In both the methods, one can add or delete cells during the process depending on the requirements. We will briefly discuss here the method of Delaunay triangulation.

Table 8.1 Connectivity matrix for 3 × 3 unstructured grid for edges

Edge No.	Node 1	Node 2
1	1	2
2	2	3
3	1	4
4	2	5
5	3	6
6	4	5
7	5	6
8	4	7
9	5	8
10	6	9
11	7	8
12	8	9

Table 8.2 Connectivity matrix for 3 × 3 unstructured grid for cells

Cell No.	Node 1	Node 2	Node 3	Node 4
1	1	4	5	2
2	2	5	6	3
3	4	7	8	5
4	5	8	9	6

It is a method of connecting a given set of points to obtain an unstructured grid consisting of triangular cells. So, what we need is a set of points distributed in the domain of interest and we have to connect them to get a triangular grid. Let us have two points P_1 and P_2 in a plane. The perpendicular bisector of the line joining the two points divide the plane into two regions, say V_1 and V_2. The region containing the point P_1 is V_1 and that containing the point P_2 is V_2. Now any point in V_1 is closer to P_1 than P_2 and vice-versa. This simple idea can be extended for a given set of points P_j in the plane, the regions V_j are the territories which can be assigned to each point P_j, such that V_j represents the space closer to P_j than any other point in the set. A *Voronoi region* is thus a convex polygon each side of which lies midway between the points it separates and is a segment of the perpendicular bisector of the line joining these points. Now if every pair of points which shares a common side of a Voronoi polygon is connected, the result is a triangulation of the set of points and the process is called Delaunay triangulation. This satisfies the following mathematical properties: (i) No point lies within a circle circumscribing a triangle, and (ii) For each triangle there is an associated vertex, and this vertex lies at the circumcentre of the triangle. Figure-8.13 shows the Voronoi regions for each point of a given set and the corresponding Delaunay triangulation. This can be described by a data structure which gives Voronoi vertex, the three points which form the triangle for that vertex, and the adjacent vertices. The data structure for Fig. 8.13 is shown in Table 8.3.

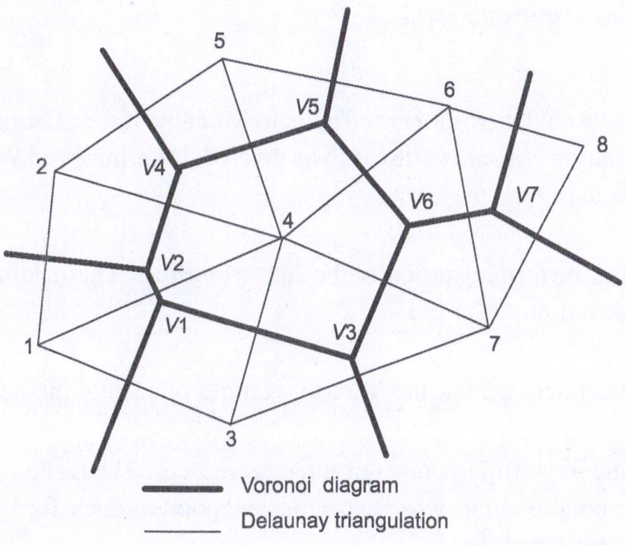

Figure 8.13 Voronoi diagram and Dalaunay triangulation for a set of points.

This data structure provides the basic infrastructure to construct the Delaunay triangulation. This procedure can be extended to three dimensions. Algorithm described in Weatherill (1990) is given below to construct a Delaunay triangulation for a two dimensional grid.

- Step 1.
  To start with define a 'convex hull' of four points, within which all the points lie and get the associated Voronoi data structure.

- Step 2.
  Introduce a new point within the convex hull.

Table 8.3 Data structure for the Fig. 8.13.

Voronoi vertex	Forming points			Neighbouring vertex		
V1	1	3	4	2	3	*
V2	1	2	4	1	4	*
V3	3	4	7	1	6	*
V4	2	4	5	2	5	*
V5	4	5	6	4	6	*
V6	4	6	7	3	5	7
V7	6	7	8	6	*	*

Note: * indicates vertex not defined.

- Step 3.
  If the new point lies within the circumcircle of any of the existing triangles, the corresponding Voronoi vertex is to be deleted. Determine the Voronoi vertices to be deleted.
- Step 4.
  Determine the forming points of the deleted vertices. These points are adjacent to the new point.
- Step 5.
  Determine the neighbouring Voronoi vertices of each of the deleted vertices.
- Step 6.
  Determine the forming points of the new vertices. The forming point include the new points, along with two adjacent points which form an edge of a neighbouring triangle.
- Step 7.
  Determine the neighbouring vertices as follows. In step 6, the forming points of a new vertex are determined. Now search through the forming points of the vertices identified in step 5. If there is a common pair of forming points, the two vertices are neighbours.
- Step 8.
  Rewrite the Voronoi data structure, removing the deleted vertices.
- Step 9.
  Go to step 2 and repeat upto step 8 for the next point.

This algorithm connects an arbitrary set of points which lie within a convex hull. The sample unstructured grid shown in Fig. 8.2 was generated using this algorithm.

8.7 MESH ADAPTATION

For a physical problem, the accuracy of the numerical solution depends largely on the quality of the grid used for discretisation. In a uniform mesh, the discretization error reduces as the number of grid points increases. Let us take a simple example of integrating a polynomial function $y = f(x)$ over an interval $x = a$ to $x = b$. If $f(x)$ is linear or a first order polynomial in x, then applying any numerical integration scheme over the full interval will give the exact integral. If it is a smooth curve, say, a second order polynimial, a few sub-intervals will produce accurate results. As the order of the polynimial increases we need smaller and smaller sub-intervals with uniform discretization for better accuracy. Extend this idea to a real situation in fluid mechanics. If the behaviour of the unknown variable is smooth enough then a uniform grid with reasonably small spacing will produce good results. But this is not the case in practice. Variables in fluid mechanics undergo rapid changes over a small distance. For inviscid flows there are shock waves, stagnation points and vortices. For viscous flows, in addition to these there are viscous dominated phenomena such as the boundary layers. To resolve these features accurately using uniform grid, number of grid points we need is practically impossible to handle even with a modern computing facility. Secondly, these features occupy a small region in the total computational field. In most of the region we do not need such a fine grid. So, to keep the number of grid points within manageable limit and at the same time resolving the complex flow field we have to think of non-uniform or stretched grid. So, our grid generation process should be interactive with the solution process. The measure of grid density will depend on the local variation of a chosen flow parameter like local pressure, local Mach number etc. What we need is a simple procedure to identify the regions of high error from a starting solution, re-generate the grid by introducing more grid points in those regions and repeat the process until we get a satisfacory resolution of the flow features. This eliminates guess work and a comparable solution can be obtained at less cost than those obtained by uniformly refining the grid over the entire flow field. Once we get the region of high error the grid can be adapted in two ways; (i) by redistridution of grid points (*moving mesh*) and (ii) by adding more of grid points in that region (*mesh enrichment*). In the first case, total number of points remain same, accuracy improves in the error prone region, but we may lose accuracy in other region from where the grid points are shifted. In the second case, number of grid points is increased and accuracy improves in the error prone area without sacrificing it in other regions but the computational cost increases.

8.7.1 Moving Mesh

Let us consider a two dimensional problem where the computation is performed in the computational (ξ, η) plane (Fig. 8.1). $\xi = \text{const.}$ and $\eta = \text{const.}$ lines correspond to non-orthogonal arcs in the physical space. Let us concentrate in ξ-direction and consider the transformation $\xi = \xi(s)$ where s is the arc length along $\eta = \text{const.}$ line in the physical space. Then the relation (Eiseman, 1985)

$$d\xi \propto w(s)ds \tag{8.84}$$

relates differential element $d\xi$ in computational plane to the elementary arc length ds in the physical plane through a non-zero weight function $w(s)$. Assuming uniform mesh in the computational plane ($d\xi = \text{const.}$), relation (8.84) satisfies *equidistribution of errors principle.* Notice that if $w = 1$, a uniform distribution of points along the arc results. Uniformity is destroyed if $w \neq 1$. This idea can be employed to get a desired distribution of points along any curvilinear coordinate direction by choosing the weight function w. Let p represent a typical flow variable whose gradient can be used as a measure for the required grid concentration, then the local value of $\partial p/\partial s$ can be used for w. So, a mathematical statement of the relationship between the computational and physical space becomes

$$d\xi \propto \left|\frac{\partial p}{\partial s}\right| ds. \tag{8.85}$$

There is no loss in generality if we assume that the weight is greater than one since the proportionality statement (8.85) has an implied scale factor. Under this assumption, relation (8.85) can be written as the sum of unity and a non-negative but finite term. A normalized form suitable for optimization and to remove singularities, can be written as (Dwyer et al., 1980)

$$\xi = \frac{\int_0^s \left(I + b\left|\frac{\partial p}{\partial s}\right|\right) ds}{\int_0^{s_{\max}} \left(I + b\left|\frac{\partial p}{\partial s}\right|\right) ds}, \tag{8.86}$$

where b is an adjustable constant used for optimization of the grid distribution. Similar expressions can be written for other directions. It is to be observed in (8.86) that for $b = 0$ a uniform distribution is obtained, while for large b equation (8.86) takes the form

$$\xi = \frac{\int_0^s |dp|}{\int_0^{s_{\max}} |dp|}. \tag{8.87}$$

Equation (8.86) can be further extended to

$$\xi = \frac{\int_0^s \left(I + b\left|\frac{\partial p}{\partial s}\right| + C\left|\frac{\partial^2 p}{\partial s^2}\right| \right) ds}{\int_0^{s_{\max}} \left(I + b\left|\frac{\partial p}{\partial s}\right| + C\left|\frac{\partial^2 p}{\partial s^2}\right| \right) ds}, \tag{8.88}$$

to take care of the situation where $\frac{\partial p}{\partial s}$ is zero but $\frac{\partial^2 p}{\partial s^2}$ is large. The constant (or a function) C controls the relative importance of second derivative influence on grid distribution. This idea of generating adaptive grids has been used in Dwyer et al. (1980) for time dependent and steady problems in multidimensional fluid dynamics and heat transfer. The concept of mesh redistribution can be implemented on both structured and unstructured meshes Weatherill, 1990).

8.7.2 Mesh Enrichment

This involves the addition of points in the region where the adaptation is required. It is difficult to add points in structured grids. Regular array of points which is the fundamental property of structured grid will no longer remain valid once we add few points in a particular region. Non-conforming nodes will be introduced and modification of solver code will be necessary after each enrichment. On unstructured grid this can be readily implemented. The addition of points involves a local reconnection of the triangular elements, and the resulting grid remains in the same form as the initial grid and the same solver can be used on the enriched grid. Once the regions of enrichment become known and the individual elements are identified there are a number of strategies for adding points. For details of enrichment procedure and application one should read Thompson et al. (1999), Weatherill (1990).

To know more about specialized grids in two and three dimensions the reader is referred to current literature Eriksson (1982), Hauser et al. (1983), Smith (1980), Taniguchi et al. (1992), Thempson et al. (1999), Weatherill (1990) for multiblock grid, unstructured grid, adaptive grid and so on for complex geometries in two and three dimensions using partial differential equations, algebraic relations, transfinite interpolation etc.

8.8 SUMMARY

The subject *grid generation* is still under active research and development. The emphasis is on the hybrid combination of different methods and algorithms to get the required quality grid suitable for the problem of interest. Nowadays algorithms are being developed for *grid less computation* (Ramesh, 2002) where, instead of the network of grid cells or mesh only a cloud of points in the computational domain is needed. But to get that cloud of points one needs to run a grid generation code. So, a grid is necessary to get the numerical solution of the governing equations of fluid flows. Ultimate quality of grid depends on how accurately a physical problem can be solved by using this.

Basic mathematical tools necessary to develop a grid generation software are presented here in a systematic way. For simplicity, these are introduced in two dimensions which can be easily extended to three dimensions. Coordinate transformations and differential equation methods are discussed in detail. Solution procedure for elliptic type partial differential equations used for grid generation are also discussed in Chapter 5. Smoothness of the grid obtained is the basic advantage of these methods. Algebraic equation methods used for geting a C-type grid using hyperbolic/parabolic functions have been introduced in detail. A function suitable for O-type grid is also introduced. The steps are presented in a mannar such that the students can write their own codes for these steps independently. A complete flow chart is also provided. Working out the exercises will develop the skill and feelings for the algebraic functions and equations. Fitting a curve on a given data points, calculation of the arc lengths and distribution of points on the curve as desired are the basic steps for algebraic grid generation. It provides lot of flexibility and the method is very fast. Cubic-spline fit and the transfinite interpolation method are also introduced. Usually hybrid method, where basic grid is generated using algebraic (or transfinite interpolation) methods and then smoothing is done using an ellptic solver is the best way to get a good grid. Fundamentals of unstructured grid generation and the ideas of moving mesh and mesh enrichmemt are also introduced. Using the basic mathematical tools skillfully to achieve the desired goal of getting a grid is more of an art than science and the quality of the grid obtained depends mostly on the skill and experience of the user.

8.9 KEY TERMS

Advancing front
Connectivity matrix
Contravariant vector
Covariant vector
Cubic spline
Delaunay triangulation
Grid adaptation
Grid topology
Mesh enrichment
Moving mesh
Multi-block grid
Structured grid
Transfinite interpolation
Unstructured grid
Voronoi region.

8.10 EXERCISE 8

1. Let y-coordinates of NACA0012 aerofoil be given by

$$y = \pm 0.6\left(0.2969\sqrt{x} - 0.126\,x - 0.3516\,x^2 + 0.2843\,x^3 - 0.1015\,x^4\right) \quad (8.89)$$

Write a computer program to get the aerofoil data. Get the leading edge x_{le} and trailing edge x_{te} (values of x for which $y = 0$) of the aerofoil. Chord of an aerofoil is defined as the distance between its trailing edge and leading edge. In this case, chord=$x_{\text{te}} - x_{\text{le}}$. Scale the data to get an aerofoil closed at the trailing edge such that $chord = 1$. Designate this data set as your NACA0012 aerofoil.

2. Define twenty one points between $x = x_{\text{le}}$ to $x = x_{\text{te}}$, including the end points with uniform distribution in x. Draw the upper surface of NACA0012 aerofoil using these points. For each (x_i, y_i) calculate θ_i using (8.49) for $i = 1, \cdots, 20$.
3. For each θ_i and (x_i, y_i) of the Exercise 2 calculate $\eta_{i\,min}$ for $i = 1, \cdots, 21$ using (8.55) or (8.56) depending on the value of $\tan\theta_i$.
4. Let $B - A = 0.02$ with $A = 0.98$, where A and B are constants for the hyperbola (8.44). Use $x = \pm 4$, $y = \pm 5$ as the outer boundary. Calculate $\eta_{i\,\text{max}}$ for all $i = 1, \cdots, 21$.
5. Let $\Delta\eta_{\text{le}} = 0.005$ and $\Delta\eta_{\text{te}} = 0.02$. Calculate all the $\Delta\eta_i$ for $i = 1, \cdots, 21$ using (8.63).
6. Change the values of $B - A$ and θ_{te} and see the role of these two parameters on the skewness of the hyperbolas.
7. Calculate the arc length of the NACA0012 aerofoil using (8.46)and (8.47) and proceed upto third order and see the results are converging.
8. On all the hyperbolas generated in Exercise 4, calculate the arc lengths from $\eta_{i\,\text{min}}$ to $\eta_{i\,\text{max}}$ using equation (8.67) for all $i = 1, \cdots, 21$.
9. Use the total arc length calculated in Exercise 8, take $s_1 = 0.005$, use geometric progression to get s_n for $n = 2, \cdots, 20$. Get the values of η_n for all s_n, $n = 2, \cdots, 20$ using (8.68). [Hint: Starting from $\eta_1 = \eta_{\text{min}}$, find η_2 using (8.68) in an iterative procedure. Proceed in the same way to get all the η's.]

10. Use the transformation (8.52) and derive the set of equations for ξ, η and ΔL_{ij} as

$$\xi = \pm\sqrt{(x-B)+\sqrt{(x-b)^2+y^2}}$$

$$\eta = \pm\sqrt{\sqrt{(x-B)^2+y^2}-(x-B)}.$$

$$\Delta L_{ij} = \int_{\eta_{i,j}}^{\eta_{i,j+1}} \sqrt{\xi^2+\eta^2}.$$

9

Inviscid Incompressible Flow

Different mathematical models of varying complexity have been studied in the different chapters of the present work. For problems with attached flows, the viscous effects are mainly confined to the boundary layer. Consequently, they have a relatively small effect on the global flow pattern and inviscid flow predictions can serve a useful role. Particularly, the pressure field may be computed quite well with such an inviscid model. For this reason and because of its relative simplicity, the inviscid flow models are important. We have confined ourselves to inviscid flows in Chapters 6, 9 and 10. Chapter 10 has been devoted to solution of inviscid compressible flow, represented by the Euler model, the full potential model, and also by the transonic small perturbation model.

At low speeds, all fluids (a liquid or a gas) behave as an incompressible fluid. We begin our study of CFD (computational fluid dynamics) with the simplest model, namely, inviscid incompressible flow.

9.1 INTRODUCTION

The problem of inviscid incompressible flow past an arbitrary body in three dimensions is of much practical interest, particularly in aircraft design, ship design or in motor-car design. If it is permitted to assume the flow field to be irrotational, as is the case with many flows of practical interest, a velocity potential exists and the governing continuity equation yields the Laplace equation for the unknown velocity potential. Such flows are called *potential flows*. Once the velocity potential is found by solving the Laplace equation under appropriate boundary condition(s), the pressure is determined from the *Bernoulli's equation,* thereby bypassing the need of solving of the nonlinear Euler equations. Thus, the velocity field may be determined independent of the pressure field. The mathematical problem here, is to solve the Laplace equation which is linear and extensively studied in the literature (Courant and Hilbent, 1953; Kellog, 1929; Helwig,1964; Prasad and Ravindran, 1985) and the boundary value problem, most often is a Neumann problem. The only difficulty that may arise is that arising through the satisfaction of the tangency boundary condition(s), particularly for arbitrary and irregular body shapes.

For axisymmetric and three-dimensional problems, exact analytical solution may be found for flow past bodies like a sphere or ellipsoid, obtained through the technique of separation of variables. In two dimensions, exact solutions of the Neumann problem for a circle or for half-space are known (Courant and Hilbert, 1953; Hellweg, 1964; Prasad and Ravindran, 1985). For flow past an arbitrary two-dimensional body shape, the problem may be reduced to that of finding a suitable *conformal transformation* of the boundary. A large number of problems have been solved by this technique. We do not discuss here, the conformal transformation technique for which standard literature may be consulted (Betz, 1964; Copson, 1950). However, it may be noted that, the conformal mapping technique is not sufficiently general, as it is restricted to problems in two dimensions only. Further, for irregular boundary shapes, often the mapping function is not readily available and requires to be evaluated numerically or approximately, (Betz 1964; Schneider, 1978; Theodorsen, 1931; Thwartes; 1960).

One may think of using any of the methods of Chapter 8 (on Grid Generation), to generate boundary-fitted coordinates so that the boundary maps along one of the coordinate axes (or coordinate planes, as the case may be) and solve the transformed equations numerically in the transformed plane. Although, this is a feasible method, it is far too laborious and inefficient compared to *the method of singularity distribution,* to be introduced in the next few sections. The method of singularity distribution was introduced in the pre-computer days, (Martenson, 1959; Prager, 1928) and significant progress has been achieved in the sixties and early seventies among which the works of Hess and Smith (1967), Woodward (1968) and Rubbert and Saaris (1972) deserve special mention. For computing incompressible inviscid flow, *these methods are numerically exact* in the sense that arbitrarily high accuracy may be achieved with sufficient refinement of the meshes and that *these methods are not restricted by approximations like thin or slender bodies or small incidence or small perturbations.* A distribution of singularities over the body surface is used and the strengths of the distributed singularities are determined by solving an integral equation. The singularities involved are *the fundamental solutions* of Laplace equation (Courant and Hilbert, 1953; Prasad and Ravindran, 1985). It may be noted, however, that if these methods be used for computing linearized subsonic or supersonic flows, as done by many authors, the methods no more remain numerically exact.

It is proved in books on Potential Theory (Courant and Hilbert, 1953; Kellog, 1929) as well as in books on fluid dynamics (Karamcheti, 1966; Lamb, 1945), that the *acyclic flow (that is, irrotational flow in a simply connected domain)* past an arbitrary body in three dimensions may be represented by a surface distribution of sources alone or doublets alone on the surface of the body. Similar results hold in two dimensions, (vander Vooran and Jong, 1970). These results provide justification of these numerical methods, often known as *panel methods*, to be discussed in the following sections.

For the sake of simplicity, we discuss here in detail steady flow in two-dimensions, past streamlined bodies like airfoils. It may be noted that for flow past a lifting airfoil the domain is no longer simply connected and use of sources alone cannot generate the exact flow. An exact flow past a lifting airfoil may be generated in various ways using distribution of singularities. For example, A.M.O.

Smith and associates used (Hess and Smith, 1967) a distribution of sources at the surface of the airfoil together with a vortex distribution. A second possibility is to use a distribution of vortices only at the airfoil surface while a third approach is to use a doublet distribution along the airfoil surface and along the wake (Vander Vooran and Jong, 1970). It may be noted that, for the steady two-dimensional lifting case, physically no wake exists, but a mathematical cut from the trailing edge to the far field boundary is required.

It is to be noted further that using vortex distribution, the condition of vanishing normal velocity at the body surface leads to a *Fredholm integral equation of the first kind.* This leads to a system of linear algebraic equations which are often *ill-conditioned.* As shown by Prager (1928) and Martensen (1959), it is possible to replace the body boundary condition by another condition which leads to a Fredholm integral equation of second kind and the difficulty of treating an ill-conditioned system may be avoided. Also, the method of Morino (1973, 1974) and similar type of methods using a Dirichlet type of boundary condition do not result in system of equations that are ill-conditioned. For three-dimensional flows, use of vortex distribution is not convenient since the unknown vorticity is a vector with two components. For this reason a doublet distribution is often preferred. Rubbert and Saaris (1972) use a combination of sources and doublets.

Another class of methods may be found in the literature, which use a distribution of singularities interior to the body surface (Basu, 1978). Although, these category of methods significantly reduce the labour involved and often deliver very good results, we do not discuss them here in view of the fact that general body shapes cannot be exactly represented by internal singularity distributions and the resulting methods belong to the category of approximate methods.

9.2 POTENTIAL FLOW PROBLEM

The problem considered is that of potential flow of an inviscid incompressible fluid past an arbitrary two-dimensional body. That is, the flow is such that it may be assumed to be two dimensional, implying thereby that, the flow variables like the velocity and pressure depend only on two space variables. The governing equation of continuity is

$$\nabla.\vec{q} = 0, \tag{9.1}$$

where $\vec{q}$ denotes the velocity vector, the density ρ being assumed to be a constant quantity. The Euler's equation of motion is

$$\frac{\partial \vec{q}}{\partial t} + (\vec{q}.\nabla)\vec{q} = -\frac{1}{\rho}\nabla p. \tag{9.2}$$

All body forces like gravity are assumed to be conservative, their potential being included in the pressure p. Equations (9.1) and (9.2) hold outside a three dimensional body Ω with boundary surface $\partial\Omega$.

Choosing the coordinate system fixed with respect to the body, the normal velocity boundary condition requires that

$$\vec{q}.\vec{n}|_{\partial\Omega} = 0, \tag{9.3}$$

$\vec{n}$ denoting the unit outward normal vector. For the present external flow problem, in which flow outside a body is considered, *a regularity condition (like vanishing of perturbation velocities)* is required at infinity.

We assume the flow field to be *irrotational.* This assumption is valid for a large number of cases of practical interest. For example, flows starting from rest by the action of conservative field of forces or by the motion of boundaries are irrotational. This assumption implies the existence of a velocity potential Φ such that

$$\vec{q} = \nabla\Phi. \tag{9.4}$$

Let

$$\vec{q} = \vec{q}_\infty + \vec{q}', \tag{9.5}$$

where $\vec{q}_\infty$ denotes the velocity field of *onset flow,* that is, the velocity field when no boundaries are present and $\vec{q}'$ denotes the *perturbation velocity field* due to the introduction of the body or the boundaries. We assume the perturbation velocity field $\vec{q}'$ to be irrotational, so that

$$\vec{q}' = \nabla\phi. \tag{9.6}$$

The perturbation field may show non-zero circulation around the body. If the circulation around the body differs from zero, the potential is made single valued by introducing a cut from the body to infinity. Since $\vec{q}_\infty$ is the velocity field of an incompressible flow, it satisfies (9.1) and so

$$\nabla.\vec{q}' = 0,$$

implying

$$\nabla.(\nabla\phi) = \nabla^2\phi = 0. \tag{9.7}$$

The boundary condition (9.3) then yields

$$\nabla\phi.\vec{n}|_{\partial\Omega} = -\vec{q}_\infty.\vec{n}. \tag{9.8}$$

The *regularity condition* at infinity requires

$$|\nabla\phi| \to 0. \tag{9.9}$$

So the mathematical problem now reduces to the solution of the Laplace equation (9.7) subject to Neumann boundary condition (9.8) and the regularity condition (9.9).

Under the above assumptions of irrotationality, the Euler's equation (9.2) may be integrated to yield the Bernoulli's equation (Karamcheti, 1966; Landau and Lifshitz, 1979; Niyogi, 1977; Oswatitsch, 1977)

$$\frac{p}{\rho} + \frac{1}{2}|\vec{q}|^2 = \frac{p_0}{\rho}, \tag{9.10}$$

where p_0 is the *stagnation pressure*, that is the value of pressure corresponding to vanishing velocity.

The pressure coefficient C_p is defined as

$$C_p = \frac{p - p_\infty}{\frac{1}{2}\rho|\vec{q}_\infty|^2} = 1 - \frac{|\vec{q}|^2}{|\vec{q}_\infty|^2}. \tag{9.11}$$

As already pointed out, in order to construct a solution of the problem stated by Eqs. (9.7) – (9.9), different types of singularity distributions may be used. One of the successful and effective methods is due to A.M.O. Smith and associates, using a distribution of sources and vortices over the boundary surface Ω.

9.3 PANEL METHODS

For solving the problem of inviscid incompressible flow past a body Ω with surface $\partial\Omega$ governed by Eqs. (9.7)–(9.9), the body surface is subdivided into a large but finite number of small surface elements, called *panels*. Appropriate kind of singularities, like source, vortex or doublets or a combination of them, as mentioned in Section 9.1, are distributed over these panels. These singularities are *fundamental solutions* of Laplace equation. In view of linearity of Laplace equation any linear combination of solutions is also a solution. These solutions satisfy the regularity condition at infinity. The problem is then reduced to finding the strength of the singularity distribution by satisfying the body boundary condition of zero normal flow at the body. This yields an integral equation for the unknown strength of the singularity. To understand this, let us study the simpler two–dimensional problem of flow past a nonlifting symmetrical airfoil using distribution of sources, as shown in Fig. 9.1, after A.M.O. Smith method.

The potential at any point P due to a distribution of sources along the surface of the airfoil is given by

$$\phi_P = \frac{1}{2\pi}\oint \ln r_{PQ}\,\sigma(Q)\,ds_Q, \tag{9.12}$$

where $\sigma(Q)$ denotes the the strength of the source distribution per unit length at the point Q on the surface of the airfoil and r_{PQ} denotes the distance between the points P and Q and ln denotes the natural logarithm.

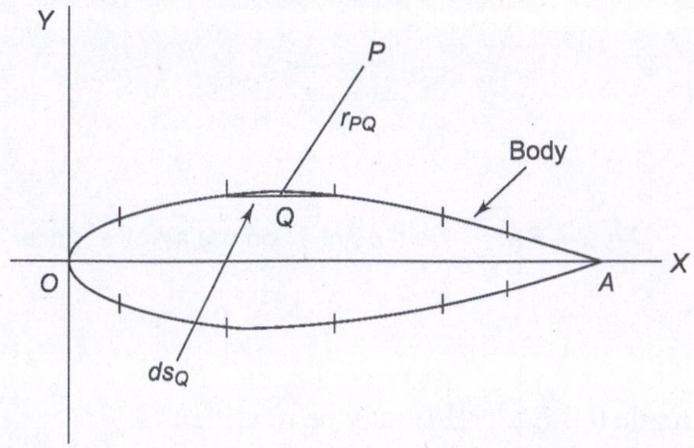

Figure 9.1 Source distribution on the surface of an airfoil; field point P outside the airfoil.

Clearly, Eq. (9.12) satisfies Laplace equation (9.7). The boundary condition Eq. (9.8) is also satisfied if

$$\frac{1}{2\pi}\frac{\partial}{\partial n_P}\oint \ln r_{PQ}\,\sigma(Q)\,ds_Q = -\vec{q}_\infty.\vec{n}_P, \tag{9.13}$$

for any point P on the surface of the airfoil, $\vec{n}_P$ denoting unit outward normal vector at the point P. The integrand $\ln r_{PQ}$ becomes singular as the point P approaches the pivotal point Q on the surface Fig. 9.2. To avoid this, the integral is divided over two parts, C_1 and C_2, where C_1 is a small neighbourhood of the point P and the remaining part of the boundary is C_2. We assume $\sigma(Q)$ to be a continuous function. The part over C_1 then yields the contribution $\frac{1}{2}\sigma(P)$ (Courant and Hilbert, 1953; Kellog, 1929). For the remaining part C_2, differentiation under the integral sign is permissible , so that Eq. (9.13) yields

$$\frac{1}{2}\sigma(P) + \frac{1}{2\pi}\int_{C_2}\frac{\partial}{\partial n_P}\ln r_{PQ}\,\sigma(Q)\,ds_Q = -\vec{q}_\infty.\vec{n}_P. \tag{9.14}$$

Since

$$\frac{\partial}{\partial n_P}(\ln r_{PQ}) = \frac{1}{r_{PQ}}\frac{\partial}{\partial n_P}r_{PQ} = -\frac{\cos\theta_P}{r_{PQ}}, \tag{9.15}$$

where θ_P denotes the angle between the vectors $\vec{n}_P$ and $\vec{PQ}$ taken from P to Q Fig. 9.2.

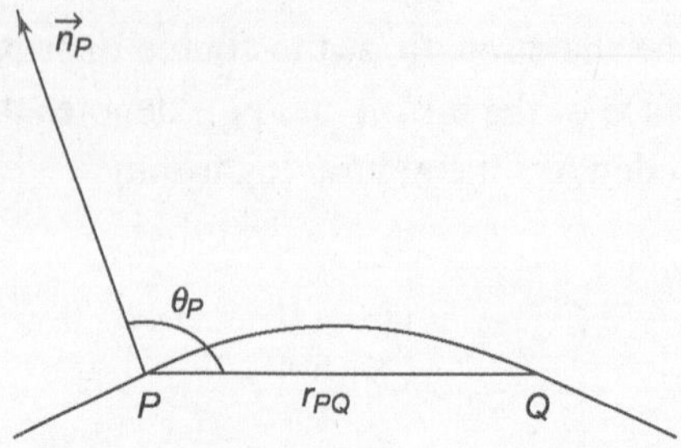

Figure 9.2 Field point P on the airfoil surface.

In terms of the angle θ_P, Eq. (9.14) may be rewritten as

$$\frac{1}{2}\sigma(P) - \frac{1}{2\pi}\oint \frac{\cos\theta_{\rm P}}{r_{\rm PQ}}\sigma({\rm Q})ds_{\rm Q} = -\vec{q}_\infty.\vec{n}_{\rm P}, \tag{9.16}$$

which is the basic equation for determination of the source strength distribution $\sigma(P)$. This is a *Fredholm integral equation of the second kind* for the unknown $\sigma(P)$. It may be noted that in this form the integrand in Eq. (9.16) remains finite as Q approaches the point P, so that the integral may be taken over the whole contour. To see this, let C be the centre of curvature of the arc PQ and M the mid point of the segment PQ. It may be observed from Fig. 9.3, that

$$\cos\theta_P = \cos(\pi - < MPC) = -\frac{PM}{MC}.$$

Also, $r_{PQ} = 2PM$, so that $\dfrac{\cos\theta_P}{r_{PQ}} = -\dfrac{\frac{PM}{MC}}{2PM} = -\dfrac{1}{2MC}$. Consequently, when Q tends to P along the surface of the airfoil, the ratio $\dfrac{\cos\theta_P}{r_{PQ}}$ approaches a finite value. The existence (Kellog, 1929) of solution of Eq. (9.16) requires the right hand side to be a continuous function. In view of the presence of $\vec{n}_P$, it means that the surface must have a continuous normal vector, so that boundaries with corners are excluded. Experience show that the above method has difficulty near concave corners, while for convex corners quite correct results may be obtained.

In the panel methods, a discrete set of control points is chosen, one point in each panel and the normal flow boundary condition is required to be satisfied at these control or pivotal points only. A linear algebraic equation is obtained corresponding to each control point, and the resulting system of algebraic equations is solved numerically by standard procedure, as already mentioned.

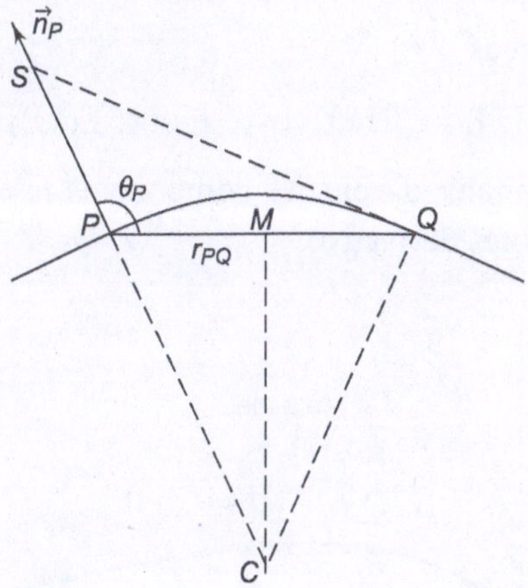

Figure 9.3 Evaluation of the integrand of Eq. (9.16) as Q tends to P.

The efficiency of the panel methods may be seen from the fact that for a three dimensional body, the integral equation is two-dimensional while for a two-dimensional problem a one-dimensional integral equation is obtained, thereby reducing the dimensionality of the problem by one. Further, for flow past bodies, the domain of the function to be found is reduced from the infinite exterior flow field to a finite domain, namely the body surface.

The A.M.O. Smith's method (Hess and Smith, 1967), for a lifting airfoil is presented in the next subsections, in some details, following Hancock and Padfield (1972).

9.3.1 AMO Smith Method for a Lifting Airfoil

We consider steady inviscid incompressible flow past an airfoil. With reference to a body fixed rectangular Cartesian coordinate system with origin at the tip of the airfoil, let the airfoil shape be given by

$$y = \begin{cases} f(x), & 0 < x < 1, \\ 0, & \text{otherwise.} \end{cases} \tag{9.17}$$

The free-stream flow velocity q_∞ is inclined at an angle α with the x-axis, taken along the chord of the airfoil, as shown in Fig. 9.4. Let $u'(x, f(x))$ and $v'(x, f(x))$ denote the perturbation velocity components on the airfoil surface. The normal flow boundary condition at the body can be written as

$$q_\infty \sin\alpha + v'(x, f(x)) = f'(x)\left[q_\infty \cos\alpha + u'(x, f(x))\right],$$

which may be rewritten as

$$\sin\alpha + v(x, f(x)) = f'(x)\,[\cos\alpha + u(x, f(x))]\,, \tag{9.18}$$

where u, v denote normalized velocity components $u = u'/q_\infty$, $v = v'/q_\infty$, and the prime on $f(x)$ denotes derivative.

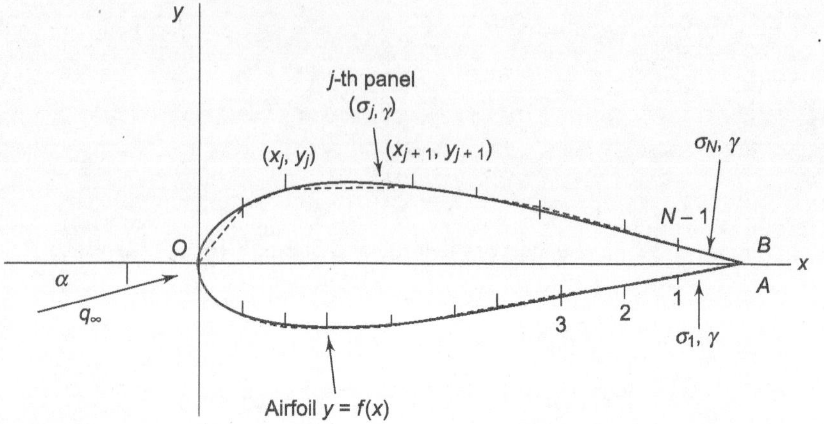

Figure 9.4 Distribution of source σ_j and constant vortex γ on the j-th panel on lifting airfoil surface.

The airfoil surface is now subdivided into N-panels by the points $1, 2, \ldots,$ $N-1$ on the airfoil surface and the consecutive points joined by straight segments, the panels being counted as $1, 2, \ldots, N$, beginning from the trailing edge A on the lower side and moving clockwise upto the trailing edge B on the upper side. Let us denote a typical element, the directed line segment from the point (x_j, y_j) to (x_{j+1}, y_{j+1}) as the j-th panel. The mid point $(\overline{x_j}, \overline{y_j})$ of this panel is chosen as the control point of the j-th panel at which point the body boundary condition is satisfied. Let the j-th panel make an angle θ_j with the positive direction of the x-axis.

A source distribution of strength σ_j on the j-th panel together with a vortex distribution of constant strength γ is placed on the panels. The strength of the source distribution is assumed to be constant over the panels but varies from panel to panel, while the vortex strength γ is assumed to be the same over all the panels. Both σ_j and γ are normalised with respect to the free-stream velocity q_∞. The normalized perturbation velocities may be expressed as

$$u_j = \sum_{i=1}^{N} A_{ji}\sigma(i) + \gamma \sum_{i=1}^{N} B_{ji}, \quad v_j = \sum_{i=1}^{N} B_{ji}\sigma_i - \gamma \sum_{i=1}^{N} A_{ji}, \tag{9.19}$$

where u_j and v_j are the normalized perturbation velocity components at the control point of the j-th panel, (that is, at the mid point of it), and A_{ji} and B_{ji} are known as *influence coefficients,* whose actual values are derived in the next subsection.

Using Eqs. (9.19) the boundary condition Eq. (9.18) for the j-th control point is satisfied if

$$\sum_{i=1}^{N}(B_{ji}-t_jA_{ji})\sigma(i)-\gamma\sum_{i=1}^{N}(A_{ji}+t_jB_{ji})=t_j\cos\alpha-\sin\alpha,\ j=1,2,\dots,N, \quad (9.20)$$

where t_j is the value of the slope $f'(x)$ at the j-th control point.

Further, *the Kutta condition* requires that there should be no flow around the sharp trailing edge, the trailing edge being a *stagnation point.* Kutta condition is stated often in other forms also, for example, the pressure must be finite and continuous at the trailing edge. This is implemented by taking the resultant velocities at the control points of the two panels on the two sides of the trailing edge to be equal. This condition delivers

$$(u_1+\cos\alpha)\cos\theta_1+(v_1+\sin\alpha)\sin\theta_1=(u_N+\cos\alpha)\cos\theta_N+(v_N+\sin\alpha)\sin\theta_N \quad (9.21)$$

the angle θ_i being the angle made by the i-th directed panel segment with the positive x-axis.

Using Eq. (9.19) in Eq. (9.21), follows on simplification

$$\sum_{i=1}^{N}(A_{1i}\cos\theta_1+B_{1i}\sin\theta_1-A_{Ni}\cos\theta_N-B_{Ni}\sin\theta_N)\sigma(i)+$$

$$\gamma\sum_{i=1}^{N}(B_{1i}\cos\theta_1-A_{1i}\sin\theta_1-B_{Ni}\cos\theta_N+A_{Ni}\sin\theta_N)$$

$$=\cos\alpha(\cos\theta_N-\cos\theta_1)+\sin\alpha(\sin\theta_N-\sin\theta_1) \quad (9.22)$$

Equations (9.20) and (9.22) constitute a system of $(N+1)$ linear algebraic equations for the $(N+1)$ unknowns $\sigma_j,\ j=1,2,\dots,N$ and γ which may be solved by standard procedures (Datta, 1995; Golub and Van Loan, 1989; Isaacson & Keller, 1966). Once the solution of these equations is obtained, the perturbation velocity components and the aerodynamic force and moment coefficients may be calculated.

9.3.2 Influence Coefficients

Let us consider a source distribution, of normalized strength σ per unit length, distributed between $-\Delta < X < \Delta$, along the X-axis (Fig. 9.5).

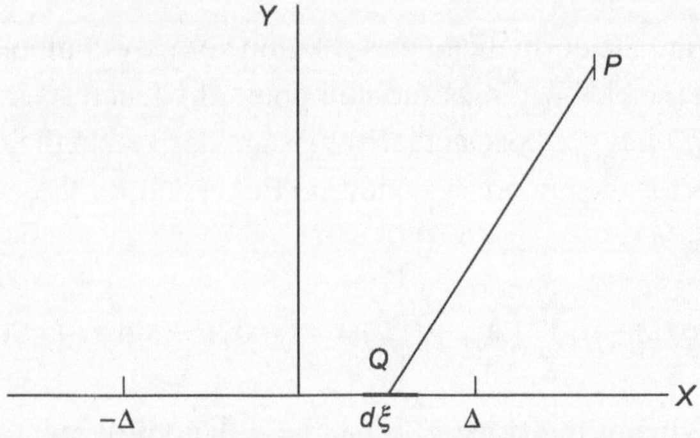

Figure 9.5 Calculation of induced velocity due to a source distribution $d\xi$ at $(\xi, 0)$ of strength σ at a field point P outside the airfoil.

Noting that the potential at any point (X, Y) due to an element of source distribution $d\xi$ of strength σ per unit length at $(\xi, 0)$ being

$$\phi = \frac{\sigma}{2\pi} \ln\left[(X-\xi)^2 + Y^2\right]^{\frac{1}{2}}, \tag{9.23}$$

the elementary velocity distributions dU, dV at any point (X, Y) may be expressed as

$$dU = \frac{\sigma}{2\pi} \frac{(X-\xi)\, d\xi}{(X-\xi)^2 + Y^2} \tag{9.24}$$

and

$$dV = \frac{\sigma}{2\pi} \frac{Y\, d\xi}{(X-\xi)^2 + Y^2}. \tag{9.25}$$

Integrating between $-\Delta$ and Δ,

$$U(X, Y) = \frac{\sigma}{2\pi} \int_{-\Delta}^{\Delta} \frac{(X-\xi)\, d\xi}{(X-\xi)^2 + Y^2} = \sigma I(X, Y, \Delta), \tag{9.26}$$

where

$$I(X, Y, \Delta) = \frac{1}{4\pi} \ln\left|\frac{(X+\Delta)^2 + Y^2}{(X-\Delta)^2 + Y^2}\right| \tag{9.27}$$

and

$$V = \frac{\sigma}{2\pi} \int_{-\Delta}^{\Delta} \frac{Y\, d\xi}{(X-\xi)^2 + Y^2} = \sigma J(X, Y, \Delta) \tag{9.28}$$

where

$$J(X, Y, \Delta) = \frac{1}{2\pi}\left[\tan^{-1}\frac{X+\Delta}{Y} - \tan^{-1}\frac{X-\Delta}{Y}\right] \tag{9.29}$$

and the principal value of the inverse tangent function, defined by $-\frac{1}{2}\pi < \tan^{-1} x < \frac{1}{2}\pi$ is to be taken. This yields the correct induced velocity distribution such that V is antisymmetric and U is symmetric about the x-axis. It may be noted that, as

$$X \to 0, Y \to 0+, \; U(0, 0+) \to 0, \; V(0, 0+) \to \frac{1}{2}\sigma,$$

$$\text{while for } X \to 0, Y \to 0-, \; U(0, 0-) = 0, \; V(0, 0-) = -\frac{1}{2}\sigma. \tag{9.30}$$

Different panels are inclined at different angles with respect to the positive x-axis of the body fixed coordinate axes ox,oy, with the origin o at the tip of the airfoil and the x-axis along the chord of the airfoil. So, a transformation of coordinates from (X, Y) to (x, y) is necessary.

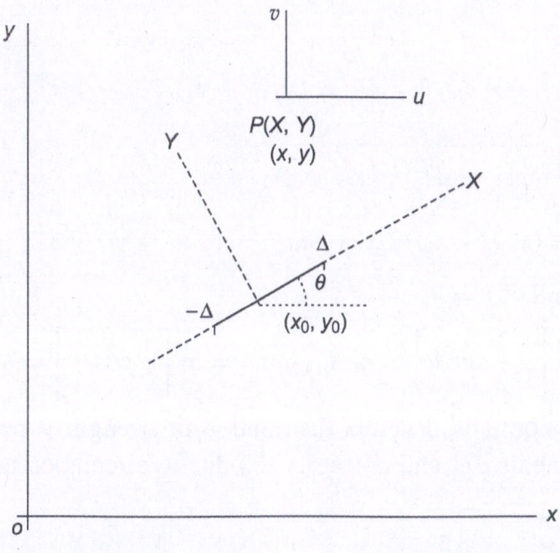

Figure 9.6 Source distribution on a panel at (x_0, y_0) inclined at an angle θ with the positive x-axis.

For a panel inclined at an angle θ with the positive x-axis, situated at (x_0, y_0), the normalized velocity components u, v are given by (Fig. 9.6)

$$\begin{aligned} u &= \sigma\,[I(X, Y, \Delta)\cos\theta - J(X, Y, \Delta)\sin\theta], \\ v &= \sigma\,[I(X, Y, \Delta)\sin\theta + J(X, Y, \Delta)\cos\theta], \end{aligned} \tag{9.31}$$

where

$$X = (x - x_0)\cos\theta + (y - y_0)\sin\theta, \quad Y = -(x - x_0)\sin\theta + (y - y_0)\cos\theta. \tag{9.32}$$

The influence of the source distribution on the i-th panel at the mid point of the j-th is given by

$$u_{ji} = \sigma_i \left[I_{ji} \cos\theta_i - J_{ji} \sin\theta_i \right] = \sigma_i A_{ji},$$

$$v_{ji} = \sigma_i \left[I_{ji} \sin\theta_i + J_{ji} \cos\theta_i \right] = \sigma_i B_{ji},\ i \neq j \tag{9.33}$$

where

$$I_{ji} = I(X_{ji}, Y_{ji}, \Delta_i), \quad J_{ji} = J(X_{ji}, Y_{ji}, \Delta_i)$$

and

$$X_{ji} = (\bar{x}_j - \bar{x}_i)\cos\theta_i + (\bar{y}_j - \bar{y}_i)\sin\theta_i$$

$$Y_{ji} = -(\bar{x}_j - \bar{x}_i)\sin\theta_i + (\bar{y}_j - \bar{y}_i)\cos\theta_i \tag{9.34}$$

where

$$\bar{x}_j = (x_{j+1} + x_j)/2, \quad \bar{y}_j = (y_{j+1} + y_j)/2, \tag{9.35}$$

$$\Delta_i = \sqrt{(x_{i+1} - x_i)^2 + (y_{i+1} - y_j)^2}/2, \tag{9.36}$$

$$\cos\theta_i = (x_{i+1} - x_i)/2\Delta, \quad \sin\theta_i = (y_{i+1} - y_i)/2\Delta \tag{9.37}$$

On the other hand, if $i = j$,

$$u_{ii} = \sigma\left[-\frac{1}{2}\sin\theta_i\right] = \sigma_i A_{ii}, \quad v_{ii} = \sigma_i\left[\frac{1}{2}\cos\theta_i\right] = \sigma_i B_{ii}. \tag{9.38}$$

Again, for a normalised vortex distribution of strength γ per unit length, the velocity components dU and dV at (X, Y) due to circulation around an element $d\xi$ are

$$\mathrm{d}U = \frac{\gamma}{2\pi}\frac{Y}{(X-\xi)^2 + Y^2}, \text{ and } dV = -\frac{\gamma}{2\pi}\frac{X-\xi}{(X-\xi)^2 + Y^2} \tag{9.39}$$

Thus, the induced velocity distributions dU and dV for strength γ are the same as dV and $-dU$ due to the strength σ given by Eqs. (9.24) and (9.25). Consequently, the velocities induced by a normalised vortex distribution are

$$u_{ji} = \gamma B_{ji}, \quad v_{ji} = -\gamma A_{ji}. \tag{9.40}$$

With these details, it is now easy to write a computer program to implement the panel method as presented above, which is left as an exercise to the reader.

9.4 PANEL METHODS (CONTINUED)

9.4.1 Mathematical Preliminaries for Morino–Kuo Method

Let D be a two-dimensional, simply connected, closed domain bounded by the closed curve S (Fig. 9.7).

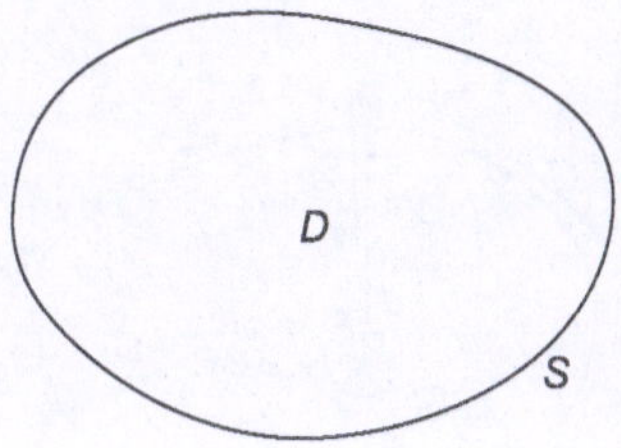

Figure 9.7 Domain for the application of Green's theorem.

If ϕ and ϕ^* are two functions satisfying the Laplace equation in D:

$$\nabla^2\phi = 0$$
$$\nabla^2\phi^* = 0$$

then Green's identity may be written as:

$$\int_S \left(\phi\frac{\partial\phi^*}{\partial n} - \phi^*\frac{\partial\phi}{\partial n}\right) dS = 0 \tag{9.41}$$

Now, let us identify ϕ as the potential due to a singularity-free flowfield in D and ϕ^* as the potential due to a two-dimensional source of unit strength, placed at a point P in D (Fig. 9.8). In order to be able to still apply Green's theorem to this situation, we have to a) exclude the singular point of the source (that is, point P), which we do by placing a circular barrier of small radius ϵ centred around it and b) keep domain D simply connected, to achive which we introduce a mathematical *cut* from S to the circumference of this circle. These artifices make domain D free of any singularities.

Surface S may now be thought to consist of several parts: S_O, the outer boundary of D; S_I, the circumference of the circle around P and S_{C1}; and S_{C2}, which make up the cut. We may now write Eq. (9.41) as:

$$\int_{S_O}\left(\phi\frac{\partial\phi^*}{\partial n}-\phi^*\frac{\partial\phi}{\partial n}\right)dS+\int_{S_I}\left(\phi\frac{\partial\phi^*}{\partial n}-\phi^*\frac{\partial\phi}{\partial n}\right)dS$$
$$+\int_{S_{C1+C2}}\left(\phi\frac{\partial\phi^*}{\partial n}-\phi^*\frac{\partial\phi}{\partial n}\right)dS=0 \tag{9.42}$$

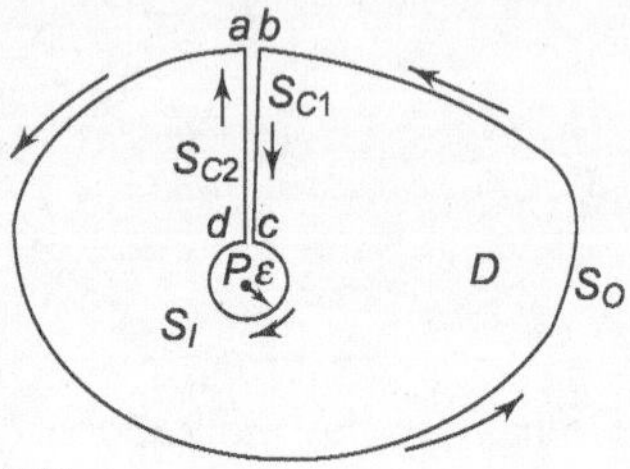

Figure 9.8 Isolating the source singularity at P.

Now, since ϕ and ϕ^* are continuous at all points on S_{C1} and S_{C2}, and at every point the normal on S_{C1} is directed against that on S_{C2}, the third integral vanishes.

Again, let the radius ϵ of the circle around P become very small. Then, the value of ϕ anywhere on this circle is very nearly its value at P, say ϕ_P, and this approximation gets better as ϵ becomes smaller. Since ϕ is continuous, $\dfrac{\partial\phi}{\partial n}$ is also nearly constant over the circle. Therefore,

$$\lim_{\epsilon\to 0}\int_{S_I}\left(\phi\frac{\partial\phi^*}{\partial n}-\phi^*\frac{\partial\phi}{\partial n}\right)dS\approx\phi_P\int_{S_I}\frac{\partial\phi^*}{\partial n}dS-\left(\frac{\partial\phi}{\partial n}\right)_P\int_{S_I}\phi^* dS \tag{9.43}$$

The first integral on the right hand side of this equation is the flux due to the source. Its value is equal to the source strength, here unity. The second integral is zero since ϕ^* is constant over the circle S_I. Hence, in the limit, the second integral of Eq. (9.42) is equal to ϕ_P. Equation (9.42) may then be written as:

$$\phi_P=\int_{S_O}\left(\phi^*\frac{\partial\phi}{\partial n}-\phi\frac{\partial\phi^*}{\partial n}\right)dS \tag{9.44}$$

In this equation, ϕ^* is the potential, at some point on S_O, due to a unit source at P; and $\dfrac{\partial\phi^*}{\partial n}$ is its derivative normal to S_O at the point in question.

We have introduced ϕ^* as the potential at some point on the boundary due to a unit source located at P. We may now turn our point of view around and say that ϕ^* is the potential *at P* due to a unit source placed somewhere on the boundary. The quantity $\dfrac{\partial\phi^*}{\partial n}$ then becomes the velocity induced at P by the unit source located on the boundary. The direction of the velocity is that of the normal to the surface at the point where the source is located.

Again, $\dfrac{\partial \phi^*}{\partial n}$ is the potential at P due to a *unit doublet* placed at the same point on the boundary as the source, with its axis directed along the normal to the boundary at that point. In a new interpretation that is vital for the subsequent development, we may *now* regard ϕ in Eq. (9.44) as the *strength* of a doublet, and $\dfrac{\partial \phi}{\partial n}$ as the *strength* of a source on S_O. Of course ϕ_P remains, as before, the value of the velocity potential at P.

Thus we see that Eq. (9.44) gives the value of ϕ at a point within D in terms of a distribution of sources and doublets on the boundary S_O. This is the importance of this equation.

9.4.2 Flow Past an Aerofoil

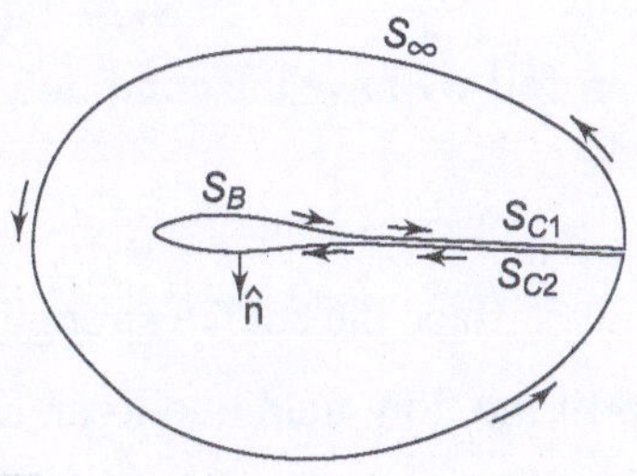

Figure 9.9 Mathematics of the flow past an aerofoil.

Let S_B represent the surface of an aerofoil immersed in a uniform stream that is bounded by a faraway surface S_∞. In order to keep the value of the velocity potential single valued and continuous, we introduce a mathematical cut $S_{C1} + S_{C2}$ extending from the trailing edge of the aerofoil to the far boundary. (The presence of circulation around a lifting aerofoil causes the value of the velocity potential to jump abruptly by an amount equal to the magnitude of the circulation each time a complete circuit around the aerofoil is made. If such circuiting is permitted, the velocity potential will be multiple valued and discontinuous, violating the condition of Green's identity. The cut prevents such circuits from being made.) Eq. (9.44), specialized to this case, becomes:

$$\phi_P = \int_{S_\infty} \left(\phi^* \frac{\partial \phi}{\partial n} - \phi \frac{\partial \phi^*}{\partial n}\right) dS + \int_{S_B} \left(\phi^* \frac{\partial \phi}{\partial n} - \phi \frac{\partial \phi^*}{\partial n}\right) dS + \int_{S_{C1+C2}} \left(\phi^* \frac{\partial \phi}{\partial n} - \phi \frac{\partial \phi^*}{\partial n}\right) dS \qquad (9.45)$$

Now, if ϕ_∞ represents the potential due to the free stream alone then, on the far boundary S_∞:

$$\phi = \phi_\infty \tag{9.46}$$

since the perturbation effect due to the aerofoil must vanish there. Also, at any point (x, y) in D, the free stream potential is:

$$\phi_\infty = V_\infty(x \cos\alpha + y \sin\alpha) \tag{9.47}$$

where V_∞ is the free stream speed far from the aerofoil and α is the angle the stream makes with the x-axis there. We may imagine that the free stream is produced by the distribution of sources and doublets on S_∞, and it is these singularities that define the velocity potential due to the free stream at point P. Then, the first integral in Eq. (9.45) becomes:

$$\int_{S_\infty} \left(\phi^* \frac{\partial\phi}{\partial n} - \phi \frac{\partial\phi^*}{\partial n}\right) dS = V_\infty(x_P \cos\alpha + y_P \sin\alpha) \tag{9.48}$$

On the surface of the aerofoil we have the condition of impenetrability:

$$\frac{\partial\phi}{\partial n} = 0 \tag{9.49}$$

This means that on the body surface the source strength is zero, and we only have a distribution of doublets. In the flow field due to an aerofoil, the quantity $\dfrac{\partial\phi}{\partial n}$ is continuous everywhere. By arguments given in the last subsection, therefore,

$$\int_{S_{C1+C2}} \phi^* \frac{\partial\phi}{\partial n} dS = 0$$

The values of ϕ itself, at corresponding points on S_{C1} and S_{C2}, are different when an aerofoil is producing lift. If $\Delta\phi$ is the difference in the values of ϕ, then:

$$\int_{S_{C1+C2}} \phi \frac{\partial\phi^*}{\partial n} dS = \int_{S_{C1}} \Delta\phi \frac{\partial\phi^*}{\partial n} dS$$

It may be noted that $\Delta\phi = \Gamma$, the circulation around the aerofoil.

Following all this discussion and simplification, Eq. (9.45) finally becomes:

$$\phi_P = V_\infty(x_P \cos\alpha + y_P \sin\alpha) - \int_{S_B} \phi \frac{\partial\phi^*}{\partial n} dS - \int_{S_{C1}} \Delta\phi \frac{\partial\phi^*}{\partial n} dS \tag{9.50}$$

Thus, for the case of an aerofoil in a free stream, the velocity potential at any point P is given in terms of yet-to-be-determined strengths of certain doublets distributed on S_B and S_{C1}.

9.4.3 A Constant-Potential Panel Method

We shall make use of Eq. (9.50) in developing a simple panel method. The method was pioneered by Morino and Kuo (1974), and developed further by Johnson and Rubbert (1975) and Maskew (1982).

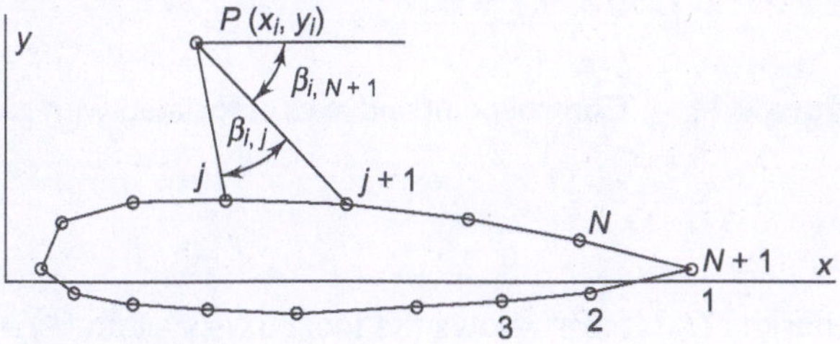

Figure 9.10 Aerofoil replaced by a scheme of panels.

9.4.4 Morino-Kuo Method

Consider that the smooth surface of the aerofoil has been replaced by an N-sided polygon inscribed in it, as shown in Fig. 9.10. The nodes are numbered sequentially, with node 1 at the trailing edge on the lower surface and node N+1 again on the trailing edge but on the upper surface. Edges or *panels* are similarly numbered. Panel j lies between nodes j and $j+1$. The central point of the method being developed is the assumption that on each edge of the polygon the value of ϕ (which is the velocity potential and, from another point of view also the strength of a doublet with axis normal to the surface) is constant.

We may now rewrite Eq. (9.50), replacing point P by its index i, as:

$$\phi_i + \sum_{j=1}^{N} \phi_j \int_{S_j} \frac{\partial \phi_{ij}^*}{\partial n} \mathrm{d}S_j + \Delta\phi \int_{S_{C1}} \frac{\partial \phi_{i,N+1}^*}{\partial n} \mathrm{d}S_j = V_\infty (x_i \cos\alpha + y_i \sin\alpha) \quad (9.51)$$

where $\frac{\partial \phi_{ij}^*}{\partial n}$ is the velocity potential induced at point i by the distributed doublet of unit strength on panel j and the quantity $\frac{\partial \phi_{i,N+1}^*}{\partial n}$ is the velocity potential induced at the point i due to the distributed doublet of unit strength on the *semi-infinite* panel coinciding with the cut S_{C1}. The exact shape of the cut is immaterial, so it is assumed to be straight. It is also to be noted that the strength of the doublet on the lower surface of the cut (S_{C1}) in Fig. 9.9 is equal to the strength of the doublet on panel 1; and the strength of the doublet on the upper surface of the cut (S_{C2}) is equal to the strength of the doublet on panel N. So we have N unknowns, ϕ_1, ϕ_2, ..., ϕ_N to determine, for which we shall have to obtain N simultaneous algebraic equations.

To this end we now associate with each panel a *control point* at which the impenetrability condition is to be satisfied. Figure 9.11 shows the control point of

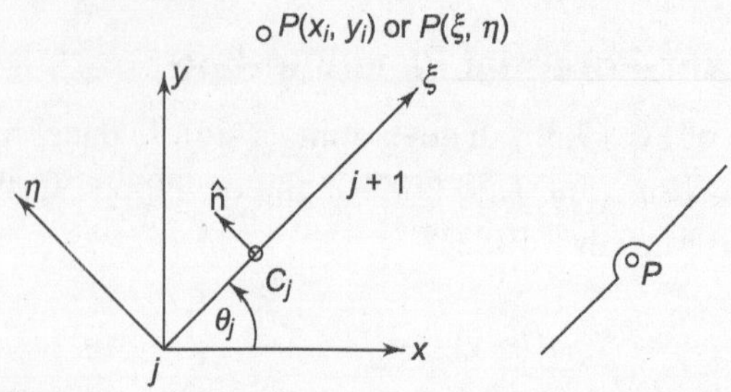

Figure 9.11 Control point and axes associated with panel j.

the jth panel, marked C_j. It also shows the local axis system (ξ, η) and its relation with the global axis system (x, y). By definition, the control point coincides with the mid-point of the panel:

$$x_{C_j} = \frac{x_j + x_{j+1}}{2}$$

$$y_{C_j} = \frac{y_j + y_{j+1}}{2}$$

The potential induced by the doublet distribution on panel j at the point P is:

$$\int_{S_j} \frac{\partial \phi_{ij}^*}{\partial n} \mathrm{d}S_j = -\frac{1}{2\pi} \int_{S_j} \frac{y_i}{(x_i - \xi)^2 + y_i^2} \mathrm{d}\xi$$

$$= \frac{\beta_{ij}}{2\pi} \tag{9.52}$$

where β_{ij} is the angle subtended by panel j at the point $P(x_i, y_i)$. It is obvious that for each point P there will be N angles β.

The angle $\beta_{i,N+1}$ becomes, simply:

$$\beta_{i,N+1} = \tan^{-1} \frac{y_i}{x_{N+1} - x_i} \tag{9.53}$$

Next we bring the point P to coincide, in turn, with the control point of each panel taking care, as before, to exclude the singularity by means of a semicircle as shown in Fig. 9.11. We then see that the angle subtended by a panel at its own control point is $\beta_{i,i} = \pi$, so that:

$$\int_{S_i} \frac{\partial \phi_{ii}^*}{\partial n} dS_i = \frac{1}{2} \tag{9.54}$$

Now we have a system of N linear algebraic equations of the form

$$\sum_{j=1}^{N} A_{ij}\phi_j = b_i, \quad i = 1, ..., N, \quad j = 1, ..., N \tag{9.55}$$

where

$$\begin{aligned} b_i &= V_\infty(x_i \cos\alpha + y_i \sin\alpha) \\ A_{ij} &= \delta_{i,j} - \frac{\beta_{i,j}}{2\pi} + \frac{\beta_{i,N+1}}{2\pi} \quad \text{when} \quad j = 1 \\ &= \delta_{i,j} - \frac{\beta_{i,j}}{2\pi} - \frac{\beta_{i,N+1}}{2\pi} \quad \text{when} \quad j = N \\ &= \delta_{i,j} - \frac{\beta_{i,j}}{2\pi} \qquad \text{otherwise} \end{aligned} \tag{9.56}$$

and $\delta_{i,j} = 1$ if $i = j$ and $\delta_{i,j} = 0$ if $i \neq j$ (Kronecker delta).

9.4.4.1 Pressure coefficient, forces, and moments Once the system of equations (9.55) is solved to obtain the velocity potential ϕ on the surface of the aerofoil, the velocity there has to be determined. This is done by numerically differentiating ϕ relative to the aerofoil arc length. One satisfactory way of doing this is to interpoalte a parabola through three neighbouring control points, with the arc length along the aerofoil surface as the independent variable and ϕ as the dependent variable and hence obtaining the derivative. Once the tangential velocity q_i at each control point is known, the pressure coefficient is got as:

$$c_{p_i} = 1 - \left(\frac{q_i}{V_\infty}\right)^2 \tag{9.57}$$

This may be integrated to obtain the lift coefficient as usual.

Another interesting way of getting the lift coefficient is to note that the circulation around the aerofoil, $\Gamma = \phi_N - \phi_1$, so that the lift force per unit span is $\rho V_\infty(\phi_N - \phi_1)$. If c is the chord of the aerofoil, then the lift coefficient per unit span is:

$$C_l = \frac{2(\phi_N - \phi_1)}{c V_\infty} \tag{9.58}$$

The lift coefficients calculated by the two methods usually agree to within a few percent.

A noticeable feature of this method, compared to most other panel methods, is that no explicit Kutta condition is necessary. This is understandable if the constant doublets on each panel are looked upon as a pair of vortices of oppsite sense, each located at one end of the panel. Then the vortices on the panels that touch the trailing edge are cancelled by the vortices at the ends of the "cut" that touch the trailing edge as well, resulting in zero vorticity there.

Figure 9.12 shows the chordwise distribution of the coefficient of pressure C_p on a NACA0012 aerofoil at an angle of attack 10 degree computed using the method described above. The aerofoil was divided into 36 panels, with most of them concentrated near the leading and trailing edges. Ideally speaking, the coefficient

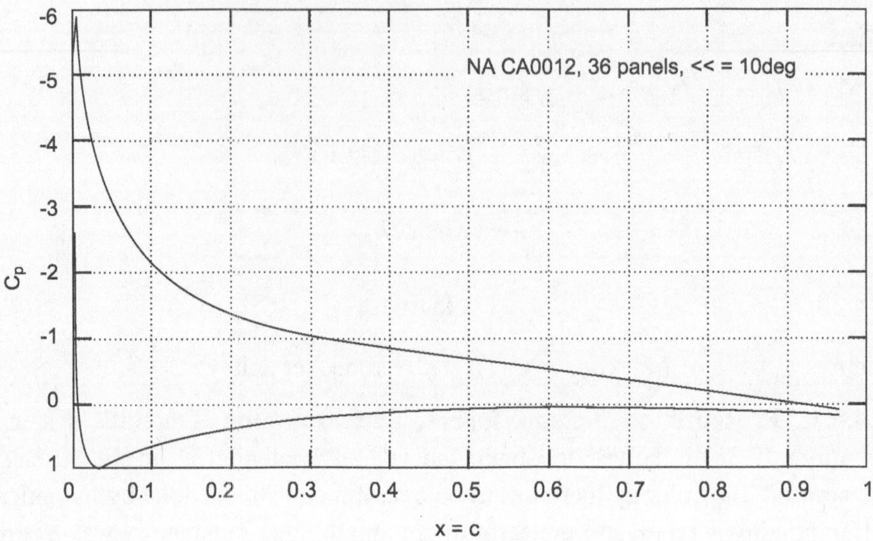

Figure 9.12 Pressure coefficient on NACA0012 at $\alpha = 10$ deg, 36 panels.

of pressure should rise to 1.0 at the trailing edge, since that is a stagnation point. As is seen here, however, this does not happen, and the rise in C_p falls short of the ideal. One reason is that the C_p is determined at the control points, and the trailing edge is not a control point. One could try to make the trailing edge panel very small, in an attempt to have a control point as near the trailing edge as possible. However, it is found from experience that neighbouring panels should not differ from each other too much in length, otherwise the system of equations may become ill-conditioned, or spurious oscillations in the variation of C_p may be introduced. These observations are, in general, true of all panel methods, exact only in the incompressible case.

An illustration Morinoprogram.c is presented in the next subsection.

9.4.5 Program 9.1: Morinoprogram.c

```
#include        <stdio.h>
#include        <stdlib.h>
#include        <math.h>

#define     FALSE           0
#define     TRUE            1

#ifndef PI
#define     PI              3.14159265358979323846
```

```
#endif

#define     TWOPI          (2.0 * PI )
#define     PIBY180        ( PI / 180.0 )

#define     INFINITY       1.0e+10

#define     PANELS         50

struct point
{   double
        x, y;
};

struct    paneltype
{   struct point
        le, te, ctl, dc;

    double
        len;
};

int
    n;                     /* actual number of panels */

struct paneltype
    panel[PANELS];

double
    alpha, chord, rhs[PANELS], infl[PANELS][PANELS];

static void
make_Geometry ( void )
{   register int
        i;

    double
        x, y, len;

    scanf ("%d %le %le", &n, &alpha, &chord);
    if ( n > PANELS )
    {   fprintf ( stderr, "Number of panels cannot exceed %d, is %d\n",
                          PANELS, n );
        fprintf ( stderr, "To increase the number of panels, change\n" );
        fprintf ( stderr, "the value of the #defined constant PANELS\n" );
```

```
        fprintf ( stderr, "and re-compile\n" );
        exit ( -1 );
    }

    printf( "# n = %d, alpha = % e, chord = % e\n", n, alpha, chord );

    for (i = 0; i < n; ++i)
        scanf ("%le %le", &panel[i].le.x, &panel[i].le.y);

    scanf ("%le %le", &panel[n-1].te.x, &panel[n-1].te.y);
    for (i = n-1; i > 0; --i)
        panel[i-1].te = panel[i].le;

    for (i = 0; i < n; ++i)
    {   panel[i].ctl.x = 0.5 * (panel[i].le.x + panel[i].te.x);
        panel[i].ctl.y = 0.5 * (panel[i].le.y + panel[i].te.y);
        y = (panel[i].te.y - panel[i].le.y);
        x = (panel[i].te.x - panel[i].le.x);
        len = sqrt (y*y + x*x);
        panel[i].len = len;
        panel[i].dc.x = x / len;
        panel[i].dc.y = y / len;
    }

/*
    printf ("Chord = % e\n", chord);
    for (i = 0; i < n; ++i)
        printf ("%d: % e % e : % e % e : % e % e : % e % e : % e\n",
                i, panel[i].le.x, panel[i].le.y, panel[i].te.x,
                panel[i].te.y, panel[i].ctl.x, panel[i].ctl.y,
                panel[i].dc.x, panel[i].dc.y, panel[i].len);
*/
}   /* make_Geometry */

static double
beta ( struct paneltype *on, struct paneltype *by )
{   double
        st, ct, dx, dy,
        xstar, ystar;

    if (on == by)
        return (PI);
    else
    {   ct = by->dc.x;
        st = by->dc.y;
```

```
        dx = on->ctl.x - by->le.x;
        dy = on->ctl.y - by->le.y;

        xstar = ct * dx + st * dy;
        ystar = ct * dy - st * dx;

        return (atan2 (ystar, xstar - by->len) - atan2 (ystar, xstar));
    }
}   /* beta */

static void
make_AMatrix ( void )
{   int
        i, j;

    double
        saveinfl[PANELS], wakeinfl;

    for (i = 0; i < n; ++i)
    {   for (j = 0; j < n; ++j)
            saveinfl[j] = infl[i][j] = beta(&panel[i], &panel[j]);

        infl[i][i] -= TWOPI;

        wakeinfl = atan2 (panel[i].ctl.y, chord - panel[i].ctl.x);
        infl[i][0] -= wakeinfl;
        infl[i][n-1] += wakeinfl;
    }
}   /* makeinfl */

static void
make_RHS ( void )
{   int
        i;

    double
        sa, ca;

    sa = sin (alpha * PIBY180);
    ca = cos (alpha * PIBY180);

    for (i = 0; i < n; ++i)
        rhs[i] = -TWOPI * (panel[i].ctl.x * ca + panel[i].ctl.y * sa);
}   /* make_RHS */
```

```
#define           MAXSIZE       PANELS

static void
LUFactorise (int N, double A[][MAXSIZE], int Pivot[MAXSIZE])
{   int
        I, J, K;

    double
        Big,
        ScaleFactor[MAXSIZE];    /* factors used in implicit scaling */

    for (K = 0; K < N; ++K)
    {   /* Find the implicit scaling factors for each row */
        for (I = K; I < N; ++I)
        {   ScaleFactor[I] = fabs(A[I][K]);

            for (J = K; J < N; ++J)
                if ((Big = fabs(A[I][J])) > ScaleFactor[I])
                    ScaleFactor[I] = Big;
        }

        /* Seek the pivotal element */
        Big = fabs(A[K][K] / ScaleFactor[K]);
        Pivot[K] = K;
        for (I = K; I < N; ++I)
        {   if ((fabs(A[I][K]) / ScaleFactor[I]) > Big)
                Pivot[K] = I;
        }

        /* Interchange the pivotal with current rows */
        if ((I = Pivot[K]) ! = K)
            for (J = K; J < N; ++J)
            {   Big = A[K][J];
                A[K][J] = A[I][J];
                A[I][J] = Big;
            }

        /* Subtract and store multipliers in the lower triangle */
        for (I = (K + 1); I < N; ++I)
        {   Big = A[I][K] = (A[I][K] / A[K][K]);
            for (J = (K + 1); J < N; ++J)
                A[I][J] -= Big * A[K][J];
        }
    }   /* End of for K */
}   /* End of LUFactorise */

static void
```

```
Backsub (int N, double A[][MAXSIZE], double *B, int Pivot[MAXSIZE])
{   int
        I, J, K;

    double
        Big, *Ptr1, *Ptr2, *PtrM;

    int
        *PtrP;

    for (K = 0, PtrP = Pivot, Ptr1 = B; K < (N-1); ++K, ++PtrP, ++Ptr1)
    {   if ((J = *PtrP) != K)
        {   Big = *(B+J);
            *(B+J) = *Ptr1;
            *Ptr1 = Big;
        }

        for (I = K + 1, Ptr2 = (B+K+1); I < N; ++I, ++Ptr2)
            *Ptr2 -= *(*(A+I)+K) * *Ptr1;
    }   /* End of for K */

    *(B+N-1) = *(B+N-1) / *(*(A+N-1)+N-1);

    for (I = N - 2, Ptr1 = (B+N-2); I >= 0; --I, --Ptr1)
    {   Big = 0.0;
        for (J = I + 1, Ptr2 = (B+I+1), PtrM = (A[I]+I+1); J < N;
             ++J, ++Ptr2, ++PtrM)
            Big = Big + *PtrM * *Ptr2;
        *Ptr1 = (*Ptr1 - Big) / *(*(A+I)+I);
    }
}   /* End of Backsub */

static void
quad ( double x1, double y1, double x2, double y2, double x3, double y3,
     double *z1, double *z2, double *z3 )

{   double
        b[MAXSIZE], a[MAXSIZE][MAXSIZE];

    int
        pivot[MAXSIZE];

    a[0][0] = x1 * x1;
    a[0][1] = x1;
    a[0][2] = 1.0;
```

```
    a[1][0] = x2 * x2;
    a[1][1] = x2;
    a[1][2] = 1.0;
    a[2][0] = x3 * x3;
    a[2][1] = x3;
    a[2][2] = 1.0;

    b[0] = y1;
    b[1] = y2;
    b[2] = y3;

    LUFactorise (3, a, pivot);

    Backsub (3, a, b, pivot);

    *z1 = b[0];
    *z2 = b[1];
    *z3 = b[2];
}   /* quad */

static void
make_Cp ( void )
{   int
        i;

    double
        a, b, c, s1, s2, s3;

    printf ("\n# Pressure coefficient\n");
    printf ("#     x               v               Cp\n");

 s2 = 0.0;
    for (i = 1; i < n-1; ++i)
    {   s1 = (i == 1 ? 0.5 * panel[0].len : s2);
        s2 = s1 + (panel[i-1].len + panel[i].len) * 0.5;
        s3 = s2 + (panel[i].len + panel[i+1].len) * 0.5;

        quad (s1, rhs[i-1], s2, rhs[i], s3, rhs[i+1], &a, &b, &c);

        c = (2.0 * a * s2 + b);
        printf ("% e  % e  % e\n", panel[i].ctl.x, c, 1.0 - c * c);
    }
}   /* make_Cp */

extern int
```

```
 main ( void )
{  int
      i, pivot[PANELS];

   make_Geometry ();

   make_AMatrix ();

   make_RHS ();

   LUFactorise (n, infl, pivot);

   Backsub (n, infl, rhs, pivot);

/***
   printf ("\nSolution\n");
   for (i = 0; i < n; ++i)
        printf ("%d: % e\n", i, rhs[i]);
***/

   make_Cp ();

   printf ("\n# Cl = % e\n", 2.0 * (rhs[n-1] - rhs[0]) / chord);

   return 0;
}  /* main */
```

9.4.6 Discretisation Error in Panel Methods

There are various sources of errors in the panel methods, associated with the satisfaction of the normal flow boundary condition. The source strength distribution is assumed to be a constant quantity over the panel. In terms of the length, say ϵ, of the largest panel, this leads to a truncation error of the order of $O(\epsilon^2)$, and clearly, relatively small panels ought to be chosen. While reasonably good results may be obtained in many cases with only about 50 or 60 panels, for a two-dimensional problem, roughly 100 panels is often sufficiently good for most purposes. The second source of error, which is a serious one , is that arising out of satisfying the body boundary condition only at the control points and not at every point of the body surface. At points on the surface, other than the control points, the normal flow velocity does not vanish in general, and *the flow is said to leak* between the pivotal points. Also, the source strengths experience a jump discontinuity at the edge of the panels where the velocity approaches infinity, introducing error. For this reason it is often said that the computed solution has significance at the control points on the body surface and outside the body surface only.

Another source of error is that irrespective of the body geometry, the panels are chosen flat. For the two dimensional case, they are straight segments while the actual body geometry is generally curved. Consequently the computed surface normals, although normal to the panels may not be the correct surface normals. For this reason, a large number of small panels have to be chosen.

Higher order panel methods have been developed as a remedy for the various discretization errors (Hess and Martin, 1974; Johnson and Rubbert, 1975). However, for most practical purposes a first-order panel method, as presented in the previous section, with a relatively large number of panels seems to be a more economic choice.

Economy may be introduced in choosing the size of the panels. Regions where stronger changes of body geometry or of flow takes place, like those near the leading and trailing edges of an airfoil, smaller panel size must be chosen, while in regions where flow changes slowly, relatively large size panels may be chosen. Further, the size of the panels should change only gradually and not abruptly.

9.5 MORE ON PANEL METHODS

The panel method presented in the previous section may be readily extended to three dimensional cases. However, the number of panels would increase heavily, and we have to deal with a large dense system of linear algebraic equations. Often, these systems are not diagonally dominant. Earlier authors (Hess and Smith, 1967) recommend direct methods, while for three dimensional cases, a higher order method followed by an iterative method is recommended (Hess and Martin, 1974).

For internal flow problems, for example, flow past a body confined in a channel or three dimensional flow around a complete aircraft or around a motor car, the use of first order panel method with constant strength source distribution over the panels, requires an enormously large number of panels. Even for a reasonable accuracy the methods turn out to be extremely laborious. For such cases, higher order methods have been recommended by several authors (see, for example, (Hess & Martin, 1974; Kraus, 1978). However, studies in Maskew (1982), Morino (1973), Morino and Kuo (1974) indicate that sufficiently good results may be obtained with these low–order panel methods in a cost effective way.

The panel method is closely related to the *boundary element method* (Brebia, 1978) which is quite convenient for internal flow problems. We do not discuss these methods, but refer the reader to literature (Brebia, 1978; Jawson and Symm, 1977).

9.6 PANEL METHODS FOR SUBSONIC AND SUPERSONIC FLOWS

Panel methods have been widely used for computing subsonic and supersonic flows past arbitrary complex configurations. However, it must be kept in mind that *all these solutions are approximate, in view of the linearization of the governing potential equation that has been presupposed.* In fact, linearization of the potential equation, assumes small perturbation, which in turn requires a thin or slender body with continuously turning tangents inclined at small angles to the flow direction everywhere. Further it requires linearization of the mass-flux density, which in turn requires either *subcritical flow* or else purely supersonic flow, the flow regimes *transonic and hypersonic* being excluded. Incidentally, a flow field is said to be subcritical if the local flow speed is smaller than the local speed of sound everywhere in the flow field.

In such cases of inviscid irrotational low speed steady *small perturbation flow,* instead of Laplace equation, the governing equation may be approximated by (Niyogi, 1977; Oswatitsch, 1956)

$$(1 - M_\infty^2)\frac{\partial^2 \phi}{\partial x^2} + \frac{\partial^2 \phi}{\partial y^2} + \frac{\partial^2 \phi}{\partial z^2} = 0, \tag{9.59}$$

where M_∞ denotes the free–stream Mach number and ϕ denotes the perturbation potential. Similarly, for purely supersonic steady irrotational flow, under the assumption of small perturbations, the governing potential equation may be linearized and reduced to the same form as Eq. (9.59), with $M_\infty^2 > 1$. For cases of subcritical subsonic flow *Prandtl–Glauert transformation* (Niyogi, 1977; Oswatitsch, 1956) may be used to reduce Eq. (9.59) to Laplace equation. For linearized supersonic flow, the panel method is directly applied to Eq. (9.59).

We wish to point it out clearly, that the governing potential equation cannot be linearized for transonic or hypersonic flows, even under the assumption of small perturbations. Thus, panel methods should not be used for computing transonic or hypersonic flow fields.

Finally, it may be mentioned that alternative formulation based on Green's Theorem of Potential Theory, (Kellog, 1929; Lamb, 1945) may be used for computing the perturbation velocity potential ϕ directly, without introducing the intermediate singularity (that is, source, vortex or doublet) distributions. Details of this approach may be found in the works of Morino (1973, 1974) and Maskew (1982). The authors propose low order panel methods for the general case of subsonic and supersonic aerodynamic flow around a lifting body having arbitrary shape and motion (that is, unstedy case) assuming small perturbations, the method being numerically exact only in the incompressible case.

9.7 SUMMARY

In many problems of practical importance involving fluid flows at high Reynolds numbers, meaningful results may be obtained by taking the flow to be inviscid and irrotational. Such flows satisfy the Laplace equation, which is linear. A solution of this equation for a given case may thus be built up from a superposition of its elementary solutions, such as sources and vortices, in conjunction with the specified boundary conditions.

Assuming that we are dealing with the flow past a body that is completely immersed in an infinite extent of fluid in all directions, the panel method begins by assuming the body surface to be composed of short linear segments, or panels. On each panel are distributed a source and a vortex of constant strength per unit length. The strength of the source is allowed to vary from panel to panel while the strength of the vortex is the same for all panels. The centre of each panel is called a control point. The velocity at the control point of a panel in a direction normal to it, induced by the distribution of singularities of yet unknown strength and the free stream, is set equal to zero in order to impose the impermeability condition. In addition, the Kutta condition at the trailing edge is imposed in any one of several ways. The resulting system of linear algebraic equations is solved for the strengths of the distributed singularities. Once these are known, the pressure on the surface, and hence the lift force (but not the drag!) may be calculated.

Panel methods can handle a wide range of configurations, including complicated shapes. The system of equations is usually well-conditioned. The presence of concave corners or edges may affect the condition adversely.

Panel methods are exact for incompressible flows. When the flow is compressible but either subcritical or completely supersonic, they can be applied provided the perturbations produced in the flow by the submerged body are small.

9.8 KEY TERMS

Acyclic flow
AMO Smith method
Bernoulli's equation
Circulation
Dirichlet problem
Doublet
Onset flow
Panel method
Potential flow
Prandtl-Galuert transformation
Pressure coefficient
Regularity condition

Fredholm integral equation
First kind
Second kind
Fundamental solution
Ill-conditioned system
Influence coefficients
Irrotational flow
Kutta condition Laplace equation
Leaking
Neumann problem
Simply connected domain
Small perturbation flow
Source
Stagnation point
Stagnation pressure
Subcritical flow
Surface singularity distribution
Inviscid irrotational flow
Vortex

9.9 EXERCISE 9

9.1 Why, do you think, it is possible to determine the lift on an aerofoil by using an inviscid flow model but not the drag?
When determining the lift on an aerofoil placed at an angle to the free stream, why is it necessary to introduce the Kutta condition? According to this condition, where is the rear stagnation point located?

9.2 Consider the problem of determining the lift on an aerofoil placed at an angle to a free stream. The surface of the aerofoil is divided into N linear panels. If you were using the AMO Smith panel method, what would be the unknown quantities whose values you have to determine using a system of linear algebraic equations? What is the number of equations you have to solve?

9.3 If the source distribution is replaced by a uniform doublet distribution of strength μ, the axes of the doublets being in the +ve Y direction. Show that the velocity potential ϕ induced at the point (X, Y) by this doublet distribution is the same as the Y-component of velocity due to the original source distribution, that is:

$$\phi(X, Y) = \frac{\mu}{2\pi}\left[\tan^{-1}\frac{X+\Delta}{Y} - \tan^{-1}\frac{X-\Delta}{Y}\right]$$

From the relations $U = \partial\phi/\partial X$ and $V = \partial\phi/\partial Y$, obtain the velocity field due to the above doublet distribution.
Show that this velocity field is the same as that due to a pair of concentrated vortices of strength μ, with one vortex located at each end of the panel. The vortex at the left end has an anti-clockwise sense while that at the right end has a clockwise sense.

9.4 Write a computer program to determine the distribution of surface pressure about a section of a circular cylinder immersed in inviscid incompressible flow. The direction of the free stream is normal to the axis of the cylinder. Since such a cylinder produces no lift, the strength of the associated vorticity distribution is zero, and only the source strengths need to be determined. The coefficient of pressure at a point on the body surface is defined as

$$C_\mathrm{p} = \frac{p - p_\infty}{\rho U_\infty^2/2}$$

where p is the static pressure at the point in question, p_∞ is the static pressure in the free stream and U_∞ is the speed of the free stream. By Bernoulli's theorem this becomes:

$$C_p = 1 - \left(\frac{u}{U_\infty}\right)^2$$

where u is the local flow speed.

Obtain results with the circumference of the cylinder divided into 8, 16, 32 and 64 panels, respectively. Compare these with the C_p values obtained analytically. The analytical expression is

$$C_\mathrm{p} = 1 - 4\sin^2\theta$$

where θ is measured from the forward stagnation point and is the angle subtended at the centre of the section of the cylinder by the arc extending from the forward stagnation point to the point where the C_p is being determined. Looking at the above comparison, what can you conclude about the effect of increasing the number of panels on a given surface?

9.5 Repeat Problem 9.4, but use uniformly distributed doublets instead of sources. What is the quality of the pressure distribution compared with that obtained using the source method for the same number of panels?

9.6 Write a computer program to compute steady incompressible nonlifting flow past an airfoil at zero incidence, using A.M.O. Smith method.

10

Inviscid Compressible Flow

Numerical solution of inviscid incompressible flow was discussed in the previous chapter. The present chapter is devoted to numerical study of inviscid compressible flow. It may be noted that linearized subsonic and supersonic flows, may be treated quite economically by the panel methods, based on *the assumptions of small perturbation,* discussed in the previous chapter. Moreover, they assume either low speed *subcritical flow* or else purely supersonic flow. It is pointed out further, that the linearized models, are quite restrictive and yield only approximate results. Their usefulness is to be seen in obtaining quick first results. In view of this, our main interest in the present chapter, is high speed flow and particularly, transonic and supersonic flow field computation. Shocks appear in such flow fields and their numerical computation is of much practical interest, although a difficult task. In the transonic speed range, flow fields with sufficiently high subsonic free-stream Mach number, become *super-critical* and the governing equations when simplified under the assumptions of small perturbation, are nolonger linear. Similarly, supersonic and hypersonic flow fields with stronger shocks cannot be correctly represented by the linearized models. Such inviscid flow fields, where the perturbations are not necessarily small, are of interest to us in this chapter.

10.1 INTRODUCTION

As pointed out above, the present chapter is devoted to numerical study of high speed flow problems. Study of super-critical or high subsonic, transonic and supersonic, even hypersonic flow fields belong to this category. High temperature effects like dissociation, ionization and thermodynamic relaxation are important in the hypersonic speed range, which is beyond the scope of the present work. Any of the methods discussed for transonic flow computation may be used for subsonic flows. Due to this, we lay more stress here on the study of transonic and supersonic flows.

Distinguishing feature of a *transonic flow field* is that both subsonic and supersonic regions are present in the flow field adjacent to one another, and that these fields are significant in determining the overall character of the flow field (Niyogi, 1982; Oswatitsch, 1977). Such flow fields occur in a wide range of aerodynamic problems like flow through nozzles, around blunt bodies moving supersonically, near airplane wings flying close to Mach number unity, around propellers and turbine blades.

Transonic speed range is one of the most efficient flight regimes, since optimum aircraft cruise performance, which is achieved at a value of free stream Mach number M_∞ that maximizes $M_\infty C_L/C_D$, is encountered in this range. Here, C_L and C_D denote the *lift and drag coefficients* respectively. For maximum maneuverability, high C_L is important which is also achieved in the same range. These features have made the analysis of transonic flow fields one of the most studied problems in fluid dynamics. However, in the transonic speed range, the flow pattern is quite complicated, because of the appearance of shock waves at the position of deceleration from supersonic to subsonic flow, as illustrated in Fig. 10.1. In general, the location of such shock waves is not known ápriori, and depends on the interaction of the free stream Mach number and the body geometry. Moreover, the appearance of a shock wave is associated with a pressure drag, called *wave drag*, which imposes a sharp positive pressure gradient on the boundary layer. This may lead to boundary layer separation, thereby increasing the drag further. The separated flow is often unstable. Furthermore, the drag arising due to the appearance of shock waves depends on the shock strength. Relatively stronger shocks often lead to wing flutter and instability. In fact, as M_∞ increases beyond the optimum performance range, such adverse transonic effects in the form of increased drag, shock-induced separation, etc. are encountered, that deteriorates performance drastically. Consequently, even today, most civilian transport aircrafts fly at Mach numbers rather less than unity and similar is the case with most military aircrafts on low flying attack missions. However, the value of M_∞ at which these performance limiting conditions occur depends strongly on the configuration. Thus one of the principal objectives of transonic configuration design is to maximize this value of M_∞ and minimize the rate of performance deterioration with M_∞ beyond this point.

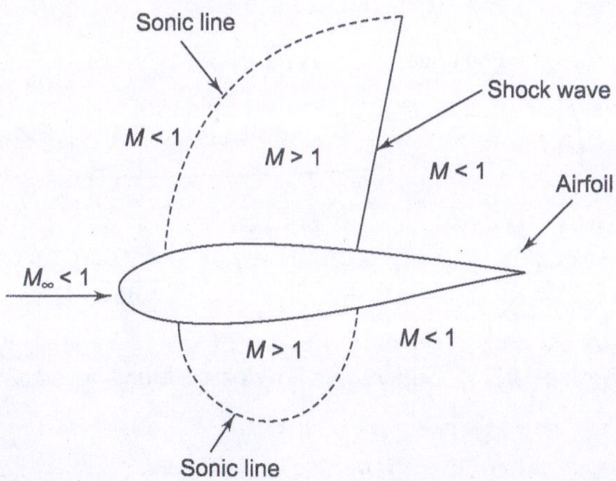

Figure 10.1 Transonic flow past an airfoil terminating in a shock.

In the present chapter, flow past airfoils and blunt bodies have also been studied where the free stream Mach number is supersonic with embedded subsonic regions in the flow field. Fighter aircrafts, whose maneuverability requirements place an extra demand on the aircraft designer, often move at such supersonic speeds. Supersonic (as well as hypersonic) flow fields are also associated with missiles, rocket nozzles, launch vehicles and so on. The knowledge of supersonic flow fields about blunt bodies is of considerable importance to the designers of these vehicles. In such cases, the flow field data are required to estimate the wave drag and to evaluate the relative and convective heat transfer rates and boundary layer effects on the body.

In the cases of flow past airfoils with $M_\infty > 1$, the onset flow remains undisturbed until the first disturbance produced by the body meets it. This leads to the formation of a *head shock (also known as a bow shock) before the airfoil leading edge*. A pair of fish-tail shocks may also be formed at the trailing edge of the airfoil. A typical supersonic flow pattern past an airfoil is shown in Fig. 10.2. However, if the leading edge is pointed, the head shock will be an *attached* one. Similarly, in the general blunt body problem, (in the case of steady flow), the flow field is usually divided into three regions, namely, a uniform supersonic flow upstream of the head shock (which is stationary relative to the body), a nonuniform subsonic region behind the shock and a nonuniform supersonic region downstream of the sonic line (Fig. 10.3). However, the shock wave may also be moving, as in unsteady problems associated with blast waves caused, for example by explosions.

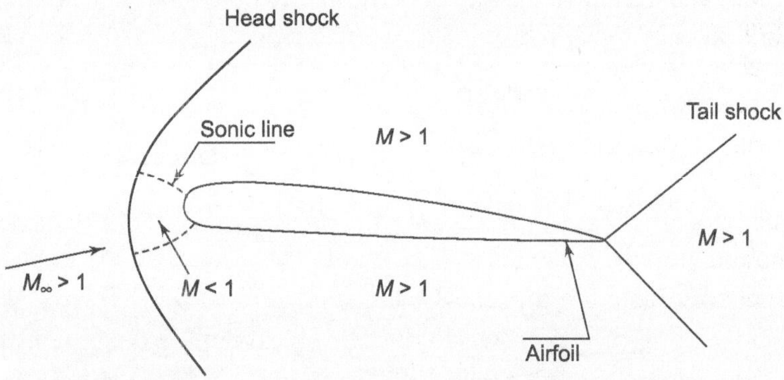

Figure 10.2 Supersonic flow past a blunt-nosed airfoil.

10.1.1 Transonic Controversy

Over, rather a long period of time, during the second world war and the subsequent quarter of a century, transonic flow fields baffled scientists and engineers by eluding quantitative study. Even in the simplest case of steady high subsonic flow past a thin profile at small incidence, where a small supersonic region is formed, embedded in an otherwise subsonic flow, the governing small perturbation equations are

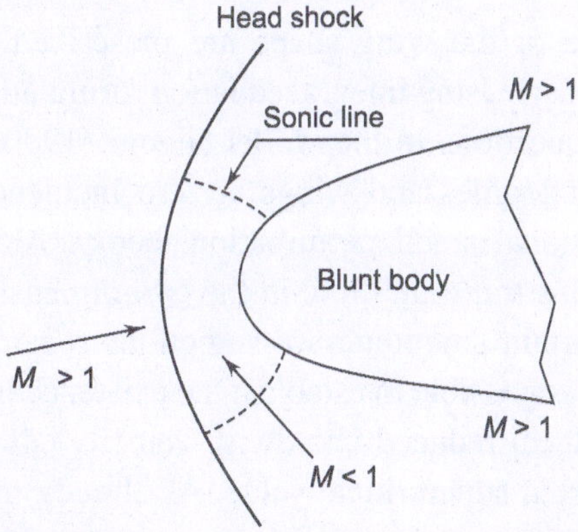

Figure 10.3 Supersonic flow past a blunt body.

nonlinear (strictly speaking, quasi-linear) and of mixed elliptic-hyperbolic type (Fig. 10.1). The line of demarcation between the subsonic and supersonic regions, that is, the sonic line, is not known *a'*priori and must be found out as a part of the solution. Experiments carried out during fifties and early sixties, showed the embedded supersonic region to terminate in a shock wave. On the other hand, exact solutions like Ringleb's solution (Niyogi, 1977; Oswatitsch, 1956) showed the existence of shock-free solution, both acceleration and deceleration through the sonic speed were continuous. It is natural then to ask whether shock-free supercritical flow past an arbitrary profile with appropriate smoothness, exists or not. This resulted in the celebrated *transonic controversy* during the fifties. Experimental results, were strongly affected by wind tunnel wall interference, so that they were not dependable, while analytical or computational methods were not sufficiently developed to resolve the controversy adequately. Morawetz (1973) established three celebrated theorems on nonexistence of shock-free neighbouring solutions for flow past profiles (Bers, 1958). One of the results may be stated as: if shock-free transonic flow past a profile with subsonic free-stream is known to exist and if the profile is changed, however little or however smoothly involving portion of an arc where the flow is supersonic, then no continuous transonic flow with the same free-stream Mach number is possible for the new profile.

Form physical point of view, it is interesting to know if a given shock-free transonic flow past a profile possesses shock-free solution for neighbouring values of the free-stream Mach number. We do not know, formally, whether this statement is true or not.

Of much theoretical interest is the question, whether super-critical (transonic) solution exists for the direct transonic problem where the free-stream Mach number and the profile shape or the wing shape are prescribed, and if so, how many such solutions may exist? Using integral equation formulation, Niyogi investigated the answer to these questions in the works Niyogi (1980, 1981, 1982 a,b,c) for steady flow past thin profiles and wings at zero incidence, under the premises of an inviscid irrotational small perturbation theory. Although, in general, the existence of continuous transonic flow in the two-dimensional case could not be established, under certain smoothness assumptions regarding the corresponding linearized solution, it is possible to establish the existence in the three-dimensional case, so long as the local reduced chordwise velocity component does not reach or exceed a certain fixed supercritical value. As already mentioned in Chapter 6, the first successful computational methods of adequate accuracy for a steady small perturbation flow are the finite difference results of Magnus and Yoshihara (1970) and Murman and Cole (1971). In a pioneering paper, Murman and Cole (1971), introduced the concept of *type-dependent differencing, which was the key to success for computing flow fields governed by mixed elliptic–hyperbolic type of equations, like the steady-state transonic flow equations.*

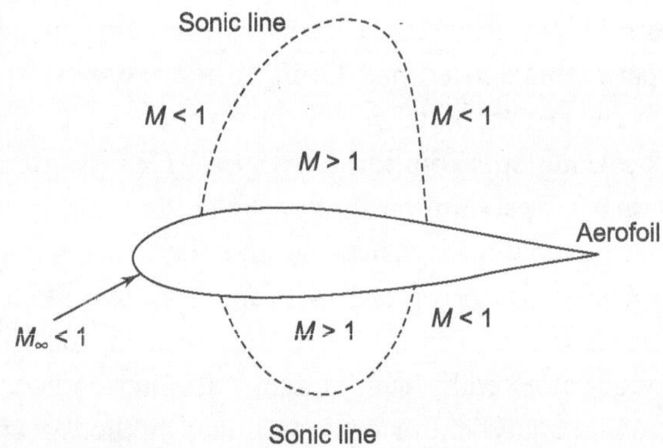

Figure 10.4 Shock-free transonic flow past an airfoil.

In this context it may be mentioned that Pearcy (1962), Nieuwland and Spee (1968) and Spee and Uijlenhoet (1968) succeeded in obtaining shock-free transonic flow past profiles of aerodynamic interest. In such cases, the transition from supersonic to subsonic speeds is smooth, that is shock-free (Fig. 10.4). For airfoil design, this type of flow is of great interest in view of the possibility of economy in cost at higher cruising speeds without suffering the drag penalty when a significant amount of supersonic region is present together with a stationary weak shock or no shock at all. However, in order to develop efficient designs, it is necessary to understand the changes in aerodynamic behaviour that occur at such transonic speed ranges.

The relatively simple problem of steady inviscid irrotational transonic flow field computation past a thin profile at small incidence has been studied in some details, in Chapter 6. The transonic small perturbation (TSP) model assumes the body to be thin, with continuously turning tangents inclined at small angles everywhere with the free-stream direction. Clearly, the small perturbation assumptions are violated at the forward and rear stagnation points.

10.2 SMALL-PERTURBATION FLOW

Flow past thin profiles or slender bodies at small angles of incidence belong to the category of small-perturbation flow. Internal flow in channels or tubes with small constrictions or bumps in the walls also belong to this category. Scientists and engineers were engaged over a long period of time with such flow fields that permit approximate treatment. A rich literature exists on *thin airfoil theory or thin wing theory* (Ashley and Landahl, 1988; Niyogi, 1977; Oswatitsch, 1956).

For purely subsonic or supersonic steady flow, approximate results of moderate accuracy may be obtained quickly using such a theory. We do not discuss these methods here but refer to the literature cited above. However, for ready reference, one or two simple results are stated here which would be useful in assessing their accuracy compared to the corresponding nonlinear inviscid flow computation.

10.2.1 Subsonic Flow Past a Thin Profile

Consider steady subsonic free-stream flow past a thin symmetric profile at zero incidence, as shown in Fig. 10.5. The profile is assumed to be thin with continuously turning tangents inclined at small angles to the free-stream. (This assumption is often violated at the leading and trailing edges of the profile.) The *thickness ratio* τ of the profile defined as the ratio $\tau = \dfrac{\text{maximum thickness}}{\text{chord length}}$ is assumed to be a small quantity.

Then the governing gasdynamic equation (7.94) may be approximated by the linear elliptic partial differential equation

$$(1 - M_\infty^2)\phi_{xx} + \phi_{yy} = 0, \quad M_\infty^2 < 1. \tag{10.1}$$

Here, ϕ is the *perturbation potential*, defined by $\phi_x = u - u_\infty$, $\phi_y = v$, u_∞ being the free-stream velocity. The *tangency boundary condition* on the surface of the profile may be shifted to the body axis, in this approximation and written for the upper part $y = h(x)$ of the profile as

$$\frac{v(x,0)}{u_\infty} = \frac{\mathrm{d}h(x)}{\mathrm{d}x}, \text{ on } y = 0+, \ 0 < x < 1, \tag{10.2}$$

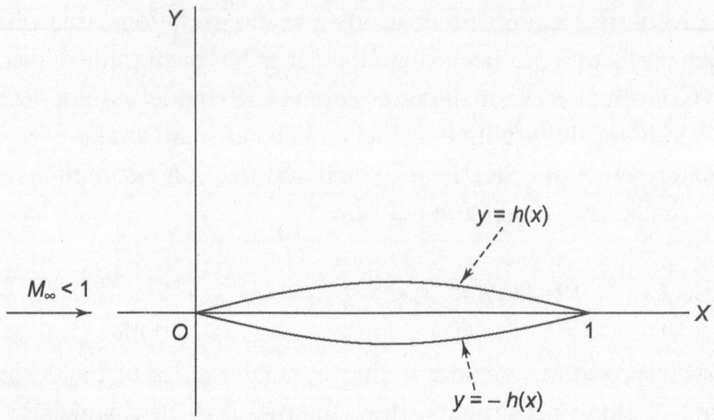

Figure 10.5 Subsonic flow past a thin symmetric profile at zero incidence.

assuming the profile to be situated on the x-axis between $0 < x < 1$. Moreover, $v = 0$, on the x-axis outside the profile. The perturbation quantities are assumed small and vanish at infinity.

Note that, Eq. (10.1) may be reduced to Laplace equation

$$\phi_{x'x'} + \phi_{y'y'} = 0, \tag{10.3}$$

by a change of variable

$$x' = x, \; y' = y\beta = y\sqrt{1 - M_\infty^2} \tag{10.4}$$

From potential theory, two types of *elementary solution* also known as fundamental solution of Eq. (10.3) are known, out of which

$$\phi(x, y) = k \ln \sqrt{(\xi - x)^2 + \beta^2(\eta - y)^2}, \tag{10.5}$$

represents a generalisation of an incompressible source ($k > 0$) or a sink ($k < 0$) situated at $\xi = x, \eta = y$. The strength of the source or sink is proportional to k. The other type is

$$\phi(x, y) = \Gamma \tan^{-1} \frac{\beta(\eta - y)}{\xi - x}, \tag{10.6}$$

which represents a *vortex* of strength proportional to Γ at $\xi = x, \eta = y$. Note that both the fundamental solutions Eqs. (10.5) and (10.6) of Laplace equation are singular at the point (x, y). Construction of solution of Eq. (10.3) by superposition of source or vortex of appropriate strength, has been discussed in details in Chapter 9, Section 9.3, where the tangency boundary condition has been satisfied on the surface of the profile. In contrast, it has been shifted approximately on the body axis, in the present discussion.

Closed form solution of the problem stated above, through Eqs. (10.1) and (10.2), together with the regularity condition at infinity, is given by Ashley and Landahl (1965), Niyogi (1977), Oswatitsch (1956) as

$$\phi(x, y) = \frac{1}{\pi\beta}\int_0^1 v_0(\xi)\ln\sqrt{(x-\xi)^2+\beta^2y^2}d\xi,$$

$$u(x, y) - u_\infty = \frac{1}{\pi\beta}\int_0^1 \frac{v_0(\xi)(x-\xi)d\xi}{(x-\xi)^2+\beta^2y^2},\ \beta = \sqrt{1-M_\infty^2}, \tag{10.7}$$

$$v(x, y) = \frac{1}{\pi}\int_0^1 v_0(\xi)\frac{\beta y}{(x-\xi)^2+\beta^2y^2}d\xi \tag{10.8}$$

Here, $v_0(x) \equiv v(x, 0+)$ is the v-component of the velocity on the upper part of the profile axis $y = 0+$, given explicitly by the simplified tangency boundary condition (10.2).

Taking the limit $y \to 0$ in (10.7) follows

$$u(x, 0) - u_\infty = \frac{1}{\pi\beta}\int_0^1 \frac{v_0(\xi)}{x-\xi}d\xi. \tag{10.9}$$

The integrand in Eq. (10.9) is singular at $\xi = x$, where *Cauchy principal value* of the integral is to be taken, defined as

$$u(x, 0) - u_\infty = \frac{1}{\pi\beta}\left\{\lim_{\epsilon\to 0}\left[\int_0^{x-\epsilon}\frac{v_0(\xi)}{x-\xi}d\xi + \int_{x+\epsilon}^1\frac{v_0(\xi)}{x-\xi}d\xi\right]\right\}. \tag{10.10}$$

The *surface pressure coefficient* C_p is given approximately as

$$C_p(x, 0) = -2\frac{u(x, 0) - u_\infty}{u_\infty}. \tag{10.11}$$

Example 10.1

Consider a symmetric parabolic arc profile

$$h(x) = \begin{cases} 2\tau(x - x^2),\ 0 < x < 1, \\ 0,\ x < 0 \text{ or } x > 1, \end{cases} \tag{10.12}$$

at zero incidence where τ is the thickness-ratio of the profile, assumed to be a small quantity.

According to the boundary condition (10.2)

$$v_0(x) \equiv v(x, 0) = u_\infty 2\tau(1 - 2x).$$

Substituting it in Eq. (10.10) and performing the integration follows

$$\frac{u(x, 0) - u_\infty}{u_\infty} = \frac{4\tau}{\pi\beta}\left[1 - \left(\frac{1}{2} - x\right)\ln\left|\frac{1-x}{x}\right|\right], \tag{10.13}$$

which is symmetrical about the line $x = \frac{1}{2}$. It approaches infinity logarithmically at the leading and trailing edges $x = 0$, and $x = 1$, respectively. The leading and trailing edges are stagnation points and there the velocity ought to be finite. This gross error in the solution, is caused by the linearization introduced in the theory. Solution Eq. (10.13) is shown in Fig. 10.6.

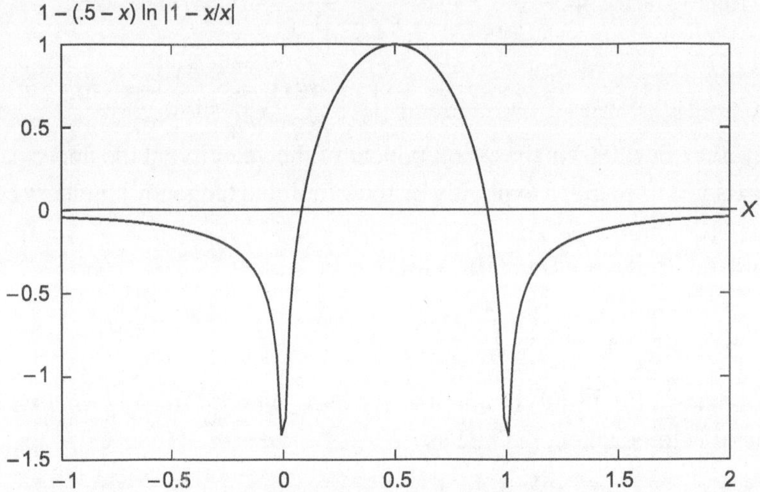

Figure 10.6 Sketch of $\frac{\pi\beta}{4\tau}\frac{u(x,0) - u_\infty}{u_\infty}$.

For an infinitely thin flat plate at small incidence α approximate closed form solution may be obtained (Ashley and Landahl, 1965; Niyogi, 1977; Oswatitsch, 1956) as

$$\frac{u_u(x,0) - u_\infty}{u_\infty} = \frac{\alpha}{\beta}\sqrt{\frac{1-x}{x}}, \text{ on the upper part, and} \tag{10.14}$$

$$\frac{u_l(x,0) - u_\infty}{u_\infty} = -\frac{\alpha}{\beta}\sqrt{\frac{1-x}{x}}, \text{ on the lower part,} \tag{10.15}$$

the suffixes u and l denoting respectively the upper and the lower parts of the profile. In deriving Eqs. (10.14) and (10.15) *Kutta-Joukowski trailing edge condition (or simply Kutta condition, as it is often referred to),* that the velocity at the trailing edge must be finite, has been used. Solution (10.14) is shown in Fig. 10.7, which shows a square-root singularity at the leading edge, arising as a consequence of the simplifying assumptions made.

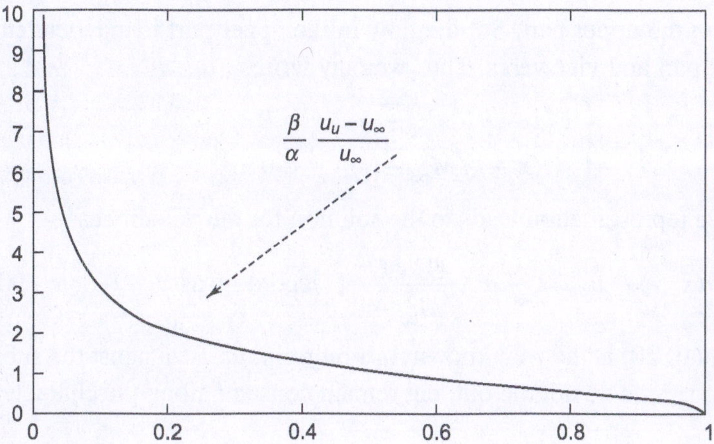

Figure 10.7 Velocity distribution for a thin profile at small incidence α.

10.2.2 Supersonic Small-Perturbation Flow

For *purely supersonic steady small perturbation flow,* the gasdynamic Equation (7.94) may be simplified to the linear hyperbolic equation

$$(M_\infty^2 - 1)\phi_{xx} - \phi_{yy} = 0, \quad M_\infty^2 > 1. \tag{10.16}$$

The simplified tangency boundary condition shifted to the body axis, given by Eq. (10.2) again remains valid in this approximation. Equation (10.16) may be reduced to the second order wave equation by a change of variable, for which the D'Alembert's solution may be written down immediately, as discussed in Chapter 4. General solution of Eq. (10.16) is

$$\phi(x, y) = F(x - y\sqrt{M_\infty^2 - 1}) + G(x + y\sqrt{M_\infty^2 - 1}), \tag{10.17}$$

where F and G are arbitrary, twice continuously differentiable functions of the arguments. The functions F and G are constant respectively along the families of characteristics

$$\xi = x - y\sqrt{M_\infty^2 - 1} = \text{const., and } \eta = x + \sqrt{M_\infty^2 - 1} = \text{const.} \tag{10.18}$$

They are known respectively as *left and right running Mach lines or characteristics.* The disturbances produced by the upper part of the profile are propagated downstream along the left running characteristics and those by the lower part are propagated downstream along the right running characteristics. Therefore, a change in the profile shape of the upper part cannot affect the flow in the lower part of the x-axis and similarly a change in the profile shape of the lower part cannot affect the flow in the upper part. So, the flow in the upper part is independent of that in the lower part and vice versa. Thus we may write

$$\phi(x, y) = \begin{cases} F(x - y\sqrt{M_\infty^2 - 1}), & y > 0, \\ G(x + \sqrt{M_\infty^2 - 1}), & y < 0 \end{cases} \tag{10.19}$$

The above representation leads to the solution for the velocity as

$$u(x, y) - u_\infty = - \pm \frac{v(x, y)}{\sqrt{M_\infty^2 - 1}}, \text{ according as } y > 0 \text{ or } y < 0. \tag{10.20}$$

Equation (10.20) is the well known *Ackeret formula.* As against the subsonic case, the perturbations do not die out, but remain constant along the characteristics.

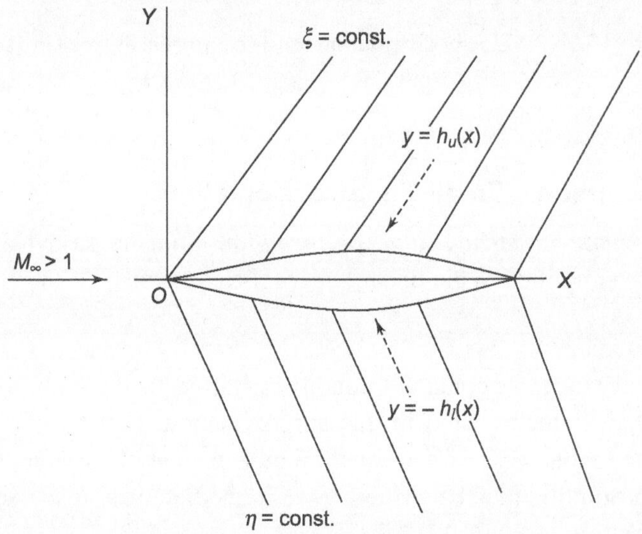

Figure 10.8 Left and right running characteristics for a thin profile in supersonic flow.

Example 10.2

Consider the symmetric thin parabolic arc profile

$$h_u(x) = h_l(x) = 2\tau(x - x^2), \; 0 \le x \le 1, \; M_\infty > 1, \tag{10.21}$$

at zero incidence. Then $\dfrac{dh_u(x)}{dx} = 2\tau(1 - 2x)$, so that on the profile we have

$$\frac{u(x, 0) - u_\infty}{u_\infty} = -\frac{2\tau}{\sqrt{M_\infty^2 - 1}}(1 - 2x), \tag{10.22}$$

which is antisymmetric about the line $x = \frac{1}{2}$, as shown in Fig. 10.9.

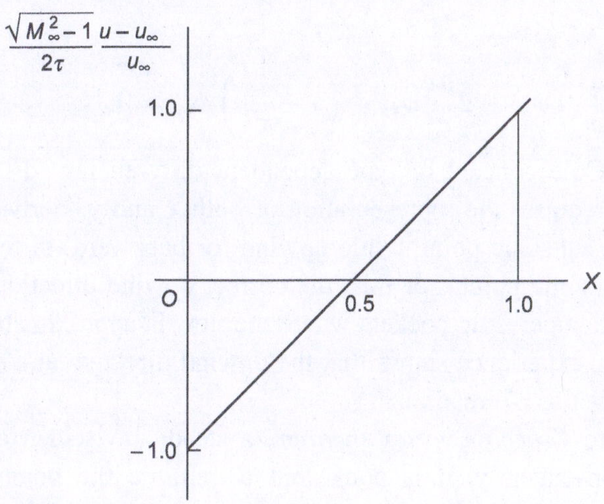

Figure 10.9 *u* perturbation on a thin airfoil in linearized supersonic flow.

After the brief discussion on linearized subsonic and supersonic flow past thin profiles, we give up the assumption of small perturbation. The subsequent discussions are concerned with high subsonic, transonic and supersonic flow where the perturbations produced need not be small. The simpler full-potential model is discussed first.

10.3 NUMERICAL SOLUTION OF THE FULL POTENTIAL EQUATION

The small perturbation assumption made in Chapter 6 and in the previous section, restricts the application of the TSP model to a small range of Mach numbers close to unity and to thin profiles and small angles of incidences. Also, the tangency boundary condition is approximately shifted to the body axis. So, there is need to develop methods to solve the exact potential equation, derived from the Euler equations assuming the flow to be irrotational. Several authors proposed methods for solving the steady inviscid transonic potential equation together with the exact boundary conditions, during the seventies, commonly known as the *full potential solution methods,* among which the works of Jameson (1974, 1975), Holst and Ballhaus (1979), Holst (1979), and Baker (1984) deserve special mention. In the following, we first discuss the solution of the non-conservative or quasi-linear form of the potential equation for the velocity potential ϕ

$$\left(1-\frac{u^2}{c^2}\right)\phi_{xx}-2\frac{uv}{c^2}\phi_{xy}+\left(1-\frac{v^2}{c^2}\right)\phi_{yy}=0. \tag{10.23}$$

The Murman-Cole concept of type-dependent differencing (discussed in detail in Chapter 6) requires the representation of both x and y -derivatives by central difference at a subsonic point, while upwind (or backward) differencing is to be used at a supersonic point. For this, the correct upwind direction is needed. For relatively small supersonic pockets where the flow is approximately aligned with the x-direction, experience shows that the upwind direction at a supersonic point may be taken as the x-direction.

According to *Crocco's vortex theorem,* a steady inviscid irrotational flow is isentropic. Consequently, it is consistent to assume the potential flow to be isentropic, as discussed in Chapter 7, Sections 7.9.1 and 7.9.2. It may be noted that so long as there is no discontinuity in the flow field, the quasi-linear form, Eq. (10.23) represents conservation of mass, momentum and energy. However, if a shock is present, entropy increases across the shock, so that the isentropic equation (10.23) cannot conserve both mass and momentum. The model preferred is that of conservation of mass, the deficiency in momentum then appears as drag, as pointed out in Jameson (1978).

At any point of the flow field, the velocity components u and v are determined by central difference formulae

$$u=\frac{\phi_{j+\frac{1}{2},k}-\phi_{j-\frac{1}{2},k}}{\Delta x},\qquad v=\frac{\phi_{j,k+\frac{1}{2}}-\phi_{j,k-\frac{1}{2}}}{\Delta y}. \tag{10.24}$$

The values at points half-way between the mesh points like $\phi_{j+\frac{1}{2},k}$ are approximated by the average values at the neighboring mesh points $\frac{1}{2}(\phi_{j+1,k}+\phi_{j,k})$. The local velocity of sound is then determined from the energy equation (7.69)

$$c^2=c_0^2-\frac{\gamma-1}{2}(u^2+v^2),$$

c_0 denoting the stagnation sound speed. At a subsonic point central difference formulae are used, as already mentioned, while at a supersonic point upwind formulae are used for the derivatives ϕ_{xx} and ϕ_{xy}, and as before, central difference is used to represent ϕ_{yy}, in all the cases. Thus at a supersonic point, the derivatives ϕ_{xx} and ϕ_{xy} are represented by

$$\frac{1}{\Delta x^2}(\phi_{j,k}-2\phi_{j-1,k}+\phi_{j-2,k})=\phi_{xx}-\Delta x\phi_{xxx}, \tag{10.25}$$

$$\frac{1}{2\Delta x\Delta y}(\phi_{j,k+1}-\phi_{j,k-1}-\phi_{j-1,k+1}+\phi_{j-1,k-1})=\phi_{xy}-\frac{\Delta x}{2}\phi_{xxy}. \tag{10.26}$$

Thus, artificial viscosity amounting to

$$\left(1-\frac{u^2}{c^2}\right)(-\Delta x\phi_{xxx})-\frac{2uv}{c^2}\left(\frac{-\Delta x}{2}\phi_{xxy}\right)=\Delta x\left[\left(\frac{u^2}{c^2}-1\right)\phi_{xxx}+\frac{uv}{c^2}\phi_{xxy}\right]$$
$$=\Delta x\left[\left(\frac{u^2}{c^2}-1\right)u_{xx}+\frac{uv}{c^2}v_{xx}\right] \quad (10.27)$$

is implicitly introduced at a supersonic point. However, when the flow is not aligned with the x-direction, a supersonic point may be reached in the flow field such that $u^2 < c^2 < u^2 + v^2$. Then, the artificial viscosity introduced becomes negative. In such cases, one of the characteristics lies ahead of the y-coordinate direction, so that the domain of dependence of that point is not correctly represented. For relatively small supersonic pockets, such a scheme may still be used, although for larger supersonic pockets, *rotated difference scheme* as introduced by Jameson (1974, 1975, 1976) must be used. We discuss here briefly, the main ideas of the *rotated difference scheme.*

10.3.1 Rotated Difference Scheme

At a point of the flow field, if s and n denote the natural coordinates, i.e. the coordinates tangential and normal to the flow direction, then the quasi-linear equation (10.23) reduce to (Niyogi, 1977; Oswatitsch, 1956)

$$(c^2 - q^2)\phi_{ss} + c^2\phi_{nn} = 0, \quad (10.28)$$

where M denotes the local Mach number $M = \frac{q}{c}$. Noting that, $\cos\theta = \frac{u}{q}$ and $\sin\theta = \frac{v}{q}$, where θ is the angle made by the local flow direction with the positive direction of x-axis, it follows that

$$\phi_{ss} = \frac{1}{q^2}\left[u^2\phi_{xx} + 2uv\phi_{xy} + v^2\phi_{yy}\right], \quad (10.29)$$

$$\phi_{nn} = \frac{1}{q^2}\left[v^2\phi_{xx} - 2uv\phi_{xy} + u^2\phi_{yy}\right]. \quad (10.30)$$

The coordinate s being aligned to the flow direction, at a supersonic point, all the derivatives appearing in the representation of ϕ_{ss} given by Eq. (10.29) are upwind differenced, while those in ϕ_{nn} are centrally differenced. At a subsonic point, all the derivatives are centrally differenced. Thus, at a supersonic point with $u > 0$, $v > 0$ the following representations for the terms in ϕ_{ss} are used

$$\phi_{xx}=\frac{1}{\Delta x^2}\left[\phi_{j,k}-2\phi_{j-1,k}+\phi_{j-2,k}\right], \quad \phi_{yy}=\frac{1}{\Delta y^2}\left[\phi_{j,k}-2\phi_{j,k-1}+\phi_{j,k-2}\right] \quad (10.31)$$

$$\phi_{xy} = \frac{1}{2\Delta x \Delta y}\left[\phi_{j,k+1} - \phi_{j,k-1} - \phi_{j-1,k+1} + \phi_{j-1,k-1}\right]. \tag{10.32}$$

The quantities on the right side of Eqs. (10.31)–(10.32) are approximations, respectively to

$$\phi_{xx} - \Delta x\phi_{xxx},\ \phi_{xy} - \frac{\Delta x}{2}\phi_{xxy} - \frac{\Delta y}{2}\phi_{xyy} \text{ and to } \phi_{yy} - \Delta y\phi_{yyy}.$$

Consequently, at a supersonic point artificial viscosity implicitly introduced amounts to

$$\left(\frac{q^2}{c^2} - 1\right)\left[\Delta x|u|(uu_{xx} + vv_{xx}) + \Delta y|v|(uu_{yy} + vv_{yy})\right], \tag{10.33}$$

which is symmetric in x and y.

10.3.2 Conservative Schemes for the Potential Equation

For numerical solution of the full-potential equation in conservation form, Jameson (1975, 1976, 1978) explicitly added artificial viscosity in order to obtain the correct directional bias in the supersonic region. The relevant conservative model has been presented in Chapter 7, Section 7.9. For the sake of convenience, we briefly repeat them here. The continuity equation representing conservation of mass is

$$\frac{\partial}{\partial x}(\rho\phi_x) + \frac{\partial}{\partial y}(\rho\phi_y) = 0, \tag{10.34}$$

and

$$\rho = \left[1 - \frac{\gamma - 1}{\gamma + 1}(\phi_x^2 + \phi_y^2)\right]^{\frac{1}{\gamma-1}} \tag{10.35}$$

where the density ρ and the velocity components ϕ_x and ϕ_y are made dimensionless by the stagnation density ρ_0 and the critical sound speed c^*, respectively.

Let us first, assume that the flow in the supersonic region is nearly parallel to the rectangular Cartesian x-axis. Then the derivatives with respect to x only require to be upwinded in the supersonic region. Following Jameson (1975), Eq. (10.34) is discretized in conservation form as

$$S_{j,k} + T_{j,k} = 0, \tag{10.36}$$

where $S_{j,k}$ is the central difference representation

$$S_{j,k} = \frac{1}{\Delta x}\left[(\rho u)_{j+\frac{1}{2},k} - (\rho u)_{j-\frac{1}{2},k}\right] + \frac{1}{\Delta y}\left[(\rho v)_{j,k+1/2} - (\rho v)_{j,k-1/2}\right], \tag{10.37}$$

and $T_{j,k}$ denotes the added artificial viscosity in conservation form

$$T_{j,k} = \frac{1}{\Delta x}\left[P_{j+\frac{1}{2},k} - P_{j-\frac{1}{2},k}\right]. \tag{10.38}$$

The artificial viscosity $P_{j+\frac{1}{2},k}$ is to be so chosen that $P = O(\Delta x)$ and it approximates the artificial viscosity corresponding to the quasi-linear form Eq. (10.23). Noting that

$$\frac{\partial}{\partial x}(\rho u) = \rho u_x - u\frac{\rho u}{c^2}u_x = \rho\left(1 - \frac{u^2}{c^2}\right)u_x,$$

we select P such that the central difference representation to P_x yields approximately the value

$$\mp\Delta x\rho\left(1 - \frac{u^2}{c^2}\right)u_{xx}, \text{ according as } u > 0 \text{ or } u < 0. \tag{10.39}$$

In order to achieve this, following Jameson (1976) let us define a switch μ as

$$\mu = \min\left[0, \rho(1 - \frac{u^2}{c^2})\right], \tag{10.40}$$

which vanishes in a subsonic flow. Then, we see that the artificial viscosity to be introduced at a supersonic point, ought to be an approximation to

$$\mp\Delta x\frac{\partial}{\partial x}(\mu u_x), \text{ according as } u > 0 \text{ or } u < 0.$$

In view of this, we choose

$$P_{j+\frac{1}{2},k} = \begin{cases} -\mu_{j,k}(u_{j+\frac{1}{2},k} - u_{j-\frac{1}{2},k}), & u_{j+\frac{1}{2},k} > 0 \\ \mu_{j+1,k}(u_{j+\frac{3}{2}} - u_{j+\frac{1}{2},k}), & u_{j+\frac{1}{2},k} < 0. \end{cases} \tag{10.41}$$

It may be observed that as $u \to c$, the switch $\mu \to 0$, so that a smooth transition through sonic values takes place.

When the supersonic flow is not aligned with the x-coordinate direction, the upwind bias of the finite difference scheme must conform to the local flow direction. The corresponding differencing may be carried out as in the case of rotated difference scheme of Jameson (1975) for the quasi-linear form, presented in the previous subsection. Noting that the conservation form Eq. (10.34) is equivalent to Eq. (10.23) multiplied by ρ, the artificial viscosity should be chosen so that

$$\frac{\partial P}{\partial x} + \frac{\partial Q}{\partial y}$$

contain terms similar to Eq. (10.33) multiplied by ρ. Noting that $\rho_x = -\frac{\rho u}{c^2}u_x$, it is possible to express the artificial viscosity in an alternative convenient form. Since,

$$\rho\left(1 - \frac{u^2}{c^2}\right)u_x = \rho\frac{u^2}{c^2}\left(\frac{c^2}{u^2} - 1\right)u_x = \left(1 - \frac{c^2}{u^2}\right)u\rho_x, \tag{10.42}$$

an equivalent form of artificial viscosity is

$$-\Delta x \frac{\partial}{\partial x}(\mu |u| \rho_x)$$

where the switching function is now redefined as

$$\mu = \max\left[0, \left(1 - \frac{c^2}{u^2}\right)\right],$$

and

$$P_{j+\frac{1}{2},k} = \begin{cases} -\mu_{j,k} u_{j,k}(\rho_{j+\frac{1}{2},k} - \rho_{j-\frac{1}{2},k}), & u_{j+\frac{1}{2},k} > 0, \\ \mu_{j+1,k} u_{j+1,k}(\rho_{j+\frac{3}{2}} - \rho_{j+\frac{1}{2},k}) & u_{j+\frac{1}{2},k} < 0, \end{cases} \tag{10.43}$$

$Q_{j,k+\frac{1}{2}}$ being defined similarly.

This form of artificial viscosity has the advantage that it may be readily extended to higher dimensions. Thus, for Eq. (10.34) the scheme may be written as

$$S_{j,k} + T_{j,k} = 0, \tag{10.44}$$

where the artificial viscosity term may be expressed as

$$T_{j,k} = \frac{1}{\Delta x}\left[P_{j+\frac{1}{2},k} - P_{j-\frac{1}{2},k}\right] + \frac{1}{\Delta y}\left[Q_{j,k+\frac{1}{2}} - Q_{j,k-\frac{1}{2}}\right], \tag{10.45}$$

$$u_{j+\frac{1}{2},k} = \frac{1}{\Delta x}\left[\phi_{j+1,k} - \phi_{j,k}\right], \tag{10.46}$$

$$v_{j+\frac{1}{2},k} = \frac{1}{4\Delta x \Delta y}\left[\phi_{j+1,k+1} + \phi_{j,k+1} - \phi_{j+1,k-1} - \phi_{j,k-1}\right], \tag{10.47}$$

with similar formulae for $v_{j,k+\frac{1}{2}}$ and $v_{j,k-\frac{1}{2}}$. The switching function is then

$$\mu = \max\left[0, \left(1 - \frac{c^2}{q^2}\right)\right] \tag{10.48}$$

The artificial viscosity added equals to

$$-\Delta x \frac{\partial}{\partial x}(\mu |u| \rho_x) - \Delta y \frac{\partial}{\partial y}(\mu |v| \rho_y), \tag{10.49}$$

and

$$P_{j+\frac{1}{2},k} = \begin{cases} -u_{j,k}\mu_{j,k}(\rho_{j+\frac{1}{2},k} - \rho_{j-\frac{1}{2},k}), \ u_{j+\frac{1}{2},k} > 0, \\ -u_{j+1,k}\mu_{j+1,k}(\rho_{j+\frac{1}{2},k} - \rho_{j+\frac{3}{2}}), \ u_{j+\frac{1}{2},k} < 0. \end{cases} \tag{10.50}$$

with similar form for $Q_{j,k+\frac{1}{2}}$.

Various other forms of artificial viscosity have been used by different authors Baker (1984), amongst which the forms due to Holst and Ballhaus (1979) and

Holst (1979), are quite neat and convenient. Holst uses Eqs. (10.44) and (10.45) and defines the artificial viscosity as

$$P_{j+\frac{1}{2},k} = -\nu_{j+\frac{1}{2},k} u_{j+\frac{1}{2},k} (\rho_{j+\frac{1}{2},k} - \rho_{j+\frac{1}{2}+r,k}), \tag{10.51}$$

where

$$r = \begin{cases} -1, & u_{j+\frac{1}{2},k} > 0, \\ 1, & u_{j+\frac{1}{2},k} < 0, \end{cases} \tag{10.52}$$

and the switching function is defined as

$$\nu = \max\left[0, CM^2(1 - \frac{1}{M^2})\right], \tag{10.53}$$

where C is a constant, chosen by the user.

Again, noting that for any mesh function $f_{j,k}$, the relation between central and upwind differences $\delta_{xx} f_{j,k} = \vec{\delta}_x \overline{\delta}_x f_{j+\frac{1}{2},k}$, hold identically, it is possible to express $S_{j,k}$ and $T_{j,k}$ in Eqs. (10.44) and (10.45) as upwind differences at points $(j + \frac{1}{2}, k)$ and $(j, k + \frac{1}{2})$. This yields on simplification, the form due to Holst and Ballhaus (1979)

$$\frac{\overline{\delta}_x(\tilde{\rho} u)_{j+\frac{1}{2},k}}{\Delta x} + \frac{\overline{\delta}_y(\overline{\rho} v)_{j,k+\frac{1}{2}}}{\Delta y} = 0, \tag{10.54}$$

where

$$\tilde{\rho}_{j+\frac{1}{2},k} = [(1 - \mu)\rho]_{j+\frac{1}{2},k} + \mu_{j+\frac{1}{2},k} \rho_{j+\frac{1}{2}+r,k}, \tag{10.55}$$

and

$$\overline{\rho}_{j,k+\frac{1}{2}} = [(1 - \mu)\rho]_{j,k+\frac{1}{2}} + \mu_{j,k+\frac{1}{2}} \rho_{j,k+\frac{1}{2}+l}, \tag{10.56}$$

$$l = \begin{cases} -1, & v_{j,k+\frac{1}{2}} > 0, \\ 1, & v_{j,k+\frac{1}{2}} < 0. \end{cases} \tag{10.57}$$

Equations (10.54) and (10.55) show that the introduction of artificial viscosity effectively retards the density, which in turn is controlled by the switching function μ defined by Eq. (10.48).

It may be noted from Eqs. (10.55) and (10.56) that here the artificial viscosity is introduced implicitly by retarding the density, unlike that in the method of Jameson, (Eqs. (10.49) and (10.50)), where it is introduced explicitly. However, the two approaches produce identical results, although the former strategy greatly simplifies the solution procedure, so that only bidiagonal or tridiagonal matrix operations are required.

For satisfying the tangency boundary condition at the body, generalized body-fitted coordinates (ξ, η) may be introduced in place of the Cartesian coordinates (x, y), as discussed in Chapter 8. The body boundary then coincides with one of the generalized coordinate directions ξ or η. The body boundary condition may then be satisfied quite conveniently and accurately. The solution procedure in the computational plane is briefly discussed in the next section.

10.4 FULL POTENTIAL SOLUTION IN GENERALISED COORDINATES

The full-potential equation in strong conservation law form, Eq. (10.34) together with Eq. (10.35) is transformed from the physical (x, y) domain to the computational (ξ, η) domain by means of the transformation

$$\xi = \xi(x, y), \quad \eta = \eta(x, y). \tag{10.58}$$

where it maintains the strong conservation law form as

$$\left(\frac{\rho U}{J}\right)_\xi + \left(\frac{\rho V}{J}\right)_\eta = 0 \tag{10.59}$$

with

$$\rho = \left[1 - \frac{\gamma - 1}{\gamma + 1}\left(U\phi_\xi + V\phi_\eta\right)\right]^{\frac{1}{\gamma-1}}. \tag{10.60}$$

Here, U, V are the contravariant velocity components along the ξ and η directions respectively given by

$$U = A_1\phi_\xi + A_2\phi_\eta, \quad V = A_2\phi_\xi + A_3\phi_\eta, \tag{10.61}$$

where

$$A_1 = \xi_x^2 + \xi_y^2, \quad A_2 = \xi_x\eta_x + \xi_y\eta_y, \quad A_3 = \eta_x^2 + \eta_y^2, \tag{10.62}$$

represent the metric quantities and

$$J = \xi_x\eta_y - \xi_y\eta_x \tag{10.63}$$

is the Jacobian of transformation. This transformation has been discussed in detail in Chapter 8.

Although the transformed Equation (10.59) is slightly more complicated than the original Cartesian form Eq. (10.34), the main advantage of the above transformation is that the boundaries associated with the physical domain are transformed to boundaries of the computational domain, chosen rectangular, so that the boundary conditions may be satisfied accurately without the need for any interpolation between grid points. This aspect is illustrated in Fig. 10.10, where the physical and computational domains for a typical transformation are shown. The inner airfoil boundary becomes the $\eta = \eta_{min}$- computational boundary while the outer physical boundary becomes the $\eta = \eta_{max}$- boundary in the computational domain. The remaining two boundaries $\xi = \xi_{min}$ and $\xi = \xi_{max}$ in the computational plane represent the lower and upper vortex sheets respectively.

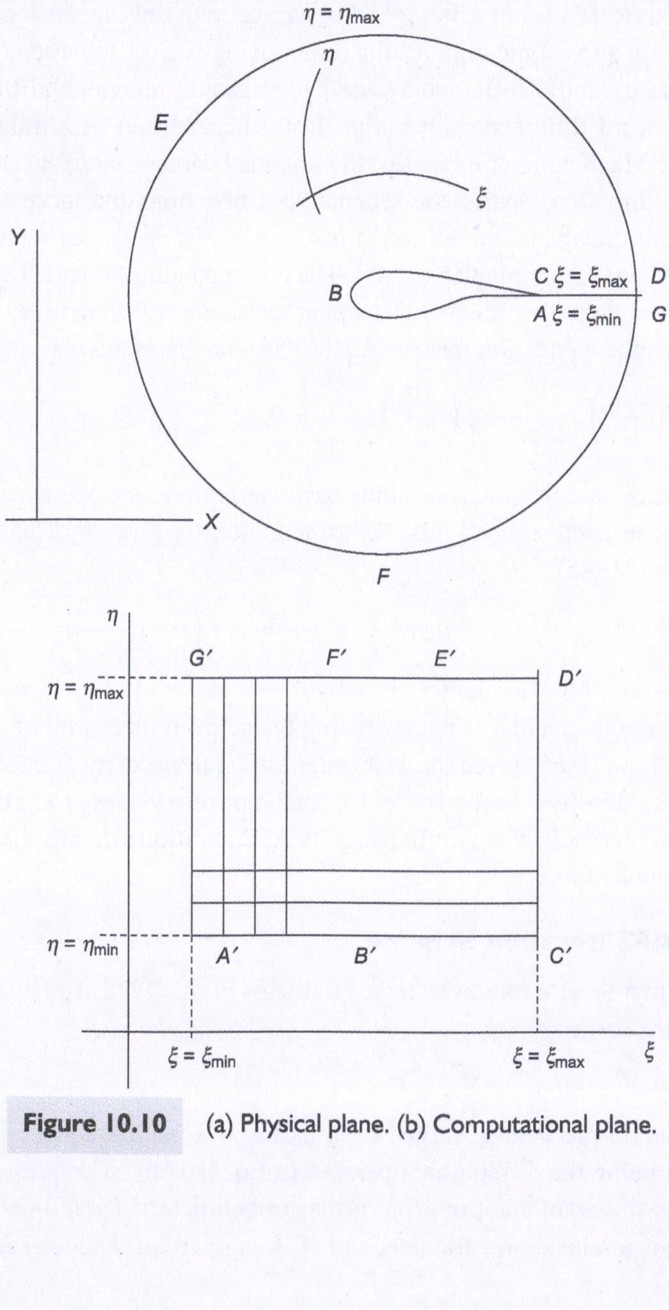

Figure 10.10 (a) Physical plane. (b) Computational plane.

10.4.1 Spatial Differencing and Artificial Viscosity

Following Holst (1979), an artificial viscosity term has been introduced implicitly in the scheme through the modification of density in the ξ direction. This provides the required upwinding for the supersonic zone and also serves the purpose of enforcing the entropy condition that discontinuous expansion shocks must be excluded.

However, Holst's scheme is second order accurate only in subsonic regions of flow, while for supersonic regions, the differencing is a combination of the second order accurate central differencing used in subsonic regions and the first order accurate upwind differencing resulting from the addition of artificial viscosity. Thus, as the Mach number increases, the scheme becomes more strongly biased in the upwind direction, that is, the scheme becomes more and more first order. To overcome this drawback, a modified form of artificial density, has been proposed by Chakrabartty and Subramanian (1985). The corresponding scheme is second order accurate throughout the flow field except at the shock wave position.

The discretized equation form of Eq. (10.59) may be written as

$$\overline{\delta}_\xi \left(\frac{\tilde{\rho} U}{J}\right)_{j+\frac{1}{2},k} + \overline{\delta}_\eta \left(\frac{\rho V}{J}\right)_{j,k+\frac{1}{2}} = 0, \tag{10.64}$$

where $\overline{\delta}_\xi$ and $\overline{\delta}_\eta$ are first order accurate backward difference operators in the ξ and η directions respectively and $\tilde{\rho}$ is the artificial density given by Chakrabartty and Subramanian (1985)

$$\tilde{\rho}_{j+\frac{1}{2},k} = \left(1 - \overline{\mu}_{i,j}\right) \rho_{j+\frac{1}{2},k} + \overline{\mu}_{i,j} \rho_{j-\frac{1}{2},k} + \overline{\lambda} \overline{\mu}_{j-1,k} \left(\rho_{j-\frac{1}{2},k} - \rho_{j-\frac{3}{2},k}\right). \tag{10.65}$$

Here, $\overline{\mu}$ is a switching parameter given by $\overline{\mu} = \max\left[0, \left(1 - \frac{1}{M^2}\right)\right]$, where M is the local Mach number, and $\overline{\lambda}$ is the switching factor from first order to second order. For $\overline{\lambda} = 0$, Eq. (10.65) gives the first order artificial viscosity due to Holst, while the artificial viscosity is second order for small nonzero values of $\overline{\lambda}$. The discretized equations may be solved very efficiently by AF2 iteration scheme, discussed in the next subsection.

10.4.2 AF2 Iteration Scheme

A general two-level iteration scheme (Ballhaus et al., 1978; Holst, 1979) can be written in the form

$$N\Delta\phi^n + \omega L\phi^n = 0 \tag{10.66}$$

for a relaxation problem governed by a partial differential equation of the form $L\phi = 0$, L being the differential operator. In Eq. (10.66), ϕ^n represents the value of ϕ at the n-th step of iteration, $\Delta\phi^n$ is the correction term $(\phi^{n+1} - \phi^n)$, $L\phi^n$ is the residual, ω is a relaxation parameter and N is an operator that determines the type of the iterative procedure. In the AF approach, N is chosen as a product of two or more factors indicated by

$$N = N_1 N_2 \dots . \tag{10.67}$$

The factors $N_1, N_2, \cdots$ must be selected such that:

1. their product is a suitable approximation to L,
2. only simple matrix operations are required, and
3. the overall algorithm is stable.

The resulting set of difference equations have been solved using the fully implicit AF2 scheme of Holst (1979). The AF2 scheme can be expressed by choosing the operator N of Eq. (10.66) for two-dimensional problem (10.64) in the form

$$\hat{\alpha} N \Delta\phi^n_{j,k} = -\left[\hat{\alpha} - \vec{\delta}_\eta \left(\frac{\rho A_3}{J}\right)_{j,k-\frac{1}{2}}\right]\left[\hat{\alpha}\overline{\delta}_\eta - \overline{\delta}_\xi \left(\frac{\tilde{\rho} A_1}{J}\right)_{j+\frac{1}{2},k} \vec{\delta}_\xi\right]\Delta\phi^n_{j,k}. \tag{10.68}$$

Here, $\hat{\alpha}$ represents an acceleration parameter. A sequence of $\hat{\alpha}$ has been used during the computations in order to reduce both high and low frequency errors in the solution. The AF2 scheme has been implemented in two steps in alternating directions as follows:

$$\text{Step 1:} \qquad \left[\hat{\alpha} - \vec{\delta}_\eta \left(\frac{\rho A_3}{J}\right)_{j,k-\frac{1}{2}}\right] f^n_{j,k} = \hat{\alpha}\omega L\phi^n_{j,k}$$

$$\text{Step 2:} \qquad \left[\hat{\alpha}\overline{\delta}_\eta - \overline{\delta}_\xi \left(\frac{\tilde{\rho} A_1}{J}\right)_{j+\frac{1}{2},k} \vec{\delta}_\xi\right]\Delta\phi^n_{j,k} = f^n_{j,k}, \tag{10.69}$$

where $f^n_{j,k}$ is an intermediate result stored at each mesh point.

10.4.3 Boundary Conditions

The exact boundary condition on the airfoil surface is that of flow tangency (that is, no flow through the airfoil surface). The implementation of this condition requires that η component of the contravariant velocity, that is, V should be zero at the airfoil surface. This boundary condition has been enforced by applying

$$\left(\frac{\rho V}{J}\right)_{j,\frac{1}{2}} = -\left(\frac{\rho V}{J}\right)_{j,\frac{3}{2}}, \tag{10.70}$$

where $k = 1$ is the airfoil surface. At the outer boundary of the computational mesh, the velocity potential and density are held fixed at the initial free stream values. However, for the lifting case, the outer boundary points must have the circulation specified, consistent with a vortex solution and updated at the end of each integration cycle. At the end of each step, the circulation is computed from the trailing edge velocity potential jump

$$\Gamma = \phi_{U_{TE}} - \phi_{L_{TE}}, \tag{10.71}$$

the suffix TE on ϕ_U and ϕ_L denoting the value evaluated at the trailing edge of the airfoil. Here, ϕ_U and ϕ_L represent the value of ϕ on the upper and lower surfaces of the airfoil respectively. As a new iteration starts, the velocity potential jump is imposed all along the vortex sheet. The correction $\Delta\hat{\phi}^n$ is differenced across the vortex sheet as

$$\Gamma^{n+1} - \Gamma^n = \Delta\hat{\phi}_U^n - \Delta\hat{\phi}_L^n. \tag{10.72}$$

However, it is difficult to impose the above jump condition since Γ^{n+1} is unknown. A suitable alternative is to estimate the value of Γ^{n+1} from

$$\Gamma^{n+1} = 3\left(\Gamma^n - \Gamma^{n-1}\right) + \Gamma^{n-2}. \tag{10.73}$$

10.4.4 Computational Results of Full-Potential Solution

The method of Holst, discussed above, is a very efficient one among the full-potential methods. Using this method, transonic flow past a NACA0012 airfoil at Mach number of $M_\infty = 0.80$ at an angle of attack of 0°, has been computed and presented in Fig. 10.11b.

For this an O-type body-fitted grid with suitable grid clustering, based on the solution to an elliptic boundary-value problem (for Poisson equations) has been generated, following Mathur and Chakrabartty (1994) (Fig. 10.11a). This is used for the finite difference solution of the full-potential equation in generalized coordinates, using second-order artificial viscosity (Chakrabarty and Subramanian, 1985). The resulting system of algebraic equations have been solved by AF2 approximate factorization scheme (Chakrabarty, 1992).

Figure 10.12 shows the results of a KORN- airfoil and its comparison with the exact solution at design conditions. The results show significant improvement with second-order artificial viscosity.

10.5 OBSERVATIONS ON THE FULL POTENTIAL MODEL

The full-potential model was studied extensively during the seventies and early eighties. This is particularly so, in view of the simplicity of the model, requiring solution of only a single second order partial differential equation. Most authors preferred the conservative model, which conserves the mass, but not the momentum, and yields a sharp shock. On the other hand, the non-conservative model does not try to conserve any of the physical quantities. The greatest objection against the

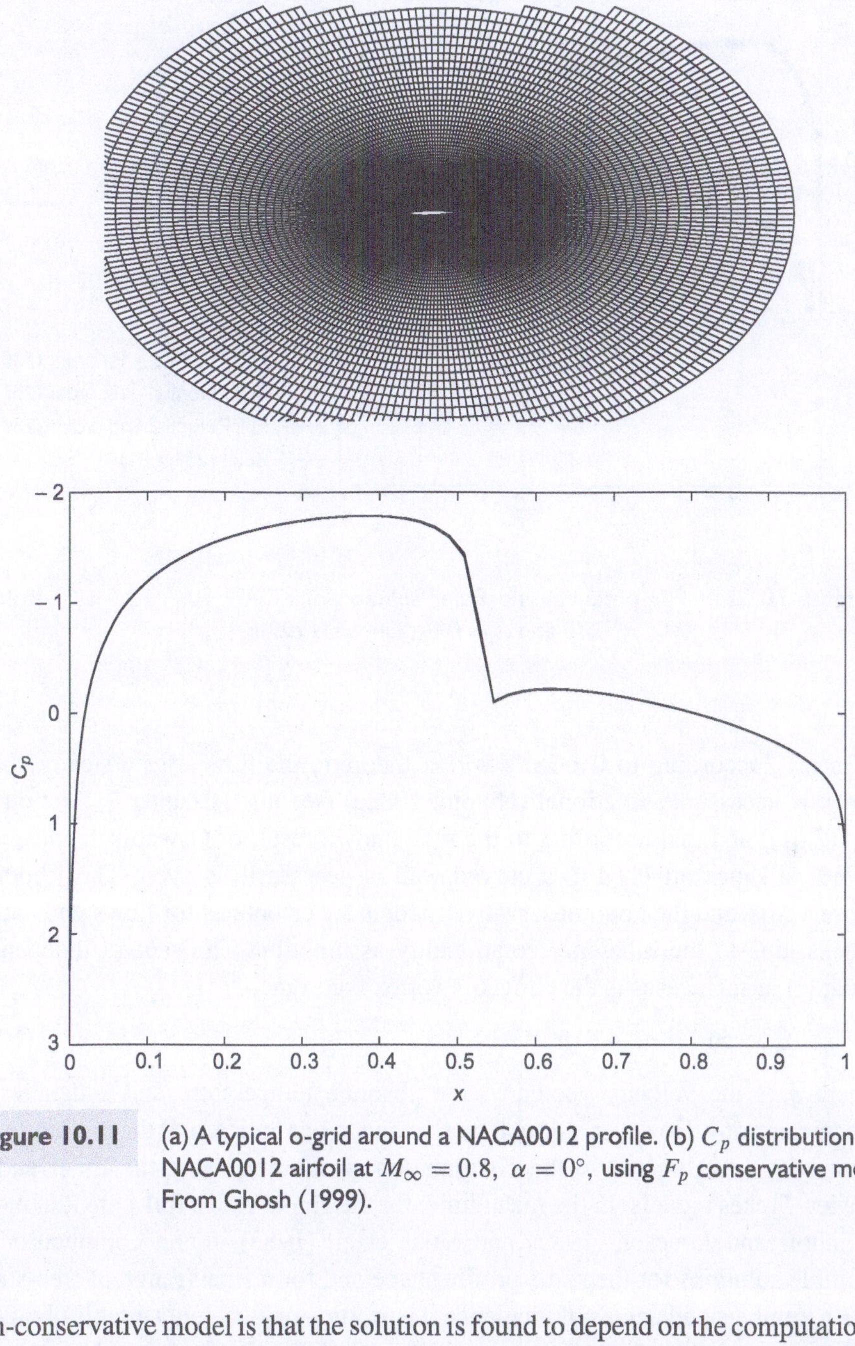

Figure 10.11 (a) A typical o-grid around a NACA0012 profile. (b) C_p distribution on a NACA0012 airfoil at $M_\infty = 0.8,\ \alpha = 0°$, using F_p conservative model. From Ghosh (1999).

non-conservative model is that the solution is found to depend on the computational grid,- a non-physical quantity (Baker, 1984). Surprisingly, comparison of solutions with that of the Euler model (see next section), show that the nonconservative solutions agree somewhat better than the corresponding conservative solutions.

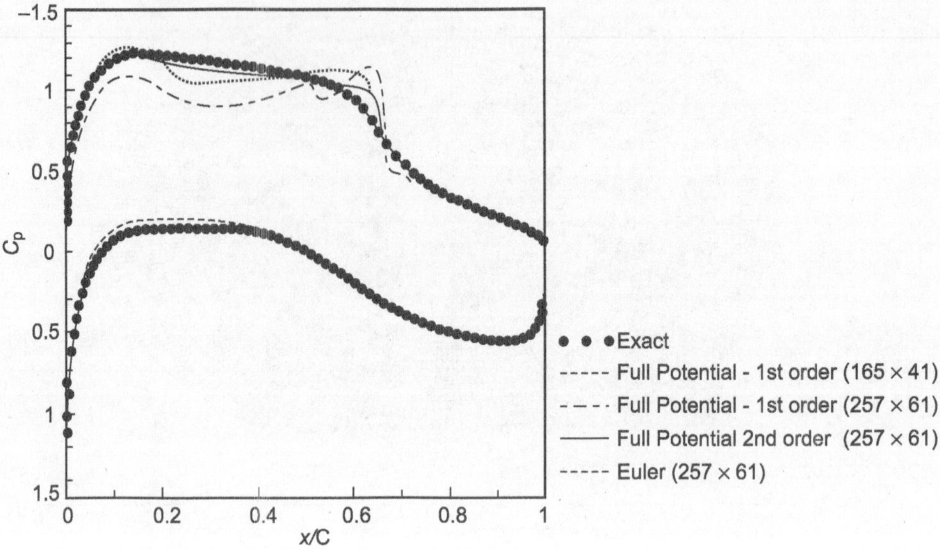

Figure 10.12 Full potential and Euler solution of KORN $70 - 10 - 13$ aerofoil for $M_\infty = 0.70$ and $\alpha = 0.0°$, Ghosh (1999).

In fact, according to Crocco's vortex theorem, the flow after a steady curved shock is necessarily rotational (Niyogi, 1982a) (see also, Chapter 7, Section 7.8, Eq. (7.61)) and that according to the well known result of Oswatitsch and Zierep (1960), a shock attached to a curved wall is necessarily curved. Thus, both the conservative and the non-conservative models are erroneous for flows with curved shocks, due to the inherent irrotationality assumption. The second independent equation in such cases is the Crocco's vortex theorem

$$\vec{q} \times \text{curl}\, \vec{q} = -T \text{ grad } s \tag{10.74}$$

where $\vec{q}$ is the velocity vector, T the absolute temperature and s denotes the specific entropy. However, no such work came to our notice, and the major upsurge of activity was directed to the computation of solution using the Euler and the Navier–Stokes models. In the mean time, for the conservative full-potential model, Steinhoff and Jameson (1982) and Salas et al. (1983), found computationally, multiple solutions for the same profile shape and for a small range of free-stream Mach numbers and angles of incidence. Depending on different suitable choices of the starting solution, the converged solutions are different. Surely, such a behaviour is not desirable. This is one of the reasons why more and more attention was paid to the Euler solutions. The next few sections, are devoted to the computational solution for the Euler model, using finite volume method.

10.6 EULER MODEL

The Euler model has been described in Chapter 7. The inviscid continuity equation, representing the conservation of mass, the Euler equations representing the conservation of momentum, the energy equation representing conservation of energy, and the equation of state, together with appropriate exact inviscid boundary conditions, constitute the Euler model. For the sake of convenience, the equations are repeated here. Numerical computation of flow fields, based on the Euler model is discussed in this section, for the two-dimensional case, using a cell–vertex finite volume method. A wellknown cell-centred finite volume method was initiated in 1981 by Jameson et.al. (1981) which proved to be a very efficient one. A number of variants of the method came up soon, among which the cell-vertex discretizations of (Chakrabarty, 1987, 1990; Hall, 1985; Ni, 1982; Rossow, 1987) are worth mentioning.

10.6.1 Governing Equations in Two Dimension

The two-dimensional Euler equations, (see Chapter 7, Section 7.8, Eq. (7.65)), may be written in integral form using vector notation as

$$\frac{\partial}{\partial t}\int_{\Omega}\vec{W}d\Omega + \int_{\partial\Omega}\bar{\bar{F}}.\vec{n}dS = 0 \tag{10.75}$$

for a fixed region Ω with boundary $\partial\Omega$. Here, $\vec{n}$ denotes the unit outward normal to $\partial\Omega$ and the flux tensor $\bar{\bar{F}}$ is given by

$$\bar{\bar{F}} = \begin{bmatrix} \rho u\vec{i}_x + \rho v\vec{i}_y \\ (\rho u^2 + p)\vec{i}_x + \rho uv\vec{i}_y \\ \rho uv\vec{i}_x + (\rho v^2 + p)\vec{i}_y \\ \rho Hu\vec{i}_x + \rho Hv\vec{i}_y \end{bmatrix} = \begin{bmatrix} \rho\vec{q} \\ \rho u\vec{q} + p\vec{i}_x \\ \rho v\vec{q} + p\vec{i}_y \\ \rho H\vec{q} \end{bmatrix}. \tag{10.76}$$

Here, $\vec{i}_x$, $\vec{i}_y$ are the Cartesian unit vectors, $\vec{q}$ is the velocity vector given by

$$\vec{q} = u\,\vec{i}_x + v\,\vec{i}_y \tag{10.77}$$

and

$$\vec{W} = \begin{bmatrix} \rho \\ \rho u \\ \rho v \\ \rho E \end{bmatrix}, \tag{10.78}$$

where p, ρ, u, v, E and H are the pressure, density, cartesian velocity components, total energy and total enthalpy per unit mass respectively. Further, for a perfect gas, pressure and specific enthalpy obey the relations

$$p = (\gamma - 1)\rho\left(E - \frac{u^2 + v^2}{2}\right) \tag{10.79}$$

and $$H = E + \frac{p}{\rho}, \tag{10.80}$$

γ being the ratio of specific heats.

10.6.2 Numerical Methods for the Euler Model

In this subsection we give a brief review of the more important numerical methods for solution of the Euler model. In parallel with the development of effective algorithms for potential flow, there were ongoing efforts to derive fast, accurate and reliable methods for solving the more accurate Euler equations that represent a valid model for inviscid rotational flows in which entropy changes may take place while mass, momentum and energy are conserved. However, the solution of the Euler equations is computationally expensive as four partial differential equations (in the two-dimensional case) are to be solved simultaneously, whereas only one equation is required to be solved in the potential flow model.

Steady state solutions are typically needed for design applications. For steady inviscid flows, there exists two alternative strategies to solve the governing Euler equations. In the first alternative, the steady equations are solved directly. The introduction of a space discretization procedure then reduces the problem to the solution of a large number of coupled nonlinear equations. These equations may be solved by a variety of iterative methods. Two alternative possibilities, in particular, are the least squares method and the Newton iteration (Giles, 1990). The steady Euler equations have also been solved by Yang et al. (1993) in a generalized Lagrangian formulation. A second-order non-oscillatory shock-capturing scheme has been developed for the purpose and applied to compute steady supersonic and hypersonic flow problems. More frequently the steady Euler equations are solved using a pseudo-transient formulation. The major problem is in achieving rapid convergence of the transient process. Viviand (1981) provides a review of the problem and possible strategies.

However, this approach is not so well developed as the second one in which the unsteady equations are integrated in time until steady state is reached. One of the reasons for this is that for the time dependent formulation, the governing system of partial differential equations is hyperbolic in supersonic as well as in subsonic regions, and there exists a highly developed theory of difference methods for hyperbolic equations. In this approach, the integration in time can be done using either explicit or implicit time stepping schemes. The explicit methods satisfy the conservation equations locally and they suffer from time step restriction imposed by the Courant–Friedrichs–Lewy (CFL) condition for stability. Particularly, when a fine grid is used, this restriction is severe and the rate of convergence is slow. The implicit schemes, on the other hand, satisfy the conservation equations globally and are not subject to time step stability limitation associated with explicit techniques.

The early standard explicit time stepping methods were constructed by Lax and Wendroff (1960), Richtmyer and Morton (1967) and subsequently modified by MacCormack (1969). In particular, the two stage predictor-corrector scheme of MacCormack has been very widely used for the solution of the unsteady Euler equations. This is the simplest known two level scheme, which is both stable and second order accurate. However, to eliminate spurious oscillations in the vicinity of shock waves, additional dissipative terms have to be introduced. The scheme was extended to multidimensional problems by MacCormack and Paullay (1972) using time splitting of the finite difference operators. The time splitting technique was used by Rizzi and Inouye (1973) for three dimensional blunt body flow along with a finite volume method.

For steady flow calculations, it is required that the final steady state should be independent of the time stepping scheme. A convenient way to obtain this is to separate space marching procedure entirely from the time marching procedure by first applying a semi-discretization. This has the additional advantage of allowing the problems of spatial discretization error, artificial dissipation and shock modelling to be studied independent of time marching stability and convergence acceleration. Such space discretization of the Euler equations (10.75) can be done conveniently in the finite volume method which is based on an integral form of the equations to be solved. Further, this is an effective method to obtain discrete approximations to conservation law equations, which preserve their conservation form. As pointed out in Chapter 4, it was shown by Lax and Wendroff (1960) that if both the governing partial differential equation and its equivalent discrete representation are cast in conservation law form, then discontinuous solutions can be computed without special treatment of the discontinuity. This in turn represents *the class of shock-capturing schemes* where the jump conditions across a discontinuity are satisfied automatically in contrast to *shock-fitting methods* for which almost, in all cases, á priori knowledge of the shock location is required and the shock waves are considered in a special manner using the Rankine–Hugoniot shock conditions explicitly. Another major advantage of the finite volume method is its flexibility to treat arbitrary geometry. Consequently, finite volume schemes have been widely used in literature for the solution of the Euler equations.

The semi-discretization procedure in finite volume method leads to a set of coupled ordinary differential equations which are to be integrated to a steady state.

When the objective is simply to reach the steady state and the details of the transient solution is not important, the time stepping scheme may be designed solely to maximize the rate of convergence. The first major choice is whether to use an explicit or an implicit method.

10.6.3 Explicit and Implicit Schemes

Explicit schemes that might be considered include linear multistep methods such as the leap-frog and Adams-Bashforth schemes, and one step mutistage methods such as the classical Runge–Kutta schemes. The two stage predictor-corrector scheme with certain modifications was used by Sells (1974) and Lerat and Sides (1982).

The one-step multistage schemes have the advantages that they require no special start up procedure and can readily be modified to give a desired stability region. They have proved extremely effective in practice as a method of solving the Euler equations. Jameson et al. (1981) used an explicit five stage Runge–Kutta type integration scheme along with a cell-centered finite volume procedure. The method of Jameson et al. (1981) was thoroughly studied by Kroll and Jain (1984, 1987), for subsonic and transonic flows around airfoils. They also analyzed different multistage schemes and their corresponding stability regions.

On the other hand, implicit schemes should yield convergence in a smaller number of time steps since in this case the time step is not constrained by a stability limitation. However, this reduction in number of time steps will be advantageous for an implicit scheme over an explicit one, only if the decrease in the number of time steps outweighs the increase in the computational effort per time step as a consequence of the need to solve coupled equations. In an implicit method, the coupled equations are solved either by an approximate factorization scheme or an iterative solution method. The main possibilities for approximate factorization are the alternating direction method and the LU decomposition method.

The alternating direction method (Peaceman & Rachford, 1955) (see Chapter 3), was given an elegant formulation for nonlinear problems by Beam and Warming (1976). The algorithm is amenable to vectorization by simultaneous solution of the tridiagonal equations along parallel coordinate lines. The method has been refined to a high level of efficiency by Pulliam and Steger (1980). The alternating direction formulation removes any restriction on the time step, at least in the two dimensional case. The use of LU decomposition method in implicit schemes can be found in the studies of Jameson and Turkel (1981). If one chooses to adopt the iterative solution technique, the principal alternatives are variants of Gauss–Seidel and Jacobi methods. Such a procedure has the advantage that it permits simultaneous overlapped calculations of the corrections at every mesh point, and is readily amenable to parallel and vector processing. A symmetric Gauss–Seidel scheme has been successfully employed in several works, e.g. (Chakravarthy, 1988). Gauss–Seidel method of iteration generally, yield a faster rate of convergence than a Jacobi method, in particular in conjunction with a flux split scheme which yields diagonal dominance. This class of schemes, however, restricts the use of vector or parallel processing.

10.6.4 Review of Acceleration Techniques

The rate of convergence of an explicit scheme can be substantially improved by using a variable time step close to the local stability limit throughout the flow field (Isaacson and Keller, 1966). This effectively increases the rate at which disturbances are propagated through the outer part of the mesh since generally the grid points are clustered near the body and expand as one moves away from the body. A similar strategy also pays with implicit schemes. Further radical improvements in the convergence rates can be realized by the multigrid time stepping technique, which extends the multigrid concept to the treatment of hyperbolic systems where the system behaviour is generally dominated by wave propagation. It is possible to accelerate the evolution of the system to steady state by using large time steps on coarse grids and interpolating the corrections back to the fine mesh. However, it remains important that the driving scheme should have the property of rapidly damping out the high frequency modes. A novel multigrid time stepping scheme was proposed by Ni in 1982. Distributed correction schemes of this type have been further developed by Hall (1985) with very good results. An alternative formulation of multigrid time stepping schemes was investigated by Jameson (1983) which corresponds to the full approximation scheme of Brandt (1977). A relatively simpler way to analyze the behaviour of multigrid time stepping schemes was proposed by Jameson (1985). Both cell centered and vertex based schemes can be devised along these lines. Hall and Salas (1985) made a comparative study of the nodal point scheme with the cell centered scheme. They concluded that cell centered schemes are more sensitive to mesh resolution than the nodal point schemes which often give better accuracy for stretched and skewed grids. Further, since cell vertices are involved in the vertex based schemes, tangency boundary conditions can be exactly satisfied, whereas an extrapolation is necessary for the cell centered approach. With properly optimized coefficients, the multistage time stepping scheme is a very efficient driver of the multigrid process. Alternating direction and LU implicit time stepping schemes, and also symmetric relaxation schemes have been explored as alternatives to the multistage procedure as a driver of the multigrid scheme (Caughy 1987; Jameson and Yoon, 1985, 1986). Very good results have been obtained by Anderson, Thomas and Whitfield (1986) using an ADI scheme with Van Leer flux splitting (Van Leer, 1982) and by Henker and Spekreijse (1984) using relaxation with Osher flux splitting (Osher and Soloman 1982). Multigrid methods have also been extended to unstructured triangular meshes (Jameson and Mavriplies, 1985). Among other techniques for convergence acceleration with the explicit methods, mention may be made of implicit residual averaging and enthalpy damping described by Jameson et al. (1981) and the distributed minimal residual (DMR) method developed by Lee and Dulikravich (1990). The DMR method is based on the idea of allowing each partial differential equation in the system to approach the converged solution at its own optimal speed and at the same time to communicate with the rest of the equations in the system.

10.6.5 Finite Volume Discretisation

For the numerical solution of the Euler equations, the time-dependent equation (10.75) in integral form has been discretized using the finite volume approach. Since the integral conservation laws allow discontinuities, this approach seems to be the most suitable for capturing shocks in the flow. Another attractive feature of finite volume method is its readiness to accommodate any type of coordinate system. For the spatial discretization of Eq. (10.75), a finite computational domain is defined by introducing far-field boundaries, which are sufficiently far away from the aerodynamic body. A structured boundary-conforming mesh is generated using curvilinear coordinates and the computational domain is subdivided into quadrilateral cells by joining the cell vertices by straight lines. However, in the finite volume formulation, the generation of a body-fitted grid and the solution process are separated since no global transformation is used. The only required data concerning the grid are the Cartesian coordinates of the four vertices of every cell in the given mesh. For Euler computations by finite volume method, both O-type grid and C-type grid (generated by an algebraic method Jain, 1983 as described in Chapter 8) have been employed. A typical C-type grid generated around a NACA0012 airfoil is shown in Fig. 10.13.

In the present method (Chakrabarty, 1990), a cell-vertex scheme has been used where the flow variables are defined at the nodal points (i, j) and for each grid point (i, j), a control volume $\Omega_{i,j}$ is formed by joining the midpoints of the four neighbouring cells with each other (Fig. 10.14). Since Eq. (10.75), is valid for any arbitrary control volume, it also holds locally for each cell $\Omega_{i,j}$. Hence,

$$\frac{\mathrm{d}}{\mathrm{d}t}\int_{\Omega i,j}\vec{W}d\Omega + \int_{\partial\Omega_{i,j}}\overline{\overline{F}}.\vec{n}\mathrm{d}S = 0 \tag{10.81}$$

where the boundary $\partial\Omega_{i,j}$ consists of four sides of the quadrilateral abcd (Fig. 10.14) and $\vec{n}$ is the unit outward normal to the surface element dS. Then, the discrete analog of Eq. (10.75) is written as

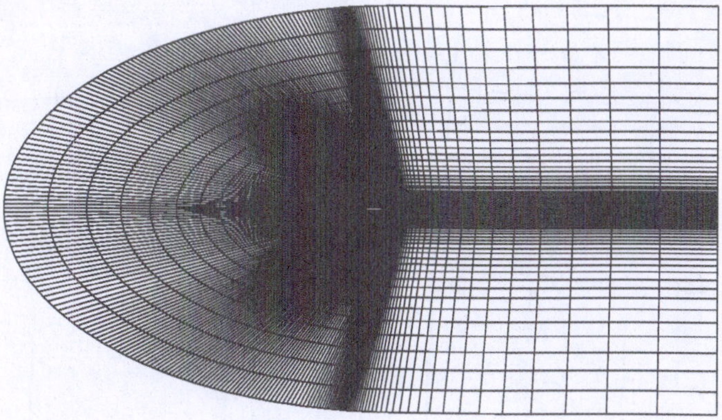

Figure 10.13 A typical C-type grid around a NACA0012 airfoil

$$V_{i,j}\left(\frac{d}{dt}\vec{W}_{i,j}\right)+\vec{Q}_{E_{i,j}}=0 \tag{10.82}$$

where, $V_{i,j}$, the volume of the cell $\Omega_{i,j}$, is computed by averaging the volumes of the four neighbouring cells Ω_1, Ω_2, Ω_3 and Ω_4 surrounding the point (i, j) and $\vec{Q}_{E_{i,j}}$ represents the net flux out of a cell and is balanced by the rate of change of $\vec{W}_{i,j}$.

In the present scheme, the fluxes are first calculated across the four neighbouring cells and then averaged to get the flux across Ω_{ij}. In particular, the flux $\vec{Q}_{E_1}$ for the cell Ω_1 may be written as

$$\vec{Q}_{E_1}=\bar{\bar{F}}.\vec{S1}_n+\bar{\bar{F}}.\vec{S2}_n+\bar{\bar{F}}.\vec{S3}_n+\bar{\bar{F}}.\vec{S4}_n, \tag{10.83}$$

where $\vec{S1}_n$, $\vec{S2}_n$ etc. are the normal vectors to the sides $S1$, $S2$ etc. of the cell Ω_1 respectively (Fig. 10.14). If SJX and SJY represent the two components of $\vec{S1}_n$ along x and y directions respectively, then the flux across the side $S1$ may be calculated as

$$\bar{\bar{F}}.\vec{S1}_n=\begin{bmatrix} \rho_{i+\frac{1}{2},j}\left(\vec{q}_{i+\frac{1}{2},j}.\vec{S1}_n\right) \\ (\rho u)_{i+\frac{1}{2},j}\left(\vec{q}_{i+\frac{1}{2},j}.\vec{S1}_n\right)+p_{i+\frac{1}{2},j}SJX \\ (\rho v)_{i+\frac{1}{2},j}\left(\vec{q}_{i+\frac{1}{2},j}.\vec{S1}_n\right)+p_{i+\frac{1}{2},j}SJY \\ (\rho H)_{i+\frac{1}{2},j}\left(\vec{q}_{i+\frac{1}{2},j}.\vec{S1}_n\right) \end{bmatrix} \tag{10.84}$$

The quantities at the midpoints are taken as simple averages of their values at the nodal points.

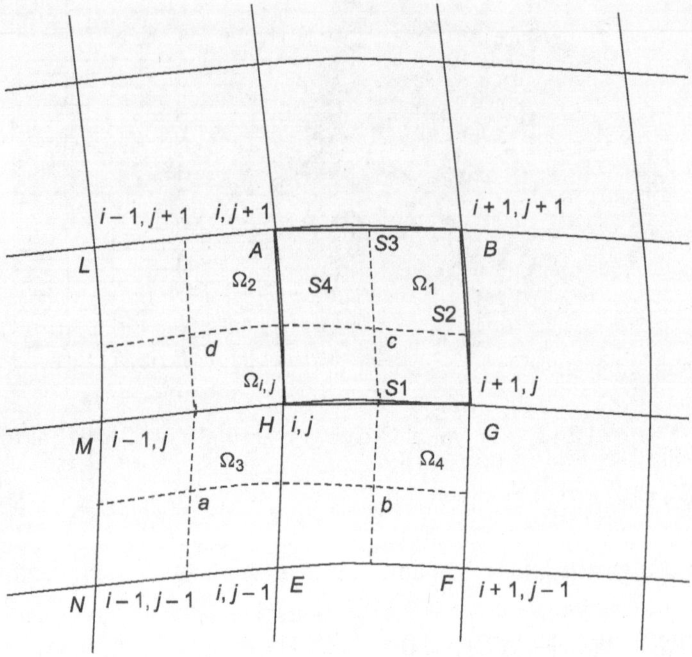

Figure 10.14 Finite volume flux computation.

The above finite volume spatial discretization is second order accurate on a Cartesian grid with constant grid sizes. For an arbitrary nonuniform grid, the accuracy of the scheme depends on the smoothness of the grid (Rossow, 1987). However, if the grid is sufficiently smooth without any abrupt change in cell shape and volume, the scheme is almost second order accurate.

10.6.6 Artificial Dissipation

The finite volume discretization Eq. (10.82), like central differencing, requires the explicit addition of artificial dissipation for numerical stability. It is necessary to add artificial viscosity like smoothing terms to damp the high frequency oscillations in the solution. In order to preserve the conservation of the scheme, dissipative terms have been introduced by adding dissipative fluxes to the semi-discrete system Eq. (10.82) as

$$V_{i,j}\left(\frac{d}{dt}\vec{W}_{i,j}\right) + \vec{Q}_{E_{i,j}} - \vec{D}_{i,j} = 0. \tag{10.85}$$

Using a blend of second and fourth order differences (Jameson et. al., 1981), the dissipative operator $\vec{D}_{i,j}$ is defined as

$$\vec{D}_{i,j} = \vec{d}_{i+\frac{1}{2},j} - \vec{d}_{i-\frac{1}{2},j} + \vec{d}_{i,j+\frac{1}{2}} - \vec{d}_{i,j-\frac{1}{2}}, \tag{10.86}$$

where the dissipative flux $\vec{d}_{i+\frac{1}{2},j}$ is expressed as

$$\vec{d}_{i+\frac{1}{2},j} = \underline{\alpha}_{i+\frac{1}{2},j} \left\{ \epsilon^{(2)}_{i+\frac{1}{2},j} \left(\vec{W}_{i+1,j} - \vec{W}_{i,j} \right) - \epsilon^{(4)}_{i+\frac{1}{2},j} \left(\vec{W}_{i+2,j} - 3\vec{W}_{i+1,j} + 3\vec{W}_{i,j} - \vec{W}_{i-1,j} \right) \right\}. \tag{10.87}$$

Here, $\underline{\alpha}_{i+\frac{1}{2},j}$ is a scaling factor chosen to give proper weightage to the dissipative terms and is given by

$$\underline{\alpha}_{i+\frac{1}{2},j} = \frac{1}{2}\left[\frac{V_{i,j}}{\Delta t^*_{i,j}} + \frac{V_{i+1,j}}{\Delta t^*_{i+1,j}} \right], \tag{10.88}$$

where $V_{i,j}$ is the cell volume and $\Delta t^*_{i,j}$ is an estimate of the time step limit Δt defined in Eq. (10.99) for a nominal Courant number of unity for the cell $\Omega_{i,j}$. (See Section 13.8 for detailed derivation.)

Further, $\epsilon^{(2)}_{i+\frac{1}{2},j}$ and $\epsilon^{(4)}_{i+\frac{1}{2},j}$ are adaptive coefficients designed to switch on enough dissipation where it is needed and are defined as

$$\epsilon^{(2)}_{i+\frac{1}{2},j} = k^{(2)} \max(\Psi_{i+1,j}, \Psi_{i,j}) \tag{10.89}$$

$$\epsilon^{(4)}_{i+\frac{1}{2},j} = \max\left(0, (k^{(4)} - \epsilon^{(2)}_{i+\frac{1}{2},j})\right) \tag{10.90}$$

where $\Psi_{i,j}$ is the pressure sensor given by

$$\Psi_{i,j} = \frac{| p_{i+1,j} - 2p_{i,j} + p_{i-1,j} |}{| p_{i+1,j} + 2p_{i,j} + p_{i-1,j} |} \tag{10.91}$$

and $k^{(2)}, k^{(4)}$ are suitable user specified constants to control the amount of dissipation.

In smooth regions of the flow, the dissipative fluxes are of third order. They provide a base level of dissipation for global damping of the oscillatory modes. However, near the shock waves where the sensor variable changes rapidly, dissipation is of first order so that the finite volume scheme behaves like a first order accurate scheme. These first order terms are needed to control spurious oscillations in the neighbourhood of shock waves. It may be noted that differences of ρH, instead of ρE has been used in the dissipative terms in the energy equation in order to admit $H =$ constant, a solution of that in the steady state.

10.7 BOUNDARY CONDITIONS

Three different types of boundary conditions occur in the present investigation.

1. *Solid body boundary condition*: At the solid wall boundary, the physical condition of zero normal velocity is applied.
2. *Condition at a coordinate cut*: An internal cut is introduced in the physical domain in order to make the computational domain simply connected. A periodicity condition has been implemented along this cut.
3. *Far-field (inflow/outflow) boundary conditions*: For the specification of far-field boundary conditions, characteristic theory has been used to provide the number and form of the boundary conditions. According to the signs of eigenvalues and one-dimensional theory of Kreiss (1970), it follows that three conditions in subsonic inflow and one condition in subsonic outflow should be specified by the free stream values. The remaining conditions should be determined from the interior solution by extrapolation. In the present investigation, following Jameson et al. (1985), the far-field boundary conditions have been implemented using the theory of Riemann invariants for a one-dimensional flow normal to the boundary. A brief resume of the theory has been presented in Chapter 7, Section (7.9.1).

   The Riemann invariants of incoming and outgoing characteristics are given by Eq. (7.79)

$$R^- = q_n - \frac{2c}{\gamma - 1}$$

$$R^+ = q_n + \frac{2c}{\gamma - 1} \tag{10.92}$$

respectively, where q_n and c represents the velocity component normal to the boundary and the speed of sound. The invariant R^- is constant along incoming characteristics and is calculated using free-stream conditions, whereas the invariant R^+, which is constant along the outgoing characteristics, is calculated by extrapolation from the interior of the field. Thus, if the subscripts ∞ and e refer to the free stream values and the values extrapolated from the interior cells adjacent to the boundary respectively, then assuming the flow is subsonic at infinity, fixed and extrapolated Riemann invariants are taken as

$$R_\infty^- = q_{n_\infty} - \frac{2c_\infty}{\gamma - 1}$$

$$R_e^+ = q_{n_e} + \frac{2c_e}{\gamma - 1}. \tag{10.93}$$

Boundary conditions for the normal velocity component and speed of sound are then constructed by adding and subtracting these invariants, that is,

$$q_n = \frac{1}{2}\left(R_e^+ + R_\infty^-\right)$$
$$c = \frac{\gamma - 1}{4}\left(R_e^+ - R_\infty^-\right). \tag{10.94}$$

At an inflow boundary, the tangential velocity component and entropy are specified as having free stream values, whereas they are extrapolated from interior at an outflow boundary. These four quantities provide a complete description of the flow in the far-field.

10.7.1 Time Stepping Scheme

The system of first-order ordinary differential equations (10.85) obtained by semi-discretisation of the Euler equation (10.75) may be written in the form

$$\frac{d}{dt}\left(\vec{W}_{i,j}\right) + \vec{P}_{i,j} = 0, \tag{10.95}$$

where the quantity $\vec{P}_{i,j}$ defined by

$$\vec{P}_{i,j} = \frac{1}{V_{i,j}}\left(\vec{Q}_{E_{i,j}} - \vec{D}_{i,j}\right) \tag{10.96}$$

is called the residual of the system (10.95). Various explicit multistage two level time stepping schemes of Runge-Kutta type have been studied by Kroll and Jain (1984, 1987) for the solution of the system of equations (10.95). In the present calculations, we employ a five stage scheme of Runge-Kutta type. At time level n, the time- stepping scheme may be written as

$$\begin{aligned}
\vec{W}_{i,j}^{(0)} &= \vec{W}_{i,j}^{n} \\
\vec{W}_{i,j}^{(1)} &= \vec{W}_{i,j}^{(0)} - \alpha_1 \Delta t \vec{P}_{i,j}^{(0)} \\
&\vdots \\
\vec{W}_{i,j}^{(5)} &= \vec{W}_{i,j}^{(0)} - \alpha_5 \Delta t \vec{P}_{i,j}^{(4)} \\
\vec{W}_{i,j}^{(n+1)} &= \vec{W}_{i,j}^{5}
\end{aligned} \tag{10.97}$$

where

$$\vec{P}_{i,j}^{(k)} = \frac{1}{V_{i,j}}\left(\vec{Q}_{E_{i,j}}^{(k)} - \vec{D}_{i,j}^{(k)}\right), \quad k = 0, 1, 2, 3, 4 \tag{10.98}$$

and the coefficients $\alpha_1, \cdots, \alpha_5$ are given by

$$\alpha_1 = \frac{1}{4}, \quad \alpha_2 = \frac{1}{6}, \quad \alpha_3 = \frac{3}{8}, \quad \alpha_4 = \frac{1}{2}, \quad \alpha_5 = 1.$$

The advantages of the present scheme are that it requires minimum of computer storage and it is second order accurate for nonlinear equations and fourth order accurate for linear equations. Further, the steady state is amenable to a variety of techniques for rapid convergence. However, as is well known, for explicit schemes the time step Δt is restricted because of stability. A necessary condition (CFL criterion) for stability of a given time stepping scheme is that the numerical domain of dependence must contain the domain of dependence of the differential equation. In the present method, a modified stability condition (Kroll and Jain 1984, 1987) given by

$$\Delta t_{i,j} \leq \underline{\lambda} V_{i,j} \left[\left| \vec{q}_{i,j}.\vec{S1}_n \right| + \left| \vec{q}_{i,j}.\vec{S4}_n \right| + c_{i,j} \left\{ \left| \vec{S1}_n \right| + \left| \vec{S4}_n \right| \right\} \right]^{-1} \tag{10.99}$$

has been used to determine the time step for the cell $\Omega_{i,j}$, where $\underline{\lambda}$ denotes the Courant number and $\vec{q}_{i,j}$ and $c_{i,j}$ represent velocity vector and speed of sound at the point (i, j) respectively. Thus the stability limit of Δt for time accurate calculations is

$$\Delta t = \min_{i,j} \Delta t_{i,j}. \tag{10.100}$$

10.7.2 Acceleration Techniques

A number of techniques may be found in literature (Jameson et al., 1981; Kroll and Jain, 1984, 1987) that are used to accelerate the convergence of the solution of unsteady Euler equations to steady state. The following three methods have been used to accelerate the convergence of the basic time stepping scheme.

1. *Local time stepping*: Since we are interested in steady state solution, the numerical solution is advanced in time using the local time step $\Delta t_{i,j}$ given by Eq. (10.99) instead of the minimum time step Δt defined by Eq. (10.100). As a consequence, the time stepping scheme can operate everywhere in the flow field at its stability limit. This procedure reduces the computational time by an order of magnitude. However, local time stepping cannot be used for time accurate calculations.
2. *Enthalpy damping*: If the enthalpy H has a constant value H_∞ in the far-field, it is constant everywhere in a steady flow. Taking advantage of this fact, one can use terms proportional to the difference between the enthalpy H and its free stream value H_∞ as forcing terms in all the equations in order to accelerate the convergence to steady state. This idea has been implemented in the present code by applying an extra step after each time step. This additional

damping step is given by

$$\vec{W}_{i,j}^{n+1} = \frac{1}{1+\sigma\left(H_{i,j}^{(r)} - H_{\infty}\right)}\vec{W}_{i,j}^{(r)} \tag{10.101}$$

except for the energy equation where

$$(\rho E)_{i,j}^{n+1} = \frac{1}{1+\sigma}\left((\rho E)_{i,j}^{(r)} - \sigma P_{i,j}^{(r)}\right). \tag{10.102}$$

Here, $\vec{W}_{i,j}^{(r)}$ denote the final values of an r-stage time stepping scheme and σ is a damping factor that is chosen empirically.

3. *Residual smoothing*: For an explicit scheme, the maximum permissible time step is restricted by the stability limit on the Courant number. It has been shown by Jameson and Baker (1983) that the Courant number and hence the stability range of an explicit multistage scheme can be increased by replacing the residuals of the scheme at each point by a weighted average of the neighbouring residuals. In order to avoid restriction on the smoothing parameter, the residual averaging is performed implicitly. In two dimensions, the implicit residual averaging has been applied in product form to replace residual $\vec{P}_{i,j}$ corresponding to unknown vector $\vec{W}_{i,j}$ by $\vec{\underline{P}}_{i,j}$ as

$$(1-\epsilon_x\delta_{xx})(1-\epsilon_y\delta_{yy})\vec{\underline{P}}_{i,j} = \vec{P}_{i,j}, \tag{10.103}$$

where δ_{xx}, δ_{yy} are the second central difference operators in x and y directions respectively and ϵ_x and ϵ_y are the smoothing parameters. Thus, at each time step and at each stage of the time stepping scheme, two sequences of tridiagonal systems have to be solved. This can be done very efficiently using LU decomposition method (Datta, 1995; Niyogi, 2003). However, numerical experiments have shown that implicit smoothing is not necessary at each stage of the time stepping scheme provided the parameters ϵ_x and ϵ_y are sufficiently increased. In the present calculations, the values of ϵ_x and ϵ_y have been fixed as unity and the smoothing has been applied only at the odd stages. The detail analysis of this technique may be obtained in the work of Kroll and Jain.

10.8 COMPUTED EXAMPLES BASED ON THE EULER MODEL

Both the full potential and Euler models described in the previous sections have been implemented for flow computation past airfoils in the transonic speed range. In order to check the accuracy and reliability of the above codes, a number of typical test cases have been investigated. The computed results were compared with experimental results (if available) or in certain cases with the exact solutions (for example, for NLR and KORN airfoils). In most cases the agreement was good. Here, we present only a few selected results. For the full potential model, the computations were carried out on an O-type grid (Fig. 10.11a), while for the Euler code both O-type grid and C-type grids (generated by an algebraic method (Jain, 1983) were used (Fig. 10.13). Grid error was eliminated by running test cases with different grid sizes and all the final results (for Euler computations) are presented on a 257 × 61 grid (with 30 points on the wake) where the far-field boundary of the computational domain has been placed at about 10–15 chords away from the airfoil.

Theoretically, the potential method should give the exact solution in the shock free cases. We have investigated NLR and KORN airfoils which are expected to give shock free solution under design conditions (Ghosh, 1999). The surface pressure (C_p) distribution around a NLR 0.1025-0.675-1.3 airfoil for free-stream Mach number $M_\infty = 0.75567$ at an angle of attack $\alpha = 0^0$ is presented in Fig. 10.15, which is designed to be shock-free under the above conditions. The shock

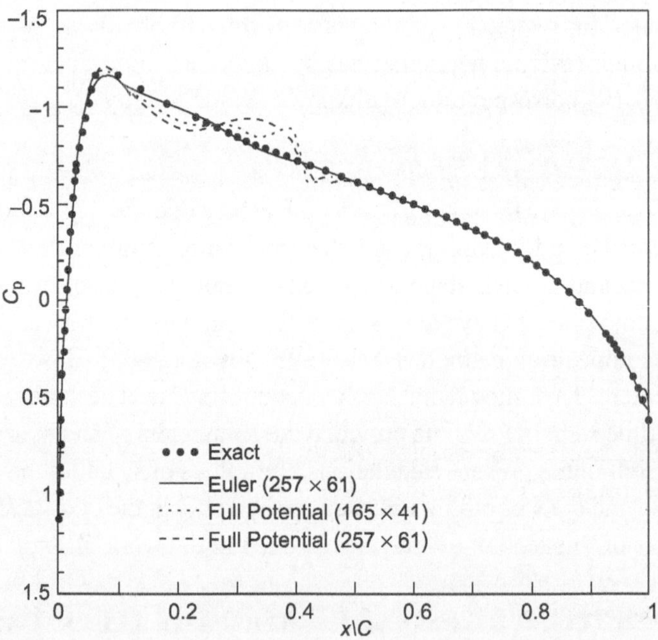

Figure 10.15 Surface pressure distribution for NLR 0.1025 – 0.675 – 1.3 airfoil at $M_\infty = 0.75567, \alpha = 0.0°$. Comparison with the exact solution. From Ghosh (1999).

free exact solution of the problem has been taken from the work of Baurdoux and Boerstoel (1968). It is to be noted that the Euler solution (with $k^{(2)} = 0, k^{(4)} = \frac{1}{16}$)

shows excellent agreement with the exact solution. However, the potential solutions with first order artificial viscosity, that is, with $\overline{\lambda} = 0$ for 165×41 grid points, show mild oscillations and is not very satisfactory. In this case, no further improvements were noticed with second order correction.

The second example considered here is that of flow past a KORN 70-10-13 airfoil with $M_\infty = 0.70$ and $\alpha = 0^0$. Figure 10.12 shows the C_p distribution for this case. Here again, it may be observed that the agreement of the potential solution (for $\overline{\lambda} = 0$) with the exact solution of Bauer et al. (1975) is satisfactory (apart from a very weak shock formation) for 165×41 grid points. The solution with 257×61 grid points shows double shock even under design condition. However, it may be noted that the potential solution with second order correction ($\overline{\lambda} = 0.01$) for 257×61 grid points shows remarkable improvement over the first order one and is smooth. Also, the Euler solution (for $k^{(2)} = 0$ and $k^{(4)} = \frac{1}{4}$) is practically shock free.

It should be noted that KORN aerofoils have open trailing edges. The results vary with respect to how we close the trailing edge to generate the grid. An improved Euler result, almost shock-free, obtained by optimising the trailing edge closure and with better grid resolution is shown in Figure 10.16.

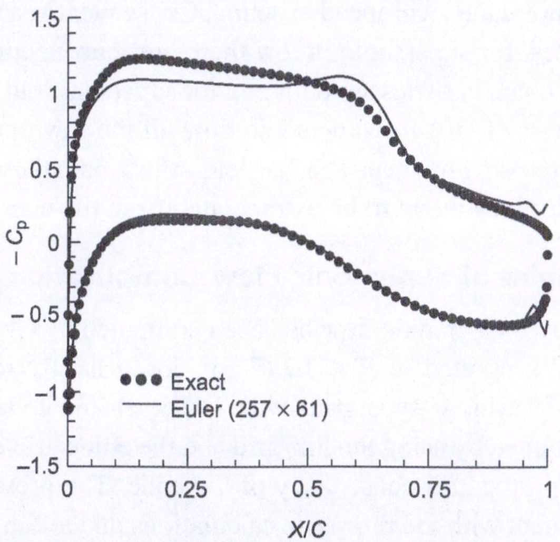

Figure 10.16 Comparison of pressure coefficient C_p obtained from the Euler solution with the exact solution for a KORN-aerofoil at $M_\infty = 0.7$ and $\alpha = 0°$.

The results of Euler computations for various airfoils are found to be quite satisfactory. Extensive numerical tests have shown that transonic flows can be well predicted using this code. However, dependence of the computed solution on the artificial viscosity parameters $k^{(2)}$, $k^{(4)}$ and more specifically on $k^{(4)}$ is noticeable. Further, the values of such parameters, for which the best solution is obtained as compared to the experimental results or the exact solutions, are different for different airfoils. This is a disadvantage of this method and, therefore, the dissipative terms should be modified to avoid such parameter dependence. This provides motivation for further study, in particular, *the TVD (total variation diminishing) and local extremum diminishing (LED) schemes* (Ghosh and Niyogi, 2000; Ghosh, 1999; Harten, 1983; Hazra, 1997; Sweby, 1985; Yee, 1989).

10.9 SUPERSONIC FLOW FIELD COMPUTATION

For computing inviscid supersonic flow only the far-field boundary conditions discussed in Section 10.7 require to be changed. However, the treatment of solid body boundary condition as well as the condition on the coordinate cut remain unaltered. For the specification of the far field boundary conditions, once again characteristic theory may be used to obtain the number and form of the boundary conditions. We know that the Jacobian matrix of unsteady two dimensional Euler equations have eigenvalues $q_n, q_n, q_n - c$ and $q_n + c$, where q_n denotes the normal velocity and c denotes the local speed of sound. Consequently, according to the sign of the eigen-values, for supersonic inflow there are four incoming characteristics while all the four characteristics are outgoing for supersonic outflow. So, following the theory of Kreiss (1970) for supersonic flow all the flow quantities have been specified as the free-streamvalues at a far field inflow boundary point while at the outflow boundaries they ought to be extrapolated from the interior points.

10.9.1 Examples of Supersonic Flow Computation

As explained above, supersonic flow has been computed by Ghosh (1999), Ghosh et al. (2000) and presented in Fig. 10.17 for flow past a NACA0012 airfoil at $M_\infty = 1.2, \alpha = 7°$ using C-type grids with 165×61 and 257×61 mesh points. The solutions obtained by using the fine grid and the crude grid do not differ much, thereby confirming the grid consistency of the code. The present solution shows very good agreement with the reference calculations due to Singh et al. (1992).

Pressure contours for a NACA0012 airfoil at supersonic free-stream Mach number $M_\infty = 4.0$ is shown in Fig. 10.18. Bow-shock ahead of the leading edge can be seen in the figure.

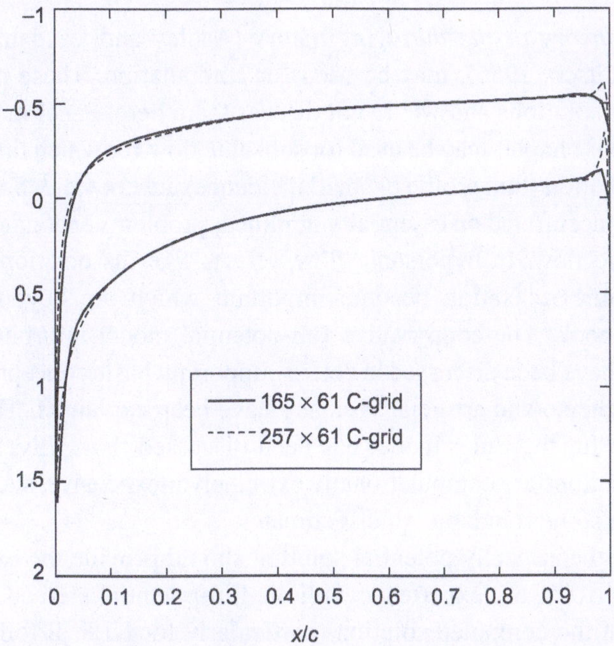

Figure 10.17 Surface pressure distribution around a NACA0012 airfoil for $M_\infty = 1.2$, $\alpha = 7^0$ using C-grid with different grid sizes. From Ghosh (1999).

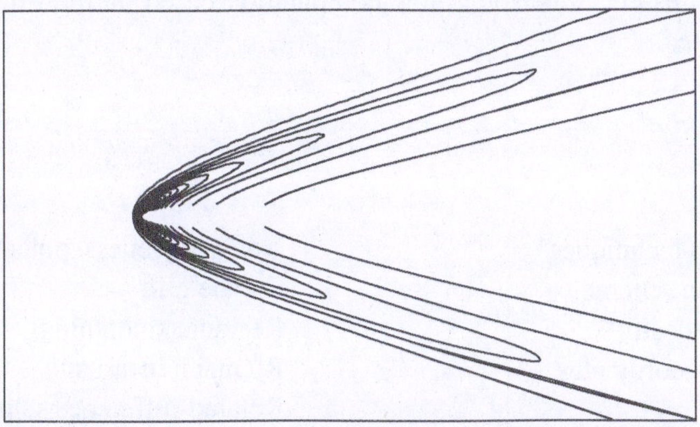

Figure 10.18 Pressure contours around a NACA0012 airfoil for $M_\infty = 4.0$, $\alpha = 0^0$ using a 165×61 O-grid. From Ghosh (1968).

10.10 SUMMARY

The more important methods for computing inviscid compressible flow past 2-D bodies in different speed ranges have been discussed in this chapter. At low subsonic flow and for purely supersonic flow past thin profiles or wings the *linearized thin airfoil theory or the thin wing theory* (Ashley and Landahl, 1965; Niyogi, 1977; Oswatitsch, 1956), may be used for computation. These methods provide approximate solutions and we do not discuss them here. Panel methods discussed in the previous chapter may be used for subsonic flow for which the assumptions of small perturbation (thin profile or small incidence) are not valid. So, the discussions have been concentrated on essentially nonlinear problems encountered in transonic or supersonic flow. In hypersonic flow, effects like dissociation, ionisation and thermodynamic relaxation become important which are beyond the scope of the present book. The conservative full-potential model in arbitrary generalised coordinates have been discussed in details. Important basic concepts like the rotated difference scheme and artificial viscosity have been explained. The finite volume computation for the Euler model has been discussed thorughly. Moreover, since potential solutions are computationally extremely inexpensive, such solutions may be useful for some rough but quick estimates.

Although theoretically potential solution should provide the exact solution for shock free airfoils, our experience with the full-potential method showed that the agreement of the computed solution (particularly for NLR airfoil) with the exact solution was not very satisfactory. In this connection, the non-uniqueness of solution of the full-potential model has been discussed and the utility of the Euler model have been stressed. Supersonic flow computation based on the Euler model has been presented.

10.11 KEY TERMS

Acceleration techniques
AF2 iteration scheme
Artificial viscosity
Body-fitted coordinates
Bow shock
Cell-vertex finite volume scheme
Consevative scheme
Crocco's vortex theorem
C-type grid
Drag coefficient
Enthalpy damping
Euler model
Full-potential method
Head shock
Lift coefficient
Local time-stepping
Non-uniqueness, potential solution
O-type grid
Residual smoothing
Riemann invariants
Rotated-difference scheme
Shock-capturing scheme
Shock-fitting scheme
Small perturbation
Sub-critical flow
Subsonic flow
Supercritical flow
Supersonic flow
Time-step limit
Time-stepping scheme
Transonic controversy
Transonic flow
Type-dependent differencing.

10.12 EXERCISE 10

10.1 For a thin symmetric profile at zero incidence in linearized subsonic flow, using distribution of sources on the profile axis derive the relations (10.7) and (10.8) for the perturbation velocities.

10.2 For the thin symmetric profile

$$h(x) = \tau(x - x^3),\ 0 < x < 1,$$

in linearized subsonic flow at zero incidence, compute the surface pressure distribution $C_p(x, 0)$. Draw a graph of the distribution. Take the thickness ratio τ of the profile as $\tau = 0.06$.

10.3 For the thin symmetric profile

$$h(x) = \tau(x - x^3),\ 0 < x < 1,$$

in linearized subsonic flow at zero incidence, compute the surface pressure distribution $C_p(x, 0)$. Draw a graph of the distribution. Take the thickness ratio τ of the profile as $\tau = 0.08$.

10.4 Using rectangular Cartesian coordinates in 3-D, with $\vec{V} = \vec{V}(u, v, w)$ show that

$$\vec{V}.\nabla(\frac{1}{2}V^2) = u^2\frac{\partial u}{\partial x} + v^2\frac{\partial v}{\partial y} + w^2\frac{\partial w}{\partial z}$$

$$+uv\left(\frac{\partial v}{\partial x} + \frac{\partial u}{\partial y}\right) + vw\left(\frac{\partial v}{\partial z} + \frac{\partial w}{\partial y}\right) + wu\left(\frac{\partial w}{\partial x} + \frac{\partial u}{\partial z}\right).\quad(10.104)$$

10.5 Solution for linearized subsonic flow past a thin infinitely long wave shaped wall

$$h(x) = \mu \sin \nu x,\ \mu = \text{ small for } M_\infty < 1,$$

was derived by Ackeret(1925) using separation of variables. Show that the pressure coefficient at the boundary is given by

$$C_p\Big|_{y=0} = -\frac{2\mu\nu}{\sqrt{1 - M_\infty^2}} \sin \nu x.$$

10.6 Derive the solution for linearized flow past the infinitely long wave shaped wall

$$h(x) = \mu \sin \nu x,\ \mu = \text{ small for } M_\infty > 1.$$

Show that the pressure coefficient is given by

$$C_{\mathrm{p}} = \frac{2\mu\nu}{\sqrt{M_\infty^2 - 1}}$$

which is constant along the characteristics $x - y\sqrt{M_\infty^2 - 1} = \text{const}$. Further, show that the wave drag C_{d} per wave length l defined by

$$C_{\mathrm{d}} = \int_0^l \frac{1}{l}\left[C_p|_{y=0}\frac{dy}{dx}\right]dx$$

is

$$C_{\mathrm{d}} = \frac{\mu^2\nu^2}{\sqrt{M_\infty^2 - 1}}.$$

10.7 In linearised subsonic flow past a thin symmetric profile at zero incidence, show that at large distances from the body, the perturbation velocities $\frac{u - u_\infty}{u_\infty}$ and $\frac{v}{u_\infty}$ decay like a dipole.

11

Boundary Layer Flow

11.1 INTRODUCTION

By the end of the nineteenth century the science of fluid dynamics had developed along two different, and rather independent, lines. One of those lines was the applied science of *hydraulics*, developed by engineers for the practical purpose of designing ships, dams and irrigation works. It was concerned with the measurement of forces on bodies immersed in flowing water, and the resistance offered to the water flowing through pipes and channels. Hydraulics was a largely empirical science and had no firm mathematical basis, but it provided answers to many questions of practical importance.

The other line was the theoretical science of *hydrodynamics*, developed by mathematicians with the aim of mathematically describing fluid motion and thereby predicting, by calculation, the very forces that the engineers were trying to measure. However, upto the time just mentioned, hydrodynamics had almost completely failed in its aim. How the new ideas based upon a combination of experiment and theory brought about a radical change in the situation will be briefly narrated in the next few paragraphs. That perspective will, hopefully, lead to a better appreciation of the importance of the concept of the *boundary layer*, which is the subject of this chapter.

Newton was perhaps the first to make a widely applicable statement about the flow of viscous fluids. He concluded from experimental observations that when a viscous fluid flows past a rigid boundary, the portion of fluid in contact with the boundary sticks to it: this is the so-called *no slip* condition; and the shear to which the fluid particles are subjected is proportional to the velocity gradient normal to the surface. (It has been subsequently found that all viscous fluids, except gases at very low pressure, satisfy the no slip condition. Many fluids, among them water and air, additionally satisfy the condition of proportionality of the shear stress with the velocity gradient. Such fluids are today called Newtonian.) More than half a century later, in the 1750s, Leonhard Euler obtained the equations of motion of an *inviscid* fluid. Shortly afterwards d'Alembert proved that an inviscid, incompressible fluid exerts *no force* on a body that is in *unaccelerated motion* relative to it: the famous "d'Alembert paradox". The "paradox" arises because this result flies in the face of the everyday experience that when a fluid, such as water, flows steadily past a submerged body, it exerts a force on the latter. The component of this force parallel to the free stream is called *drag*.

Many decades again were to pass before Navier, in 1822, obtained the first mathematically correct equations of motion of a viscous fluid. Mathematicians like Cauchy and Poisson improved these equations, which were given their final form by Stokes in 1845. Today these equations are called the Navier–Stokes equations.

They are non-linear, coupled partial differential equations that are not amenable to analytical solution unless drastic simplifying assumptions are made. Stokes himself applied them to calculate the resistance faced by a sphere moving *very slowly* in a viscous fluid (so-called "creeping flow"). The assumption of very slow motion means that viscous forces predominate inertia forces during the motion that the latter may be completely ignored, rendering the equations linear. (Stokes's result was used by Harvey and Millikan in the famous "Millikan oil drop experiment" to determine the charge of an electron.)

Further attempts to apply the Navier–Stokes equations to determine fluid forces in more familiar circumstances than Stokes's sphere were thwarted by the mathematically complicated nature of these equations. One might think it reasonable to simplify the Navier–Stokes equations by neglecting viscous effects altogether, at least for fluids with small viscosities. However, doing this leads to the Euler equations mentioned earlier, and from there to the d'Alembert paradox! Also, there is ample experimental evidence which shows that viscosity plays a crucial role, far beyond what its magnitude might lead one to imagine, in the flow of even slightly viscous fluids such as water and air. It is not therefore permissible to ignore viscosity altogether in studying the steady motion of a viscous fluid, no matter how small this viscosity might be.

In sum, therefore, when, at the beginning of the twentieth century (1903), the Wright brothers made their historic aircraft flight, this was the state of the science of fluid mechanics: there was hydraulics, which was practical but depended heavily on empiricism; and there was hydrodynamics, which was mathematically elegant but powerless to tackle problems arising out of everyday fluid-dynamic phenomena, including many of those associated with flight. And the mathematical complexity of the Navier–Stokes equations made the chasm between the two sub-disciplines of fluid mechanics appear almost unbridgable.

It was Ludwig Prandtl who, with his insight into the flow of viscous fluids obtained through careful experiments, in 1904 built the bridge that spanned this chasm between hydraulics and hydrodynamics. This allowed the latter to make rapid strides towards being able to determine the resistance faced by bodies moving through viscous fluids in many (although by no means all) situations of practical importance. Prandtl observed that the effect of viscosity on the fairly rapid flow of a slightly viscous fluid such as water is confined to a *relatively narrow region* close to a boundary, and that outside this region the flow is nearly like that of an inviscid fluid. This narrow region, where viscous effects are important, Prandtl called the *boundary layer*. Outside the boundary layer, in the "outer flow", the fluid could be considered inviscid without introducing any great inaccuracies.

Prandtl's observation, that in practically important situations involving the flow of slightly viscous fluids such as water and air, viscous effects were largely confined to a narrow region near a rigid surface, allowed the Navier–Stokes equations to be simplified sufficiently that they could be solved for a fairly wide range of situations. This had a very profound effect on the development of hydrodynamics and its new counterpart, aerodynamics. In the rest of this chapter we shall explore exactly how the concept of the boundary layer has led to the simplifications just described and their consequences in terms of solving the Navier–Stokes equations.

11.2 THE BOUNDARY LAYER: PHYSICAL CONSIDERATIONS

Figure (11.1) shows a slightly curved, rigid and impermeable, surface BB over which is flowing a fluid of small viscosity. (The vague terms "slightly curved" and "small viscosity" will be made more precise by and by.) Since this surface bounds the flowfield on one side it will, in the subsequent text, sometimes be called the *boundary*. The free stream is, by and large, parallel to BB. A curvilinear coordinate system o-ξ-η is associated with BB, where ξ is the coordinate along the surface and η is the coordinate normal to it. Viscous effects are important within the boundary layer, which is a thin layer of fluid close to the boundary. In Fig. (11.1) the outer edge of the boundary is shown with a broken line nearly parallel to BB. LL is a typical streamline within the boundary layer. Outside the boundary layer the flow is considered to be inviscid.

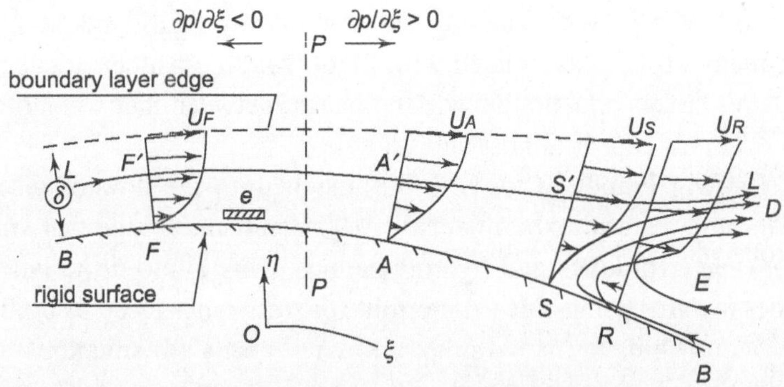

Figure 11.1 Boundary layer due to flow over a slightly curved surface

The pressure in the flowing fluid near the surface is changing: to the left of the broken line PP, the pressure gradient is negative ($\partial p/\partial\xi < 0$) or *favourable*, while to the right of this line the pressure gradient is positive ($\partial p/\partial\xi > 0$) or *adverse*. Gradual changes in the pressure gradient are typical for flows over curved surfaces, such as aerofoils. Flows with shocks may have sudden and significant changes in the pressure gradient in the region of a shock. The adjectives favourable and adverse are applied, physically speaking, to pressure gradients aiding and opposing the flow, respectively. The changing pressure along the stream causes the velocity, U to change, too. Outside the boundary layer the relation between pressure and velocity is given by the Bernoulli equation. In Fig. (11.1), $U_F > U_A > U_S > U_R$.

Let u and v be the streamwise and normal components, respectively, of the flow velocity at any point in the flow field. At the surface BB, $v = 0$, since the surface is, by assumption, impermeable. This condition would be true even if the fluid were inviscid. The fluid being viscous, those of its particles that are in contact with the surface BB stick to it, so that $u = 0$ at the surface. Therefore, *at the surface BB the fluid is at rest.*

Now consider the flowfield in the neighbourhood of the point F on BB. Moving away from F along a direction normal to BB, we find that the streamwise component of velocity, u, starts increasing. The rate of increase is rapid at first but gradually decreases until, at some short distance from the surface, u becomes almost constant with a magnitude equal to U_F. Beyond this point the flow may be regarded as inviscid. The distance, δ, from the surface at which the streamwise flow velocity "almost" attains a constant value is called the *boundary layer thickness*. This definition is only qualitative and, indeed, the boundary layer thickness cannot be exactly defined since, strictly speaking, the streamwise flow speed never attains its full inviscid value. More will be said subsequently about definite measures of the boundary layer thickness.

The *velocity profile*, that is, how the streamwise velocity grows as one moves away from the surface, is an important characteristic of the boundary layer. As has been said, the rate of growth of velocity is largest near the boundary and gradually decreases to zero as one moves away from it. According to Newton's law of viscosity, the shear stress τ on an element of the fluid is given by:

$$\tau = \mu \frac{\partial u}{\partial \eta} \tag{11.1}$$

therefore τ is largest at the boundary and decreases as we move away from it. It is the shear stress exerted on the surface of an immersed body (which is also a boundary of the flow) by the fluid that is partly responsible for the drag on the body. This component is called the *skin friction drag*. There is another major component of drag that will become apparent in due course.

Figure (11.2) shows a small element of fluid (such as the element e in Fig. (11.1)), of dimensions $d\xi \times d\eta$, aligned along the ξ–η axes. Since the curvature of the boundary has been assumed to be small, the ξ coordinate is approximately linear in this small region. As shown in the figure, the pressure gradient $\partial p/\partial \xi$

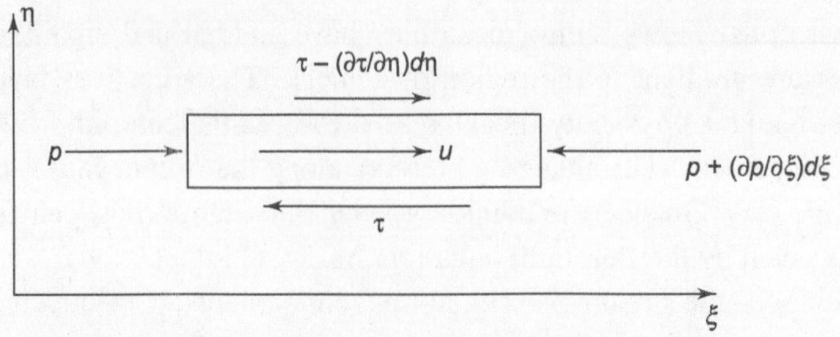

Figure 11.2 Forces on a fluid element in a boundary layer.

is positive, or adverse. Applying Newton's second law of motion to this element gives:

$$(\rho\, d\xi\, d\eta)\,\frac{du}{dt} = -\frac{\partial \tau}{\partial \eta}\, d\xi\, d\eta - \frac{\partial p}{\partial \xi}\, d\xi\, d\eta$$

where ρ is the density of the fluid; or

$$\frac{du}{dt} = -\frac{1}{\rho}\left(\frac{\partial \tau}{\partial \eta} + \frac{\partial p}{\partial \xi}\right) \tag{11.2}$$

As is apparent from this equation, the shear stress, which always opposes the motion, tends to retard the fluid element. So does the pressure gradient when it is adverse. If, on the other hand, the pressure gradient is favourable and sufficiently large, it can overcome the retarding shear stress and cause the fluid element to accelerate. Read with Eq. (11.1), the equation further tells us that fluid particles near the boundary are retarded more than others because the shear stress there is large.

Figure (11.2) shows also that the fluid particle shown there exerts a retarding force on the particle just above it, and that this retarding force decreases in magnitude as one moves away from the boundary.

Keeping these points in mind let us follow the fluid particle e starting from the point F in Fig. (11.1). As it moves to the right, the particle encounters a gradually reducing favourable pressure gradient, which means that its rate of acceleration decreases. Upon crossing the line PP that marks the beginning of the region of adverse pressure gradient, it has certainly begun to retard. In retarding, it also tends to slow down the particle immediately above it, and that particle tends to slow down the particle above it and so on. The influence of this slowing down is apparent in the velocity profile at A. There, the streamwise velocity u grows less rapidly with η than it does at F, which is situated in a region of favourable pressure gradient. Close to A the velocity profile is less full that it is at F and may even be slightly concave.

The result of this slower growth in the streamwise velocity as we move away from the surface at A is that, in order to satisfy the requirement of continuity, the fluid now has to flow across a greater depth than it did before. Thus, the fluid that flowed through the cross section FF' at F now has to flow through a cross section $AA' > FF'$. This effectively pushes the streamline LL away from the surface. This is the *displacement effect* of the boundary layer. Again, due to the slower growth of the streamwise velocity with η, we now have to travel farther away from the surface in a direction normal to it before u becomes almost independent of the distance from it. In other words, the boundary layer thickness at A is greater than what it is at F. Also, since the velocity gradient at the wall is smaller at A than it is at F, Eq. (11.1) implies that $\tau_A < \tau_F$. In sum, in a region of adverse pressure gradient it is generally true that:

- the boundary layer is thicker and the streamlines are farther away from the surface, and
- the shear stress at the boundary is less than in a region of favourable pressure gradient.

11.2.1 Separation of the Boundary Layer from the Surface

While its momentum will allow the fluid particle e to penetrate some distance into a region of adverse pressure gradient, a point will ultimately be reached where all this momentum is used up in overcoming the twin retarding agents of shear and adverse pressure and the particle is brought to rest. In Fig. (11.1) this occurs when e arrives at a point just above S. Several new things happen here.

Firstly, the shear stress at the wall vanishes. We know that fluid particles *at* the boundary are at rest, all along the boundary. Now we see that, at S the fluid particle close to, but not on, the boundary is also at rest. The velocity gradient at and around S is, therefore, zero and so is the shear stress.

Secondly, when the fluid particle e comes to rest, it experiences no shear stress but it is still acted upon by the pressure which is greater on the right than on the left. Therefore, e feels a force trying to push it leftward, but it cannot move in that direction because particles following it are pushing on it from that direction. The only way it can move is *up*, that is away from the surface, and this it does along the *dividing streamline* SD. The boundary layer is then said to have *separated* from the boundary. *Beyond this point we may no longer meaningfully speak of a "boundary layer".*

In the vicinity of S, therefore, there is a remarkable change in the flow pattern. Streamlines which were closely following the surface all along from F, and being only gradually pushed outward as the boundary layer thickened, suddenly turn sharply upward on coming close to S. No fluid particles from the left are able to cross the dividing streamline SD. And in the vacuous region to the right of SD, fluid flows in from the right towards S, turns sharply around and then flows outwards again. This is a region of circulating, or eddying, flow (region E). The viscosity of the fluid sets neighbouring fluid particles also into eddying motion, which then break away into the free stream to form the *wake*. The large eddies formed in the zone of separated flow lead to a modification of the pressure distribution on the boundary, causing a net force in the direction of the free stream. This streamwise force arising out of the modified pressure distribution is the other major component of drag, and is called the *pressure drag*. The first component, it may be recalled, is the skin friction drag, caused by the frictional shear stress at the boundary. Together, the skin friction drag and the pressure drag make up the *boundary layer drag* or *from drag*.

It may be mentioned that bodies for which the skin friction drag is significantly larger than the pressure drag are said to be *streamlined*; bodies for which the converse is true are called *bluff*. Long slender bodies, only slightly inclined to the stream, are streamline shapes; examples are aerofoil sections, aircraft wings and fuselages. Squat bodies, such as spheres, or cylinders placed with their axes normal to the stream, are bluff. Ducts that diverge rapidly also behave like bluff bodies to fluids flowing through them. It is the combination of shape and attitude to the stream that determines if a body is streamline or bluff: an aircraft wing greatly inclined to the flow loses its streamline character and becomes a bluff shape.

11.2.2 Turbulence

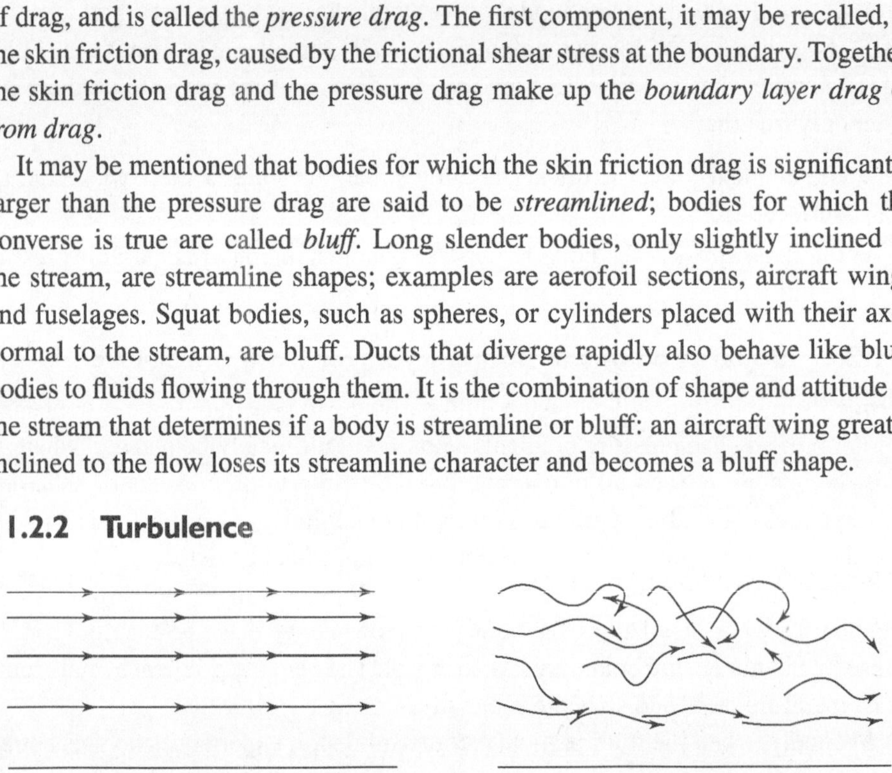

(a) Laminarflow (b) Turbulentflow

Figure 11.3 Streamlines in laminar and turbulent flow.

The flow of a fluid at low speeds (strictly speaking, at low Reynolds numbers) is *laminar*, that is, the streamlines are smooth curves with regular and predictable shapes (Fig. 11.3). The individual fluid particle velocities are in the direction of the overall or mean flow at any section of the flowfield. This means that there is almost no flow in a direction normal to a rigid surface bounding the flowfield. In a laminar flow, viscosity is caused by the transfer of momentum normal to the mean flow, brought about by the random movement of individual molecules.

Beyond a certain *critical Reynolds number* a remarkable, and fairly sudden, change comes about in the flow pattern. The previously regular arrangement of streamlines is replaced by one of chaotic motion. Although the mean flow remains as before, individual streamlines are no longer the smooth curves they were in laminar flow. Instead, they are sinuous and the motion of individual fluid particles bears no apparent relation to the mean motion, either in magnitude or in direction. This kind of flow is termed *turbulent*.

Laminar flow is provoked into turbulence when conditions are such that the effect of disturbances, which are forever present in a flow, do not die down but amplify instead. Presence of turbulence in the free stream, roughness or protrusions on the boundary, acoustic disturbances or an adverse pressure gradient are all capable of hastening the transition from laminar to turbulent flow. The determining factor in the transition process is, however, the Reynolds number. Below a certain low value of the Reynolds number it is almost impossible to provoke turbulence: the disturbances all die down. Above a certain high value of the Reynolds number it is almost impossible to prevent transition to turbulence: the slightest disturbance will cause this transition. In the interim range the factors listed above determine whether the flow will remain laminar or become turbulent.

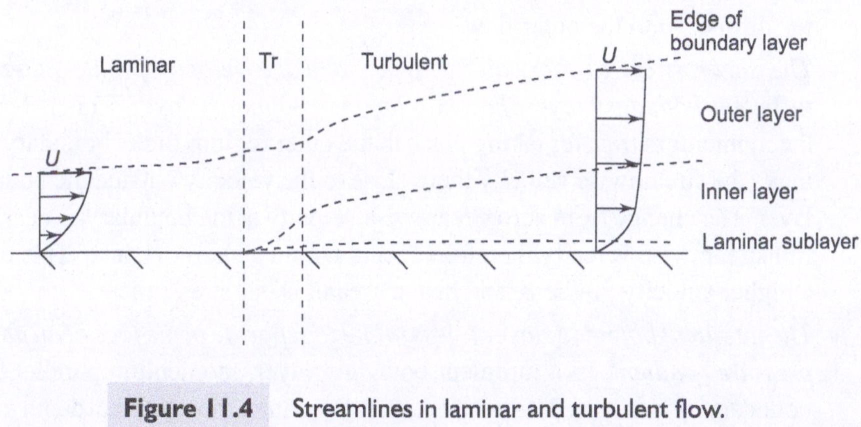

Figure 11.4 Streamlines in laminar and turbulent flow.

Figure (11.4) shows the boundary layer over a flat plate making the transition from laminarity to turbulence. *Tr* is the region where the transition occurs: the so-called *transition zone*. At sufficiently high Reynolds numbers this region is small enough that we may speak of the *transition point*.

The structure of the turbulent boundary layer is more complex than that of the laminar boundary layer (Duncan et al., 1970). Across the thickness of such a layer, three distinct zones are observed. Very close to the surface is a region of laminar flow, called the *laminar sub-layer*. There the rigid boundary suppresses the random movement of fluid particles normal to the mean flow, forcing the flow to be laminar. Beyond the laminar sub-layer and extending up to about 40% of the thickness of the boundary layer is the *inner layer*. This is where the most intense turbulent activity takes place. Here are found eddies of a wide range of sizes, and velocity fluctuations over a range of frequencies. In this region the fluctuating components of velocity are large and may reach magnitudes up to 10% of the mean velocity. The larger eddies transfer kinetic energy to the smaller ones; and the very smallest eddies dissipate their kinetic energy through viscous action. Beyond the inner layer, and extending up to the edge of the boundary layer, is the *outer layer*. This region is characterised by large eddies and relatively low frequencies of velocity fluctuation. There is large-scale transfer of streamwise momentum between this layer and the outer flow. As a result the mean flow velocity across this layer is more uniform than it is in the outer region of a laminar boundary layer.

The major differences between the characteristics of laminar and turbulent boundary layers are as follows:

- *The thickness of the turbulent boundary layer is significantly greater than that of the laminar boundary layer*. This happens because the large-scale transport of streamwise momentum normal to the mean flow causes the boundary layer to "diffuse" into the outer flow.
- *The shear stress at the wall is higher, and the velocity profile fuller in a turbulent boundary layer than in a laminar boundary layer*. This is because the momentum transfer taking place in the outer regions of the boundary layer make the streamwise velocity there close to the velocity outside the boundary layer. The change from zero streamwise velocity at the boundary to nearly the full streamwise velocity therefore occurs within a short distance. This causes a higher velocity gradient, and hence greater shear stress, there.
- *The turbulent boundary layer is less prone to separate in the face of an adverse pressure gradient*. In a turbulent boundary layer, momentum parallel to the boundary is convected from the outer flow into most of the depth of the boundary layer. This transfer of momentum, which occurs on a scale much larger than in a laminar boundary layer, helps the fluid particles to overcome the retardation due to shear and adverse pressure gradient and thereby delays separation of the boundary layer.

In a turbulent boundary layer, shear stresses over and above those due to viscosity are caused by the transfer of momentum normal to the mean flow, brought about by the random movement of fairly large chunks of fluid normal to the boundary. These stresses are known as *Reynolds stresses*. The momentum transferred in this way is *much larger* than is possible by individual molecules, hence the effective viscosity in turbulent flow is much greater than in laminar flow. The apparent additional viscosity is known as *eddy viscosity* and, in the inner region of a turbulent boundary layer, can greatly exceed the real viscosity.

11.2.3 Measures of Boundary Layer Thickness

It has been remarked before that, while it is a useful physical concept, the boundary layer thickness δ cannot be precisely defined. This is because, strictly speaking, the flow velocity at the edge of the boundary layer *never* becomes exactly equal to the velocity of the outer flow, no matter how large we consider the thickness of the boundary layer to be. As a matter of convention, δ is taken to be that distance normal to the surface where the velocity $u = 0.99U$, U being the speed in inviscid flow. This definition gives an idea of the thickness of the boundary layer but has an element of arbitrariness about it (the number 0.99).

There are other lengths, related to δ, which may be more precisely defined and which are also physically meaningful (Duncan et al., 1970). It has already been remarked that as the flow within the boundary layer slows down (due to the retarding effect of viscous shear stress and also perhaps the adverse pressure gradient), the requirement of continuity causes the streamlines to be pushed outward. The *displacement thickness* δ^* is that distance by which the outer potential flow at any streamwise position is pushed outward as a consequence of the decrease in velocity in the boundary layer.

Figure (11.5) shows the velocity profile at streamwise station ξ in a typical boundary layer. BB is the edge of the boundary layer. Let η_∞ be a distance along η which is much larger than the boundary layer thickness. The volume of fluid flowing through the section from $\eta = 0$ to $\eta = \eta_\infty$ is:

$$Q = \int_{\eta=0}^{\eta_\infty} u(\eta)d\eta \tag{11.3}$$

Now, if the flow were completely inviscid and flowing with an average speed U, the same volume Q could flow through a section of depth $\eta_\infty - \delta^*$, that is:

$$Q = \int_{\eta=\delta^*}^{\eta_\infty} Ud\eta \tag{11.4}$$

Equality of the two integrals above amounts, graphically, to the two hatched areas

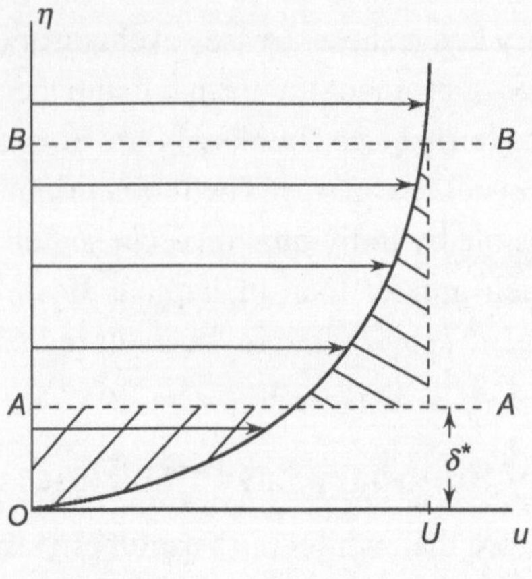

Figure 11.5 The displacement thickness.

in Fig. (11.5) being equal. From the above two equations we obtain:

$$\delta^* = \int_{\eta=0}^{\eta_\infty} \left(1 - \frac{u(\eta)}{U}\right) d\eta \tag{11.5}$$

An alternative, and highly fruitful way (as will turn out!) of looking at the displacement thickness is that to the outer, inviscid, flow the boundary of the flowfield is situated not at $\eta = 0$ but at $\eta = \delta^*$.

Similarly, the *momentum thickness* θ is defined as the depth of fluid flowing at the speed of the inviscid flow such that the momentum carried by it is equal to the defect in momentum caused by the presence of the boundary layer:

$$\theta = \int_{\eta=0}^{\eta_\infty} \frac{u(\eta)}{U}\left(1 - \frac{u(\eta)}{U}\right) d\eta \tag{11.6}$$

The lengths δ^* and θ are physically meaningful quantities precisely defined and are widely used to characterise a boundary layer.

11.3 THE BOUNDARY LAYER EQUATIONS

We shall restrict ourselves here to the study and computation of boundary layers in steady, two-dimensional incompressible flows. This does not mean that the effects of compressibility or three-dimensionality are not important. With modern high-speed aircraft, quite the opposite is in fact true. However, we are constrained to adopt the said restrictions because the generalisation to three dimensions and/or compressible flows significantly increases the order of complexity of the problem, which is hard to accommodate in an introductory exposition.

11.3.1 Assumptions of the Boundary Layer Theory

Although we have mentioned these before, here we again set forth the assumptions upon which the boundary layer theory stands.

- The fluid whose flow is being studied must be only slightly viscous. A more precise way of saying this is that the flow must be at a sufficiently large Reynolds number. In such cases, the effects of viscosity are confined to a thin boundary layer, outside which the flow may be regarded as inviscid.
- The boundary over which the boundary layer is being studied must present a streamline shape to the flow: that is, the boundary layer should remain attached to the boundary.
- The boundary layer is *thin*, that is, its thickness δ (as determined by some appropriate measure) is much smaller than some length L which is characteristic of the dimension of the boundary in the streamwise direction:

$$\delta \ll L \tag{11.7}$$

  Therefore, if L is a quantity of order unity ($L \sim O(1)$), then δ is an infinitesimal of order higher than unity. For example, the flow of air over the wing of an aircraft is typically at a Reynolds numbers of several millions. Then, the boundary layer over a wing having a chord a few metres long is only a few centimetres thick. This is true as long as the boundary layer is attached.

  To take some examples of streamwise length scales, in studying the flow past an aerofoil section, L could be the chord of the aerofoil; similarly, for the case of the flow past a slender fuselage held at a small angle to the stream, it could be the length of the fuselage. Not all problems have meaningful length scales, however. In the case of the flow past a semi-infinite flat plate placed parallel to the stream there is no length scale in the streamwise direction.

11.3.2 The Boundary Layer Equations for Laminar Flow

Steady, two-dimensional, incompressible viscous flow is governed by the equations of continuity:

$$\frac{\partial u}{\partial \xi} + \frac{\partial v}{\partial \eta} = 0 \tag{11.8}$$

and momentum (the Navier–Stokes equations) along the ξ and η directions:

$$u\frac{\partial u}{\partial \xi}+v\frac{\partial u}{\partial \eta}=-\frac{1}{\rho}\frac{\partial p}{\partial \xi}+\nu\left(\frac{\partial^2 u}{\partial \xi^2}+\frac{\partial^2 u}{\partial \eta^2}\right) \tag{11.9}$$

$$u\frac{\partial v}{\partial \xi}+v\frac{\partial v}{\partial \eta}=-\frac{1}{\rho}\frac{\partial p}{\partial \eta}+\nu\left(\frac{\partial^2 v}{\partial \xi^2}+\frac{\partial^2 v}{\partial \eta^2}\right) \tag{11.10}$$

where ξ, the coordinate along the (perhaps slightly curved) boundary and η, the coordinate normal to the boundary are the independent variables; and the velocities u and v along ξ and η, respectively, and the pressure p are the dependent variables. The kinematic viscosity, ν, is assumed to be constant.

In the streamwise direction, the magnitude of velocity change over a length L is usually not greater than U, the speed of the inviscid flow. Therefore, the partial derivative $\partial u/\partial \xi \sim O(1)$. Now since $\partial u/\partial \xi \sim O(1)$, it follows that $\partial v/\partial \eta \sim O(1)$ since, from Eq. (11.8), $\partial v/\partial \eta = -\partial u/\partial \xi$. But the change in η within the boundary layer is $O(\delta)$, hence the change in v within the boundary layer is also $O(\delta)$. But at the boundary, $v = 0$, hence its absolute value at the edge of the boundary layer is $O(\delta)$. To summarise:

$$\frac{\partial u}{\partial \xi}, \frac{\partial v}{\partial \eta} \sim O(1)$$

$$v \sim O(\delta)$$

within the boundary layer.

Again, outside the boundary layer where the Bernoulli equation applies, we may say that $O(\Delta p) = O(\rho U \Delta U) = O(1)$. Hence, in the streamwise direction pressure changes are of order unity.

11.3.2.1 Non-dimensionalisation of the governing equations It is clear that the length scales along the ξ and η directions are of different orders of magnitude, as are the changes in the velocities u and v within the boundary layer. We shall therefore non-dimensionalise the governing equations with the aim of making the various terms of the order of unity as far as possible. To do this, we introduce the following non-dimensional variables:

$$\overline{\xi}=\xi/L, \quad \overline{\eta}=\eta/\delta$$

$$\overline{u}=u/U, \quad \overline{v}=vL/U\delta$$

$$\overline{p}=p/\rho U^2$$

Then the Eqs. (11.9) and (11.10) non-dimensionalized become:

$$\overline{u}\frac{\partial \overline{u}}{\partial \overline{\xi}} + \overline{v}\frac{\partial \overline{u}}{\partial \overline{\eta}} = -\frac{\partial \overline{p}}{\partial \overline{\xi}} + \frac{1}{Re}\left[\frac{\partial^2 \overline{u}}{\partial \overline{\xi}^2} + \left(\frac{L}{\delta}\right)^2 \frac{\partial^2 \overline{u}}{\partial \overline{\eta}^2}\right] \tag{11.11}$$

$$\overline{u}\frac{\partial \overline{v}}{\partial \overline{\xi}} + \overline{v}\frac{\partial \overline{v}}{\partial \overline{\eta}} = -\left(\frac{L}{\delta}\right)^2 \frac{\partial \overline{p}}{\partial \overline{\eta}} + \frac{1}{Re}\left[\frac{\partial^2 \overline{v}}{\partial \overline{\xi}^2} + \left(\frac{L}{\delta}\right)^2 \frac{\partial^2 \overline{v}}{\partial \overline{\eta}^2}\right] \tag{11.12}$$

respectively, where $Re = UL/\nu$ is the Reynolds number.

11.3.2.2 Order of magnitude analysis Considering Eq. (11.11) first, we see that all terms on the left hand side are $O(1)$, as is the pressure term on the right. Looking at the viscous terms, both $\partial^2\overline{u}/\partial\overline{\xi}^2$ and $\partial^2\overline{u}/\partial\overline{\eta}^2$ are $O(1)$. However, due to the presence of the factor $(L/\delta)^2$, the second term within brackets is much larger than the first. We may therefore *ignore* the term $\partial^2\overline{u}/\partial\overline{\xi}^2$.

The assumption of a thin boundary layer will hold only if the viscous term is no greater in magnitude than the inertial and pressure terms. This *requires* that the coefficient $L^2/(\delta^2 Re)$ be $O(1)$ or, in other words:

$$\frac{\delta}{L} \sim O\left(\frac{1}{\sqrt{Re}}\right) \tag{11.13}$$

This is what we mean by the statement that at high Reynolds numbers the boundary layer is "thin". Incidentally, this shows that *the laminar boundary layer grows inversely as the square root of the Reynolds number*, which is a very important result. The simplified equation then becomes:

$$\overline{u}\frac{\partial \overline{u}}{\partial \overline{\xi}} + \overline{v}\frac{\partial \overline{u}}{\partial \overline{\eta}} = -\frac{\partial \overline{p}}{\partial \overline{\xi}} + \frac{1}{Re}\left(\frac{L}{\delta}\right)^2 \frac{\partial^2 \overline{u}}{\partial \overline{\eta}^2} \tag{11.14}$$

Now looking at Eq. (11.12), we see that the magnitude of the pressure term dominates over all the others. Therefore, we may ignore all terms of this equation in comparison with the pressure term, which leads to:

$$\frac{\partial \overline{p}}{\partial \overline{\eta}} = 0 \tag{11.15}$$

This equation tells us that the pressure across the boundary layer is constant, that is, the pressure at the boundary is the same as it is at the edge of the boundary layer. This implies that we may determine the pressure at the edge of the boundary layer by solving the inviscid flow there, and is *central* to the simplification afforded by the boundary layer theory to the momentum equations of a viscous fluid.

When the boundary is more than just 'slightly' curved, Eq. (11.15) needs modification. Figure (11.6) shows a fluid particle, of density ρ and dimensions $d\xi \times d\eta$, negotiating a significantly curved surface. The required centripetal force f must be provided by the difference in pressure between the two faces parallel to

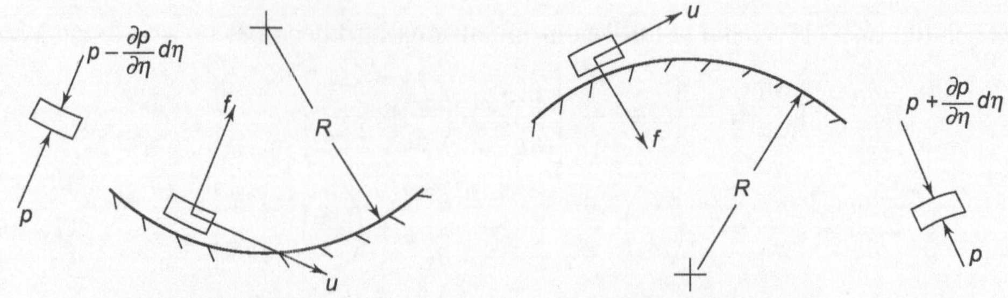

Figure 11.6 Effect of curvature upon the normal pressure gradient.

the boundary:

$$f = -\frac{\partial p}{\partial \eta}\, d\eta\, d\xi$$

for the surface of positive curvature (Fig. (11.6a)). But f must be equal to the mass of the fluid particle times its centripetal acceleration:

$$\rho\, d\xi\, d\eta\, \frac{u^2}{R} = -\frac{\partial p}{\partial \eta}\, d\eta\, d\xi$$

from which, with our system of non-dimensionalisation, we obtain:

$$\frac{\partial \overline{p}}{\partial \overline{\eta}} = -\frac{\delta}{R} \tag{11.16}$$

An identical result obtains for the surface of negative curvature (where R must, of course, be considered negative). *Now* if we define a 'slightly curved' boundary as one whose radius of curvature is much larger than the thickness of the boundary layer on it, then Eq. (11.16) reduces to Eq. (11.15). When the boundary is more than slightly curved, we use Eq. (11.16). The interested reader may refer to Cebeci and Bradshaw (1988) for a more rigorous analysis.

11.3.2.3 Obtaining the laminar boundary layer equations In dimensional form, the boundary layer equations (11.14)–(11.16) become:

$$u\frac{\partial u}{\partial \xi} + v\frac{\partial u}{\partial \eta} = -\frac{1}{\rho}\frac{\partial p}{\partial \xi} + \nu\frac{\partial^2 u}{\partial \eta^2} \tag{11.17}$$

and

$$\frac{\partial p}{\partial \eta} = -\frac{u^2}{R} \tag{11.18}$$

or

$$\frac{\partial p}{\partial \eta} = 0 \tag{11.19}$$

We shall use this dimensional form of the equations for calculations on laminar boundary layers.

11.3.3 The Boundary Layer Equations for Turbulent Flow

As described earlier, turbulent flow is characterised by random and rapid variations of flow velocity superposed on a steady mean flow. This leads to a profound change in the behaviour of the boundary layer and in the magnitudes of the shearing stresses, to describe which of the equations of the boundary layer have to be derived again. The following development is adapted from the lucid exposition by Duncan et al. (1970).

The components of flow velocity in turbulent motion may be represented as:

$$u = u_s + u', \qquad v = v_s + v' \tag{11.20}$$

where the primed quantities vary randomly with time while the quantities with suffix s are independent of time. At any time t, the mean of these velocities over a period T that is much larger than the time taken by the slowest of the random fluctuations is:

$$\overline{u}(t) = \frac{1}{2T}\int_{\tau=t-T}^{\tau=t+T} u\,\mathrm{d}\tau = u_\mathrm{s}, \qquad \overline{v}(t) = \frac{1}{2T}\int_{\tau=t-T}^{\tau=t+T} v\,\mathrm{d}\tau = v_\mathrm{s}$$

(where the overbar now denotes the time average) which follows from the fact that

$$\int_{\tau=t-T}^{\tau=t+T} u'\mathrm{d}\tau = \int_{\tau=t-T}^{\tau=t+T} v'\mathrm{d}\tau = 0$$

due to the completely random nature of u' and v'.

Let us consider a plane of height Δh and depth unity normal to the ξ direction, which is the direction of the mean flow. The flux through this plane is $\rho(u_s + u')\Delta h$ and the components of momentum transferred across it in unit time are:

$$\rho(u_\mathrm{s} + u')^2, \qquad \rho(u_\mathrm{s} + u')(v_\mathrm{s} + v')$$

having the time averaged values:

$$\rho u_\mathrm{s}^2 + \rho\overline{u'^2}, \qquad \rho u_\mathrm{s} v_\mathrm{s} + \rho\overline{u'v'} \tag{11.21}$$

The term $\rho\overline{u'^2}$, arising from the fluctuating velocity, represents a normal stress acting on the plane in the negative ξ direction; similarly $u'v'$ represents a shearing stress in the plane, acting in the negative η direction:

$$\tau'_{\xi\xi} = -\rho\overline{u'^2}, \qquad \tau'_{\xi\eta} = -\rho\overline{u'v'} \tag{11.22}$$

These, and $\tau'_{\eta\eta} = -\rho\overline{v'^2}$, are the Reynolds stresses mentioned earlier, and are all of similar magnitude.

Of the two stress components $\tau'_{\xi\xi}$ and $\tau'_{\xi\eta}$, the first one obviously is not zero, since u'^2 is always positive. As it turns out, the second term is not zero either. When a fluid particle moves outward from close to the boundary ($v' > 0$) then it arrives in a region where the mean u is greater than the region where it came from; hence the particle itself *very likely* has a $u' < 0$ at its new position, making $u'v'$ negative. Similarly, when a fluid particle moves in towards the boundary ($v' > 0$), it arrives at its new location very likely with a $u' > 0$ and $u'v'$ is again negative. Therefore, the quantity $\overline{u'v'}$ is negative. The quantities u' and v' are *strongly correlated* in the sense that they almost certainly have opposite signs.

Replacing u and v by $u_s + u'$ and $v_s + v'$, respectively, in the continuity equation Eq. (11.8), we get:

$$\frac{\partial(u_s + u')}{\partial\xi} + \frac{\partial(v_s + v')}{\partial\eta} = 0 \tag{11.23}$$

The time-average of this equation is:

$$\frac{\partial u_s}{\partial\xi} + \frac{\partial v_s}{\partial\eta} = 0 \tag{11.24}$$

which means also that:

$$\frac{\partial u'}{\partial\xi} + \frac{\partial v'}{\partial\eta} = 0 \tag{11.25}$$

If we make similar substitutions in the momentum equations (11.9) and (11.10), simultaneously replacing p with $p_s + p'$ gives:

$$(u_s + u')\frac{\partial(u_s + u')}{\partial\xi} + (v_s + v')\frac{\partial(u_s + u')}{\partial\eta} = -\frac{1}{\rho}\frac{\partial(p_s + p')}{\partial\xi} + \nu\left(\frac{\partial^2(u_s + u')}{\partial\xi^2} + \frac{\partial^2(u_s + u')}{\partial\eta^2}\right)$$

$$(u_s + u')\frac{\partial(v_s + v')}{\partial\xi} + (v_s + v')\frac{\partial(v_s + v')}{\partial\eta} = -\frac{1}{\rho}\frac{\partial(p_s + p')}{\partial\eta} + \nu\left(\frac{\partial^2(v_s + v')}{\partial\xi^2} + \frac{\partial^2(v_s + v')}{\partial\eta^2}\right)$$

Taking the time averages of these equations, noting that $\overline{u'}$, $\overline{v'}$ and all their spatial derivatives are zero and using Eq. (11.25), we obtain:

$$u_s\frac{\partial u_s}{\partial\xi} + v_s\frac{\partial u_s}{\partial\eta} = -\frac{1}{\rho}\frac{\partial p_s}{\partial\xi} + \left(\nu\frac{\partial^2 u_s}{\partial\xi^2} - \frac{\partial\overline{u'^2}}{\partial\xi}\right) + \left(\nu\frac{\partial^2 u_s}{\partial\eta^2} - \frac{\partial\overline{u'v'}}{\partial\eta}\right) \tag{11.26}$$

$$u_s \frac{\partial v_s}{\partial \xi} + v_s \frac{\partial v_s}{\partial \eta} = -\frac{1}{\rho}\frac{\partial p_s}{\partial \eta} + \left(\nu \frac{\partial^2 v_s}{\partial \xi^2} - \frac{\partial \overline{u'v'}}{\partial \xi}\right) + \left(\nu \frac{\partial^2 v_s}{\partial \eta^2} - \frac{\partial \overline{v'^2}}{\partial \eta}\right) \tag{11.27}$$

These equations are very similar to Eqs. (11.9) and (11.10), except that the viscous terms are augmented by the Reynolds stresses. Therefore, an order of magnitude analysis, as already performed for the laminar case, may be expected to give similar results if we assume that the thickness of the turbulent boundary layer, and hence the length scale along η, is much smaller than the length scale along ξ. As a consequence of this assumption, ξ-derivatives of quantities of similar magnitudes are much smaller than η-derivatives of the same quantities. The ξ-momentum equation simplified then becomes:

$$u_s \frac{\partial u_s}{\partial \xi} + v_s \frac{\partial u_s}{\partial \eta} = -\frac{1}{\rho}\frac{\partial p_s}{\partial \xi} + \left(\nu \frac{\partial^2 u_s}{\partial \eta^2} - \frac{\partial \overline{u'v'}}{\partial \eta}\right) \tag{11.28}$$

and the η-momentum equation gives:

$$\frac{\partial p_s}{\partial \eta} = 0 \tag{11.29}$$

11.3.4 Handling the Reynolds Stresses: Turbulence Modelling

The similarity of Eq. (11.28) to Eq. (11.17) is not quite complete, however, due to the presence of the Reynolds stress term $\partial \overline{u'v'}/\partial \eta$. This term is an additional unknown but we cannot find an additional equation from the laws of fluid mechanics to make the system of equations complete. The presence of this term therefore makes calculations on the turbulent boundary layer different from, and more complex than, calculations on the laminar boundary layer.

In order to determine the Reynolds stress term it becomes necessary to resort to *ad hoc* assumptions about its behaviour. These assumptions are based partly on theoretical analysis and partly on experimental results, and are collectively known as *turbulence modelling*. In all of them the aim is to express the Reynolds stress as a function of the mean flow.

Turbulence models of various degrees of sophistication and complexity have been proposed. Many of them require the solving of partial differential equations in addition to the equations of motion, while some others are algebraic. The reason why there are so many different models is that no two of them give quite the same result when applied to a given problem, so that some of them work better in a given situation than others. Moreover, each model involves the setting of a number of tuning parameters ('knobs and handles') so that the task of choosing and applying a turbulence model to a given problem is an art. In this chapter we shall use an algebraic turbulence model originally due to Prandtl.

11.3.5 Mathematical Nature of the Boundary Layer Equation (Boundary Conditions)

The boundary layer equations, both for laminar and turbulent flow (Eqs. (11.17) and (11.28), respectively), are partial differential equations with ξ and η as the independent variables and u, or u_s, as the dependent variable. (The turbulent boundary layer equation involves the Reynolds stress which, strictly speaking, is an unknown quantity. But with a suitable turbulence model this becomes known in terms of the mean flow speed u_s.) The highest derivative of u is of the second order in η and of the first order in ξ. Also, there are no mixed derivatives. Hence the boundary layer equation is *parabolic* in nature. In this equation ξ is the *time-like* coordinate, meaning that conditions at any point ξ_2 do not influence those at another point ξ_1 if $\xi_1 < \xi_2$. This behaviour is in contrast to that of the Navier–Stokes equations which, being of elliptic character, allow conditions anywhere in the flowfield to influence the flow at all points in that flowfield.

The parabolic nature of the boundary layer equation, together with the fact that there is only one such equation, gives a significant computational advantage over the Navier–Stokes equations which are elliptic and two in number.

Parabolic partial differential equations form what are called *initial-boundary value problems*. In this particular case, an *initial* condition needs to be prescribed at some point on the ξ axis, say at $\xi = 0$: this is the starting velocity profile. In addition, a pair of *boundary* conditions need to be prescribed, at $\eta = 0$ and $\eta = \delta$, respectively. Mathematically:

$$u(0, \eta) = U(0)f(\eta), \quad 0 \leq \eta \leq \delta \tag{11.30}$$

is the initial condition, $U(0)f(\eta)$ being the initial velocity profile, and

$$u(\xi, 0) = 0 \tag{11.31}$$

$$u(\xi, \delta) = U(\xi) \tag{11.32}$$

are the boundary conditions. $U(\xi)$ is the speed of the inviscid flow at the edge of the boundary layer. $U(\xi)$ is assumed to be known, being obtained from an inviscid computation independent of the boundary layer computation.

11.4 COMPUTATIONS ON THE LAMINAR BOUNDARY LAYER

11.4.1 Objectives

The primary aim of calculating a solution to the boundary layer equations is to be able to determine the drag exerted by the fluid on the rigid boundary past which it flows. An additional important piece of information obtainable is the point of separation, if any, of the boundary layer. As described earlier, the drag consists of two clearly identifiable components: one due to the skin friction and the other due to the modification in the pressure distribution that is brought about by the presence of the boundary layer. The skin friction drag predominates over the pressure drag for 'streamline' shapes, while the converse is true for 'bluff' shapes.

The boundary layer calculation also gives the displacement thickness. This quantity is not only a definite measure of the boundary layer thickness, but serves a greater purpose. Suppose, starting from a distribution of pressure on the boundary due purely to inviscid flow, we obtain a solution of the boundary layer equations and from there the displacement thickness. The displacement thickness may be taken to be the distance by which the outer inviscid flow considers the boundary to be 'pushed out' as a result of the displacement effect of the boundary layer. Then, having obtained the displacement thickness from a first boundary layer calculation, we may 'move' the physical boundary outward by this amount and recalculate the inviscid flow and hence the pressure. Obviously, this means that the boundary layer calculation has to be repeated. With the new value of the displacement thickness so obtained, the boundary may be "moved" to a new position and the inviscid calculation repeated. This cycle of inviscid-viscous iterative calculation may be repeated until there is no significant change in the value of the displacement thickness. The pressure then prevailing at the edge of the boundary layer is the pressure in the 'real' viscous flow, and may be integrated to obtain the pressure drag. This technique of alternating inviscid calculation and viscous correction is known as *viscous-inviscid interaction.*

11.4.2 Similarity Transformation and the Falkner–Skan Equation

The solution of the equation for the boundary layer over a semi-infinite, thin, flat plate placed parallel to the flow was first obtained by Blasius in 1908 (Duncan et al., 1970). In obtaining the solution, Blasius argued that no characteristic length could be associated with the streamwise direction in this particular problem because of the semi-infinite nature of the flat plate. In the absence of a characteristic length, the velocity profiles at all streamwise stations ought to be *similar*. That is,

$$\frac{u(\xi, \eta)}{U(\xi)} = f(\zeta) \tag{11.33}$$

where ζ is a non-dimensional variable consisting of some suitable combination of ξ and η and $f(\zeta)$ is a function whose form is to be determined. This problem has since been known as the Blasius problem.

In 1931 Falkner and Skan showed that similar solutions existed in all cases where the flow velocity just outside the boundary layer (that is, the velocity of the

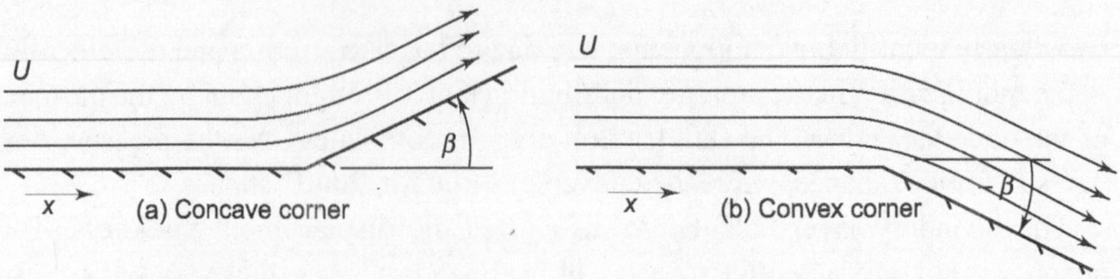

Figure 11.7 Potential flow past corners concave and convex.

"inviscid" flow) was of the type:

$$U(\xi) = k\xi^m \tag{11.34}$$

where m is a dimensionless, real constant and k is a constant having suitable dimensions. For $m > 0$ this describes the inviscid flow near a concave corner of angle β (Fig. 11.7a), with

$$m = \frac{\beta}{\pi - \beta} \tag{11.35}$$

The corner itself is a stagnation point. The Blasius problem corresponds to $\beta = m = 0$. Similarly the "Hiemenz problem", or "stagnation point flow", in which the freestream impinges at right angles on an infinite flat plate, corresponds to $\beta = \pi/2$ or $m = 1$. Again, $m < 0$ corresponds to the flow past a convex corner of angle β (that is, $\beta < 0$, Fig. 11.7b). In this case the corner is a singular point, where the flow velocity is infinite.

Let now

$$\zeta = \frac{\eta}{\delta(\xi)}$$

where δ is some suitable measure of the boundary layer thickness. We use here Eq. (11.13) taking it to be an equality, replace L with ξ, and write:

$$\delta(\xi) = \frac{\xi}{\sqrt{Re_\xi}}$$

where $Re_\xi = \xi U(\xi)/\nu$. Then:

$$\zeta = \eta\sqrt{\frac{U(\xi)}{\nu\xi}}$$

But we have assumed an outer, inviscid flow of the form:

$$U(\xi) = k\xi^m$$

Therefore:

$$\zeta = \sqrt{\frac{k}{\nu}} \cdot \eta \xi^{(m-1)/2} \tag{11.36}$$

Next we introduce a non-dimensional stream function $f(\zeta)$, related to the dimensional stream function $\psi(\xi, \eta)$ as follows:

$$\begin{aligned}\psi(\xi, \eta) &= \sqrt{\nu k} \cdot \xi^{(m+1)/2} f(\zeta) \\ &= \nu \frac{\xi}{\eta} \zeta f(\zeta)\end{aligned} \tag{11.37}$$

With the definitions:

$$u = \frac{\partial \psi}{\partial \eta}, \quad v = -\frac{\partial \psi}{\partial \xi}$$

we then obtain:

$$\begin{aligned}u &= U(\xi) f' \\ v &= -\frac{\nu}{\eta} \zeta \left[\frac{m+1}{2} f + \frac{m-1}{2} \zeta f' \right]\end{aligned} \tag{11.38}$$

where the $'$ denotes differentiation with respect to ζ.
Continuing in this way, we further obtain:

$$\begin{aligned}\frac{\partial u}{\partial \xi} &= \frac{U(\xi)}{\xi} \left[m f' + \frac{m-1}{2} \zeta f'' \right] \\ \frac{\partial u}{\partial \eta} &= U(\xi) \frac{\zeta}{\eta} f'' \\ \frac{\partial^2 u}{\partial \eta^2} &= \frac{U^2(\xi)}{\nu \xi} f'''\end{aligned} \tag{11.39}$$

In the outer, inviscid flow, by Bernoulli's theorem:

$$\frac{p(\xi)}{\rho} + \frac{U^2(\xi)}{2} = \text{constant}$$

hence:

$$-\frac{1}{\rho} \frac{\mathrm{d}p}{\mathrm{d}\xi} = U(\xi) \frac{\mathrm{d}U(\xi)}{\mathrm{d}\xi} \tag{11.40}$$

Using Eqs. (11.38)–(11.40) in the laminar boundary layer equation Eq. (11.17) and simplifying, we finally obtain:

$$f''' + \frac{m+1}{2} f f'' + m(1 - f'^2) = 0 \tag{11.41}$$

which is the Falkner–Skan equation.

At the rigid boundary, $u(\xi, 0) = v(\xi, 0) = 0$; also, $u(\xi, \eta \to \infty) = U(\xi)$. Hence, from the pair of equations (11.38) we get the following boundary conditions for Eq. (11.41):

$$f(0) = f'(0) = 0$$
$$f'(\zeta \to \infty) = 1 \tag{11.42}$$

11.4.3 Laminar Boundary Layer on a Flat Plate

Flow past a semi-infinite flat plate aligned with the stream corresponds to $m = 0$, for which Eq. (11.41) reduces to:

$$2f''' + ff'' = 0 \tag{11.43}$$

to which the boundary conditions set forth in Eq. (11.42) apply. This is the Blasius equation.

11.4.3.1 Solution by the "shooting method" One way of solving the two-point boundary value problem embodied in the Blasius equation is to guess an initial value of $f''(0)$, obtain the solution of the resulting initial value problem and, by repeated adjustment of the initial guess, to try and satisfy the far boundary condition $f'(\zeta \to \infty) = 0$. This is called the *shooting method*, since it is reminiscent of the method used by artillery to hit a target by repeated adjustments of the elevation angle of the barrel (Niyogi, 2003).

Needless to say, an initial guess close to the actual value greatly helps the shooting method to converge rapidly. Many approximate equations for the velocity profile in a laminar flow past a semi-infinite flat plate are known (Duncan et al., 1970). One such equation is:

$$\frac{u(\eta)}{U_0} = \frac{3}{2}\frac{\eta}{\delta} - \frac{1}{2}\left(\frac{\eta}{\delta}\right)^2$$

so that:

$$f'(\zeta) = \frac{3\zeta}{2} - \frac{\zeta^2}{2}$$

from which:

$$f''(0) = \frac{3}{2}$$

Now, if we define:

$$f(\zeta) = f_1(\zeta)$$
$$f'(\zeta) = f_1'(\zeta) = f_2(\zeta)$$
$$f''(\zeta) = f_2'(\zeta) = f_3(\zeta)$$

Then we may write Eq. (11.43) as:

$$f_1' = f_2$$
$$f_2' = f_3$$
$$f_3' = -0.5 f_1 f_3 \tag{11.44}$$

with the initial conditions:

$$f_1(0) = f_2(0) = 0, \quad f_3(0) = 1.5 \tag{11.45}$$

Now if these equations are integrated numerically using any suitable method (such as the fourth order Runge Kutta method; see Niyogi (2003), it is found that f_2 attains a value of approximately 2.73 at $\zeta \approx 3.5$, and remains constant thereafter. However, this value of $f_2(\zeta \to \infty)$ is way above the desired value of 1. So we have to find a means of zeroing in on the proper value of $f_3(\zeta = 0)$ such that the far boundary condition $f_2(\zeta \to \infty) = 1$ is satisfied.

Since the equation set (11.44) is homogeneous, integrating it with the initial condition $f_3 = 0$ would lead to each of the three variables f_1, f_2 and f_3 being identically zero for all ζ. Hence the right initial value of f_3 should lie between 0 and 1.5. Let the desired value of $f_3(\zeta = 0)$ be ϕ. Then:

$$f_2(\zeta \to \infty) = g(\phi) \tag{11.46}$$

where $g(\phi)$ is some function of ϕ whose form we do not know but whose value for a given ϕ we do know. Seeking the right value of ϕ to satisfy the far boundary condition therefore becomes a problem of obtaining a root of the algebraic equation

$$g(\phi) - 1 = G(\phi) = 0 \tag{11.47}$$

Since we know that the desired value of ϕ lies between 0 and 1.5, we may readily use any suitable algorithm for refining the approximate value of the root. Here is how the iterations go, to five places of decimals, using the *regula falsi* algorithm (Niyogi, 2003) for solving the algebraic equation (11.47):

ϕ	$G(\phi)$	$g(\phi) = f_2(\zeta \to \infty)$
0	−1	0
1.5	1.73266	2.73266
0.54892	0.39808	1.39808
0.39262	0.11817	1.11817
0.35113	0.03793	1.03793
0.33830	0.01249	1.01249
...		
0.33206	0.00000	1.00000

Hence, the desired initial value of f_3 that will make $f_2(\zeta \to \infty) = 1$ is 0.33206.

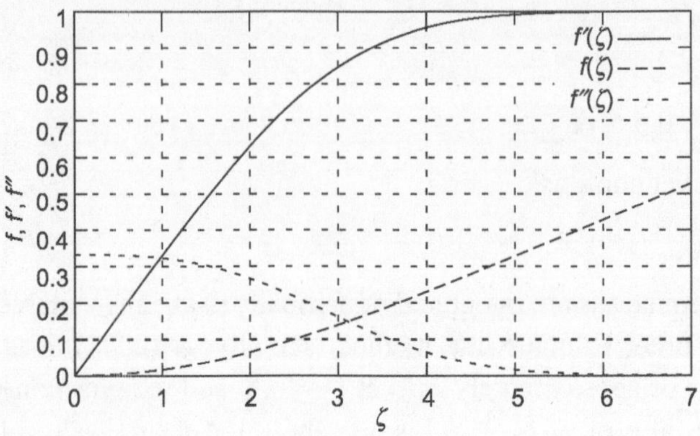

Figure 11.8 Relating to the boundary layer on a flat plate.

The graphical data is as follows:

η	f	f'	f''
0.00000e+00	0.00000e+00	0.00000e+00	3.32058e-01
5.00000e–01	4.14929e–03	1.65886e–01	3.30912e–01
1.00000e+00	1.65572e–02	3.29781e–01	3.23008e–01
1.50000e+00	3.70139e–02	4.86790e–01	3.02581e–01
2.00000e+00	6.50026e–02	6.29767e–01	2.66752e–01
2.50000e+00	9.96313e–02	7.51261e–01	2.17412e–01
3.00000e+00	1.39681e–01	8.46046e–01	1.61360e–01
3.50000e+00	1.83770e–01	9.13042e–01	1.07773e–01
4.00000e+00	2.30575e–01	9.55520e–01	6.42341e–02
4.50000e+00	2.79014e–01	9.79516e–01	3.39808e–02
5.00000e+00	3.28328e–01	9.91543e–01	1.59067e–02
5.50000e+00	3.78058e–01	9.96880e–01	6.57856e–03
6.00000e+00	4.27963e–01	9.98974e–01	2.40202e–03
6.50000e+00	4.77933e–01	9.99700e–01	7.74103e–04
7.00000e+00	5.27925e–01	9.99923e–01	2.20167e–04

11.4.3.2 Displacement thickness and skin friction coefficient The aim of boundary layer calculation is ultimately to determine the drag on the submerged body, and perhaps also the displacement thickness of the boundary layer. The shearing stress at the wall, at station ξ, is:

$$\tau_w(\xi) = \frac{\partial u}{\partial \eta}$$

which, using the second of Eqs. (11.40) and (11.36) becomes:

$$\tau_w(\xi) = \frac{\rho \nu U}{\sqrt{\nu}} \sqrt{k \zeta^{m-1}} f'' \tag{11.48}$$

Now, the *local skin friction coefficient* on one side of the plate at station ξ is defined as:

$$c_f(\xi) = \frac{\tau_w(\xi)}{\rho U^2/2}$$

which, using Eq. (11.48) and noting from the table above that $f''(0) = 0.332$, gives:

$$c_f(\xi) = \frac{0.664}{\sqrt{Re_\xi}} \tag{11.49}$$

where $Re_\xi = U\xi/\nu$ is the *local Reynolds number*.
The drag coefficient of a length c of the plate, measured from the leading edge, is:

$$C_D = \frac{2}{c} \int_0^c c_f(\xi) \mathrm{d}\xi$$

where the factor 2 arises because of contributions from both sides of the plate. With Eq. (11.49) this becomes:

$$C_D = \frac{2.66}{\sqrt{Re}} \tag{11.50}$$

where $Re = Uc/\nu$.
The displacement thickness was defined in Eq. (11.5) as:

$$\delta^* = \int_0^\infty \left(1 - \frac{u(\eta)}{U}\right) \mathrm{d}\eta$$

With u defined as in the first of Eqs. (11.38) and $\eta = \zeta \cdot \xi/\sqrt{Re_\xi}$, this integral becomes:

$$\frac{\delta^*}{\xi} = \frac{1}{\sqrt{Re_\xi}} \int_0^\infty (1 - f')\, \mathrm{d}\zeta$$

which may be numerically evaluated using the values of f' given in the table to obtain:

$$\frac{\delta^*}{\xi} = \frac{1.72}{\sqrt{Re_\xi}} \tag{11.51}$$

These results show that in a laminar boundary layer the local skin friction coefficient and displacement thickness are both inversely proportional to the square root of the local Reynolds number.
To get an estimate of the drag coefficient and the displacement thickness, we calculate these values for $c = 1$ m and $\eta = 1$ m, respectively, assuming our plate

to be immersed in an air stream flowing past it at $U = 100$ m/s. At standard atmospheric conditions the kinematic viscosity of air is 1.455×10^{-5} m^2/s. This gives a Reynolds number value of $Re = Re_\xi = 6.87 \times 10^6$. Therefore, $C_D = 1.01 \times 10^{-3}$ and $\delta^* = 6.56 \times 10^{-4}$ m. It is noteworthy that the boundary layer thickness 1 m from the plate leading edge is of the order of 1 mm.

11.4.3.3 What of the displacement effect? It was explained earlier (Section 11.2.3) how the slowing down of the fluid within the boundary layer causes the outer, inviscid, flow to be pushed out, or displaced, by an amount equal to the boundary layer displacement thickness δ^*. As a result of this displacement, the outer flow changes has to be recalculated. A change in the outer flow causes the streamwise pressure gradient $\frac{\partial p}{\partial \xi}$ in Eq. (11.17) to change, too. Accordingly, the boundary layer computation has to be done again. This iterative process then has to be repeated until a convergent state is reached, and the converged result may be taken to describe the actual flow situation accurately.
Strictly speaking, then, we should be doing this iterative procedure for the flat plate problem. However, extensive comparisons of results of the type obtained in the last section with experimental results show such close agreement Duncan et al., (1970) that it is not necessary to carry out this iterative process for the problem of the semi-infinite flat plate placed parallel to the flow.

11.4.4 Non-Similar Solutions of the Boundary Layer Equation

The cases for which no similar solutions to the boundary layer equations exist are by far in the majority. Then the pair of partial differential equations (11.17) may be solved directly, or they may first be transformed by means of a Falkner–Skan type transformation into a single third order partial differential equation before solution (Cebeci and Cousteix (1999), pp. 82–83). Here we shall take the latter approach. As before, let:

$$\zeta = \frac{\eta}{\delta(\xi)} = \eta\sqrt{\frac{U(\xi)}{\nu\xi}}$$

be the similarity variable. Now, however, the non-dimensional stream function f is a function of *both* ξ and η, so that:

$$\psi(\xi, \eta) = \sqrt{U(\xi)\nu\xi}\, f(\xi, \eta) \tag{11.52}$$

With the definitions:

$$u = \frac{\partial \psi}{\partial \eta}, \quad v = -\frac{\partial \psi}{\partial \xi}$$

the continuity equation is identically satisfied. Then, following steps similar to those before we obtain, from the second of the boundary layer equations:

$$\frac{\partial^3 f}{\partial \eta^3} = \frac{\rho U \delta}{\mu}\left[\frac{\delta}{U}\frac{\mathrm{d}U}{\mathrm{d}\xi}\left\{\left(\frac{\partial f}{\partial \eta}\right)^2 - 1 - f\frac{\partial f}{\partial \eta}\right\} - \frac{\mathrm{d}\delta}{\mathrm{d}\xi} f \frac{\partial^2 f}{\partial \eta^2} + \delta\left(\frac{\partial f}{\partial \eta}\frac{\partial^2 f}{\partial \xi \partial \eta} - \frac{\partial f}{\partial \xi}\frac{\partial^2 f}{\partial \eta^2}\right)\right] \qquad (11.53)$$

with the boundary conditions:

$$f(0) = \frac{\partial f}{\partial \eta}_{|\eta=0} = 0$$

$$\frac{\partial f}{\partial \eta}_{|\eta\to\infty} = 1$$

In contrast to the Falkner–Skan equation, this is a partial differential equation in ξ, η. Its solution using a technique known as the Keller Box Scheme is described next.

11.4.5 The Keller Box Scheme

The Keller box scheme is a two-point scheme for a parabolic system of first-order partial differential equations. The first-order equations are approximated on an arbitrary rectangular net (Fig. 11.9) using central difference for the derivatives and averages for the non-derivative quantities. This scheme is second-order accurate in both independent variables. Due to its using only two neighbouring points for writing the finite difference equations, it can work on variable width meshes without its accuracy being affected.

In order to use the Keller box scheme, we have to represent the third-order equation (11.53) as a system of three first-order equations by introducing auxiliary variables, thus:

$$\frac{\partial f}{\partial \eta} = g$$

$$\frac{\partial g}{\partial \eta} = h \qquad (11.54)$$

$$\frac{\partial h}{\partial \eta} = \frac{\rho U \delta}{\mu}\left[\frac{\delta}{U}\frac{dU}{d\xi}(g^2 - 1 - fg) - \frac{d\delta}{d\xi} f h + \delta\left(g\frac{\partial g}{\partial \xi} - h\frac{\partial f}{\partial \xi}\right)\right]$$

According to the box scheme, the first of these equations is discretised thus:

$$\frac{f_j^n - f_{j-1}^n}{h_j} = \frac{g_j^n + g_{j-1}^n}{2}$$

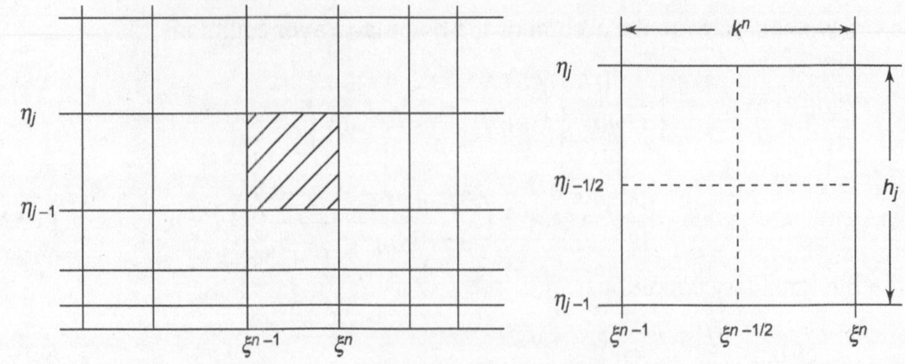

Figure 11.9 Nomenclature for the Keller box scheme.

The other two equations are similarly treated and, after some algebraic manipulations, a non-linear block tridiagonal system of algebraic equations is obtained. At a given ξ station these may be solved using a Newton linearisation procedure. The values of f, g and h so obtained serve as inputs for the next ξ station. In this way the solution is carried forward in a marching procedure. Details are available in Cebeci and Cousteix (1999 pp. 84–89).

11.5 TURBULENT BOUNDARY LAYERS

At sufficiently high Reynolds numbers the boundary layer becomes turbulent, as has already been described. The Reynolds-averaged equations for the turbulent boundary layer involve the turbulent shear stresses (the so-called Reynolds' stresses) which, being statistical in nature, are not describable by the equations of fluid dynamics. Various kinds of turbulence models attempt to relate these stresses to the mean flow quantities. The nature and application of some of these turbulence models are described in Chapters 12 and 13.

11.6 SUMMARY

For flows at high Reynolds numbers, the effect of viscosity is confined to a narrow region, called the boundary layer, near the boundary. Within this region the effects of viscosity, namely shear stresses and loss of mechanical energy of the fluid, are pronounced. Outside this region the flow is essentially inviscid.

The dimension of the boundary layer normal to the boundary is considerably smaller than the characteristic length of the flow. This fact leads to a simplification of the

Navier–Stokes equations. In particular, it is found that the pressure does not vary appreciably across the depth of the boundary layer.

When the external pressure obeys a power law, then a similarity transformation makes it possible to turn the partial differential equations governing the boundary layer into an ordinary differential equation with a similarity variable as the independent variable. In this situation it becomes relatively easy to get a solution. When no similarity transformation is possible, the partial differential equations may either be solved directly or, using a similarity-type transformation, the two partial differential equations may be transformed into a single partial differential equation of the third order and then solved. The latter approach is the more straightforward.

11.7 KEY TERMS

Adverse pressure gradient
Boundary layer
Favourable pressure gradient
Laminar flow
Turbulent flow
Reynolds number
Prandtl's hypothesis
Falkner–Skan equation
shooting method
Keller box scheme.

11.8 EXERCISE 11

1. Flow in a convergent channel. If the velocity in the outer flow is of the form

$$U = -\frac{u_1}{\xi}, \quad u_1 < 0$$

this represents flow in a converging channel with flat walls (Schlichting and Gerstin (2000), pp. 152–153).

Introduce (a) the similarity variable:

$$\zeta = \frac{\eta}{\xi}\sqrt{\frac{u_1}{\nu}} = \tan\theta\sqrt{\frac{Ur}{\nu}}$$

where θ is the polar angle measured counter-clockwise from the η-axis and r is the radial distance measured from the apex of the channel, and (b) the dimensionless stream function:

$$f(\zeta) = -\frac{\sqrt{\nu u_1}}{\psi(\xi, \eta)}$$

and show that Eq. (11.17) transforms into:

$$f''' - f'^2 + 1 = 0$$

with the boundary conditions:

$$f'(\zeta = 0) = 0,$$
$$f'(\zeta \to \infty) = 1,$$
$$f''(\zeta \to \infty) = 0.$$

Obtain a solution of this equation using the shooting method, and compare with the analytical solution given in Schlichting and Gerstin (2000), pp.152–153.

2. Block-tridiagonal system of equations. A 3 × 3 *block tridiagonal* system of equations looks like this:

$$\begin{bmatrix} [D]_1 & [U]_2 & [0] & [0] & [0] & \cdots & [0] \\ [L]_1 & [D]_2 & [U]_3 & [0] & [0] & \cdots & [0] \\ [0] & [L]_2 & [D]_3 & [U]_4 & [0] & \cdots & [0] \\ [0] & [0] & \cdots & \cdots & \cdots & \cdots & [0] \\ [0] & [0] & \cdots & \cdots & [L]_{n-2} & [D]_{n-1} & [U]_n \\ [0] & [0] & [0] & \cdots & \cdots & [L]_{n-1} & [D]_n \end{bmatrix} \begin{Bmatrix} [x]_1 \\ [x]_2 \\ [x]_3 \\ \vdots \\ [x]_{n-1} \\ [x]_n \end{Bmatrix} = \begin{Bmatrix} [b]_1 \\ [b]_2 \\ [b]_3 \\ \vdots \\ [b]_{n-1} \\ [b]_n \end{Bmatrix}$$

where $[D]_i$, $[U]_i$ and $[L]_i$ are 3 × 3 matrices and [0] is the 3 × 3 null matrix; and $\{x\}_i$ and $\{b\}_i$ are 3 × 1 vectors.

Such a system of equations may be solved efficiently by extending the Thomas algorithm for a simple tri diagonal system. The extension is done by replacing each scalar in the original algorithm with a 3 × 3 matrix or 3 × 1 vector, as appropriate. Scalar additions and multiplications are replaced by matrix additions and multiplications, and the operation of taking the reciprocal of a scalar is replaced by that of taking the inverse of a 3 × 3 matrix.

Write a computer program to solve a 3 × 3 block tridiagonal system using the extended Thomas algorithm.

3. Computer program for the Keller box scheme. Write a computer program to implement the Keller box scheme. The detailed development of the scheme is available in Cebeci and Cousteix (1999), pp. 84–89.

(The block-tridiagonal solver of the last exercise will form the core of this computer program.)

12

Viscous Incompressible Flow

In the previous chapter, methods for incompressible viscous flow computation based on boundary layer model were discussed. In that model, the Navier–Stokes equations have been approximated using the boundary-layer theory of Prandtl (Prandtl, 1904; Schneider, 1978). Computation of fluid flow problems based on the complete Navier–Stokes equations (N–S Eqs.) is the subject matter of the present chapter and the next one. In the present chapter we assume the flow to be incompressible while the next chapter is devoted to compressible viscous flow. Computation of laminar flow has been discussed first. The latter sections briefly discuss methods for turbulent flow computation based on incompressible *Reynolds averaged Navier–Stokes (RANS)* equations.

12.1 INTRODUCTION

N–S Equations describe the flow of viscous fluid most accurately. These equations are known for more than one and a half century. As already mentioned, these are a system of nonlinear partial differential equations. Except very few special cases, no mathematical method is available to solve them. So, scientists and engineers looked for approximate solutions which gave rise to the various approximate models. Although, strictly speaking, all fluids are viscous, under certain conditions it is possible to introduce the approximate model of inviscid fluid. At low speeds all fluid behave as an incompressible fluid. Computational methods developed for computing solutions of compressible Navier–Stokes equations are, in general, not applicable to problems for incompressible flow. The more important methods for incompressible flow are presented in this chapter.

Several approximate models like the *thin-layer model* or the *parabolised Navier–Stokes equations* commonly classified under the category of the *reduced Navier–Stokes equations* (Fletcher, 1988), have been investigated during the seventies and early eighties. None of these models are satisfactory for problems with regions of *reversed flow* or for problems with *large areas of separation.* For many problems there is no obvious dominant flow direction. For accurate treatment of such problems it becomes necessary to solve the N–S Equations. without introducing any approximation in the equations or in the boundary and initial conditions. Moreover, since the N–S Equations may be solved for laminar flow in a reasonable amount of time (at least in the 2-D cases) on easily available computational equipments like the personal computers, the modern trend is to go for the solution of the N–S Equations, without introducing any approximation in the equations or in the boundary conditions. For this reason, we do not discuss any of the above approximate models.

In this chapter we assume the flow to be incompressible and governed by the N–S Equations. At sufficiently low Reynolds number, the flow remains *laminar.* Methods for computational solution of N–S Equations under laminar conditions has reached a state of maturity and it is now possible to solve such problems accurately, with confidence. However, at higher Reynolds number the flow grows *turbulent*. Study of turbulent flow is rather a difficult area, analytically as well as computationally. Enormous effort has been made all over the world for over a century, in order to develop satisfactory methods for predicting solutions of problems involving turbulent flow. In spite of best efforts, the level of progress is only mediocre. Even today, one has to rely heavily on empirical data for methods of turbulent flow computation.

It is important to note that most of the common flow problems of practical interest are turbulent. For example, flow past an aircraft, a ship, an automobile or a tower are turbulent. We encounter turbulent flow in heat exchangers, nuclear reactors, pumps, gas and steam turbines. Turbulence plays an important role in flow of rivers. Atmospheric turbulence is important in weather prediction. These examples indicate the widespread need for satisfactory methods for computing turbulent flows.

Over the past two decades several computational methods have come up for computing turbulent flows. The most accurate among these, namely *Direct Numerical Simulation* needs vast computing power and is limited at present to the study of only the simplest of 2-D problems. Another important development is the *Large Eddy Simulation* model. These models are beyond the scope of the present work. Interested readers are referred to Biswas and Eswaran (2002), Gatski (1996) and Hartel (1996).

In view of its great need in applications, one looks for approximate simple computational methods for predicting turbulent flows. Most frequently, we are interested in the mean motion. The equations, obtained by averaging the N–S Equations over a small period of time, provide for the governing equations of the mean motion. However, these equations contain the unknown *Reynolds stress* terms. Requisite number of equations are not available to determine them. This is the well-known *closure problem.* For relating the Reynolds stress terms to the mean motion, one has to make *ad-hoc* assumptions or rely on empirical data or on empirical or semi-empirical methods. Development and study of such models is known as *turbulence modelling.* It may be noted that all such models involve empirical information. Thus, using some suitable turbulence model for determining the Reynolds stress terms, a closed system of equations is obtained, from which approximate solutions may be computed. The simplest of such turbulence models are the algebraic *eddy viscosity models,* in which the effect of turbulence is taken care of by introducing a *turbulent kinematic viscosity* ν_T. Among other turbulence models, the two-equation K–ϵ model (Launder and Spalding, 1974), K denoting the turbulent kinetic energy and ϵ the rate of dissipation of turbulent energy, has found wide application. We discuss turbulence modelling briefly in this chapter. Only simple algebraic models and the K–ϵ model have been discussed. For other models the reader is referred to Biswas and swaran (2002), Launder and Spalding (1972), Majumdar (1991), Rodi (1980).

12.2 INCOMPRESSIBLE FLOW COMPUTATION

The computational methods developed for solving viscous fluid flow problems may be classified under (a) methods for compressible flow and (b) methods for incompressible flow. The methods developed for incompressible flow are generally characterised by the use of pressure as a main dependent variable. Also, frequently staggered grid arrangement is used for such computations in order to avoid decoupling of the pressure field. Such decoupling is known as the *checker-board effect.* The schemes for compressible flow, on the other hand use density as a primary variable and extract pressure from the Bernoulli's equation and the equation of state. Most of such methods are based on nonstaggered grids in which all the unknown dependent variables are stored at the same location. These schemes could not be used for incompressible flows or for flows at very low Mach numbers since in these cases the pressure density coupling becomes very weak. Hence, study of unsteady flow fields using pressure based methods gained importance for incompressible flow.

The main obstacle to choose the pressure as a working variable is the need to devise a mechanism by which the continuity and the momentum equations could be linked together. The difficulty stems from the presence of the pressure in the momentum equations in the form of source terms, but the pressure does not appear in the continuity equation. If we regard the momentum equations as equations determining the velocities, then we see that the pressure possesses no equation of its own. This difficulty is circumvented by deriving a pressure equation from a joint manipulation of the continuity and the momentum equations. The pressure equation may be derived either by using the parent differential equations or else by utilizing the finite difference analogues of these equations directly.

For computing incompressible laminar flow, different basic formulations exist depending on the choice of the unknowns constituting the dependent variables. We shall discuss here only flows in 2-D. The basic unknowns may be the *primitive variables*, namely the two rectangular Cartesian velocity components u, v and the pressure p. For incompressible laminar flow, not much energy change occurs. Generally the energy equation is not coupled with the continuity and the momentum equations and need not be considered. If required, this equation may be considered separately. A second formulation is the *stream-function vorticity approach* while a third formulation is based on the *vorticity velocity.* Among others, there are the *stream-function only* approach, etc. We discuss here two approaches, namely those based on the primitive variables and the stream-function vorticity formulations that found wide applications. For other methods the reader may consult references Fletcher (1988) and Peyret and Taylor (1983). The primitive variable approach may be readily extended to 3-D. Its main difficulty is in specification of boundary condition on pressure. The main difficulty with the stream-function vorticity approach is that the extension to 3-D is not straight forward, as no stream-function exists in 3-D. Moreover, determination of vorticity at a boundary may not be easy. An inconvenience of the stream-function vorticity approach is that the pressure is not explicitly obtained. Consequently, additional calculations are required for its determination.

We consider two-dimensional incompressible fluid flow, without external body forces, heat or energy supply. Governing continuity and momentum equations (N–S Equations) in rectangular Cartesian coordinates in nondimensional form, as discussed in Chapter 7, are

$$\text{Continuity: } \frac{\partial u}{\partial x} + \frac{\partial v}{\partial y} = 0, \tag{12.1}$$

$$\text{x-momentum: } \frac{\partial u}{\partial t} + u\frac{\partial u}{\partial x} + v\frac{\partial u}{\partial y} = -\frac{\partial p}{\partial x} + \frac{1}{Re}\left(\frac{\partial^2 u}{\partial x^2} + \frac{\partial^2 u}{\partial y^2}\right) \tag{12.2}$$

$$\text{y-momentum: } \frac{\partial v}{\partial t} + u\frac{\partial v}{\partial x} + v\frac{\partial v}{\partial y} = -\frac{\partial p}{\partial y} + \frac{1}{Re}\left(\frac{\partial^2 v}{\partial x^2} + \frac{\partial^2 v}{\partial y^2}\right). \tag{12.3}$$

Here, u, v denote the velocity components along x and y-axial directions, p the pressure, the density being absorbed in the Reynolds number Re. It is to be noted that Eqs. (12.2) and (12.3) are not in conservative form. The corresponding conservative equations, that conserve momentum, are

$$\text{x-momentum: } \frac{\partial u}{\partial t} + \frac{\partial u^2}{\partial x} + \frac{\partial uv}{\partial y} = -\frac{\partial p}{\partial x} + \frac{1}{Re}\left(\frac{\partial^2 u}{\partial x^2} + \frac{\partial^2 u}{\partial y^2}\right) \tag{12.4}$$

$$\text{y-momentum: } \frac{\partial v}{\partial t} + \frac{\partial uv}{\partial x} + \frac{\partial v^2}{\partial y} = -\frac{\partial p}{\partial y} + \frac{1}{Re}\left(\frac{\partial^2 v}{\partial x^2} + \frac{\partial^2 v}{\partial y^2}\right). \tag{12.5}$$

A typical initial boundary-value problem associated with the above N–S Equations (12.1)–(12.3) or (12.1), (12.4) and (12.5) may be stated as follows:

To find the velocity vector $\vec{V}$ and the pressure p which are solutions of Eqs. (12.1–12.3) in a bounded domain Ω with boundary C, such that on the boundary

C, $\vec{V}$ satisfies the condition

$$\vec{V} = \vec{V}_C, \tag{12.6}$$

and at initial time $t = 0$, the velocity vector $\vec{V}$ satisfies the condition

$$\vec{V} = \vec{V}_0. \tag{12.7}$$

Moreover, the boundary value $\vec{V}$ must satisfy the integral relation

$$\int_{\Omega} \vec{V}_C \cdot \vec{N} \mathrm{d}s = 0, \tag{12.8}$$

$\vec{N}$ denoting the unit normal vector to Ω. Over and above, the initial value $\vec{V}_0$ must satisfy

$$\nabla \cdot \vec{V}_0 = 0. \tag{12.9}$$

The above boundary and initial conditions are directly applicable to the primitive variables approach. For other approaches, appropriate boundary and initial conditions are to be derived from these conditions.

We wish to observe that for the velocity–pressure formulation it is essential that the condition $\nabla \cdot \vec{V} = 0$, be satisfied at all time, for example, *at each time step of an explicit scheme for numerical computation.* This is an additional constraint, requiring solution of an algebraic system of equations at each time step. It may be noted that no boundary condition for the pressure is required.

In the following, we discuss the stream-function vorticity approach first.

12.3 STREAM-FUNCTION VORTICITY APPROACH

Let us eliminate pressure p from the momentum equations (12.2) and (12.3). Differentiating both sides of (12.2) with respect to y and Eq. (12.3) with respect to x and subtracting, pressure can be eliminated. We wish to express the resulting equation in terms of the vorticity ζ.

The vorticity vector $\vec{\zeta}$ is defined as

$$\vec{\zeta} = \nabla \times \vec{V}, \tag{12.10}$$

where the ∇-operator is defined as the *vector*

$$\nabla \equiv i_x \frac{\partial}{\partial x} + i_y \frac{\partial}{\partial y}, \tag{12.11}$$

i_x, i_y being unit vectors along rectangular Cartesian coordinate directions x and y, and the velocity vector

$$\vec{V} = i_x u + i_y v. \tag{12.12}$$

Let ζ denote the magnitude of the vorticity vector $\vec{\zeta}$, given by

$$\zeta = |\vec{\zeta}| = \frac{\partial v}{\partial x} - \frac{\partial u}{\partial y}. \tag{12.13}$$

Using continuity equation (12.3) we obtain the *vorticity transport equation* in terms of ζ as

$$\zeta_t + u\zeta_x + v\zeta_y = \frac{1}{Re}\left[\zeta_{xx} + \zeta_{yy}\right]. \tag{12.14}$$

Introducing the stream-function ψ, defined by

$$\frac{\partial \psi}{\partial y} = u, \quad \frac{\partial \psi}{\partial x} = -v, \tag{12.15}$$

in (12.13) yields

$$\psi_{xx} + \psi_{yy} = -\zeta. \tag{12.16}$$

Equations (12.14)–(12.16) are the *basic equations of the stream-function vorticity formulation.*

One-dimensional analogue of the vorticity transport equation (12.14), namely

$$\zeta_t + u\zeta_x = \mu\zeta_{xx}, \tag{12.17}$$

where u is a known quantity and μ is a nonnegative constant, is the *parabolic type model convection–diffusion equation,* discussed in Chapter 4, Section 4.10. All the methods discussed there, may now be applied to solve the vorticity transport equation (12.14).

The vorticity transport equation (12.14) is in *non conservative form.* Using continuity equation (12.1), it may be expressed in the *conservative form*

$$\frac{\partial \zeta}{\partial t} + \frac{\partial (u\zeta)}{\partial x} + \frac{\partial (v\zeta)}{\partial y} = \frac{1}{Re}\left(\frac{\partial^2 \zeta}{\partial x^2} + \frac{\partial^2 \zeta}{\partial y^2}\right). \tag{12.18}$$

In terms of velocity vector $\vec{V} = i_x u + i_y v$, Eq. (12.18) may be rewritten as

$$\frac{\partial \zeta}{\partial t} + \nabla \cdot (\vec{V}\zeta) = \frac{1}{Re}\nabla^2 \zeta. \tag{12.19}$$

It is well known that equations in conservative form generally yield more accurate and dependable results compared to that of nonconservative form (Roache, 1972).

We note that pressure is not explicitly available as solution of the stream-function vorticity equations (12.16) and (12.18). In order to determine the pressure, one has to solve a Poisson equation for the pressure, commonly known as the *pressure Poisson equation,* which is derived in the next subsection.

12.3.1 Pressure Poisson Equation

Differentiating (12.2) with respect to x and (12.3) with respect to y and adding yields

$$\frac{\partial}{\partial t}(u_x + v_y) + u_x^2 + v_y^2 + u\left(u_{xx} + v_{xy}\right) + v\left(u_{xy} + v_{yy}\right) + 2v_x u_y$$
$$= -\nabla^2 p + \frac{1}{Re}\nabla^2\left(\frac{\partial u}{\partial x} + \frac{\partial v}{\partial y}\right). \quad (12.20)$$

Note that, by virtue of the continuity equation (12.1), we have

$$u_{xx} + v_{xy} = \frac{\partial}{\partial x}\left(u_x + v_y\right) = 0,$$

$$u_{xy} + v_{yy} = \frac{\partial}{\partial y}\left(u_x + v_y\right) = 0,$$

and $$u_x^2 + v_y^2 = u_x(-v_y) + v_y(-u_x) = -2u_x v_y.$$

Then, Eq. (12.20) changes over to the Poisson equation for pressure

$$\nabla^2 p = 2\left(u_x v_y - v_x u_y\right). \quad (12.21)$$

Introducing in (12.21) the stream-function ψ according to Eq. (12.15), yields the pressure Poisson equation

$$\nabla^2 p = G, \quad (12.22)$$

where

$$G = 2\left[\psi_{xx}\psi_{yy} - \psi_{xy}^2\right] \quad (12.23)$$

For numerical computation, second-order central difference representation may be used for the right hand side expression G. Then,

$$G = 2\left[\frac{\psi_{j+1,k} - 2\psi_{j,k} + \psi_{j-1,k}}{\Delta x^2}\,\frac{\psi_{j,k+1} - 2\psi_{j,k} + \psi_{j,k-1}}{\Delta x^2}\right.$$
$$\left. - \left(\frac{\psi_{j+1,k+1} - \psi_{j+1,k-1} - \psi_{j-1,k+1} + \psi_{j-1,k-1}}{4\Delta x\,\Delta y}\right)^2\right]. \quad (12.24)$$

12.3.2 Boundary Conditions for Stream-Function and Vorticity

The basic equations for incompressible viscous flow in 2-D of the stream-function vorticity formulation are the vorticity transport equation (12.14)

$$\zeta_t + u\zeta_x + v\zeta_y = \frac{1}{Re}\left[\zeta_{xx} + \zeta_{yy}\right]. \quad (12.25)$$

and the Poisson equation for the stream-function

$$\psi_{xx} + \psi_{yy} = -\zeta \tag{12.26}$$

where the velocity components are

$$\frac{\partial \psi}{\partial y} = u, \quad \frac{\partial \psi}{\partial x} = -v, \tag{12.27}$$

the vorticity ζ being defined as

$$\zeta = \frac{\partial v}{\partial x} - \frac{\partial u}{\partial y}. \tag{12.28}$$

Equation (12.25) is a second-order partial differential equation of parabolic type for the unknown vorticity ζ. Starting from an initial distribution of vorticity $\zeta = \zeta_0(x, y, t_0)$ at $t = t_0$ the solution marches forward step by step in the t-direction. For computing the solution, values of ζ must be known on the boundary C of the computational domain. However, boundary values of ζ are not prescribed and these values must be derived from the prescribed boundary conditions. Typical boundary conditions have been presented in Eqs.(12.6)–(12.9). Appropriate boundary conditions for ψ and ζ may be derived from them.

The stream-function ψ may be determined from Poisson equation (12.26), once the vorticity ζ is known. This equation is of elliptic type. Methods of solving such equations have been discussed in Chapter 5. For example, one may use the ADI method or the approximate factorisation methods or the multigrid methods for its numerical solution. The incompressible unsteady N–S Equations are a system of partial differential equations of mixed parabolic–elliptic type. The stream-function vorticity approach, separates them into two equations, namely the parabolic-type vorticity transport equation (12.25) for ζ and the elliptic-type Poisson equation (12.26) for the stream-function.

The prescription of boundary and initial conditions are guided by the type of the governing partial differential equations. Parabolic-type equations require an initial condition from which the solution marches forward in time, while for an elliptic-type equation no initial conditions may be prescribed. It has to be a boundary-value problem of Dirichlet or Neumann type or a combination. We presently discuss in details the boundary and initial conditions to be prescribed for well-posed problems.

Equation (12.26) is a Poisson equation for the stream-function ψ. While solving for ψ the vorticity distribution ζ appearing on the right side of (12.26) is known. Equation (12.26) is of elliptic type and requires two conditions for ψ to be prescribed on the boundary of the computational domain. If the velocity components be prescribed, in view of (12.27) the derivatives of ψ are known on the boundary. In order to make the value of ψ unique, we may require to fix the value of ψ at one point on the boundary. Consequently, the two boundary conditions may be taken to be of the form

$$\psi|_C = \alpha, \quad \text{and} \quad \left.\frac{\partial \psi}{\partial n}\right|_C = \beta, \tag{12.29}$$

n denoting direction normal to the boundary C.

For flow past an immersed body, necessary far-field boundary conditions like

$$\zeta = 0, \tag{12.30}$$

is often used instead of the second condition $\frac{\partial \psi}{\partial n}|_C = \beta$ in (12.29).

The boundary, through which the flow enters the computational domain is called the *inflow boundary* while the boundary through which it leaves the domain is called the *outflow boundary.* At inflow and at outflow boundaries it is appropriate to prescribe values of all but one of the dependent variables (Gustaffson, 1978). Since, far away from an immersed body the viscous terms are negligibly small, the flow behaves as an inviscid fluid and only one boundary condition may be prescribed at an outflow boundary.

As illustration the more common boundary conditions are now explained by means of the following well-known examples, discussed earlier by many authors like Anderson et al. (1984), Fletcher (1988), Peyrect and Taylor (1983), Roache (1972), Shankar and Deshpande (2000).

Example 12.1

The two-dimensional driven cavity: An example of internal flow with walls on all the four sides is the two-dimensional driven cavity problem. Consider a rectangular cavity with walls AB, BC, CD and AD, on the four sides, as shown in Fig. 12.1, filled with incompressible viscous fluid. The lid of the cavity AD moves to the right with uniform speed $u = 1$, parallel to itself, which sets the fluid inside in motion. To study the resulting motion. This problem has been used as a test case for comparing different numerical methods for solving the incompressible N–S Equations

We observe that walls AB, BC, CD are part of a streamline $\psi = \text{const.}$. For convenience, the constant is chosen as zero. The no-slip boundary condition on the walls are $u = 0$, and $v = 0$. On the walls AB and CD, $v_y = 0$, so that $\zeta = -\psi_{xx}$, while on BC and AD, $v_x = 0$, so that $\zeta = -\psi_{yy}$. The boundary conditions are shown in Fig. 12.1.

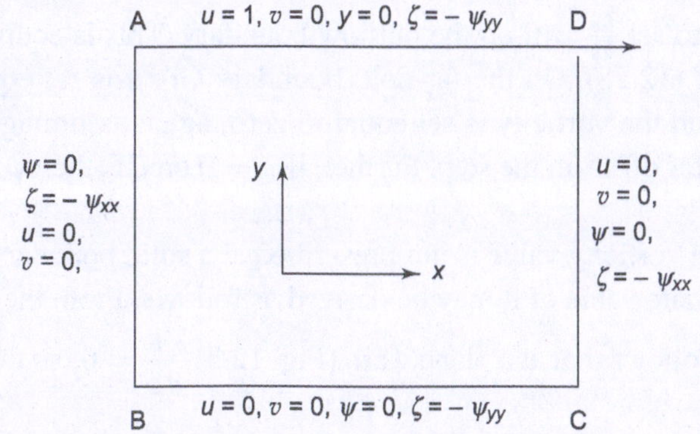

Figure 12.1 Boundary conditions for stream-function and vorticity for the driven cavity problem.

Example 12.2

Flow past a backward-facing step: An example, frequently discussed in literature is that of incompressible viscous flow past a backward-facing step, ABCDE as shown in Fig. 12.2. The flow separates at the corner B and reattaches at D on the wall CE, forming a region of recirculating flow behind the step, around BCD. Here, GA and FE are inflow and outflow boundaries respectively. GF is the far-field boundary.

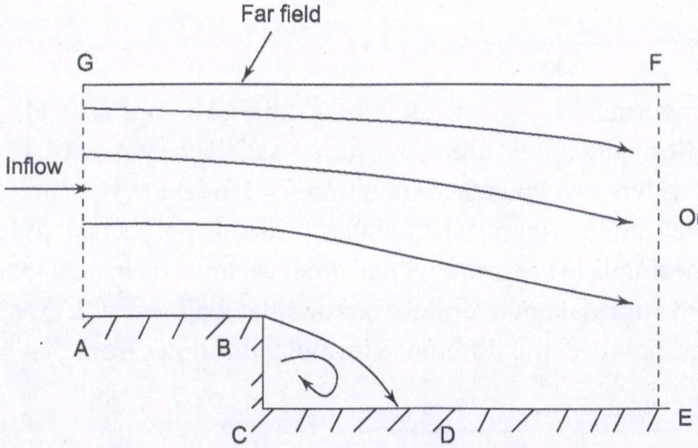

Figure 12.2 Flow past a backward-facing step.

On the inflow boundary GA, u is specified, which in view of (12.27) implies that ψ is known at the inflow boundary. The outflow boundary FE is assumed to be sufficiently far away from the step. Among various alternatives, it is generally recommended to set $\frac{\partial v}{\partial x} = 0$ on the outflow boundary. This is equivalent to $\frac{\partial^2 \psi}{\partial y^2} = -\zeta$, in view of (12.28). On the far-field boundary GF, u is set equal to the free-stream value and the vorticity is set equal to zero, again assuming the far-field to be sufficiently far off from the step. Further, if $v = 0$ on GF, then ψ is fixed and set equal to ψ_G.

Note that the vorticity value is not prescribed at a solid boundary.

An approximate value of it may be derived as follows. From the relations $\psi_y = u,\ \psi_x = -v$, follows that at a plane wall, (Fig. 12.3) $\dfrac{\partial v}{\partial x} = 0$, so that $\zeta = \frac{\partial v}{\partial x} - \frac{\partial u}{\partial y}$ yields

$$\zeta|_{\text{wall}} = -\frac{\partial u}{\partial y}|_{\text{wall}} = -\psi_{yy}|_{\text{wall}}. \tag{12.31}$$

Let us consider a flat wall at which we take the index $k = 1$. Expanding by Taylor series, we see that

$$\begin{aligned}\psi_{j,2} &= \psi_{j,1} + \Delta y \frac{\partial \psi}{\partial y} + \frac{\Delta y^2}{2}\frac{\partial^2 \psi}{\partial y^2} + O(\Delta y^2)\\ &= \psi_{j,1} + \frac{\Delta y^2}{2}\psi_{yy}|_{\text{wall}} + O(\Delta y^2)\\ &= \psi_{j,1} - \frac{\Delta y^2}{2}\zeta|_{\text{wall}} + O(\Delta y^2),\end{aligned}$$

which yields on simplification, yields

$$\zeta_{\text{wall}} = -\frac{2(\psi_{j,2} - \psi_{j,1})}{\Delta y^2} + O(\Delta y) \tag{12.32}$$

This first order formula is known for a long time (Thompson et al., 1985) and has been used in many applications delivering satisfactory results. However, for more accurate results, the boundary condition for ζ ought to be represented more accurately. Since at interior mesh points we are using second order accurate formulae, it is desirable to use also a second order accurate formula for the boundary value of ζ. Such a formula may be derived for a flat wall (Fig. 12.3) as follows:

Using a three-point central difference formula, it follows from $\zeta_{j,k} = -\frac{\partial^2 \psi}{\partial y^2}|_{j,k}$

$$\zeta_{j,k} = -\frac{\psi_{j,k-1} - 2\psi_{j,k} + \psi_{j,k+1}}{\Delta y^2} + O(\Delta y^2) \tag{12.33}$$

assuming the wall to be situated at $k = 1$. Now, consider the third order representation for $\frac{\partial \psi}{\partial y}$

$$g_j = \frac{\partial \psi}{\partial y}|_{j,k} = \frac{1}{6\Delta y}\left[-2\psi_{j,0} - 3\psi_{j,1} + 6\psi_{j,2} - \psi_{j,3}\right] + O(\Delta y^3)(12.34)$$

We note that the mesh point $(j, 0)$ lies outside the computational domain. So, eliminating $\psi_{j,0}$ from (12.33) and (12.34) yields on simplification

$$\zeta_{j,1} = \frac{1}{\Delta y^2}\left[-7\psi_{j,1} + 8\psi_{j,2} - \psi_{j,3}\right] - \frac{3g_j}{\Delta y} + O(\Delta y^2). \qquad (12.35)$$

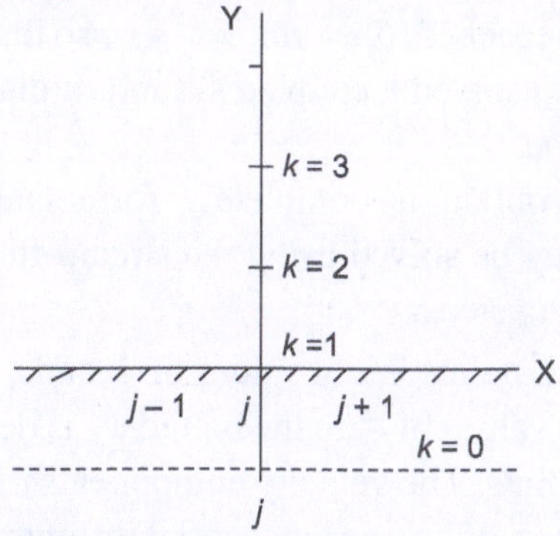

Figure 12.3 Boundary value of vorticity at a flat wall

This formula was given by Jensen (1959) and used among others by Ghia et al. (1982). According to Gupta and Manohar (1979), the second order representation of the boundary value of vorticity, generally leads to more accurate results. Some caution is necessary, however, in using a higher order formula, since often one encounters instability in such cases, particularly at higher Reynolds numbers.

12.3.3 Method of Solution

The basic equations of the stream-function vorticity approach are Eqs. (12.25)–(12.28). For the conservative formulation, Eq. (12.18) is used instead of Eq. (12.25). More common procedures for determining the boundary conditions for vorticity and stream function have been discussed in the previous subsection. Note that in view of the presence of the unknown velocity components Eqs. (12.25) or (12.18) is a nonlinear partial differential equation for the unknown ζ. Once ζ is known at all mesh points of the computational domain and the boundary values of ζ are known ψ may be computed at all internal mesh points by solving the Poisson equation (12.26). Subsequently, differentiating ψ the velocity components may be determined. It may be noted further, that only the vorticity transport equation (12.25) (or 12.18) depends explicitly on the time t. For unsteady problems, the vorticity distribution changes at each time step. Consequently, the changed stream-function field must be determined at each time step. Since (12.25) is nonlinear, in general, an iterative algorithm is necessary for its solution. Thus, at each time step the ζ and ψ fields are updated either sequentially or else as a coupled system. Mallinson and Vahl Davis (1973) used a sequential algorithm, and employed a pseudo-transient approach to solve Eq. (12.26) by rewriting it as

$$\psi_t - \left[\psi_{xx} + \psi_{yy} + \zeta\right] = 0. \tag{12.36}$$

When the steady-state is reached, the term $\psi_t \to 0$, so that Eq. (12.22) is regained. Rubin and Khosla (1981) solved a coupled system of discretised equations, using a *strongly implicit method.*

Once the fluid flow problem is completely formulated in the computational domain, the problem may be solved using the stream-function vorticity approach according to the following steps:

1. At all internal mesh points initial values of ψ and ζ are prescribed (might be guess values). The values of ζ on the boundary is determined from the values at internal mesh points. The velocity components u and v at a mesh point are determined by means of three-point central difference formula using (12.27). Note that, in view of the presence of the unknown velocity components u, v, Equation (12.25) or (12.18) is a nonlinear partial differential equation in ζ. Once ζ is known at all mesh points and the boundary values of ζ are known, ψ may be computed at all internal mesh points by solving the Poisson Equation (12.26). Only then differentiating ψ the velocity components may be determined.
2. At any time t, the solution of vorticity transport equation is advanced to $t + \Delta t$, using any of the explicit or implicit methods discussed in Chapter 4, Section 4.10. In general, a second order accurate method is preferable to the first order methods. In particular, the ADI (alternating direction implicit) method is a popular choice in view of its efficiency, stability and accuracy.
3. Poisson equation (12.26) is then solved by any of the methods discussed in Chapter 5, at the interior points of the computational domain. The approximate factorisation scheme or the multigrid method may be used for this purpose, which are very efficient.
4. In view of Eq. (12.27), the velocity components may now be computed using three-point central difference formulae.

5. The boundary values of the vorticity are then updated, using the interior values of the velocity components and using, for example, the second order formula (12.35), or the first order formula (12.32).
6. Convergence is then tested. If all the flow variables at all the internal mesh points do not change more than some prescribed tolerance from the corresponding values of the previous time step, the solution has converged. The pressure Poisson equation (12.22) may now be solved. If only the final steady-state solution is of interest, the pressure Poisson equation is solved only once, after convergence has been attained.

   On the other hand, in the case of no convergence, go to step 2 and repeat the steps.

The method described through the above six steps is a sequential algorithm, since while solving for ζ^{n+1} at time step $n+1$, it uses the velocity component values of the previous time step, that is, it uses the available values u^n, v^n. This is a process of *local linearisation,* frequently adopted to solve nonlinear equations. Some loss of accuracy is often associated with such a linearisation. On the other hand, one may decide to solve a coupled system where all the unknowns are at the time level $n+1$; that is, we have to solve simultaneously the system of equations for ζ^{n+1}, ψ^{n+1}, u^{n+1}, v^{n+1}. Such a solution procedure, followed by Ghia et al. (1982) for computing driven flow in a square cavity, yields very accurate results and does not generally suffer from any instability problem at relatively high Reynolds number.

Rubin and Khosla (1981) solve the conservative form of the vorticity transport equation (12.18) and (12.26) as a coupled system. In order to obtain a diagonally dominant discretised system for relatively large values of Reynolds number, the convection term $\dfrac{\partial u\zeta}{\partial x}$ is discretised as

$$\frac{\partial u\zeta}{\partial x} = \mu_x \frac{(u\zeta)^{n+1}_{j+1,k} - (u\zeta)^{n+1}_{j,k}}{\Delta x} + (1-\mu_x)\frac{(u\zeta)^n_{j,k} - (u\zeta)^n_{j-1,k}}{\Delta x}$$
$$+\frac{1}{2}\Delta x(1-2\mu_x)\frac{(u\zeta)^n_{j+1,k} - 2(u\zeta)^n_{j,k} + (u\zeta)^n_{j-1,k}}{\Delta x^2}, \quad (12.37)$$

where

$$\mu_x = \begin{cases} 0, & \text{if } u_{j,k} \geq 0, \\ 1, & \text{if } u_{j,k} < 0. \end{cases}$$

The above scheme was put forward by Khosla and Rubin (1974). It is an upwind scheme at the implicit level $n+1$, which reduces to a three-point central difference scheme under steady-state conditions. Ghia et al. (1982) combine the Rubin and Khosla formulation with multigrid to compute flow in a square driven cavity for

Reynolds number upto 10,000 on a 257 × 257 uniform grid. The reader is referred to this work for details of a very efficient method.

Some fluid flow problems may be unsteady and no steady-state solution might exist. On the other hand, for many fluid flow problems one might be interested in the steady-state solution only. In such a case, the vorticity transport equation (12.25) changes over to

$$u\zeta_x + v\zeta_y = \frac{1}{Re}\left[\zeta_{xx} + \zeta_{yy}\right], \tag{12.38}$$

which is of elliptic type. Several authors solved directly the steady-state equation instead of solving the unsteady transport equation till a steady-state has been reached. However, the unsteady approach turns out to be more efficient than solving the steady-state equation, which for relatively high Reynolds number grow extremely slow.

There exists many other approaches for solving the equations of stream-function vorticity approach. Reader is referred to Peyret and Taylor (1983), Roache (1972) and Fletcher (1988), for detailed account of these methods.

12.4 PRIMITIVE VARIABLES APPROACH

The more popular methods for solving the N–S Equations (12.1)–(12.3) or the conservative form equations (12.1), (12.4) and (12.5) using the primitive variables employ *staggered grids,* that do not store all the dependent variables at the same node on the computational mesh. In connection with their MAC (Marker and Cell) method, Harlow and Welch (1965) introduced the concept of storing the velocity components at locations staggered with respect to those at which the pressure is stored. Subsequently, many schemes and well-known codes have been developed that are based on staggered grids, for example, SMAC (simplified MAC) by Amsden and Harlow (1970), SIMPLE (semi-implicit method for pressure linked equations) by Patankar and Spalding (1972) SIMPLER (SIMPLE revised) by Patankar (1980), SIMPLEC by Van Doormal and Raithby (1984), PISO by Issa (1985), Isaacson and Keller (1966), QUICK (1979) (quadratic upwind interpolation for convective kinetics) (Leonard, 1979) and many other variants of the original MAC method. The detailed development and review may be found in the monograph by Patankar (1980). The projection method of Chorin (1968), the method of fractional steps by Yanenko (1971), Temam (1969), Peyret and Taylor (1983) are popular methods for the solution of unsteady incompressible N–S Equations.

One of the main advantages in using the staggered grids is that the pressure gradient terms in the momentum equations could be centered about the velocity which they drive without resulting in the decoupling of adjacent pressure nodes, commonly referred to as *the checker-board effect* in the pressure field. The boundary conditions on pressure are not required when the velocities are prescribed on the boundary, since the domain boundaries can be chosen to fall on the velocity nodes. This is not possible when nonstaggered meshes are employed for discretising the governing equations of fluid flow.

12.4.1 Discretisation Using Staggered Grid

As mentioned above, in a staggered grid, all the dependent variables are not stored at the same location or mesh point, but are stored at different locations. We discuss here the staggered grid arrangement put forward by Harlow and Welch (1965) in introducing their MAC method, which has been subsequently used by many other authors.

Figure 12.4 shows such a typical cell, $ENWS$, centered at j, k. The pressure is stored at the cell-centre, its value being denoted by $p_{j,k}$. The u-component of the velocity is stored at the middle point E of the eastern cell-face, denoted by $u_{j+\frac{1}{2},k}$ and the v-component of the velocity is stored at the middle point N of the northern cell-face denoted by $v_{j,k+\frac{1}{2}}$. Portions of the neighbouring cells are also shown in the same figure.

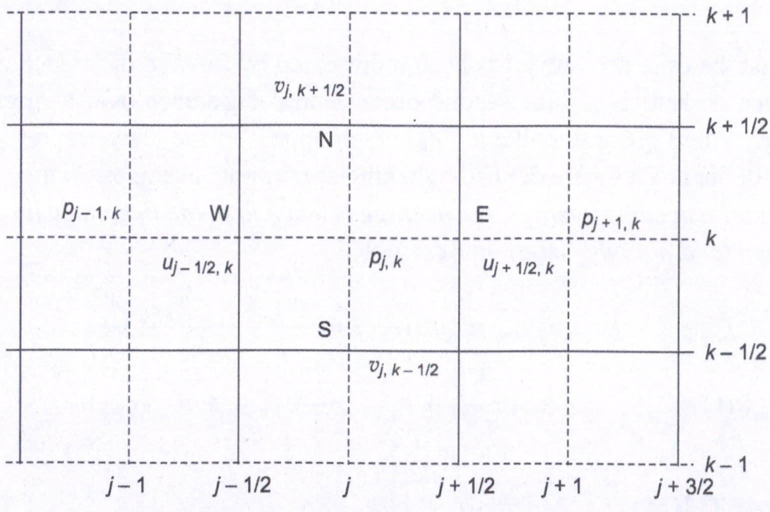

Figure 12.4 Discretisation using a staggered grid.

Using the above figure, the discretisations of the various terms in the N–S Equations are defined as follows:

$$\left.\frac{\partial u}{\partial t}\right|_{j+\frac{1}{2},k} = \frac{u^{n+1}_{j+\frac{1}{2},k} - u^{n}_{j+\frac{1}{2},k}}{\Delta t} + O(\Delta t),$$

$$\left.\frac{\partial u}{\partial x}\right|_{j+\frac{1}{2},k} = \frac{u_{j+1,k} - u_{j,k}}{\Delta x} + O(\Delta x^2),$$

$$\left.\frac{\partial u^2}{\partial x}\right|_{j+\frac{1}{2},k} = \frac{u^2_{j+1,k} - u^2_{j,k}}{\Delta x} + O(\Delta x^2),$$

$$\left.\frac{\partial uv}{\partial y}\right|_{j+\frac{1}{2},k} = \frac{(uv)_{j+\frac{1}{2},k+\frac{1}{2}} - (uv)_{j+\frac{1}{2},k-\frac{1}{2}}}{\Delta y} + O(\Delta y^2), \tag{12.39}$$

$$\left.\frac{\partial^2 u}{\partial x^2}\right|_{j+\frac{1}{2},k} = \frac{u_{j+\frac{3}{2},k} - 2u_{j+\frac{1}{2},k} + u_{j-\frac{1}{2},k}}{\Delta x^2} + O(\Delta x^2),$$

$$\left.\frac{\partial^2 u}{\partial y^2}\right|_{j+\frac{1}{2},k} = \frac{u_{j+\frac{1}{2},k+1} - 2u_{j+\frac{1}{2},k} + u_{j+\frac{1}{2},k-1}}{\Delta y^2} + O(\Delta y^2),$$

$$\left.\frac{\partial p}{\partial x}\right|_{j+\frac{1}{2},k} = \frac{p_{j+1,k} - p_{j,k}}{\Delta x} + O(\Delta x^2), \tag{12.40}$$

Note that the time derivative has been represented by forward difference while all the space derivatives by the second order central difference over a single mesh spacing. A few quantities like $u_{j+1,k}$ which appear in the above representations are not defined. These are defined by taking appropriate averages. *In the case of a product term needing average, the average is taken first and then requisite product is formed (and not otherwise).* For example,

$$u^2_{j+1,k} = (uu)_{j+1,k} = \frac{(u_{j+\frac{1}{2},k} + u_{j+\frac{3}{2},k})^2}{4}, \tag{12.41}$$

$$(uv)_{j+\frac{1}{2},k+\frac{1}{2}} = \frac{1}{2}(u_{j+\frac{1}{2},k+1} + u_{j+\frac{1}{2},k})\frac{1}{2}(v_{j,k+\frac{1}{2}} + v_{j+1,k+\frac{1}{2}}). \tag{12.42}$$

12.5 THE MAC METHOD

The MAC (Marker and Cell) method, put forward by Harlow and Welch (1965) is a pioneering work in computing solution of incompressible laminar N–S Equations. It was originally devoted to free-surface flow computation; however, the method is equally effective in other incompressible laminar flow computation. Discretisation on a staggered grid is a basic feature of the method. The momentum equations are discretised using FTCS (forward-time-central-space) scheme, the time derivative represented by first-order forward representation and all the space derivatives represented by second-order central difference. We note that for inviscid flow the FTCS scheme is unconditionally unstable; however, for viscous flow it is conditionally stable, as explained in Section 4.10 in Chapter 4.

The MAC method connects the momentum equations with the continuity equation by deriving a Poisson equation for the pressure, so that the pressure gets an equation of its own, commonly referred to as the *pressure Poisson equation.* This equation may be derived either by manipulating the original N–S Equations or else by manipulating the discretised equations. The MAC method introduces the concept of massless *marker particles* which are advected with the flow but do not participate in momentum or energy transfer. Their use is in determining the position and shape of the free-surface. We do not discuss this aspect here; interested reader may refer to the original paper (Harlow and Wetch, 1965).

Discretisation of typical terms of the N–S Equations on a staggered grid has been presented in the previous subsection. The discretised form of the N–S Equations (12.4), (12.5) in conservative form are

$$\frac{u^{n+1}_{j+\frac{1}{2},k}-u^{n+1}_{j+\frac{1}{2},k}}{\Delta t}+\frac{u^2_{j+1,k}-u^2_{j,k}}{\Delta x}+\frac{(uv)_{j+\frac{1}{2},k+\frac{1}{2}}-(uv)_{j+\frac{1}{2},k-\frac{1}{2}}}{\Delta y}=-\frac{p^{n+1}_{j+1,k}-p^{n+1}_{j,k}}{\Delta x}$$
$$+\frac{1}{Re}\left[\frac{u_{j+\frac{3}{2},k}-2u_{j+\frac{1}{2},k}+u_{j-\frac{1}{2},k}}{\Delta x^2}+\frac{u_{j+\frac{1}{2},k+1}-2u_{j+\frac{1}{2},k}+u_{j+\frac{1}{2},k-1}}{\Delta y^2}\right] \tag{12.43}$$

$$\frac{v^{n+1}_{j,k+\frac{1}{2}}-v^{n+1}_{j,k+\frac{1}{2}}}{\Delta t}+\frac{(uv)_{j+\frac{1}{2},k+\frac{1}{2}}-(uv)_{j+\frac{1}{2},k-\frac{1}{2}}}{\Delta x}+\frac{v^2_{j,k+1}-v^2_{j,k}}{\Delta y}=-\frac{p^{n+1}_{j+1,k}-p^{n+1}_{j,k}}{\Delta y}$$
$$+\frac{1}{Re}\left[\frac{v_{j+1,k+\frac{1}{2}}-2v_{j,k+\frac{1}{2}}+v_{j-1,k+\frac{1}{2}}}{\Delta x^2}+\frac{v_{j,k+\frac{3}{2}}-2v_{j,k+\frac{1}{2}}+v_{j,k-\frac{1}{2}}}{\Delta y^2}\right] \tag{12.44}$$

It may be noted that the superscripts n have been omitted for quantities at time-level n and only the quantities at time-level $n+1$ have been shown with superscript $n+1$ in the above equations.

Solving for $u^{n+1}_{j+\frac{1}{2},k}$ and $v^{n+1}_{j,k+\frac{1}{2}}$ we obtain from Eqs. (12.43) and (12.44)

$$u^{n+1}_{j+\frac{1}{2},k}=F^n_{j+\frac{1}{2},k}-\frac{\Delta t}{\Delta x}\left[p^{n+1}_{j+1,k}-p^{n+1}_{j,k}\right], \tag{12.45}$$

$$v^{n+1}_{j,k+\frac{1}{2}}=G^n_{j,k+\frac{1}{2}}-\frac{\Delta t}{\Delta y}\left[p^{n+1}_{j,k+1}-p^{n+1}_{j,k}\right], \tag{12.46}$$

where the mesh functions $F^n_{j+\frac{1}{2},k}$ and $G^n_{j,k+\frac{1}{2}}$ may be written in compact form in terms of the central difference operator δ as

$$F_{j+\frac{1}{2},k}=u_{j+\frac{1}{2},k}-\frac{\Delta t}{\Delta x}\delta_x u_{j+\frac{1}{2},k}-\frac{\Delta t}{\Delta y}\delta_y(uv)_{j+\frac{1}{2},k}+\frac{1}{Re}\left[\frac{\delta_{xx}}{\Delta x^2}+\frac{\delta_{yy}}{\delta y^2}\right]u_{j+\frac{1}{2},k}, \quad (12.47)$$

$$G_{j,k+\frac{1}{2}}=v_{j,k+\frac{1}{2}}-\frac{\Delta t}{\Delta x}\delta_x(uv)_{j,k+\frac{1}{2}}-\frac{\Delta t}{\Delta y}\delta_y(v^2)_{j,k+\frac{1}{2}}+\frac{1}{Re}\left[\frac{\delta_{xx}}{\Delta x^2}+\frac{\delta_{yy}}{\delta y^2}\right]v_{j,k+\frac{1}{2}}. \quad (12.48)$$

It may be noted that the pressure is implicitly present in these equations at time-level n. The continuity equation is discretised in terms of *dilation* $D_{j,k}$ for the cell centered at (j, k) as

$$D^{n+1}_{j,k} = 0, \text{ where}$$

$$D_{j,k} = \frac{u^{n+1}_{j+\frac{1}{2},k} - u^{n+1}_{j-\frac{1}{2},k}}{\Delta x} + \frac{v^{n+1}_{j,k+\frac{1}{2}} - v^{n+1}_{j,k-\frac{1}{2}}}{\Delta y}. \quad (12.49)$$

12.5.1 The Pressure Poisson Equation

The differntial form of the pressure Poisson equation may be established from the momentum and the continuity equations as

$$\nabla^2 p = -\frac{\partial^2 u^2}{\partial x^2} - 2\frac{\partial^2(uv)}{\partial x \partial y} - \frac{\partial^2 v^2}{\partial y^2} - \frac{\partial D}{\partial t}$$
$$+\frac{1}{Re}\left(\frac{\partial^2 D}{\partial x^2} + \frac{\partial^2 D}{\partial y^2}\right) \equiv S_P, \text{ (say)} \quad (12.50)$$

where

$$D = \frac{\partial u}{\partial x} + \frac{\partial v}{\partial y}. \quad (12.51)$$

Derivation of this equation is left as an exercise to the reader. We derive instead the discrete version of the equation directly.

Substituting for $u^{n+1}_{j+\frac{1}{2},k}$, $u^{n+1}_{j-\frac{1}{2},k}$, $v^{n+1}_{j,k+\frac{1}{2}}$ and $v^{n+1}_{j,k-\frac{1}{2}}$ from Eqs. (12.45) and (12.46), yields on simplification, the discrete form of the pressure Poisson equation

$$\frac{p^{n+1}_{j+1,k} - 2p^{n+1}_{j,k} + p^{n+1}_{j-1,k}}{\Delta x^2} + \frac{p^{n+1}_{j,k+1} - 2p^{n+1}_{j,k} + p^{n+1}_{j,k-1}}{\Delta y^2}$$
$$= \frac{1}{\Delta t}\left[\frac{F^n_{j+\frac{1}{2},k} - F^n_{j-\frac{1}{2},k}}{\Delta x} + \frac{G^n_{j,k+\frac{1}{2}} - G^n_{j,k-\frac{1}{2}}}{\Delta y}\right] \quad (12.52)$$

The right hand side expression in Eq. (12.52) may be simplified and put to the convenient form

$$=\frac{D^n_{j,k}}{\Delta t}-\left[\frac{\delta_{xx}}{\Delta x^2}u^2_{j,k}+2\frac{\delta_x\delta_y}{\Delta x\Delta y}(uv)_{j,k}+\frac{\delta_{yy}}{\Delta y^2}v^2_{j,k}\right]-\frac{1}{Re}\left(\frac{\delta_{xx}}{\Delta x^2}+\frac{\delta_{yy}}{\Delta y^2}\right)D^n_{j,k}, \quad (12.53)$$

where the central difference operators over single mesh spacings, $\delta_x, \delta_{xx}, \delta_y, \delta_{yy}$ etc. are defined as

$$\delta_x u_{j,k} = u_{j+\frac{1}{2},k} - u_{j-\frac{1}{2},k}, \ \delta_y u_{j,k} = u_{j,k+\frac{1}{2}} - u_{j,k-\frac{1}{2}},$$

$$\delta_{xx} u_{j,k} = u_{j+\frac{1}{2},k} - 2u_{j,k} + u_{j-\frac{1}{2},k}, \ \delta_{yy} u_{j,k} = u_{j,k+\frac{1}{2}} - 2u_{j,k} + u_{j,k-\frac{1}{2}}. \quad (12.54)$$

The pressure Poisson equation (12.50) has to be solved at each time step by standard iterative or direct procedures discussed in Chapter 5, relaxation method being a popular choice. Since this equation is of elliptic type, boundary conditions on the boundary of the computational domain is needed for this. However, boundary condition for pressure is not required by the MAC method or its variants. The boundary of the computational domain is suitably chosen such that the boundary passes over the velocity nodes (Peyret and Taylor, 1983). Subsequent developments of MAC method try to reduce the computational cost by solving a *pressure-correction formula* instead of it.

Once the pressure is thus determined, the velocity components $u^{n+1}_{j+\frac{1}{2},k}$, $v^{n+1}_{j,k+\frac{1}{2}}$, may be computed from Eqs. (12.45) and (12.46). Since the velocity components are determined explicitly, time-step restrictions are involved, determined from linear stability analysis, by locally freezing the nonlinear coefficients. (Note that the von Neumann stability analysis requires the partial differential equation to be linear and is not applicable to nonlinear equations.)

12.5.2 Stability Restriction

For numerical stability, certain restrictions are imposed on the mesh sizes Δx, Δy and on Δt. We observe that the time step is governed by two restrictions. The first one is related to the convection of the fluid, requiring that the fluid cannot move through one cell in one time step. More precisely, linear stability gives

$$\Delta t_1 \le \text{Min}\left[\frac{\Delta x}{|u|}, \frac{\Delta y}{|v|}\right]_{j,k} \quad (12.55)$$

Secondly, the momentum must not diffuse more than one cell in one time step. Based on Hirt's stability analysis related to viscous effects, this yields

$$\Delta t_2 \le \text{Min}\left[\frac{1}{2}Re\frac{\Delta x^2\Delta y^2}{\Delta x^2+\Delta y^2}\right]_{j,k} \quad (12.56)$$

The time step to be actually used in computations, is then taken as

$$\Delta t = \text{FAC} \times [\text{Min}(\Delta t_1, \Delta t_2)]_{j,k}, \tag{12.57}$$

where FAC is a safety-factor, used to ensure stability.

12.5.3 Boundary Conditions in Primitive Variables

Along a no-slip wall, we require both u and v-velocity component values u_w and v_w at the wall, indicated by the suffix w must be zero, that is $u_w = 0 = v_w$.

At an inflow boundary, the u-velocity component is prescribed, for example,

$$u(y) = U_0. \tag{12.58}$$

Often, at inflow the v-component is set as $v = 0$.

The boundary conditions on u and v are evaluated using appropriate averaging where needed. At a no-slip wall along $k = w$ (in Fig. 12.4) yields

$$u_w = 0,\ u_{j+\frac{1}{2},w} = 0,\ u_{j-\frac{1}{2},w} = 0,$$

$$v_w = 0,\ v_{j,w} = \frac{1}{2}(v_{j,w+\frac{1}{2}} + v_{j,w-\frac{1}{2}}),\ \text{or } v_{j,w-\frac{1}{2}} = -v_{j,w+\frac{1}{2}}. \tag{12.59}$$

Note that $k = w - \frac{1}{2}$ is inside the wall; this condition ensures $v_{j,w} = 0$.

12.6 SOLUTION SCHEME

The solution procedure begins with initialising the velocity field $(u_{j,k}, v_{j,k})$. This may be done either from the result of the previous cycle or else from prescribed initial conditions. The pressure Poisson equation may be solved by standard schemes like the SOR (successive over relaxation). However, as the number of mesh points increase, SOR iterations grow slower and a very large number of iteration steps would be required to reach a satisfactory level of convergence. It is important to note that, the continuity equation must be satisfied sufficiently accurately because otherwise the solution of momentum equations may develop nonlinear instability. In order to reduce the computational time for each cycle, subsequent authors, like Mukherjea (1990), Layek (1996), Layek et al. (1996), Maikap et al. (2003a, 2003b), Mahapatra (2002) in their work kept the number of SOR iterations limited, say to 100 iteration steps. Convergence of the pressure solution cannot be expected with such a small number of iterations. Therefore the velocity field obtained after solving the momentum equations using already known inaccurate pressure may not satisfy the equation of continuity to adequate accuracy. This necessitates a corrector stage. In this stage pressure and subsequently velocities are corrected to get a more accurate velocity field in terms of conservation of mass.

12.6.1 Variants of the MAC Method

We wish to note that all the above methods mentioned in Section 12.4, originate from the MAC method. In the SIMPLE algorithm the dependent variables are located on a staggered grid and a decoupled solution strategy is used. First an approximate pressure field is employed to solve the momentum equations. A *pressure-correction equation* is then derived by combining the finite difference approximations of continuity and momentum equations, Biswas (2003). The pressure–correction equation is solved by alternate time sweeps, and both the velocities and the pressure are updated to reflect the pressure corrections. Typically, all changes are underrelaxed in order to achieve numerical stability. Modifications of SIMPLE to improve the pressure–velocity coupling have been suggested by Raithby and Schneider (1979). It may be noted that in deriving the pressure-correction equation, the diffusive and the convective terms in the momentum equations have been dropped (Patankar, 1980). Among others, it may be mentioned that for computation of viscous flow in complex configurations using collocated variable arrangement, a pressure-based method have been put forward by Majumdar (1986) and Rodi et al. (1989).

In convective flow situations, discretisation of the convective terms of the Navier–Stokes equations need particular care. The diffusive terms, in general, present no difficulty. Three-point second-order accurate central difference representation is widely used for these terms. Derivatives of pressure are also discretised using central difference formulae. On the other hand, proper treatment of the convective terms is not so straightforward. This is so, because central differencing of the convective terms generally lead to instability, particularly, at higher Reynolds numbers when viscous domination is significantly reduced.

12.6.2 Treatment of Convective Terms

In the present scheme (Layek, 1996; Maikap et al., 2003a; Mukherjea, 1990) as it was mentioned previously, the convective terms are differenced with a combination of central differencing and second order upwind differencing schemes. The convecting velocities ($u_{\mathrm{r}}, u_{\mathrm{l}}, u_{\mathrm{t}}, u_{\mathrm{b}}, v_{\mathrm{t}}, v_{\mathrm{b}}$) at the interface of the control volume are calculated by the same interpolation formula for both the schemes. Thus for the u-momentum equation the different symbols are defined by

$$u_{\mathrm{r}} = (u_{ij}^n + u_{i+1j}^n)/2, \quad u_{\mathrm{l}} = (u_{ij}^n + u_{i-1j}^n)/2,$$

$$u_{\mathrm{t}} = (u_{ij}^n + u_{ij+1}^n)/2, \quad u_{\mathrm{b}} = (u_{ij}^n + u_{ij-1}^n)/2,$$

$$v_{\mathrm{t}} = (v_{ij}^n + v_{i+1j}^n)/2, \quad v_{\mathrm{b}} = (v_{ij-1}^n + v_{i+1j-1}^n)/2.$$

Here suffices r, l, t and b correspond to right, left, top and bottom middle positions of the control volume respectively. Let ϕ be the momentum flux. In the second order upwind formulation the choice of taking the momentum flux passing through the interface of the control volume depends on the sign of the convecting velocities at that interface (Roache, 1972; Biswas, 2003). Thus the momentum fluxes for u-momentum are given by

$$\begin{aligned}
\phi_{u\mathrm{r}} &= u_{ij}^n \quad \text{if} \quad u_\mathrm{r} \geq 0,\\
\phi_{u\mathrm{r}} &= u_{i+1j}^n \quad \text{if} \quad u_\mathrm{r} < 0,\\
\phi_{u\mathrm{l}} &= u_{i-1j}^n \quad \text{if} \quad u_\mathrm{l} \geq 0,\\
\phi_{u\mathrm{l}} &= u_{ij}^n \quad \text{if} \quad u_\mathrm{l} < 0.
\end{aligned} \tag{12.60}$$

Suffix u of ϕ represents the quantity for u-momentum. Similarly,

$$\begin{aligned}
\phi_{u\mathrm{t}} &= u_{ij}^n \quad \text{if} \quad v_\mathrm{t} \geq 0,\\
\phi_{u\mathrm{t}} &= u_{ij+1}^n \quad \text{if} \quad v_\mathrm{t} < 0,\\
\phi_{u\mathrm{b}} &= u_{ij-1}^n \quad \text{if} \quad v_\mathrm{b} \geq 0,\\
\phi_{u\mathrm{b}} &= u_{ij}^n \quad \text{if} \quad v_\mathrm{b} < 0.
\end{aligned} \tag{12.61}$$

Hence finite difference form of different convective terms in the u-momentum equation are given by

$$\begin{aligned}
\frac{\partial u^2}{\partial x} &= (1-\beta)\frac{u_\mathrm{r}^2 - u_l^2}{\delta x} + \beta\frac{u_\mathrm{r}\phi_{u\mathrm{r}} - u_\mathrm{l}\phi_{u\mathrm{l}}}{\delta x},\\
\frac{\partial u^2}{\partial y} &= (1-\beta)\frac{u_\mathrm{t}^2 - u_b^2}{\delta y} + \beta\frac{u_\mathrm{t}\phi_{u\mathrm{t}} - u_\mathrm{b}\phi_{u\mathrm{b}}}{\delta y},\\
\frac{\partial uv}{\partial y} &= (1-\beta)\frac{u_\mathrm{t}v_\mathrm{t} - u_\mathrm{b}v_\mathrm{b}}{\delta y} + \beta\frac{v_\mathrm{t}\phi_{u\mathrm{t}} - v_b\phi_{u\mathrm{b}}}{\delta y},
\end{aligned} \tag{12.62}$$

where β is a combination factor which is determined from the stability criteria. With $\beta = 0$ the scheme becomes central differencing and with $\beta = 1$ it is a second order upwind differencing scheme.

12.7 CASE STUDY: SEPARATED FLOW IN A CONSTRICTED CHANNEL

Flow in constricted channels occur in many fluidic devices (orifices, valves) and have wide applications in engineering. Numerical solutions have been obtained by many authors like Shyy and Sun (1993), Huang and Seymour (1995). It is well known that the flow through a symmetric sudden expansion in a channel becomes asymmetric about the central plane as the Reynolds number is increased, as investigated in Durst et al. (1974), Cherdron et al. (1978), Sobey (1985), Fearn et al. (1990), Durst et al. (1993). Complete N–S Equations have to be solved numerically in order to study the flow separation phenomena. A large number of works appeared using the stream-function vorticity approach as also the primitive variables approach. Methods based on the stream-function vorticity formulation generally lead to faster numerical computation than that of the primitive variable formulation. However, in general, the former approach provides poor analysis of flow in the near wake region, as observed by Braza et al. (1986).

The primitive variable approach based on staggered grid have been used in many recent works on constricted flow in channels or in tubes. We mention here the works of Lee (1990), Layek (1996), Layek et al. (1996), Mahapatra et al. (2002), Maikap et al. (2003a). As illustration of computational method using the primitive variables approach, following Maikap et al. (2003b) we present here briefly the problem and results of computation of viscous incompressible flow through a channel with asymmetric double constriction without assuming centreline symmetry.

12.7.1 The Problem and Method of Solution

We consider an infinitely long parallel plate channel having asymmetric double constrictions on both the plates (Fig. 12.5) filled with homogeneous Newtonian incompressible viscous fluid. Initially the fluid is at rest everywhere within the channel except at infinite distance upstream of the double asymmetric constriction where a parabolic velocity profile is prescribed.

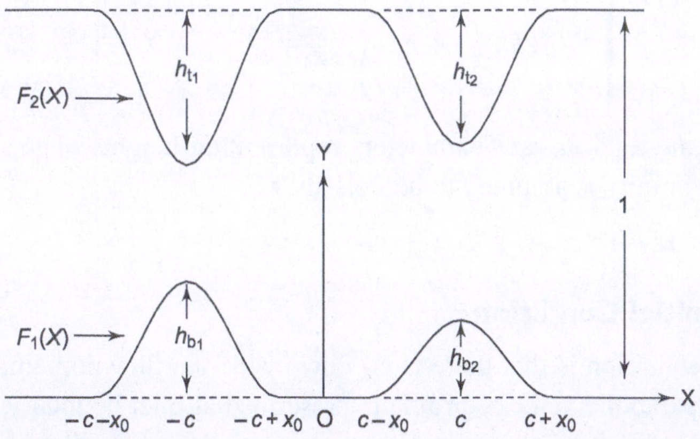

Figure 12.5 Geometry of the asymmetric double constriction.

A co-ordinate transformation has been employed to map the infinite irregular domain to a finite rectangular computational domain. Pressure Poisson equation has been solved and a pressure-velocity correction scheme has been invoked. Time step is determined using restricted stability criteria at each time iteration. This two-stage algorithm achieves the convergence criteria for the mass conservation equation quite efficiently. The correction formulae do not require boundary conditions.

12.7.2 Boundary Conditions

The streamwise and transverse velocity components should be zero at the rigid walls (no-slip condition). At the inlet and outlet boundaries the Poiseuille flow is prescribed. On the upper ($Y = F_2(X)$) and lower ($Y = F_1(X)$) walls of the channel, the no-slip boundary conditions are

$$u = 0, \quad v = 0 \quad \text{at} \quad Y = F_2(X), F_1(X). \tag{12.63}$$

The functions $F_1(X)$ and $F_2(X)$, which represent the shapes of the lower and upper walls of the double asymmetric constricted channel (as shown in Fig. 12.5), are:

$$F_1(X) = \begin{cases} 0 & : \quad -\infty < X \le -c - X_0 \\ \frac{1}{2} h_{b_1}\left(1 + \cos\left(\frac{\pi(X+c)}{X_0}\right)\right) & : \quad |X + c| \le X_0 \\ 0 & : \quad |X| \le c - X_0 \\ \frac{1}{2} h_{b_2}\left(1 + \cos\left(\frac{\pi(X-c)}{X_0}\right)\right) & : \quad |X - c| \le X_0 \\ 0 & : \quad c + X_0 \le X < \infty \end{cases},$$

$$F_2(X) = \begin{cases} 1 & : \quad -\infty < X \le -c - X_0 \\ 1 - \frac{1}{2} h_{t_1}\left(1 + \cos\left(\frac{\pi(X+c)}{X_0}\right)\right) & : \quad |X + c| \le X_0 \\ 1 & : \quad |X| \le c - X_0 \\ 1 - \frac{1}{2} h_{t_2}\left(1 + \cos\left(\frac{\pi(X-c)}{X_0}\right)\right) & : \quad |X - c| \le X_0 \\ 1 & : \quad c + X_0 \le X < \infty \end{cases} \tag{12.64}$$

Here h_{t_1},h_{t_2},h_{b_1} and h_{b_2} are parameters representing heights of the contractions. The flow at infinity is assumed to be Poiseuille,

$$u = Y - Y^2 \quad \text{at} \quad X = \pm\infty. \tag{12.65}$$

12.7.3 Initial Condition

The initial condition is that there is no flow inside the flow domain, while on the other hand parabolic velocity profile is prescribed at inlet boundary. Physically it represents that the flow is approaching the constriction gradually.

12.7.4 Co-ordinate Transformation

As explained in detail in Maikap et al. (2003a), a transformation has been introduced to map the infinite irregular physical domain to a finite rectangular computational domain. The co-ordinate transformation used in this study is

$$x = \tanh(kX), \quad y = \frac{Y - F_1(X)}{F_2(X) - F_1(X)}, \tag{12.66}$$

where k is a parameter, that controls the grid distribution in an efficient manner. The grids in the physical plane are dense near the origin due to the nature of the function $\tanh(kX)$. The transformation defined in (12.66) transforms the curved upper boundary $Y = F_2(X)$ into the straight line $y = 1$, the curved lower boundary $Y = F_1(X)$ into the straight line $y = 0$ and the outflow and inflow boundaries at $X = \pm\infty$ into $x = \pm 1$.

12.7.5 Numerical Solution

A computer code has been developed to solve numerically, the governing equations, under appropriate boundary and initial conditions employing a finite difference technique on staggered grid, as explained above and in details in Maikap et al. (2003a). This code is capable of predicting laminar separated flow which, by virtue of their complex structure, cannot be treated analytically.

12.7.5.1 Type of grid used Grid distributions in the physical plane for the asymmetric double constriction are nonuniformly spaced. The transformation defined by Eq. (12.66) enables one to space grid points uniformly in the computational plane. In the new system of co-ordinates (x, y), a uniformly spaced rectangular grid is superimposed on $[-1, 1] \times [0, 1]$ computational domain. All the computations are carried out in this rectangular domain.

12.7.6 Results and Discussion

In order to discuss flow characteristics numerical results have been obtained for two-dimensional asymmetric double constricting passage corresponding to different cases given in the following Table.

Table 12.1 Table: Geometries of constriction and Reynolds number studied

Case	h_{t_1}	h_{t_2}	h_{b_1}	h_{b_2}	X_0	c	*Re* range for calculation	Computed separation *Re*	
								upper wall	lower wall
M_1	0.4	0.4	0.2	0.2	2.0	3.0	50–5000	330	1500
M_3	0.4	0.0	0.0	0.2	2.0	3.0	100–5000	650	3200
M_7	0.2	0.2	0.4	0.4	2.0	2.0	50–5000	1500	330

From the above table it is evident that the flow separates at the upper and lower walls at different Reynolds numbers which is expected in view of the asymmetric geometry. In the case M_1 the flow separates at the upper wall at a smaller Reynolds number compared to that at the lower wall of the channel. This is so because the height of the constrictions at the upper wall is higher than that at the lower wall. In the case M_3 there are single constrictions asymmetrically situated, one at the upper wall and the other on the lower wall. Figure 12.6 represents u-velocity profiles at Reynolds number $Re = 2000$ at different stations for the cases M_1. It may be seen from Fig. 12.6 that there is separation at the upper wall in the valley of the two constrictions. In the downstream region of the second constriction, the flow first separates at both the upper and lower walls, then reattaches and again at a further downstream region separates at the lower wall.

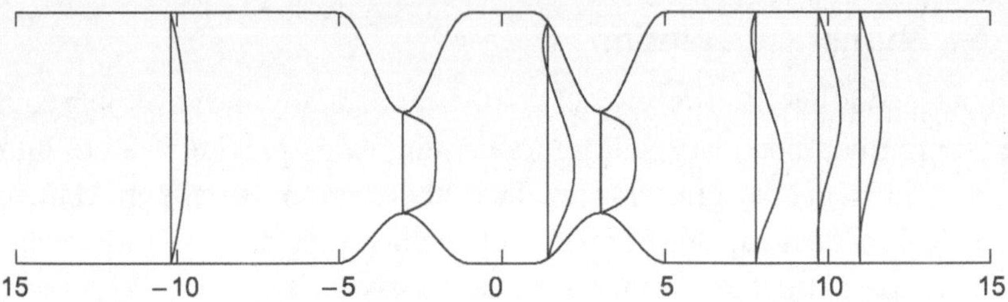

Figure 12.6 u-velocity profiles at $Re = 2000$, M_1. From Maikap et al. (2003b).

12.7.6.1 Wall vorticity The nondimensional wall vorticity is of particular interest due to its relation to the value of shear stress acting on the solid wall. Figure 12.7 presents wall vorticity for the case M_3, at $Re = 5000$. The magnitude of the wall (upper and lower) vorticity values increase rapidly when the flow approaches the constriction reaching the peak value near the maximum constricted area. At high Reynolds numbers, the peak value of wall vorticity is found to occur slightly upstream of the narrowest cross section. At a location downstream of this peak value, the magnitude of the wall (upper and lower) vorticity values decrease rapidly. Wall vorticity values change sign for higher Reynolds number. The negative value of upper wall vorticity and positive value of lower wall vorticity indicate the occurrence of *separation involving circulation with back flow near the wall.* The length of reattachment (the distance between separation and reattachment points) for both the upper and lower walls may be determined from these wall vorticity curves.

Figure 12.7 shows the distribution of wall vorticity at $Re = 5000$ for the case M_3. In this case there is only one constriction on the upper wall and another at the lower wall, at different locations of the channel. In this case an interesting phenomenon is observed. We note that, at the position where the constriction at the upper wall is situated, there is no corresponding constriction at the lower wall. Inspite of this, the wall vorticity at the lower wall does not maintain a constant value. Moreover, the wall vorticity changes in such a way as if there were a constriction at the lower wall. This change in lower wall vorticity value takes place only due to the influence and interaction of the constriction at the upper wall. It may be noted that in the downstream region there is a constriction at the lower wall, but no corresponding constriction at the upper wall. Even then the upper wall vorticity increases and attains a maximum value and then decreases to reach the constant value for the undisturbed flow. All these happen due to the influence of the constriction at the lower wall.

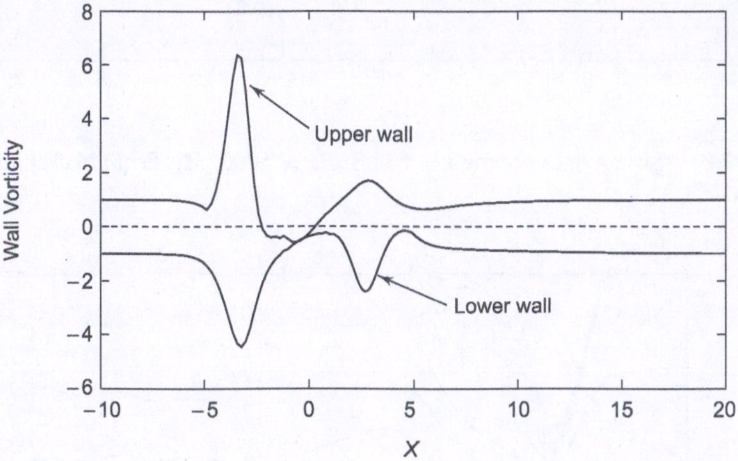

Figure 12.7 Wall Vorticity distribution at $Re = 5000$, M_3. From Maikap et al. (2003b).

12.7.6.2 Streamlines and vorticity contours The streamlines for the constriction case M_1 at Reynolds number $Re = 5000$ are shown in Figure 12.8 at nondimensional time $t = 634.3$. The corresponding isovorticities are shown in Fig. 12.9 for time $t = 894.8$. Large areas of recirculating zone may be seen in the figures downstream of the constrictions in both the walls. Fig. 12.8 shows formation of small eddies near the wall, large separation regions and formation of small eddies trapped in larger eddies. This is found to occur in the valley of the two constrictions as also downstream of the second constriction. They show the formation of small eddies at the walls which diffuse with time and again reappear. The process of formation of eddies is continuous in the valley of the constrictions and downstream of the second constriction. Significant perturbation in the streamlines may be noticed even quite far downstream of the second constriction.

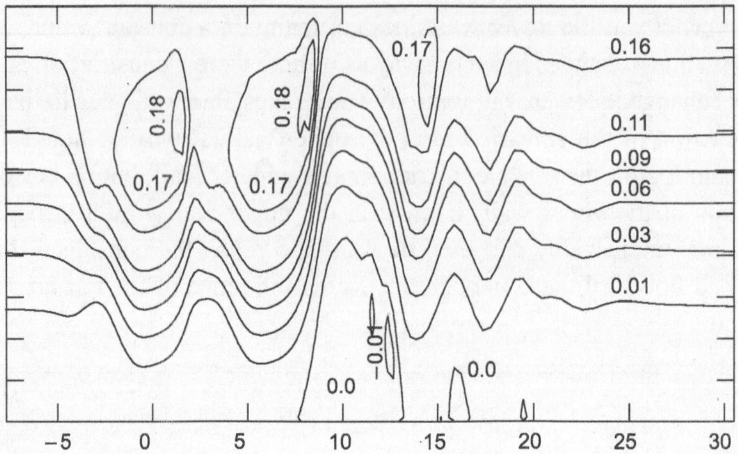

Figure 12.8 Streamlines at time $t = 634.3$, $Re = 5000$, M_1. From Maikap et al. (2003b)

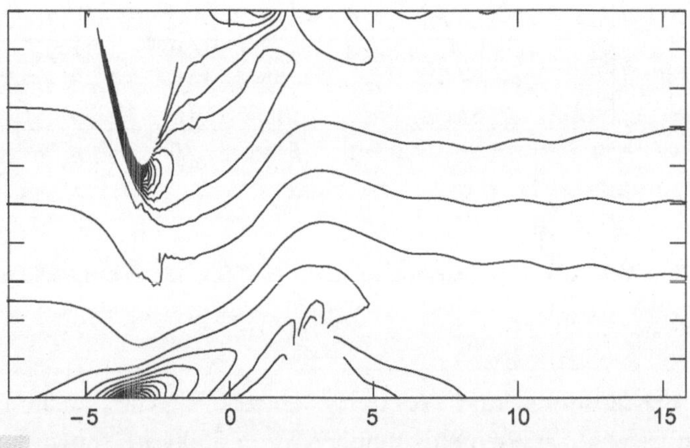

Figure 12.9 Vorticity contour at time $t = 894.8$, $Re = 5000$for M_3. Form Maikap et al. (2003b).

12.7.7 Conclusion: Case Study

Numerical solution for unsteady flow of a viscous incompressible fluid through a long channel with asymmetric double constriction has been investigated for several Reynolds number ranges corresponding to different constriction shape geometries. Flow unsteadiness depends on the length and height of the constrictions and on the Reynolds number. Wall vorticity and the separation length changes with time. The flow separates in the valley of the two constrictions as also downstream of the second constriction. *Secondary separation* may be seen in Fig.12.10, for the case M_7. In the Reynolds number range investigated, no eddy appeared on the upstream side of the contrictions and that for sufficiently large time, the flow approaches steady flow.

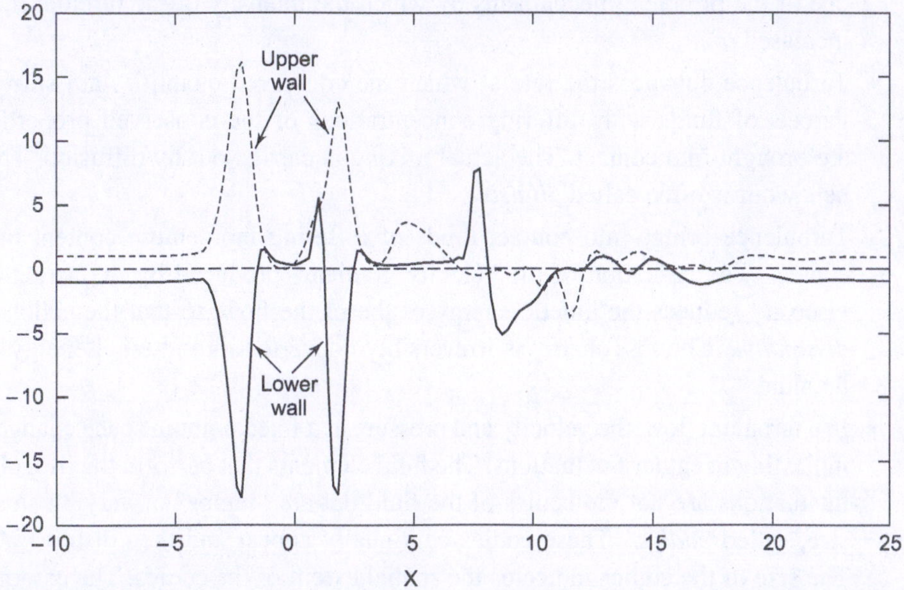

Figure 12.10 Distribution of vorticity on the upper and lower walls at Re = 4000, the case M_7. From Maikap et al. (2003b).

12.8 TURBULENT FLOW

For computation of a turbulent flow field, it is helpful to understand the physical nature of the flow. This is discussed in the following subsection.

12.8.1 Physical Characteristics of Turbulent Flow

As already mentioned, at higher Reynolds numbers the flow grows turbulent. From experimental study and dimensional analysis important characteristic features of a turbulent flow have come to our knowledge (Biswas and Eswaran, 2002; Landau and Lifshitz, 1989; Majumdar, 1991; Schlichting and Gersten, 2000) which we note in the following:

- Turbulent flows are highly unsteady.

- They are three-dimensional. The time-averaged velocity in a turbulent flow field may be a function of two space coordinates, but the instantaneous field seems to be random.
- Much vorticity is found to be present in the field. Stretching of the vortices is one of the principal mechanisms by which the intensity of the turbulence is increased.
- Turbulence enhances the rate at which the conserved quantities are stirred. Parcels of fluids with differing concentrations of the conserved properties are brought into contact. The actual mixing is carried out by diffusion. This behaviour is often called *diffusive.*
- Turbulence brings into contact fluids of differing momentum content into contact. The reduction of the velocity gradients produced by the action of viscosity reduces the kinetic energy of the of the flow, so that the action is *dissipative.* The lost energy is irreversibly converted into internal energy of the fluid.
- In a turbulent flow, the velocity and pressure at a fixed point in space changes, indicating irregular fluctuations. The fluid elements that perform the irregular fluctuations are not molecules of the fluid but are "lumps" of varying small size, called *eddies.* These eddies continually appear and then disintegrate. The size of the eddies indicates the spatial extent of the eddies. The external conditions of the flow determine the size of the eddies.

12.8.2 Incompressible Reynolds Averaged Navier–Stokes Equations

As already discussed, for turbulent flow computation, we have to take recourse to *turbulence modelling,* which starts with the time-averaged Navier–Stokes equations. In Chapter 7, we indicated how the averaging is done. For incompressible flow, we derive here the equations in detail, since these are the basic equations of turbulence modelling.

Turbulent motion being a random motion in which changes occur so rapidly that it is not possible to follow in detail the motion of the fluid, Osborne Reynold (1894) conceived *turbulent motion as a simple superposition of a fluctuating motion on a mean motion.* According to Reynolds' idea that each flow variable, like the velocity components (u, v, w) at a point in turbulent motion may be written as

$$u = \overline{u} + u', \quad v = \overline{v} + v' \quad w = \overline{w} + w', \quad p = \overline{p} + p' \tag{12.67}$$

where u', v', w', p' denote the fluctuating components. The time-averaging has been already defined in Eq. (7.49). We repeat it here for ready reference.

A time-averaged quantity $\overline{g}$ is defined as

$$\overline{g} = \frac{1}{\Delta t}\int_{t_0}^{t_0+\Delta t} g\,dt, \tag{12.68}$$

Here, Δt should be large compared to the period of random fluctuations associated with turbulence, but small with respect to any typical slow variations in the flow field associated with ordinary unsteady flows.

12.8.2.1 Properties of averaging The operations of averaging is introduced axiomatically, by imposing the following conditions (Reynolds conditions):

$$\overline{a+b} = \overline{a} + \overline{b};$$

$$\overline{Ka} = K\overline{a}, \; K = \text{ constant }, \quad \overline{K} = K.$$

$$\overline{\frac{\partial a}{\partial t}} = \frac{\partial \overline{a}}{\partial t}, \; \overline{\frac{\partial a}{\partial x^\alpha}} = \frac{\partial \overline{a}}{\partial x^\alpha}, \; x^\alpha = x, \text{ or } y \text{ or } z;$$

$$\overline{\overline{a}b} = \overline{a}\,\overline{b}. \tag{12.69}$$

Putting $b = 1$ in the last condition gives
$\overline{\overline{a}} = \overline{a}$. So, $\overline{a'} = \overline{a - \overline{a}} = \overline{a} - \overline{\overline{a}} = \overline{a} - \overline{a} = 0.$

Putting $b = b'$ and $b = \overline{b}$ respectively in Eq. (12.69) we see that

$$\overline{\overline{a}b'} = \overline{a}\,\overline{b'} = 0; \text{ and } \overline{\overline{a}\,\overline{b}} = \overline{a}\,\overline{b}; \tag{12.70}$$

Consequently,

$$\begin{aligned}\overline{ab} &= \overline{(\overline{a} + a')(\overline{b} + b')} = \overline{\overline{a}\,\overline{b} + \overline{a}b' + a'\overline{b} + a'b'} \\ &= \overline{\overline{a}\,\overline{b}} + \overline{\overline{a}b'} + \overline{a'\overline{b}} + \overline{a'b'} = \overline{a}\,\overline{b} + 0 + 0 + \overline{a'b'} \\ &= \overline{a}\,\overline{b} + \overline{a'b'}.\end{aligned} \tag{12.71}$$

It may be noted that $\overline{a'b'}$ does not vanish identically.

The basic equations for incompressible Navier–Stokes equations have been presented in Chapter 7, Section 7.4.2. For the sake of convenience, they are repeated here. The continuity equation is

$$\frac{\partial u}{\partial x} + \frac{\partial v}{\partial y} + \frac{\partial w}{\partial z} = 0, \tag{12.72}$$

The momentum equations along x, y and z directions, in conservative form, are

$$\rho\left[\frac{\partial u}{\partial t} + \frac{\partial u^2}{\partial x} + \frac{\partial uv}{\partial y} + \frac{\partial uw}{\partial z}\right] = -\frac{\partial p}{\partial x} + \left(\frac{\partial \tau_{xx}}{\partial x} + \frac{\partial \tau_{xy}}{\partial y} + \frac{\partial \tau_{xz}}{\partial z}\right) + f_x, \tag{12.73}$$

$$\rho\left[\frac{\partial v}{\partial t} + \frac{\partial vu}{\partial x} + \frac{\partial v^2}{\partial y} + \frac{\partial vw}{\partial z}\right] = -\frac{\partial p}{\partial y} + \left(\frac{\partial \tau_{yx}}{\partial x} + \frac{\partial \tau_{yy}}{\partial y} + \frac{\partial \tau_{yz}}{\partial z}\right) + f_y, \tag{12.74}$$

$$\rho\left[\frac{\partial w}{\partial t}+\frac{\partial wu}{\partial x}+\frac{\partial wv}{\partial y}+\frac{\partial w^2}{\partial z}\right]=-\frac{\partial p}{\partial z}+\left(\frac{\partial \tau_{zx}}{\partial x}+\frac{\partial \tau_{zy}}{\partial y}+\frac{\partial \tau_{zz}}{\partial z}\right)+f_z, \quad (12.75)$$

where the viscous stress tensor is given by

$$\tau_{xx}=\sigma_x=2\mu\frac{\partial u}{\partial x},\ \tau_{yy}=\sigma_y=2\mu\frac{\partial v}{\partial y},\ \tau_{zz}=\sigma_z=2\mu\frac{\partial w}{\partial z},\ \tau_{xy}=\tau_{yx}=\mu\left(\frac{\partial u}{\partial y}+\frac{\partial v}{\partial x}\right),$$

$$\tau_{xz}=\tau_{zx}=\mu\left(\frac{\partial u}{\partial z}+\frac{\partial w}{\partial x}\right),\ \tau_{yz}=\tau_{zy}=\mu\left(\frac{\partial v}{\partial z}+\frac{\partial w}{\partial y}\right). \quad (12.76)$$

Substitution Eqs. (12.67) are introduced in the above N–S Equations and time-average is taken, using relations (12.69–12.71). Noting that,

$$\overline{\rho\frac{\partial u}{\partial t}}=\rho\frac{\partial \bar{u}}{\partial t},\ \text{since}\ \frac{\partial \overline{u'}}{\partial t}=0,$$

$$\frac{\overline{\partial(\rho uu)}}{\partial x}=\frac{\partial}{\partial x}(\rho\bar{u}^2),\quad \overline{\frac{\partial}{\partial y}(\rho uv)}=\frac{\partial}{\partial y}(\rho\bar{u}\bar{v})+\frac{\partial}{\partial y}\overline{(\rho u'v')},$$

and so on, follows on simplification the time-averaged equation for the x-momentum (12.73)

$$\rho\left[\frac{\partial \bar{u}}{\partial t}+\frac{\partial \bar{u}^2}{\partial x}+\frac{\partial(\bar{u}\bar{v})}{\partial y}+\frac{\partial(\bar{u}\bar{w})}{\partial z}\right]$$
$$=-\frac{\partial \bar{p}}{\partial x}+\left\{\frac{\partial \tau_{xx}}{\partial x}+\frac{\partial \tau_{xy}}{\partial y}+\frac{\partial \tau_{xz}}{\partial z}\right\}$$
$$+\left[\frac{\partial}{\partial x}(-\overline{\rho u'^2})+\frac{\partial}{\partial y}(-\overline{\rho u'v'})+\frac{\partial}{\partial z}(-\overline{\rho u'w'})\right], \quad (12.77)$$

and two similar equations. The continuity equation becomes

$$\frac{\partial \bar{u}}{\partial x}+\frac{\partial \bar{v}}{\partial y}+\frac{\partial \bar{w}}{\partial z}=0, \quad (12.78)$$

which is of the same form as Eq. (12.72), and may be obtained from this equation by replacing the velocity components (u, v, w) by the corresponding mean velocities $(\bar{u}, \bar{v}, \bar{w})$.

Apart from the mean velocity fluctuation product terms, Eq. (12.77) and the other two similar equations mentioned, are of the same form as Eq. (12.73) (and the corresponding momentum equations (12.74)–(12.75)). We conclude that the effect of turbulence is, as if, additional stresses $\overline{\overline{\tau^t}}$ with components

$$\tau^t_{xx}=-\overline{\rho u'^2},\ \tau^t_{yy}=-\overline{\rho v'^2},\ \tau^t_{zz}=-\overline{\rho w'^2},$$

$$\tau^t_{yz}=-\overline{\rho v'w'},\ \tau^t_{zx}=-\overline{\rho w'u'},\ \tau^t_{xy}=-\overline{\rho u'v'}, \quad (12.79)$$

are introduced into the fluid, while the existing flow parameters are replaced by their mean values. These additional stresses, arising on account of turbulent fluctuations, are known as *Reynolds stresses* or *eddy stresses.* Thus, we may think that the stress tensor $\overline{\overline{\tau}}$ of the mean motion is the sum of laminar viscous stress tensor $\overline{\overline{\tau^l}}$ and an additional stress tensor $\overline{\overline{\tau^t}}$, defined in Eq. (12.79), arising due to the turbulent fluctuations:

$$\overline{\overline{\tau}} = \overline{\overline{\tau^l}} + \overline{\overline{\tau^t}}. \tag{12.80}$$

the components of $\overline{\overline{\tau^l}}$ being identical with the original stress tensor $\overline{\overline{\tau}}$ defined in Eq. (12.76). For determining the mean motion, we thus have the continuity equation (12.78), the x-momentum equation (12.77) and two other similar momentum equations for the y and z directions—a total of four equations for determining ten unknowns, namely u, v, w, p and the six Reynolds stress components, expressed in terms of *correlations between fluctuating velocity components.* The number of equations is smaller than the number of unknowns. This difficulty is known as the *closure problem.*

Since the eddy stresses are unknown functions, the above system can be closed only if certain assumptions are made about their values in terms of the mean values $\overline{u}, \overline{v}, \overline{w}$. This leads to *turbulence modelling* consisting of *assumptions and empirical results,* which express the Reynolds stress components, *comprising of correlations between fluctuating velocity components,* in terms of flow quantities of the mean motion.

A turbulent flow is said to be two-dimensional if the mean motion is two-dimensional. Incompressible RANS equations for two-dimensional flow obtained from Eq. (12.77) are

$$\rho\left[\frac{\partial \overline{u}}{\partial t} + \frac{\partial \overline{u}^2}{\partial x} + \frac{\partial (\overline{u}\overline{v})}{\partial y}\right] = -\frac{\partial \overline{p}}{\partial x} + \left\{\frac{\partial \tau_{xx}}{\partial x} + \frac{\partial \tau_{xy}}{\partial y}\right\} + \left[\frac{\partial}{\partial x}(-\rho\overline{u'^2}) + \frac{\partial}{\partial y}(-\rho\overline{u'v'})\right], \tag{12.81}$$

and

$$\rho\left[\frac{\partial \overline{v}}{\partial t} + \frac{\partial \overline{v}\overline{u}}{\partial x} + \frac{\partial (\overline{v}^2)}{\partial y}\right] = -\frac{\partial \overline{p}}{\partial y} + \left\{\frac{\partial \tau_{yx}}{\partial x} + \frac{\partial \tau_{yy}}{\partial y}\right\} + \left[\frac{\partial}{\partial x}(-\rho\overline{u'v'}) + \frac{\partial}{\partial y}(-\rho\overline{v'^2})\right], \tag{12.82}$$

the continuity equation for the mean flow being given by

$$\frac{\partial \overline{u}}{\partial x} + \frac{\partial \overline{v}}{\partial y} = 0. \tag{12.83}$$

12.8.3 Closure Problem and Turbulence Modelling

We note here a few salient points regarding the closure problem and modelling discussed in detail in Majumdar (1991).

- Reynolds averaged Navier–Stokes (RANS) equations are solved for computing the mean flow.
- At each point of the flow field, turbulence velocity correlations, $\overline{\rho u'^2}$, $\overline{\rho u'v'}, \cdots$ are calculated using additional model equations, called *turbulence models*.
- Exact transport differential equations may be established for each Reynolds stress component. However, the equations for determining the second-order correlations have third-order correlations as unknown, which again, require modelling.
- Closure is possible *only through empirical information* relating fluctuating velocity correlations as a function of the mean flow parameters.
- It may be noted further, that the turbulence models are always semi-empirical; even, the model transport equations contain empirical constants.
- Empirical information may be in the form of *purely algebraic relations* connecting Reynolds stresses to the mean flow quantities or their gradients. On the other hand, they may be in the form of *differential equations* for transport of turbulence parameters, like the turbulence kinetic energy or their dissipation, which determine the Reynolds stresses.
- The empirical information depend heavily on experimental results and are of only limited validity. Usually it is not possible to extrapolate them to other geometries or other flow arrangements.
- For computing the time-averaged velocity, pressure or temperature the RANS equations are solved along with the turbulence model equations. The governing coupled non-linear partial differential equations are solved now-a-days by computational methods like the finite volume (or finite element) methods. We may also use other methods, like the finite difference method. However, the finite volume or the finite element methods are more efficient as they need lesser amount of computation.

12.8.4 Boussinesq Hypothesis

Boussinesq (1877) conceived that the effect of turbulence is only to increase the viscosity of the fluid and introduced the hypothesis that the edddy stresses are given in terms of the mean motion by

$$\tau_{xx}^{t} = 2\rho\nu_t \frac{\partial \overline{u}}{\partial x}, \ldots, \quad \tau_{xy}^{t} = \rho\nu_t \left(\frac{\partial \overline{v}}{\partial x} + \frac{\partial \overline{u}}{\partial y}\right), \ldots \tag{12.84}$$

The quantity ν_t is called the *coefficient of eddy kinematic viscosity.* It may be noted that ν_t is not a physical constant characteristic of the fluid. It varies from point to point and is a property of the flow.

For water $\nu = 0.018,\ \nu_t = 5.3$ for slow turbulence, and it is enormously large for large-scale motion. For air, $\nu = 0.122$ and $\nu_t = O(10^5)$ in atmospheric turbulence.

Observation of turbulent flows show certain transport features that have similarities with laminar flows. For example, momentum, heat or mass generally flow from regions of higher velocity, temperature or concentration to regions with lower values. Moreover, the rate of transport is enhanced by increasing the velocity fluctuations (energy scale) or by increasing the size of the eddies (length scale). These facts may motivate one to think that the *eddy viscosity concept* of Boussinesq (1877) might be useful for modelling turbulent flows.

One drawback of such a model may be noticed immediately. The convection effects originating due to the fluctuating velocities are expressed as diffusion effect in models using the eddy-viscosity hypothesis. Apparently, there is hardly any physical justification for the models based on the concept of eddy viscosity. In spite of it, these models are very popular and are successful in a large number of simple turbulent flow problems.

12.8.5 Eddy Viscosity Models

The eddy viscosity models are classified under Majumdar (1991)

- *Zero equation models,* in which no transport differential equation is solved for the turbulence parameters. Well-known classical examples are
  1. Mixing length theory of Prandtl (1925);
  2. Vorticity transfer theory of Taylor;
  3. Similarity hypothesis of Von Karman (1930).
- A popular and widely used model is that due to Cebeci–Smith (1974). Other zero-equation models, useful mainly for compressible turbulent flow computation are
  1. Baldwin–Lomax model (1978);

2. Johnson and King model (1985).

- *One-equation models,* in which transport differential equation is solved for one turbulence parameter, of which an example is the equation for turbulent kinetic energy due to Prandtl (1945), Kolmogorov (1942).
- *Two-equation models,* that solve two transport differential equations, one each for the velocity scale ($V = \sqrt{K}$) and the other is a length-scale related variable. The most widely tested and popular two-equation model is the K–ϵ model put forward by Jones and Launder (1972).

12.8.6 Zero-Equation Models

In these models, the Reynolds shear stress terms in the momentum equations are modelled by algebraic eddy viscosity or mixing-length formulations. For example, for a 2-D incompressible flow, the eddy kinematic viscosity ν_t is defined by

$$-\rho\overline{u'v'} = \rho\nu_t\frac{\partial\bar{u}}{\partial y} \tag{12.85}$$

In the mixing-length approach, the Reynolds shear stress is approximated by

$$-\rho\overline{u'v'} = \rho l^2\left(\frac{\partial\bar{u}}{\partial y}\right)^2, \tag{12.86}$$

l denoting the mixing-length. Several formulations have been suggested to specify the variation of eddy viscosity or mixing length distributions. We mention briefly the Cebeci–Smith model (1974) which has been widely used. It is a two-layer model and assumes that for wall boundary layer flows, the inner and outer layers may be described in terms of eddy viscosities. A modified version of it, suitable for compressible flows have been proposed by Baldwin–Lomax (1978), and discussed in detail, in Chapter 13, Section 13.4.1, where various applications with computational results may be found.

12.8.7 K–ϵ Model

Zero-equation models are useful and reasonably good for wall boundary-layer flows, although they lack generality. The two-equation models have less limitations. The K–ϵ model due to Jones and Launder (1977), is based on eddy-viscosity concept, ν_t being given by

$$\nu_t = C_\mu\frac{K^2}{\epsilon} \tag{12.87}$$

Here, C_μ is a constant and K and ϵ are obtained from differential equations representing transport of turbulence kinetic energy K and the rate of dissiption ϵ. The transport equations are given by

$$\overline{u}\frac{\partial K}{\partial x} + \overline{v}\frac{\partial K}{\partial y} = \frac{\partial}{\partial y}\left(\frac{\nu_t}{\sigma_k}\frac{\partial K}{\partial y}\right) + \nu_t\left(\frac{\partial \overline{u}}{\partial y}\right)^2 - \epsilon, \tag{12.88}$$

$$\overline{u}\frac{\partial \epsilon}{\partial x} + \overline{v}\frac{\partial \epsilon}{\partial y} = \frac{\partial}{\partial y}\left(\frac{\nu_t}{\sigma_c}\frac{\partial \epsilon}{\partial y}\right) + C_{\epsilon 1}\frac{\epsilon}{K}\nu_t\left(\frac{\partial \overline{u}}{\partial y}\right)^2 - C_{\epsilon 2}\frac{\epsilon^2}{K} \tag{12.89}$$

The three terms on the right of Eq. (12.88) respectively represent turbulent diffusion, turbulent energy production and viscous dissipation. The value of the parameters in this equation are given by

$$C_\mu = 0.09,\ C_{\epsilon 1} = 1.44,\ C_{\epsilon 2} = 1.92,\ \sigma_k = 1.0, \sigma_\epsilon = 1.3. \tag{12.90}$$

These equations apply only to free shear flows. For wall boundary layer flows, they require modifications (Biswas and Eswaran, 2002; Majumdar, 1991).

12.9 SUMMARY

Incompressible viscous flow has been studied in this chapter. For computation of 2D laminar flow, the sream-function vorticity method and the MAC method have been discussed. The MAC method may be readily extended to the study of 3D flows. Flow in a 2D channel with constriction has been discussed in detail, as a case study. Incompressible RANS equations have been derived. Turbulence modelling has been discussed briefly.

12.10 KEY TERMS

Backward facing step problem
Boundary conditions, stream-function vorticity
MAC method
Boussinesq hypothesis
Closure problem
Conservative form
Constricted channel flow
Convective terms-treatment
Dilation
DNS (direct numerical simulation)
Discretisation on staggered grid
Driven cavity problem
Eddies
Large eddy simulation model
MAC (Marker and Cell) method
Mixing-length,
Non-conservative form
Pressure Poisson equation
Pressure-correction formula
Primitive variables formulation
Reversed flow
Reynolds averaged Navier–Sokes equations
Reynolds stresses
Second upwinding
Secondary separation
Flow separation

Eddy stresses
Eddy viscosity
Eddy viscosity models
Incompressible Navier–Stokes equations
Incompressible viscous flow
Laminar flow
Stability restriction
Staggered grid
Stream-function vorticity method
Turbulent flow
Turbulence modelling
Vorticity transport equation.

12.11 EXERCISE 12

12.1 Establish the differential form of the pressure Poisson equation using primitive variables

$$\nabla^2 p = -\frac{\partial^2 u^2}{\partial x^2} - 2\frac{\partial^2 (uv)}{\partial x \partial y} - \frac{\partial^2 v^2}{\partial y^2} - \frac{\partial D}{\partial t} + \frac{1}{Re}\left(\frac{\partial^2 D}{\partial x^2} + \frac{\partial^2 D}{\partial y^2}\right), \quad (12.91)$$

where the dilation term D is

$$D = \frac{\partial u}{\partial x} + \frac{\partial u}{\partial y}. \quad (12.92)$$

12.2 Compute solution of the driven square cavity problem, in the unit square in the first quadrant using the stream-function vorticity formulation, with boundary conditions stated as in Fig. 12.1. Take $\Delta x = \Delta y = 0.025$.

12.3 Compute solution of problem 12.2 using primitive variables approach. Solve the pressure Poisson equation at each time step correct to 5-D, using SOR-iteration scheme. Write a computer program for this and obtain results.

12.4 Compute solution of the backward facing step, as described in Fig. 12.2 using the stream-function vorticity formulation. Take the far-field boundary at a distance of 2-units from the inflow- wall, the step being of unit height. Write a computer program and execute it to get results.

12.5 Compute solution of problem 12.4 using the primitive variables approach.

12.6 Using the continuity equation for 3-D incompressible flow, show that the nonconservative equation

$$\frac{\mathrm{d}f}{\mathrm{d}t} = \frac{\partial f}{\partial t} + u\frac{\partial f}{\partial x} + v\frac{\partial f}{\partial y} + w\frac{\partial f}{\partial z} = \frac{\partial f}{\partial t} + \vec{V}.\vec{\nabla} f, \quad (12.93)$$

may be rewritten as

$$\frac{\mathrm{d}f}{\mathrm{d}t} = \frac{\partial f}{\partial t} + \frac{\partial (fu)}{\partial x} + \frac{\partial (fv)}{\partial y} + \frac{\partial (fw)}{\partial z} = \frac{\partial f}{\partial t} + \vec{\nabla}.(f\vec{V}), \quad (12.94)$$

Hence, derive the conservative form of the 3-D N–S Equations (7.36)–(7.38).

13

Viscous Compressible Flow

We have seen in Chapter 7 that the domain of viscous compressible flow, in general, satisfies the assumption of continuous media and can be described by the conservation laws of mass, momentum and energy forming a set of equations called Navier–Stokes equations. Discussion on *direct numerical simulation* (DNS) of viscous compressible flow past bodies of practical interest is beyond the scope of this book. Instead, we will restrict ourselves to solution of the *Reynolds averaged Navier–Stokes* (RANS) equations with turbulence modelling. In this chapter, the solution methodology of the most general form of the governing differential equations describing the compressible viscous flows alongwith some practical examples will be discussed.

13.1 INTRODUCTION

The subject, computational fluid dynamics (CFD), dealing with basic fluid flow problems with associated transport of heat and mass, encountered in variety of engineering applications has been developed extensively during the last two and a half decades. For compressible flow, the subject CFD is an amalgamation of the subjects like, fluid mechanics, thermodynamics, numerical analysis and computer science. Mathematical theory of differential equations with vector and tensor algebra forms the basis for the development of the subject. It has been seen in the last chapter that under the assumption of incompressibility Navier–Stokes equations were reduced to simpler form. It should be noted that for incompressible flows under the assumption of constant density, the subject thermodynamics has no role to play. For compressible flow, the Navier–Stokes equations generally mean laws of conservations of mass, momentum and energy and to make the system complete, the density and temperature variations along with the thermodynamic equation of state are to be considered. The appearence of shock waves, sonic line, flow separation etc. further complicates compressible flows in comparison to its incompressible counterpart. It has been widely accepted that the unsteady Navier–Stokes equations for compressible flow as discussed in Chapter 7 are valid equations to describe the motion of a continuous media under the assumption that *mean-free-path* of the fluid molecules is much less than the scales of interest of the fluid flow. In that case, these equations are valid for both laminar and turbulent flows.

Nowadays, it is possible to get an workstation or even a personal computer (PC) which is adequate for some approximate simulation of two and three dimensional viscous flows past aerofoils, wings, missiles, complete aircraft and even for flow through inlets and engines. This development was possible due to simultaneous advancement of computer technology alongwith the large scale development of numerical algorithms. However, more efforts are necessary to achieve an efficient and reliable solution for flow past bodies of complex geometry of practical interest.

13.2 DYNAMIC SIMILARITY

Let us introduce some dynamic *similarity paprameters* used frequently in the study of viscous compressible fluid mechanics. The motion of a fluid is said to be completely determined if the fluid properties like density ρ, pressure p, temperature τ, coefficient of viscosity μ and velocity vector $\vec{U}$ are known as functions of time and position. Usually it is the presence of an obstacle in a uniform motion which generates the flow field and so, the flow field depends on the shape and size of the obstacle. It is desirable to find a relation between the flow patterns generated by geometrically similar bodies. If we keep the shape of the obstacle fixed (geometrical similarity) and change the size by varying a charactericstic length scale l (say, the chord of an aerofoil or diameter of a circular cylinder), the flow pattern will remain unaltered and the flow quantities can be obtained by changing the fundamental units of length, mass and time (Lighthill, 1963). This property permits us to test a model of an aircraft (say) in wind-tunnel to know what the results will be for a full scale aircraft in a free flight. So, the similarity observed in the flows generated by geometrically similar bodies is known as *dynamic similarity*.

Introducing some reference units of lengh L, mass M and time T, the new set of scaled variables (primed) becomes:

$$l' = \frac{l}{L}, \quad U' = \frac{U}{LT^{-1}}, \quad \rho' = \frac{\rho}{ML^{-3}}, \quad \mu' = \frac{\mu}{ML^{-1}T^{-1}}. \tag{13.1}$$

Which follows

$$\frac{\rho' U' l'}{\mu'} = \frac{\rho U l}{\mu}. \tag{13.2}$$

What we see in Eq. (13.2) is that the two flows having different flow variables can be made similar if the group of flow variables representing a dimensionless quantity remains invariant. Conditions for dynamically similar flows can also be obtained from fluid dynamic equations. Introducing a reference length L, a reference velocity U_∞ and a reference density ρ_∞, other non-dimensional variables can be derived as

$$x' = \frac{x}{L},\ y' = \frac{y}{L},\ z' = \frac{z}{L},\ T' = \frac{U_\infty T}{L},\ U' = \frac{U}{U_\infty},\ V' = \frac{V}{U_\infty},\ W' = \frac{W}{U_\infty},$$

$$\rho' = \frac{\rho}{\rho_\infty},\ p' = \frac{p - p_\infty}{\rho U_\infty^2},\ \vec{f}\,' = \frac{\vec{f} L}{\rho U_\infty^2},\ \bar{\bar{\tau}}' = \frac{\bar{\bar{\tau}} L}{\mu U_\infty},\ \text{etc.} \tag{13.3}$$

It can be shown that the governing equations (7.13–7.15) will remain unchanged in dimensionless form. Equations (7.18)and (7.19)will become

$$\bar{\bar{\sigma}} = -p\bar{\bar{I}} + \frac{1}{Re}\bar{\bar{\tau}} \text{ and } \bar{\bar{\tau}} = \lambda \text{ div } \vec{V}\bar{\bar{I}} + 2\mu \text{ def } \vec{V} \tag{13.4}$$

respectively. Where $Re = \frac{\rho U l}{\mu}$ represents the same group of variables as in Eq. (7.2) and is called the *Reynolds number*, introduced in chapter-7 as the ratio of *inertial force* to *viscous force*. Eq. (7.20) gives for a perfect gas

$$\vec{q} = -\frac{\gamma}{Re P_r} k \text{ grad } e, \tag{13.5}$$

where

$$P_r = \frac{\mu c_p}{k} = \frac{\mu/\rho}{k/c_p\rho} = \frac{\nu}{k/c_p\rho} = \frac{\text{kinematic viscosity}}{\text{thermal diffusivity}} \tag{13.6}$$

is the *Prandtl number*, a measure of the relative importance of heat conduction and viscosity of the fluid and c_p is the specific heat of the fluid at constant pressure.

Apart from these, another important parameter is *Mach number* M, a ratio of the local fluid velocity U to the local velocity of sound $a = \frac{\gamma p}{\rho}$. This parameter, $M = U/a$, is considered as a measure of compressibility. For a flow to be considered incompressible, it must be (i) slowly varying, (ii) having very low Mach number, (iii) having small difference in temperature between obstacle and stream and (iv) having a length scale small compared to the scale height of the atmosphere (Lighthill, 1963). That is the parameters $\frac{\omega l}{a}, \frac{U}{a}, \frac{T_{\text{wall}} - T_\infty}{T_\infty}, \frac{gl}{a^2}$ are small (with g = acceleration due to gravity). A flow with small Mach number implies the ratio of the variation of the density to the variation of velocity is small. The flow is called subsonic or supersonic depending on the value of M is less than or greater than one respectively.

Measured or computed values of pressure (load) distribution on the surface of the obstacle or the aerodynamic forces acting on the obstacle are generally expressed in terms of dimensionless coefficients, like pressure coefficient C_P, drag coefficient C_D, lift coefficient C_L, moment coefficient C_M etc., such that these quantities will remain independent of the choice of units of length, density, velocity and coefficient of viscosity depending only on the Reynolds number for the steady case. These coefficients are obtained by normalizing the quantities by some reference values like pressure by free stream static pressure, forces by the product of free stream static pressure and some reference area etc.

13.3 RANS (REYNOLDS AVERAGED COMPRESSIBLE NAVIER–STOKES) EQUATIONS

Although the Navier–Stokes equations remain valid for turbulent flows, the very rapid variations in space and time which characterize such flows can not be taken into account by the most refined discretization of space and time allowable on the most powerful computers, at least for realistic applications, taking into account the limitations in memory size and computing speeds. The variation of the scales of interest in turbulent flow is random and very large. It typically consists of turbulent eddies (lump of fluid in swirling motion) of varying sizes. Large "scales" (say L) are of the same order as for the mean flow and these contain the energy of the turbulence. Kolmogorov scales, the smallest scales (say l), are responsible for dissipating the turbulence energy. Both the scales can be obtained from dimensional analysis (Malan, 1999). An eddy of every size plays an important role in the turbulent transport mechanism. Qualitatively, the length scale l is proportional to $Re^{-3/4}$ times of L (Reynolds, 1990). So, as the Reynolds number increases, the smallest scales in turbulent flow get smaller and so do the significant scales that need to be resolved. For a driven cavity flow it has been shown by Deshpande and Milton (1990) that for a grid spacing of $\frac{1}{1000}$ against that of $\frac{1}{80}$, increased computational effort due to increased number of grid points amounts to $\left(\frac{1000}{80}\right)^3 \approx 2000$ times of what was needed originally for the spacing of $\frac{1}{80}$.

Even the present day computers are inadequate with respect to speed and storage capacity to obtain solutions of practical problems. At present, the determination of turbulent flows by global approach is based on the Reynolds' time averaged Navier–Stokes (RANS) equations (or simplified forms of these equations) supplemented with algebraic relations or partial differential equations, which constitute a turbulence model, so as to obtain a closed system of equations. Some methods and algorithms which are being successfully used for compressible flows will be discussed in this chapter for two dimensional flows. The extension of these to three dimensions are straight forward. Reynolds averaged Navier–Stokes (RANS) equations have been derived in Chapter 7 and can be written by dropping the over bars and tildes for covenience and with the same definitions of the variables as:

$$\frac{\partial \rho}{\partial t} + \frac{\partial}{\partial x_j}(\rho u_j) = 0 \tag{13.7}$$

$$\frac{\partial}{\partial t}(\rho u_i) + \frac{\partial}{\partial x_j}(\rho u_i u_j) = -\frac{\partial p}{\partial x_i} + \frac{\partial}{\partial x_j}\left(\tau_{ij} - \overline{\rho u_i'' u_j''}\right) \tag{13.8}$$

$$\frac{\partial}{\partial t}(\rho H) + \frac{\partial}{\partial x_j}(\rho u_j H) = \frac{\partial p}{\partial t} + \frac{\partial}{\partial x_j}\left(u_i \tau_{ij} + \overline{u_i'' \tau_{ij}} - \overline{\rho u_j'' H''} + k\frac{\partial T}{\partial x_j}\right) \tag{13.9}$$

where H denotes the total enthalpy per unit mass given by

$$H = E + \frac{p}{\rho}. \tag{13.10}$$

Equations (13.7)–(13.9) are called RANS equations. They have the same general form as the instantaneous Navier–Stokes equations with the velocities and scalar variables now representing mass- and time-averaged variables respectively with additional terms $\overline{\rho u_i'' u_j''}$ and $\overline{u_i'' \tau_{ij}} - \overline{\rho u_j'' H''}$. These additional terms represent the effect of turbulence. Now, the stress tensor is augmented by additional stress called *Reynolds stress tensor* and the heat flux vector is augmented by additional *turbulent heat flux*. Due to the presence of these two extra unknowns the above equations are not closed and some modelling is required to get a closed and solvable system.

For classical aerodynamics problems, Navier–Stokes equations constitute a very accurate mathematical model to be used directly for laminar flows. Considerable advancement in the computation of viscous compressible flows using RANS equations with turbulence models has been achieved recently due to the availability of high speed computers and the substantial improvement in the efficiency and accuracy of the numerical algorithms.

13.4 TURBULENCE MODELLING

We have seen now that the Navier–Stokes equations are valid governing equations for compressible turbulent flow. To solve these equations numerically, the so called *direct numerical simulation* (DNS) is prohibitively expensive to resolve the wide range of length and time scales of turbulence. Next approach is *large eddy symulation* (LES), where, the large scale eddies are treated directly like DNS but the small eddies, where the length scale is less than the computational mesh size are resolved by turbulence models. This is a compromise between DNS and solving RANS equations using some turbulence models. For simple flows like flow over a flat plate or the jet flow where either the boundaries are very simple or there is no solid boundary, a reasonable solution can be obtained using DNS or LES methods.

In turbulence modelling, what it requires is that the Reynolds stresses in Eq. (13.8) are to be related to the mean velocity gradients or mean strain rate tensor *(Boussinesq approximation)* as:

$$-\overline{\rho u_i'' u_j''} = \mu_t \left(\frac{\partial u_i}{\partial x_j} + \frac{\partial u_j}{\partial x_i} \right) - \frac{2}{3} \left(\rho k + \mu_t \frac{\partial u_i}{\partial x_i} \right) \delta_{ij}. \tag{13.11}$$

Although it is an approximation (Hinze, 1975), this approach works surprisingly well for many practical cases. μ_t is called turbulent *eddy viscosity*. Turbulent kinematic viscosity ($\nu_t \equiv \mu_t/\rho$), has the unit $[L^2/s]$. Effective viscosity μ becomes $\mu = \mu_l + \mu_t$, where μ_l is the *laminar viscosity*. Similarly, turbulent heat flux term in Eq. (13.9) can be written as (Majumdar, 1998):

$$-\overline{\rho u_j'' H''} = -\frac{\mu_t}{P_{rt}} \frac{\partial H}{\partial x_j}, \tag{13.12}$$

with the laminar heat flux $= \left(-\frac{\mu_l}{P_{rl}}\frac{\partial H}{\partial x_j}\right)$. Effective Prandtl number, P_r now becomes $P_r = P_{rl} + P_{rt}$ where the suffices l and t represent the laminar and turbulent values of the parameters and k is the coefficient of thermal conductivity. Typical values for P_{rl} and P_{rt} are 0.72 and 0.95 respectively.

13.4.1 Algebraic Turbulence Models

The algebraic or zero equation model based on eddy viscosity approach was originally proposed by Cebeci and Smith (1974) and was later modified by Baldwin and Lomax (1978). Eddy viscosity μ_t can be related to the mean strain using Prandtl's mixing-length approach. No additional differential equations are to be solved in these approaches. Other advantages of these models are:

1. they need minimum computer time and storage;
2. they are easy to implement and
3. they give reasonably accurate results for attached and separated flows past simple geometries.

Flows to be considered here are either fully turbulent or the transition point is to be prescribed from experimental data. Advantage of Baldwin–Lomax model over that of Cebeci–Smith is that the computation of the boundary layer edge has been avoided, which makes the model very popular for computation of compressible flows. In both these models turbulent eddy viscosity μ_t has been calculated in two layers called $\mu_t\mid_{\text{inner}}$ and $\mu_t\mid_{\text{outer}}$. The inner and outer layers are defined by $y \le y_{\text{crossover}}$ and $y > y_{\text{crossover}}$ respectively, where y is the normal distance from the wall and $y_{\text{crossover}}$ is the smallest value of y at which $\mu_t\mid_{\text{inner}}$ equals $\mu_t\mid_{\text{outer}}$. The inner formulation for both the models is same namely,

$$\mu_t\mid_{\text{inner}} = \rho l^2 \mid \omega \mid, \tag{13.13}$$

where the length scale, $l = ky[(1 - \exp(-y^+/A^+)]$ with Van Driest's damping factor $[(1 - \exp(-y^+/A^+)]$, $\mid \omega \mid$ is the magnitude of the vorticity given by $\left(\frac{\partial u}{\partial y} - \frac{\partial v}{\partial x}\right)$ for two dimensional flow, law-of-the-wall coordinate $y^+ = \frac{\sqrt{\rho_w \tau_w}}{\mu_w} y$ and $A^+ = 26$. Cebeci–Smith used ρ instead of ρ_w, suffix w representing the value of the quantity at the wall.

The outer formulation as per Cebeci–Smith model is:

$$\mu_t\mid_{\text{outer}} = \rho \alpha u_e \delta_i \gamma \tag{13.14}$$

with Klebanoff's intermittency factor

$$\gamma = (1 + 5.5(y/\delta)^6)^{-1}, \tag{13.15}$$

where, the parameters are defined as: Clauser's constant $\alpha = 0.0168$, boundary layer thickness δ defined as the distance from the wall where $u/u_e = 0.99$, incompressible displacement thickness $\delta_i = \int_0^{\delta} \left(1 - \frac{\rho u}{\rho_e u_e}\right) dy$, Von Karman constant $k = 0.4$. Suffix e defines the value of the quantity at the edge of the boundary layer. As we see now, in order to get the outer eddy viscosity, the location of the boundary layer edge is to be defined first, then calculate its position and then the flow variables are to be calculated at that point. To do this with numerical accuracy is a difficult task. To avoid this, Baldwin and Lomax (1978) proposed the following modification in the outer formulation:

$$\mu_t \mid_{\text{outer}} = \alpha C_{cp} \rho F_{\text{wake}} F_{\text{KLEB}}, \tag{13.16}$$

where $F_{\text{wake}} = \min\{y_{\max} F_{\max}, C_{\text{wk}}\, y_{\max} (u_{\text{dif}})^2 / F_{\max}\}$. $y_{\max}$ and $F_{\max}$ are to be determined from the function

$$F(y) = y \mid \omega \mid [1 - \exp(-y^+/A^+)] \tag{13.17}$$

with the following definition

$$F_{\text{KLEB}} = [1 + 5.5(\frac{C_{\text{KLEB}}\, y}{y_{\max}})^6]^{-1}. \tag{13.18}$$

u_{dif} is the difference between the maximum and minimum of total velocity and τ_w is the shear stress at the wall. The constants used are $k = 0.4$, $\alpha = 0.0168$, $C_{CP} = 1.6$, $C_{wk} = 0.25$ and $C_{\text{KLEB}} = 0.3$.

These models are designed for structured tetrahedral or hexahedral grid in two or three dimensional problems respectively. The algorithm requires marching outwards along a grid line from the wall. The models discussed above are well tested to give reasonably accurate results for attached flows with weak shocks. For strong shocks with shock-induced separation at the foot of the shock, these models in general predict stronger shocks. This problem was studied by Chakrabartty and Dhanalakshmi (1995) using various modification of the models depending on how the eddy viscosity can be calculated between the wall and the edge of the bubble. They observed that the separation bubble and the skin-friction distribution can be predicted well by all the four models proposed by them. A sample result is shown in Fig. 13.1, where it is shown that the pressure coefficient C_p predicted by the Baldwin–Lomax turbulent model compares well with experimental data on the lower surface and upto about 60% on the upper surface. Predicted shock is stronger and its position is far downstream, whereas, the comparison of the local skin-friction coefficient C_f with experiment is good and the bubble at the foot of the shock (portion where C_f is negative) has been predicted well.

As pointed out by Bradshaw (1994), for transonic flows, the normal pressure gradient is not negligible inside the boundary layer and so does not obey the boundary layer assumption, which has been used only through the turbulence models. Velocity gradient produced due to this, in principle, leads to extra production of turbulence through the product of mean velocity gradient and turbulent shear stress. Prediction of correct wall shear stress merely implies correct prediction of mean velocity profiles and not the velocity gradients. In rapidly growing flows near separation where normal pressure gradient affects the mean velocity gradient one cannot expect an acceptable pressure distribution. So, for better results in both skin-friction and pressure distribution, turbulence models should be developed for Navier–Stokes computations without any boundary-layer type of approximations.

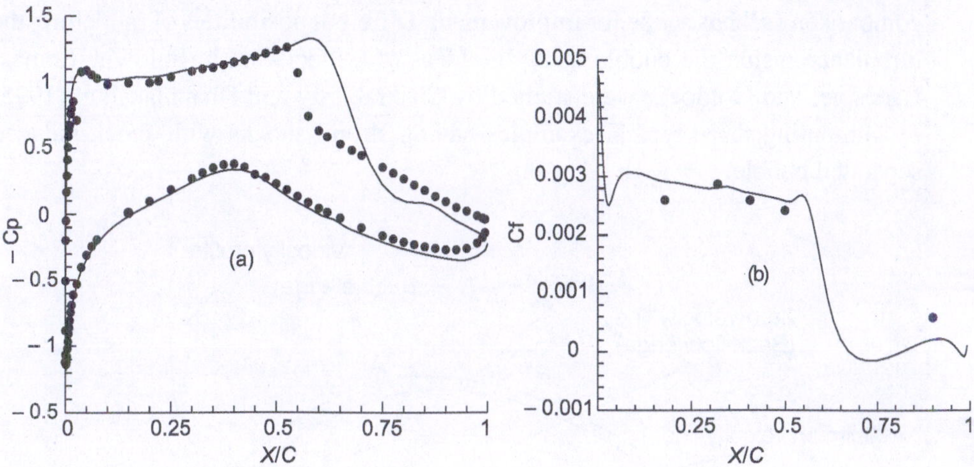

Figure 13.1 (a) Surface pressure ($-C_p$) and (b) local skin-friction (C_f) distribution for RAE-2822 aerofoil at $M_\infty = 0.75$, $\alpha = 2.81°$, $Re_\infty = 6.2 \times 10^6$, Grid size= 257×61, ••• Expt. (Cook et al., 1979), ___ RANS computation with Baldwin–Lomax turbulence model.

13.4.2 Other Models

As the boundary layer becomes separated, the above models are no longer relevant. Goldberg (1986) attempted to address this problem in a rigorous manner and proposed a model to treat the separated region. His model is based on the assumption and observation that (i) the stress scale is given by the maximum shear stress in the separated layer, not by the wall stress as proposed by Baldwin and Lomax and (ii) the shear stress has qualitatively the same turbulent structure when it is detached as when it is attached and the length scale is the height of the separated region.

Radespiel (1989) also used a similar idea by replacing the distance from the wall by the distance from the minimum velocity line (see Fig. 13.2 for nomenclature) and the shear stress at the wall by its maximum value to prevent vanishing eddy viscosity at the separation point. He however did not treat the region between the wall and the minimum velocity line. He obtained a pressure distribution with stronger shock downstream of the experimental shock position and skin-friction distribution shows separation but no reattachment on the aerofoil. However by using Johnson and King (1984) model, which is not simply an eddy viscosity model but also contains features of a Reynolds stress model, the results improved both in shock position and skin-friction distribution with reattachment near the trailing edge. Aftershock comparison still has scope for improvement. Different posibilities of modelling the turbulence inside the bubble using the ideas of Cebeci–Smith, Baldwin–Lomax, Radespiel and Goldberg were studied by Chakrabartty and Dhanalakshmi (1995) by computing three typical examples having strong shocks with shock-induced separated bubble.

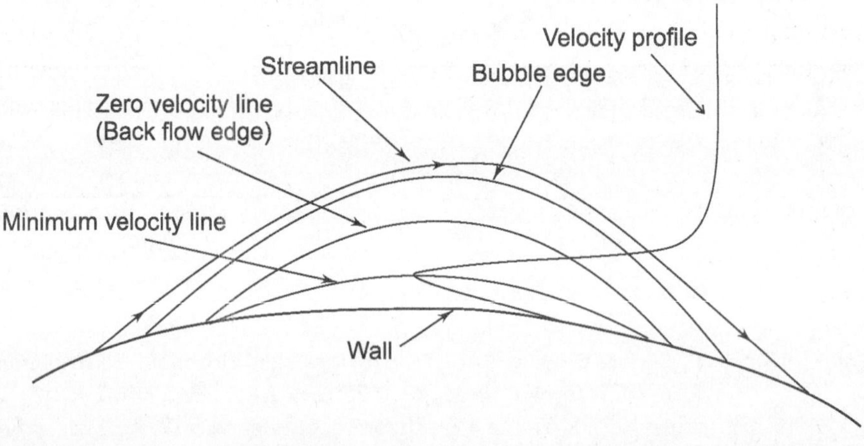

Figure 13.2 Schematic view of separated flow bubble and its nomenclature.

Some typical turbulence models widely used in commercial CFD codes like Spalart–Allmaras one-equation model (Spallart and Allmaras, 1992), some variants of $k - \epsilon$ type two-equation model and Reynolds stress transport model are described by Malan (1999) and Majumdar (1998). Among these, Spalart–Allmaras model is gaining popularity and is being used widely in compressible flow computations. It is an one equation model where the turbulent kinetic energy k is obtained by solving one equation and a length scale is prescribed using a mixing length approach. One advantage in this model is that the calculation of the length scale related to local shear layer thickness is not necessary and so it is highly suited for multigrid computation or computation with unstructured grid. Implementation of Reynolds stress transport model is difficult particularly for three dimensional flows. Different variants of $k - \epsilon$ model are being used widely for incompressible flow problems. Their application in compressible flow problems is not so popular. For massively separated turbulent flows, dominated by the leading edge vortex separation for flow past low aspect ratio delta wings at high angles of attack, Degani and Schiff (1986) introduced modification to the basic Baldwin–Lomax model to ensure that the scales used to determine the outer eddy viscosity are correct relative to the boundary layer and not to the vortex core. This was used by Chakrabartty, et. al. (1998) to analyse vortex flow over a cropped delta wing with round leading edge.

13.5 BOUNDARY CONDITIONS

Boundary conditions for Euler and RANS equations have been discussed in detail in Chapter 7. Here, we shall discuss the implementation of some of the boundary conditions used frequently. These implementation procedures are common to both Euler and RANS equations. At the solid wall: no boundary condition is required for density and it can be calculated directly using the continuity equation. For velocity components, boundary condition for the viscous flow is the *no slip* condition, which means that the relative velocity of the fluid with respect to the wall is zero. This can be implemented directly for the body-fitted coordinate system using finite difference or vertex-based/nodal-point finite volume discretization, where the grid points lie on the surface and the flow variables are defined at the grid points. Difficulty arises in Cartesian coordinate system where the grid points do not lie on the surface, in that case some extrapolation procedure has to be followed. Similar difficulty arises for cell-centred finite volume approach, where the grid points are on the surface but flow variables are at the centre of the cell. Here also some extrapolation formula or one sided finite difference formula is to be used. Another approach is to use image cells below the solid wall boundary as shown in Fig. 13.3, and assign the values of ρu and ρv in the image cells as the negative of their values in the first cell above the wall. Temperature at the wall is either prescribed ($T = T_{\mathrm{w}}$) or the normal derivative of temperature (adiabatic wall condition) $\dfrac{\partial T}{\partial n} = 0$. This can be treated as an equivalent condition for internal energy ($e = c_{\mathrm{v}}T$). For inviscid flow, the flow should be tangential or the normal velocity should be zero. This can be implemented in the following manner. The normal velocity q_n at any point on the wall is given by

$$q_n = ul + vm, \tag{13.19}$$

where u, v are the Cartesian velocity components in the x, y directions and l, m are the respective direction cosines of the normal $\vec{n}$ at that point. These components are replaced by the corrected velocity components u_c, v_c which enforce the zero normal velocity. These are obtained as

$$u_c = u - (ul + vm)l, \quad v_c = v - (ul + vm)m. \tag{13.20}$$

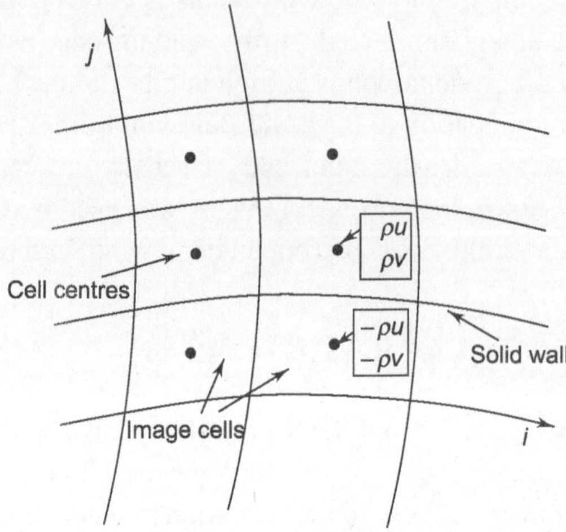

Figure 13.3 Solid wall boundary condition for cell-centred finite volume scheme.

Different types of internal cut boundaries arise depending on whether the overlapping region of the two connecting domains consists of single cell-layer or double cell-layer. In a multiblock grid, two block-boundaries can be linked by overlapping one or two grid-cell layers. Figure 13.4 shows how two blocks are connected through one layer, where last cell of block-1(between $i = I - 1$ and $i = I$) is overlapped with the first cell (between $i = 1$ and $i = 2$) of block-2.

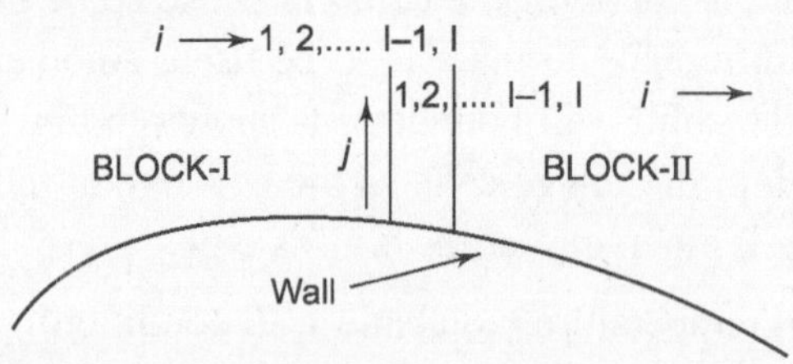

Figure 13.4 Cut with one overlapping grid layer.

Cut with two overlapping grid-layers appears even in a single block grid, from the trailing edge of the aerofoil to downstream boundary of the grid. Figure 13.5

explains the two overlapping grid layers after the trailing edge of an arofoil in C-type grid.

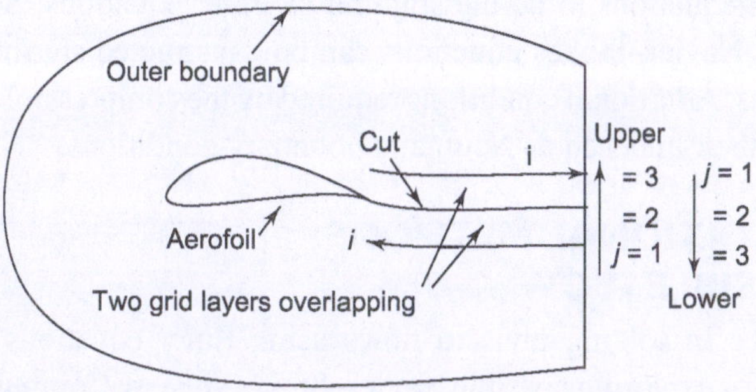

Figure 13.5 Cut with two overlapping grid layer.

Periodicity boundary condition which is similar to two-layer-cut condition is explained in Fig. 13.6. for internal flow through cascades.

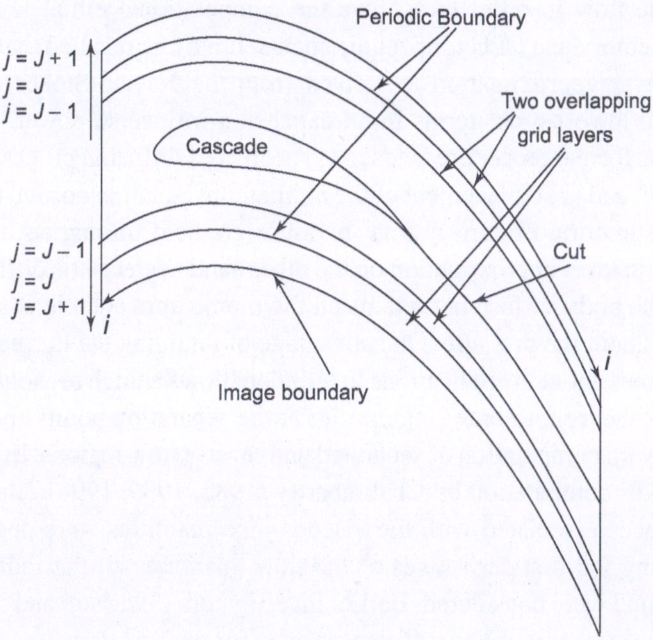

Figure 13.6 Periodic boundaries with two overlapping grid layer.

Far field boundary conditions for Euler and Navier–Stokes equations have been discussed in Chapter 7. It can be noted once again that at open boundaries with smooth flow the viscosity and heat conduction terms are neglible in comparison with convective terms. In this case, one can expect that the limit solutions of Navier–Stokes equations to be the solution of Euler equations. So, the boundary conditions for Navier–Stokes equations can be constructed starting with those of Euler equations. Additional conditions required by the compressible Navier–Stokes equations are then imposed as Neumann boundary conditions.

13.6 BASIC COMPUTATIONAL METHODS FOR COMPRESSIBLE FLOW

Past experience in solving inviscid flows using Euler equations provides many opportunities for exploring possible Navier–Stokes solvers. Computation of viscous flow past bodies of practical interest using Navier–Stokes equations representing the conservation of mass, momentum and energy needs a robust and efficient algorithm to deal with the coupled non-linear set of equations and is to be supplemented by the availability of high speed computers with large memory.

Classical boundary-layer or thin-layer approximations (Chakrabartty, 1989), 1990; Radespiel, 1989; Swanson and Turkel, 1985) have frequently been used to model the flow in order to achieve the computational efficiency in terms of computer memory and CPU time requirements. It may be recalled that the classical boundary layer approximation is derived from the Navier–Stokes equations by retaining the lowest order terms in an expansion in inverse powers of Reynolds number. This formal procedure leads to (1) neglect of diffusion process parallel to a body surface, and (2) replacement of the momentum equation normal to the surface with the assumption of zero normal pressure gradient throughout the boundary layer. The thin layer approximation on the other hand neglects the diffusion process parallel to the body surface but retains all the momentum equations and makes no assumption about the pressure. One advantage of retaining the normal momentum equation comes about in applications to turbulent flows at high reynolds number. Its use removes the troublesome singularities at the separation points and permits the straight forward computation of separated and reverse flow regions. In the thin layer Navier–Stokes computation by Chakrabartty (1987, 1989, 1990) the streamwise like differences associated with the viscous flux quantities were neglected while for computing the first derivatives of the flow quantities all the four sides of the control volume were considered. Earlier investigators (Swanson and Turkel, 1985) neglected the streamwise like differences even for the first derivatives.

For the discretization in space, it is advantageous to use body fitted/curvilinear coordinte system, where one family of grid lines coincides with the body surface. Numerical treatment of the boundary conditions are greatly simplified in this approach. Now, in the arbitrary curvilinear meshes, RANS equations can be discretized either by converting the equations into a generalized curvilinear coorinate system (ξ, η), where the computational domain becomes rectangular or by using the equations in their premitive variable form in Cartesian coordinate system and use the mesh system independently. Generally, the first approach is suitable for the finite difference methods, where the discretization in uniform rectangular mesh in computational domain is more accurate but it increases the complexity in the transformed governing equations. An implicit method originally developed by Beam and Warming (1976) for a conservative hyperbolic system was later extended to solve complete RANS equations (Beam and Warming, 1978). One variation of this scheme presented by Hollanders and Viviand (1980) where three time levels $n-1$, n and $n+1$ of $\vec{W}$ has been used will be discussed here. Time dependent RANS equations (13.7–13.9) can be written in Cartesian coordinates for two dimensional flows in conservative form as

$$\frac{\partial \vec{W}}{\partial t} + \frac{\partial \overline{\overline{F}}}{\partial x} + \frac{\partial \overline{\overline{G}}}{\partial y} = 0. \tag{13.21}$$

Where $\vec{W}$ is the vector of conserved variables and $\overline{\overline{F}}$ and $\overline{\overline{G}}$ are the total fluxes in x- and y-directions respectively. Let us discretize the time derivative term by a three point second order implicit finite difference formula as

$$\frac{\partial \vec{W}^{n+1}}{\partial t} = (3\vec{W}^{n+1} - 4\vec{W}^{n} + \vec{W}^{n-1})/(2\Delta t) + O(\Delta t^2). \tag{13.22}$$

Substituting this, the Eq. (13.21) can be written in discretized form at $(n+1)^{th}$ time level (i.e. $t = (n+1)\Delta t$) as

$$\vec{W}^{n+1} + \frac{2\Delta t}{3}\left[\frac{\partial \overline{\overline{F}}^{n+1}}{\partial x} + \frac{\partial \overline{\overline{G}}^{n+1}}{\partial y}\right] = \frac{4\vec{W}^{n}}{3} - \frac{\vec{W}^{n-1}}{3} + O(\Delta t^3) \tag{13.23}$$

where $\overline{\overline{F}}^{n+1}$ and $\overline{\overline{G}}^{n+1}$ mean $\overline{\overline{F}}(W^{n+1})$ and $\overline{\overline{G}}(W^{n+1})$ respectively. Inviscid parts of the fluxes are functions of the conservative variables alone, whereas the viscous parts contain the derivatives of the conservative variables also. Using the taylor series expansion of second order like

$$\overline{\overline{F}}_{\mathrm{E}}^{n+1} = \overline{\overline{F}}_{\mathrm{E}}^{n} + A^n(\vec{W}^{n+1} - \vec{W}^{n}) + O(\Delta t^2) \tag{13.24}$$

and

$$\overline{\overline{G}}_{\mathrm{E}}^{n+1} = \overline{\overline{G}}_{\mathrm{E}}^{n} + B^n(\vec{W}^{n+1} - \vec{W}^{n}) + O(\Delta t^2) \tag{13.25}$$

for the inviscid fluxes $\overline{\overline{F}}_{\mathrm{E}}$ and $\overline{\overline{G}}_{\mathrm{E}}$ with $A = \dfrac{\partial \overline{\overline{F}}_{\mathrm{E}}}{\partial \vec{W}}$ and $B = \dfrac{\partial \overline{\overline{G}}_{\mathrm{E}}}{\partial \vec{W}}$ and also for the splitted viscous fluxes in x- and y-directions such as $\overline{\overline{F}}_{V} = \overline{\overline{F}}_{Vx} + \overline{\overline{F}}_{Vy}$ and

$\overline{\overline{G}}_V = \overline{\overline{G}}_{Vx} + \overline{\overline{G}}_{Vy}$ the following equations in delta form can be obtained.

$$\left(I + \frac{2\Delta t}{3}\left[\delta_x(A - M + P_x)^n - \delta_{xx}P^n + \delta_y(B - N + Q_y)^n - \delta_{yy}Q^n\right]\right)\Delta\vec{W}^n$$

$$= -\frac{2\Delta t}{3}(\delta_x\overline{\overline{F}}^n + \delta_y\overline{\overline{G}}^n) + \frac{\Delta\vec{W}^{n-1}}{3} + \frac{2\Delta t}{3}(\delta_x\Delta F_{Vy}^{n-1} + \delta_y\Delta G_{Vx}^{n-1}) + O(\Delta t^3) \quad (13.26)$$

where $M = \frac{\partial\overline{\overline{F}}_{Vx}}{\partial\vec{W}}$, $N = \frac{\partial\overline{\overline{G}}_{Vy}}{\partial\vec{W}}$, $P = \frac{\partial\overline{\overline{F}}_{Vx}}{\partial\vec{W}_x}$, $Q = \frac{\partial\overline{\overline{G}}_{Vy}}{\partial\vec{W}_y}$. δ_x, δ_{xx}, δ_y and δ_{yy} are difference operators, I is the unit matrix, $\Delta\varphi^n = \varphi^{n+1} - \varphi^n$ for any quantity φ.

The operator in the left hand side of (13.26) can be factorized approximately and the following two step implicit (alternating direction implicit (ADI)), non-iterative, finite difference scheme in x- and y-direction respectively can be obtained as:

$$\left(I + \frac{2\Delta t}{3}\left[\delta_x(A - M + P_x)^n - \delta_{xx}P^n\right]\right)\Delta\vec{W}^* = \text{RHS of (13.26)} \quad (13.27)$$

and

$$\left(I + \frac{2\Delta t}{3}\left[\delta_y(B - N + Q_y)^n - \delta_{yy}Q^n\right]\right)\Delta\vec{W}^n = \Delta\vec{W}^*. \quad (13.28)$$

Using the standard finite difference approximations for δ_x, δ_y, δ_{xx} etc. the above two equations can be solved for $\Delta\vec{W}^n$ by inverting two matrices of tridiagonal form one in each direction using standard methods of matrix inversion. The unknown solution vector $\vec{W}$ at $(n+1)^{th}$ time step can be obtained from $\vec{W}^{n+1} = \vec{W}^n + \Delta\vec{W}^n$.

The above scheme has been extended to nodal point finite volume discretisation by Dutta (1995) for computing transonic flow past aerofoils.

13.7 FINITE VOLUME COMPUTATION IN 2D

In the finite volume formulation, the generation of a body fitted grid using curvilinear coordinates and the solution procedure are separated since no global transformation is used. The only required data concerning the grid are the Cartesian coordinates for the vertices of every cell in the given mesh. For a two dimensional problem, the computational domain is divided into quadrilateral cells by joining the cell vertices by straight lines. Computation of face area and cell volume have been discussed in Chapter 2.

The capability of finite volume method to treat the arbitrary geometry is well known. Basic strategy applied by Swanson and Turkel (1985) for spatial discretisation is the cell centred scheme originally developed for Euler equations by Jamson et al. (1981), in which the flow quantities are associated with the centre of a cell in the computational mesh. On the other hand, cell vertex or nodal point schemes proposed by Ni (1982) and Hall (1985) have been implemented by Rossow (1987) for Euler equations, where the flow quantities are ascribed at the cell vertices. Rossow (1987) showed that for a skewed grid zeroth order error can occur for cell centred scheme whereas nodal point schemes remain at least first order accurate. The scheme proposed by Hall (1985) has been used by Chakrabarty (1987, 1988, 1989) for the computation of thin layer Navier–Stokes equations and that of Ni (1982) has been used by Davis et al. (1986) for analysis of viscous flow through cascades. Surface boundary conditions can be satisfied exactly at the vertices along the body surface, and the pressure on the wall can be computed directly by the nodal point scheme, whereas an extrapolation is necessary if one uses the cell centred scheme. Müller (1986) used the implicit central difference scheme of Beam and Warming (1976) to compute laminar transonic flow past aerofoils at medium Reynolds numbers. Müller and Rizzi (1980) developed a cell-centred finite volume space discretization of the viscous fluxes for compressible Navier–Stokes equations and implemented explicit Runge-Kutta scheme (Jameson et al., 1981) for time integration. The two dimensional unsteady Navier–Stokes equations, neglecting body forces and heat sources can be written in integral form as

$$\frac{\partial}{\partial t}\iiint\limits_{V} \vec{W}\,dV + \iint\limits_{S} \bar{\bar{H}}_E \cdot \vec{n}\,ds + \iint\limits_{S} \bar{\bar{H}}_V \cdot \vec{n}\,ds = 0. \tag{13.29}$$

Here the integrals are over a surface S with normal $\vec{n}$ enclosing a volume V and ds and dv are the elementary surface and volume elements respectively. Let ρ, (u, v), p, E, T and H denote the density, Cartesian velocity components, pressure, total internal energy, absolute temperature and total enthalpy respectively. The vector $\vec{W}$ of the conservative variables/unit volume, the Eulerian flux $\bar{\bar{H}}_E$, and viscous flux $\bar{\bar{H}}_V$, per unit area per unit time, can be written as

$$\vec{W} = \begin{bmatrix} \rho \\ \rho u \\ \rho v \\ \rho E \end{bmatrix}, \quad \bar{\bar{H}}_E = \begin{bmatrix} \rho\vec{q} \\ \rho u\vec{q} + p\vec{e}_x \\ \rho v\vec{q} + p\vec{e}_y \\ \rho H\vec{q} \end{bmatrix},$$

and

$$\bar{\bar{H}}_V = \begin{bmatrix} 0 \\ \bar{\bar{\tau}} \cdot \vec{e}_x \\ \bar{\bar{\tau}} \cdot \vec{e}_y \\ \bar{\bar{\tau}} \cdot \vec{q} - \vec{Q} \end{bmatrix},$$

where

$$\vec{q} = u\vec{e}_x + v\vec{e}_y,\ \bar{\bar{\tau}} = \begin{bmatrix} \sigma_x & \tau_{xy} \\ \tau_{yx} & \sigma_y \end{bmatrix}$$

The components of the normal stress tensors $\sigma_x,\ \sigma_y$, shear stress tensor $\bar{\bar{\tau}}$, and heat flux Q can be expressed for a Newtonian fluid as

$$\sigma_x = -\lambda\left(\frac{\partial u}{\partial x} + \frac{\partial v}{\partial y}\right) - 2\mu\frac{\partial u}{\partial x},$$

$$\sigma_y = -\lambda\left(\frac{\partial u}{\partial x} + \frac{\partial v}{\partial y}\right) - 2\mu\frac{\partial u}{\partial y},$$

$$\tau_{xy} = \tau_{yx} = -\mu\left(\frac{\partial u}{\partial y} + \frac{\partial v}{\partial x}\right),$$

$$\vec{Q} = k\nabla T = k\left(\frac{\partial T}{\partial x}\vec{e}_x + \frac{\partial T}{\partial y}\vec{e}_y\right),$$

$$E = \frac{RT}{\gamma - 1} + \frac{1}{2}\left(u^2 + v^2\right),$$

and

$$H = E + p/\rho.$$

$\vec{e}_x, \vec{e}_y$ are the unit vectors of the Cartesian coordinate system (x, y), and k is the coefficient of thermal conductivity. For the perfect gas, $p = \rho RT$, where $R = C_\mathrm{p} - C_\mathrm{v}$. The specific heats C_p and C_v at constant pressure and volume respectively are constants. The quantities μ and λ are the first and second coefficients of viscosity respectively. λ can be taken to be equal to $(-2/3\mu)$ (*Stokes hypothesis*). Sutherland's law (see Chapter 7) relates μ and T. The coefficient of thermal conductivity k can be evaluated by using a fixed value of Prandtl number.

According to the nodal point (or vertex-based) finite volume scheme the flow variables are defined at the nodal points (i, j) and the fluxes are to be calculated across a control volume Ω_{ij}, (*abcd* in Fig. 13.7 formed by joining the midpoints of the cell boundaries. Since equation (13.29) is valid for any control volume, it also holds locally for each cell. Hence

$$\frac{\partial}{\partial t}\int\limits_{\Omega_{ij}} \vec{W}\mathrm{d}\Omega + \int\limits_{\partial\Omega_{ij}} \bar{\bar{H}}\cdot\vec{n}\mathrm{d}s = 0, \tag{13.30}$$

with $\bar{\bar{H}} = \bar{\bar{H}}_\mathrm{E} + \bar{\bar{H}}_\mathrm{V}$. Where $\partial\Omega_{ij}$ is the boundary of the cell Ω_{ij} consisting of four sides of the quadrilateral $abcd$, and $\vec{n}$ is the outer normal to the surface element ds. The volume V_{ij} of the cell Ω_{ij} can be computed by averaging the volumes of the four neighbouring cells Ω_1, Ω_2, Ω_3 and Ω_4 sorrounding the point (i, j). The discrete analog of (13.30) is written as

$$V_{ij}\left(\frac{d}{dt}\vec{W}_{ij}\right)+\vec{Q}_{ij}=0, \tag{13.31}$$

where $\vec{Q}_{ij}$ represents the net flux out of a cell and is balanced by the rate of change of W_{ij}. The total flux $\vec{Q}_{ij}$ can be eavaluated as the sum of the Euler flux $\vec{Q}_{Eij}$ and the viscous flux $\vec{Q}_{Vij}$. In the present approach the Euler fluxes are first calculated across the neighbouring cells and then the average is taken to get the flux across Ω_{ij} as

$$\vec{Q}_{\mathrm{E}ij}=\frac{1}{4}\sum_{n=1}^{4}\vec{Q}_{\mathrm{E}n}, \tag{13.32}$$

where the subscript n refers to the face number of the cell, $\vec{Q}_{\mathrm{E}n}$ is the net inviscid flux across the cell. The value of the conserved variables representing a particular cell face can be obtained by averaging its values at the two end points. Net inviscid flux $\vec{Q}_{\mathrm{E}n}$ across the cell is the algebraic sum of the fluxes across the four sides of the cell. Now, say, for the cell Ω_1 the flux Q_1 can be written as

$$Q_1=\overline{\overline{H}}_{\mathrm{E}}\cdot\vec{S}_{1n}+\overline{\overline{H}}_{\mathrm{E}}\cdot\vec{S}_{2n}+\overline{\overline{H}}_{\mathrm{E}}\cdot\vec{S}_{3n}+\overline{\overline{H}}_{\mathrm{E}}\cdot\vec{S}_{4n}, \tag{13.33}$$

where $\vec{S}_{1n}$, $\vec{S}_{2n}$, etc. are the normal vectors to the sides S_1, S_2, etc. of the cell Ω_1 respectively. Let $\vec{e}_x$ and $\vec{e}_y$ be the unit vectors along the coordinate directions x and y respectively. Let us define (SIX, SIY) and (SJX, SJY) as the x and y components of the normal vectors of the face whose normal is along i and j directions respectively. The total flux (Euler and viscous) across the side S_1 can be calculated in the following way.

$$\overline{\overline{H}}\cdot\vec{S}_{1n}=\begin{bmatrix}\rho_{i+1/2,j}(\vec{q}_{i+1/2,j}\cdot\vec{S}_{1n})\\(\rho u)_{i+1/2,j}(\vec{q}_{i+1/2,j}\cdot\vec{S}_{1n})+p_{i+1/2,j}SJX_{i,j}+\overline{\overline{\tau}}\cdot SJX_{i,j}\\(\rho v)_{i+1/2,j}(\vec{q}_{i+1/2,j}\cdot\vec{S}_{1n})+p_{i+1/2,j}SJY_{i,j}+\overline{\overline{\tau}}\cdot SJY_{i,j}\\(\rho H)_{i+1/2,j}(\vec{q}_{i+1/2,j}\cdot\vec{S}_{1n})+\overline{\overline{\tau}}\cdot\vec{q}\cdot\vec{S}_{1n}-\vec{Q}\cdot\vec{S}_{1n}\end{bmatrix} \tag{13.34}$$

The quantities at the midpoints are the simple averages of their values at the nodal points. Now from the definition of $\overline{\overline{\tau}}$ and $\vec{Q}$ the following quantities

$$\begin{aligned}&\overline{\overline{\tau}}\cdot SJX=(\sigma_x\vec{e}_x+\tau_{yx}\vec{e}_y)\cdot(SJX\vec{e}_x+SJY\vec{e}_y)=\sigma_x SJX+\tau_{yx}SJY\\&\overline{\overline{\tau}}\cdot SJY=\tau_{xy}SJX+\sigma_y SJY\\&\overline{\overline{\tau}}\cdot\vec{q}\cdot\vec{S}_{1n}=(u\sigma_x+v\tau_{xy})\cdot SJX+(u\tau_{yx}+v\sigma_y)\cdot SJY\end{aligned}$$

and

$$\vec{Q}\cdot\vec{S}_{1n} = k\frac{\partial T}{\partial x}\cdot SJX + k\frac{\partial T}{\partial y}\cdot SJY$$

have to be evaluated at $(i+1/2, j)$ and these will contribute to the viscous part of the total flux. Euler fluxes can be calculated directly from the flow variables and the geometrical surface vectors stored at each point (i, j), whereas for viscous fluxes the first derivatives of the flow variables are to be used. Similar expressions can be derived for the other three sides. For viscous fluxes we need the derivatives at the midpoint of the grid surfaces. To calculate the derivatives φ_x, φ_y, etc at $(i+1/2, j)$,

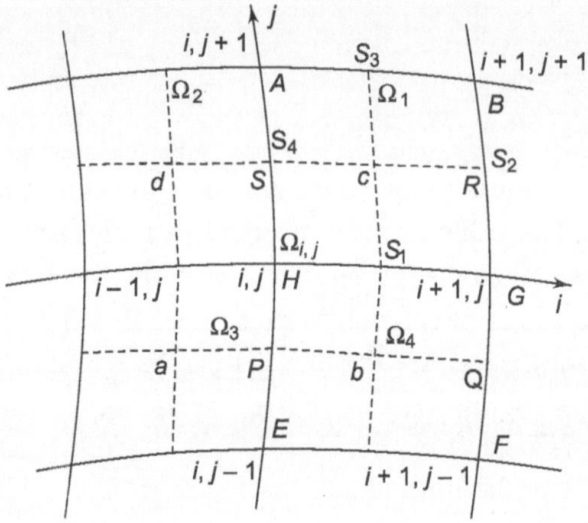

Figure 13.7 Finite volume mesh for nodal point scheme.

where φ represents any of the flow variables, let us consider the volume $PQRS$ as the control volume (Fig. 13.7) with volume V enclosed by the surface S. Then by Green's theorem

$$\iiint_V \frac{\partial\varphi}{\partial x}\mathrm{d}V = \iint_S \varphi\cos\,(\vec{n}, x)\,\mathrm{d}S, \tag{13.35}$$

where $\mathrm{d}V$ and $\mathrm{d}S$ are the elementary volume and surface element respectively. Assuming the derivatives are constant over the volume and φ is constant on the surface, these can be represented by their values at the midpoints. So,

$$\frac{\partial\varphi}{\partial x}\Big|_{i+1/2,j} = \frac{1}{V}\{\varphi_1\cos\,(\vec{n}_1, x)PQ + \varphi_2\cos\,(\vec{n}_2, x)QR$$
$$+\varphi_3\cos\,(\vec{n}_3, x)RS + \varphi_4\cos\,(\vec{n}_4, x)SP\} \tag{13.36}$$

where

$$\varphi_1 = \tfrac{1}{4}\{\varphi_{i,j} + \varphi_{i,j-1} + \varphi_{i+1,j} + \varphi_{i+1,j-1}\}$$
$$\varphi_2 = \varphi_{i+1,j}$$
$$\varphi_3 = \tfrac{1}{4}\{\varphi_{i,j} + \varphi_{i+1,j} + \varphi_{i+1,j+1} + \varphi_{i,j+1}\}$$
$$\varphi_4 = \varphi_{i,j},$$

and the product of the area and the x-projection of the nornal vectors,

$$\cos(\vec{n}, x) \cdot \mathrm{PQ} = 1/2[\mathrm{SJX}_{EF} + \mathrm{SJX}_{GH}]. \tag{13.37}$$

Similar expressions can be written for the other sides.

Finaly, let us denote the present scheme as scheme-A, which can be summarised as:

(i) Calculate the first derivatives of all the flow variables at $(i + 1/2, j)$ using the full control volume $PQRS$ for all (i, j) using the Green's theorem.

(ii) Calculate the stress tensors and the heat flux terms.

(iii) Find the difference of flux quantities across two faces AB and HG to get the viscous flux over the cell $HGBA$, for thin layer type of approximation (neglecting streamwise-like differences) (Chakrabartty, 1989,1990). Swanson and Turkel (1985), neglected this even in the first step.

(iv) Add these with the corresponding Euler flux for the same cell.

(v) From the four neighbouring cells of the point (i, j), take the average to get the flux across Ω_{ij} sorrounding the point (i, j).

In step (iii), if the fluxes across all the four faces are calculated and their algebraic sum is taken to get the required flux without any approximation, it needs very high computing effort in terms of CPU time and memory. This scheme is suitable for computing two and three dimensional flow with thin layer approximation. For complex flow field it introduces error while neglecting streamwise-like differences. Another alternative way of calculating the viscous fluxes, name it as scheme B, also introduced by Chakrabartty (1990a), can be summarised as:

(i) From the premitive values at the points H, G, B and A, calculate the fluxes across the faces HG, GB, BA and AH (Fig. 13.7). Use Green's theorem to get the first derivatives at the centroid of the cell Ω_1. Similarly get the first derivatives at the centroids of the cells Ω_2, Ω_3 and Ω_4 sorrounding the point (i, j).

(ii) Calculate σ_x, τ_{xy}, etc. at the centroids of the four neighbouring cells Ω_1, Ω_2, Ω_3 and Ω_4.

(iii) Knowing σ_x, τ_{xy}, etc. at a, b, c and d, calculate the fluxes across ab, bc, cd and da for full RANS equations or only across ab and cd for thin layer approximation.

(iv) Use Green's theorem to get the second derivatives from the net viscous fluxes across the cell Ω_{ij}.

(v) Add the viscous corrections to the corresponding Euler corrections to update the flow variables.

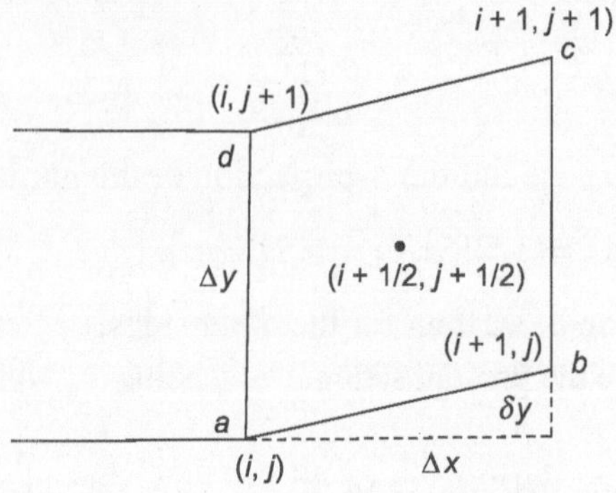

Figure 13.8 Control volume for first derivatives of scheme-B.

Total change of the flow variables over a single time step due to the inviscid terms are included in the distribution formula through a piecewise integration of the time rate derivative of the Euler equations around spatially translated control volumes (defined by *abcd*). One extra step is now added to the numerical approach to include the viscous terms. The changes in time of the flow variables due to the viscous terms have been calculated from an integration of the shear stress and heat conduction terms around the same control volume (i.e. *abcd*) that have been used to calculate the second order time rate changes and added to the inviscid time rate of changes just prior to updating of the flow variables. The domain of influence of the viscous operator remains consistent with that of the inviscid operator.

Ultimately for Navier–Stokes solution we have to compute two differentiations using Green's theorem in finite volume method. These two schemes differ with respect to the point at which we compute the first derivatives and with respect to the different control volumes. Geometrical idea of scheme B is somewhat similar to that used by Davis et al. (1986).

It is to be noted that whether we use the thin layer approximation or not, we have to calculate the first derivatives at the centroids of each cell which consume the major computational effort for viscous correction. So, in step (iii) above, calculations of fluxes across two or four faces do not make much difference in the total computational effort necessary for viscous flow computations. In earlier works Chakrabarty (1989, 1990), Swanson and Turkel (1985) for thin layer approximations, first derivatives are to be calculated at each $(i + 1/2, j)$ points but for full RANS, these are to be calculated at each $(i + 1/2, j)$ and $(i, j + 1/2)$. To make it clear let us assume that in a given mesh system there are M and N number of cells in i and j-directions respectively. Then in scheme B for full RANS equations as well as for thin layer approximation first derivatives are to be calculated on MN number of points. Whereas according to the earlier schemes like scheme A first derivatives are to be calculated on $M \cdot (N + 1)$ number of points for thin layer approximation and on $M \cdot (N + 1) + N \cdot (M + 1)$ number of points for full RANS equations. So, to compute full RANS equations, scheme B is more economical than those used in earlier works. This scheme has been extended to three dimension by Chakrabartty and his colleagues (1996a, 1996b, 1998, 2001, 2002, 2003a, 2003b, 2003c) and applied to various practical problems in aeronautics and space.

The simulation of viscous flows requires the approximation of the diffusion terms involving gradients of the flow variables for the surface integrals (see Eq. (13.29)). One possibility is to choose an approximation which follows the idea of finite differences. Therein a local transformation and chain rule differentiation are applied although the remainder of the numerical solution is based on a finite volume solution. The other approach uses the integral theorem of Green (13.35) which is more consistent with the philosophy of finite volume formulation.

To get the derivatives and their order of accuracy, let us consider the scheme B. Here we need the derivatives of the flow variables at the point $(i + 1/2, j + 1/2)$. Consider the skewed grid as in Fig. 13.8. In this case,

$$SJX_{ab} = \delta y, \quad SJX_{cd} = -\delta y, \quad SIX_{bc} = \Delta y, \quad SIX_{da} = -\Delta y, \quad V = \Delta x \cdot \Delta y.$$

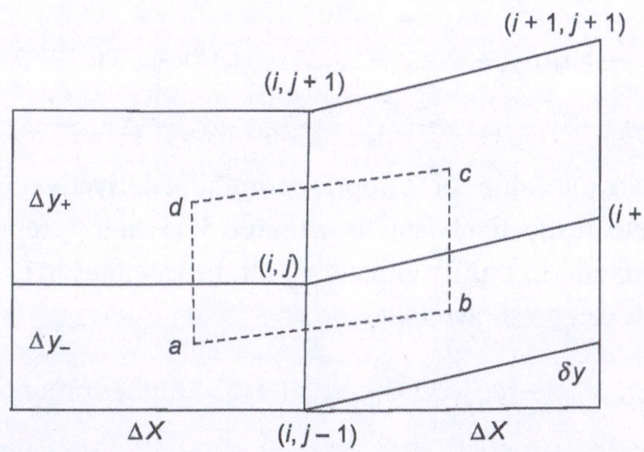

Figure 13.9 Control volume for second derivatives in scheme B.

Using Green's theorem for the control volume $abcd$ the derivatives can be expressed as

$$\frac{\partial \varphi}{\partial x}\Big|_{i+1/2,j+1/2} = \frac{1}{2V}[\varphi_{i,j}(\delta y - \Delta y) + \varphi_{i+1,j}(\Delta y + \delta y) + \varphi_{i+1,j+1}(\Delta y - \delta y) + \varphi_{i,j+1}(-\delta y - \Delta y)], \qquad (13.38)$$

where the representative values of φ on the face have been obtained by averaging those at the two vertices. Now it can be shown by expanding all the φ values in (13.38) by Taylor series about $(i + 1/2, j + 1/2)$ that

$$\frac{\partial \varphi}{\partial x}\Big|_{i+1/2,j+1/2} \approx \varphi_x \big|_{i+1/2,j+1/2} + (\text{terms with } \varphi_{xxx}, \text{ etc.}) \qquad (13.39)$$

which gives second order accurate first derivatives.

The derivatives at the second stage referred in step (iv) of scheme B can be obtained as follows. Here we need the gradients at the point (i, j) with input (values of the first derivatives) stored at $(i + 1/2, j + 1/2)$ points as shown in Fig. 13.9.The surface vectors along the sides of $abcd$ are the averages of those along the four similar surfaces around it. So,

$$SJX_{ab} = \frac{1}{2}\delta y, \quad SJX_{cd} = -\frac{1}{2}\delta y, \quad SIX_{bc} = \frac{1}{2}(\Delta y_+ + \Delta y_-),$$

$$SIX_{da} = -\frac{1}{2}(\Delta y_+ + \Delta y_-), \quad \text{and } V = \frac{1}{2}\Delta x(\Delta y_+ + \Delta y_-).$$

Therefore, the derivative at point (i, j) can be expressed as

$$\frac{\partial \varphi}{\partial x}\Big|_{i,j} = \frac{1}{4V}[\varphi_a(\delta y - \Delta y_+ - \Delta y_-) + \varphi_b(\delta y + \Delta y_+ + \Delta y_-) + \varphi_c(-\delta y + \Delta y_+ + \Delta y_-) + \varphi_d(-\delta y - \Delta y_+ - \Delta y_-)], \qquad (13.40)$$

where φ_a, φ_b, etc. are the values of φ (representing first derivatives of the variables) at the points a, b, etc. respectively and the average of φ_a and φ_b represents its value along the surface ab and so on. By expanding all the φ values in (13.40) by Taylor series about (i, j) it can be shown that

$$\frac{\partial \varphi}{\partial x}\Big|_{i,j} \approx \varphi_x \big|_{i,j} + \frac{1}{8}[\varphi_{xy}(\Delta y_+ - \Delta y_- + \delta y)] + \text{higher order terms}, \qquad (13.41)$$

which shows that it gives at least first order accurate derivatives for sretched and skewed grid. It should be noted here that for smooth Cartesian grid this gives second order accurate derivatives. Similar results can be achieved for the derivatives in other directions also.

13.8 SOLUTION PROCEDURE

Unsteady RANS equations (13.29) are solved numerically in a computational domain. A semi-discretisation is used which completely separates the discretisation of space and time derivatives. The resulting system of ordinary differential equations in time is solved using explicit five stage Runge-Kutta time stepping scheme. Although natural damping terms (dissipation) are present in the equations, artificial dissipation terms are properly blended not to overwhelm the natural viscosity but just to make the algorithm numerically stable. To accelerate the rate of convergence local time stepping, enthalpy damping and residual smoothing can be applied. For viscous flows, total enthalpy may be taken as constant if the surfaces are adiabatic and both the laminar and turbulent Prandtl numbers are unity. Details of these steps have been discussed in Chapter 10.

For viscous flow computation the time step limit $\Delta t_{i,j}$ at the point $(i.j)$, can be derived by satisfying the Courant, Friedrichs and Lewy (CFL) stability criterion for advection-diffusion equation as follows (Payret and Taylor, 1983):

Let us consider the two dimensional advection diffusion equation

$$\frac{\partial f}{\partial t} + A\frac{\partial f}{\partial x} + B\frac{\partial f}{\partial y} - \nu\nabla^2 f = 0. \tag{13.42}$$

Where $f(x, y)$ is any unknown function, A, B and $\nu(> 0)$ are assumed to be constants. Direct explicit scheme to solve (13.42) using finite difference discretisation is

$$\frac{1}{\Delta t}(f_{i,j}^{n+1} - f_{i,j}^{n}) + A\Delta_x^0 f_{i,j}^n + B\Delta_y^0 f_{i,j}^n - \nu(\Delta_{xx} + \Delta_{yy})f_{i,j}^n = 0. \tag{13.43}$$

Where $f_{i,j}^n \equiv f(i\Delta x,\ j\Delta y,\ n\Delta t)$ with i, j, n are integers and

$$\Delta_x^0 f_{i,j} = \frac{1}{2\Delta x}(f_{i+1,j} - f_{i-1,j})$$

$$\Delta_y^0 f_{i,j} = \frac{1}{2\Delta y}(f_{i,j+1} - f_{i,j-1})$$

$$\Delta_{xx} f_{i,j} = \frac{1}{2\Delta x^2}(f_{i+1,j} - 2f_{i,j} + f_{i-1,j})$$

$$\Delta_{yy} f_{i,j} = \frac{1}{2\Delta y^2}(f_{i,j+1} - 2f_{i,j} + f_{i,j+1}).$$

Scheme (13.43) which is first order accurate in time and second order accurate in space can be written as

$$f_{i,j}^{n+1} = f_{i,j}^{n} - \frac{A\Delta t}{2\Delta x}(f_{i+1,j}^{n} - f_{i-1,j}^{n}) - \frac{B\Delta t}{2\Delta y}(f_{i,j+1}^{n} - f_{i,j-1}^{n})$$
$$+\frac{\nu\Delta t}{\Delta x^2}(f_{i+1,j}^{n} - 2f_{i,j}^{n} + f_{i-1,j}^{n}) + \frac{\nu\Delta t}{\Delta y^2}(f_{i,j+1}^{n} - 2f_{i,j}^{n} + f_{i,j-1}^{n}). \quad (13.44)$$

For stability analysis in Von-Neumann sense, let

$$f_{j,k}^{n} = \sum_{p=-\infty}^{\infty}\sum_{q=-\infty}^{\infty} \rho^{n}(p,q)e^{ipj\Delta x}e^{iqk\Delta y},$$

and

$$\alpha = \left(1 - \frac{2\nu\Delta t}{\Delta x^2} - \frac{2\nu\Delta t}{\Delta y^2}\right),$$
$$\beta = \left(-\frac{A\Delta t}{2\Delta x} + \frac{\nu\Delta t}{\Delta x^2}\right), \quad \gamma = \left(\frac{A\Delta t}{2\Delta x} + \frac{\nu\Delta t}{\Delta x^2}\right),$$
$$\delta = \left(-\frac{B\Delta t}{2\Delta y} + \frac{\nu\Delta t}{\Delta y^2}\right), \quad \phi = \left(\frac{B\Delta t}{2\Delta y} + \frac{\nu\Delta t}{\Delta y^2}\right).$$

Then we get

$$\rho^{n+1} = \alpha\rho^{n} + \beta\rho^{n}e^{ip\Delta x} + \gamma\rho^{n}e^{-ip\Delta x} + \delta\rho^{n}e^{iq\Delta y} + \phi\rho^{n}e^{iq\Delta y}$$

or

$$\frac{\rho^{n+1}}{\rho^{n}} = g = \alpha + \beta(\cos\xi + i\sin\xi) + \gamma(\cos\xi - i\sin\xi)$$
$$+\,\delta(\cos\eta + i\sin\eta) + \phi(\cos\eta - i\sin\eta) \quad (13.45)$$

where $\xi = p\Delta x, \quad \eta = q\Delta y$. It can be verified that $\alpha + \beta + \gamma + \delta + \phi = 1$ or $\alpha = 1 - (\beta + \gamma) - (\delta + \phi)$.

So, (13.45) becomes

$$g = 1 - (\beta+\gamma) + (\beta+\gamma)\cos\xi + i(\beta-\gamma)\sin\xi$$
$$+[-(\delta+\phi) + (\delta+\phi)\cos\eta + i(\delta-\phi)\sin\eta].$$

or

$$|\, g\,| \leq |\, 1 - (\beta+\gamma) + (\beta+\gamma)\cos\xi + i(\beta-\gamma)\sin\xi\,|$$
$$+\,|\, -(\delta+\phi) + (\delta+\phi)\cos\eta + i(\delta-\phi)\sin\eta\,|$$
$$\leq \sqrt{BX^2+CX+1} + |\, -(\delta+\phi)+(\delta+\phi)\cos\eta + i(\delta-\phi)\sin\eta\,| \quad (13.46)$$

where

$$X = 1/2(1 - \cos\xi), \quad B = 16\beta\gamma,$$
$$C = 4[(\beta - \gamma)^2 - (\beta + \gamma)].$$

The square of the second part of right hand side of (13.46) is

$$4\delta\phi(1 - \cos\eta)^2 + 2(\delta - \phi)^2(1 - \cos\eta) = DY^2 + EY.$$

Where

$$Y = 1/2(1 - \cos\eta), \quad D = 16\delta\phi, \quad E = 4(\delta - \phi)^2.$$

So,

$$\begin{aligned} |\, g\,|^2 &\le BX^2 + CX + 1 + DY^2 + EY + 2\sqrt{(BX^2 + CX + 1)(DY^2 + EY)} \\ &\le 1 - [X(-BX - C) + Y(-DY - E) - 2\sqrt{(BX^2 + CX + 1)(DY^2 + EY)}]. \end{aligned}$$

For $0 \le X \le 1$ and $0 \le Y \le 1$, the condition $|\, g^2 \,| \le 1$ becomes

$$-B - C - D - E - 2\sqrt{(B + C + 1)(D + E)} \ge 0. \tag{13.47}$$

and

$$C + E \le 0. \tag{13.48}$$

For this case, condition (13.48) becomes

$$4[(\beta - \gamma)^2 - (\beta + \gamma)] + 4(\delta - \gamma)^2 \le 0$$

or

$$\frac{A^2\Delta t^2}{\Delta x^2} - \frac{2\nu\Delta t}{\Delta x^2} + \frac{B^2\Delta t^2}{\Delta y^2} \le 0$$

or

$$\left[\frac{A^2}{\Delta x^2} + \frac{B^2}{\Delta y^2}\right]\Delta t \le \frac{2\nu}{\Delta x^2}. \tag{13.49}$$

The condition (13.47) gives

$$\begin{aligned} &-16\beta\gamma - 4(\beta - \gamma)^2 + 4(\beta + \gamma) - 16\delta\phi - 4(\delta - \phi)^2 \\ &-2\sqrt{16\beta\gamma + 4(\beta - \gamma)^2 - 4(\beta + \gamma) + 1}\sqrt{16\delta\phi + 4(\delta - \phi)^2} \ge 0. \end{aligned}$$

Which becomes after simplification

$$\Delta t \le \frac{\Delta x^2 \Delta y^2}{(\Delta x^2 + \Delta y^2)2\nu}. \tag{13.50}$$

For $\Delta x = \Delta y$, the conditions (13.49) and (13.50) become

$$(A^2 + B^2)\Delta t \le 2\nu \tag{13.51}$$

and

$$\frac{\nu \Delta t}{\Delta x^2} \le \frac{1}{4}. \tag{13.52}$$

Similarly for the upwind scheme (Peyret and Taylor, 1983), CFL condition can be written for a particular cell of sides Δx and Δy as

$$\Delta t \;\le\; \lambda \frac{\Delta x^2 \Delta y^2}{2\nu(\Delta x^2 + \Delta y^2) + \Delta x \Delta y(\mid A \mid \Delta y + \mid B \mid \Delta x)} \tag{13.53}$$

Here λ is the Courant number of the scheme and $\mid A \mid$, $\mid B \mid$ are the spectral radii of the Jacobians and ν is the dissipation coefficient. Replacing $\mid A \mid$ and $\mid B \mid$ by the corresponding eigenvalues of the Jacobians of the Euler equations the time step limit takes the form (Chakrabartty, 1991; Chakrabartty and Dhanalakshmi, 1996)

$$\Delta t \;=\; \lambda \frac{\Delta t_1 \Delta t_2}{\Delta t_1 + \Delta t_2} \tag{13.54}$$

where Δt_1 and Δt_2 are the time step limits for convective and diffusive terms respectively. In particular

$$\Delta t_1 \;=\; \frac{\Delta t_i \Delta t_j}{\Delta t_i + \Delta t_j},$$

$$\Delta t_i \;=\; \frac{V}{\mid u\, SIX + v\, SIY \mid + c\sqrt{SIX^2 + SIY^2}},$$

$$\Delta t_j \;=\; \frac{V}{\mid u\, SJX + v\, SJY \mid + c\sqrt{SJX^2 + SJY^2}},$$

$$\Delta t_2 \;=\; \frac{V^2}{2\nu(SIX^2 + SIY^2 + SJX^2 + SJY^2)}$$

where V is the volume of the cell and SIX, SJX and SIY, SJY are the x and y components of the normal vector multiplied by the surface area of the surface whose normal is along i and j directions respectively. ν represents the coefficient of kinematic viscosity, u and v are the velocity components in x and y directions respectively and c is the velocity of sound.

13.9 COMPUTATIONAL RESULTS

The nodal point finite volume methods described in the preceding sections have been extended to three dimensions and applied to many practical problems by Chakrabartty and his group. Some of the simple two dimensional problems will be discussed here like; (i) laminar flow over a flat plate, (ii) laminar and turbulent viscous flow past aerofoils, (iii) internal flow through nozzle, (iv) turbulent flow through cascades, etc. The reference lengths for the definition of free stream Reynolds number for these cases are the length of the flat plate, the chord of the aerfoil, and half of the height of the nozzle throat respectively. In all the computations, the viscous fluxes have been computed only at the first stage of the Runge-Kutta scheme and frozen for the remaining stages. About 30% of the computer time has been saved in this way without affecting the accuracy of the final solution. Many more two and three dimensional cases of practical interest have been computed by Chakrabartty and his group and are available in Chakrabartty et al. (1999), Chakrabartty et al. (2002, 2003a, 2003b, 2003c).

It is important to note here that the pressure coefficient, lift coefficient etc. can be predicted well even by inviscid flow computation using Euler equations. Presence of vorticity in the flow can be taken care of by the Euler equations, but there is no nechanism in the Euler equations to create vorticity in the absence of shock waves and geometrical singularities such as corners and sharp leading and trailing edges. It is the shear flow which is responsible for the formation of boundary layer, wake and the complex phenomena of flow separations. To simulate these accurately, we need to compute viscous flows using Navier–Stokes equations. Therefore, the validation of the viscous flow computations should be made against experiments or against some exact solutions for simple problems (if available) by comparing local skin friction coefficient, drag and moment coefficients, etc. which are actually dominated by viscous flows.

13.9.1 Flow Over a Flat Plate

For the flat plate case, the computation started from a free stream boundary about twice the length ahead of the leading edge and the upper boundary was fixed at a distance of one (i.e. the length of the plate). On these two boundaries the free stream conditions and the far field inflow/outflow conditions were applied. The positions of these two outer boundaries were fixed where no further significant changes in the solution occur. Symmetry, wall and exit boundary conditions were applied on the boundary between the free stream and the leading edge of the plate, on the plate and on the downstream boundary respectively. In the flow direction the grid spacing has been taken to be uniform, whereas in the normal direction a geometrical stretching has been applied (see Chapter 8). The minimum cell height Δy_{min} near the plate has been adjusted such that 15–20 cells lie inside the boundary layer. The results presented here are for the grid size (121 × 51), where 80 cells are in the free stream zone (from free stream to the leading edge of the plate), 40 cells are on the plate and 50 cells are in normal direction. Coordinates are chosen such that x-axis lies along the plate and y-axis is along the normal direction with origin at the leading edge of the plate. In Fig. 13.10, the skin friction coefficients C_{floc} $(= \tau_{\text{w}}/(\frac{1}{2}\rho U_\infty^2))$, where τ_{w} is the shear stress on the wall, ρ the density and

U_∞ is the free stream velocity) have been plotted against local Reynolds number R_{ex} $(= \frac{\rho U_\infty x}{\mu})$. The results compared very well with classical Blasius solution (Jones & Watson, 1963) throughout the plate. Computed value of total skin-friction drag $C_{DF} = 0.0179$ compares well with that (0.0188) obtained by Blasius. Velocity

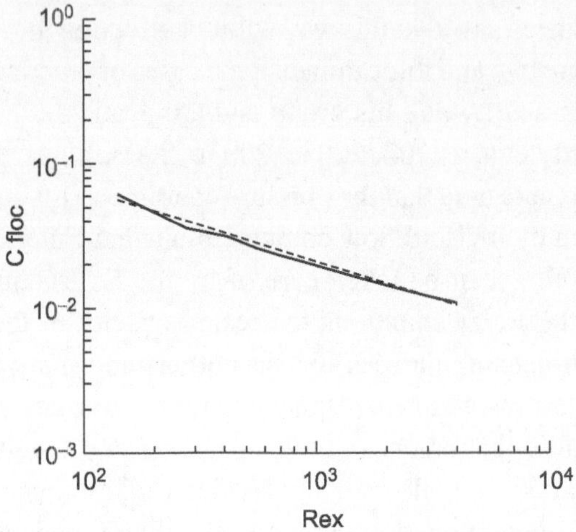

Figure 13.10 Comparison of local skin-friction distribution on a flat plate, at $M_\infty =$ 0.75; ____ Present, - - - Blasius (Jones and Watson, 1963).

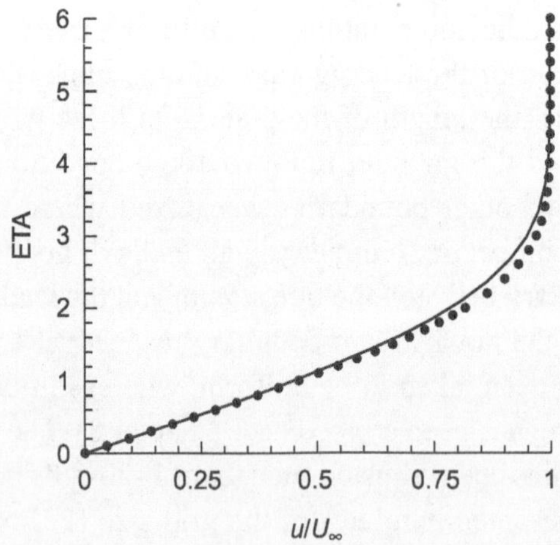

Figure 13.11 Comparison of velocity profile on a flat plate, at $M_\infty = 0.75$; ____ Present, • • • Blasius (Jones and Watson, 1963).

profile u/U_∞ versus ETA, ($\eta = \sqrt{0.5 \cdot R_{ex} \cdot y/x}$) has been compared with that obtained by Blasius at a representative station ($x = 0.67$) has been compared in Fig. 13.11. It shows very good comparison with that obtained by Blasius. Temperature profile ($ETA \sim T/T_\infty$) and the convergence history (ln $\frac{d\rho}{dt} \sim$ no. of iterations) are shown in Figs. 13.12 and Fig. 13.13 respectively.

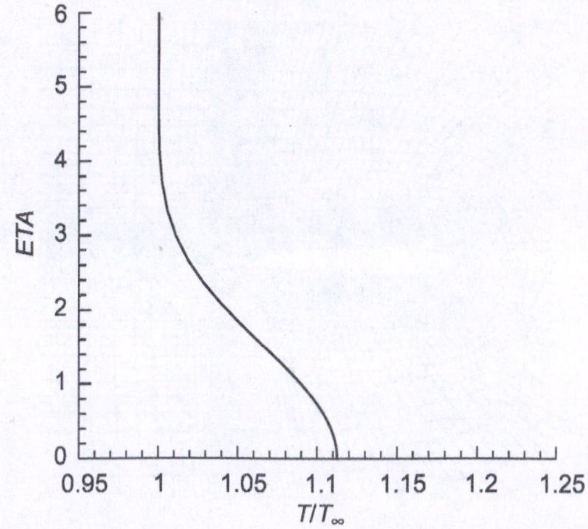

Figure 13.12 Temperature profile on a flat plate, at $M_\infty = 0.75$.

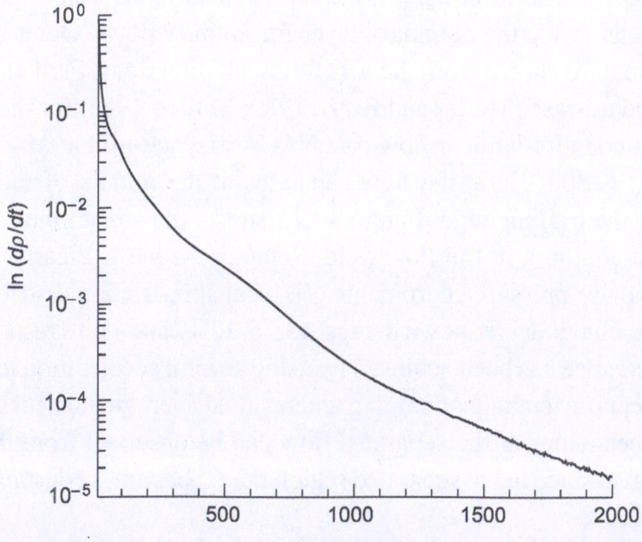

Figure 13.13 Convergence behaviour for computation of flow over a flat plate at $M_\infty = 0.75$.

13.9.2 Viscous Flow Past NACA0012 Aerofoil

A typical C-type (165 × 61) grid generated by algebraic grid generation procedure discussed in Chapter 8 for computing flow past NACA0012 aeroil has been shown in Fig. 13.14. A C-type grid can simulate the wake better than an O-type grid and so, for viscous flow computation a C-type grid is preferable. The minimum height

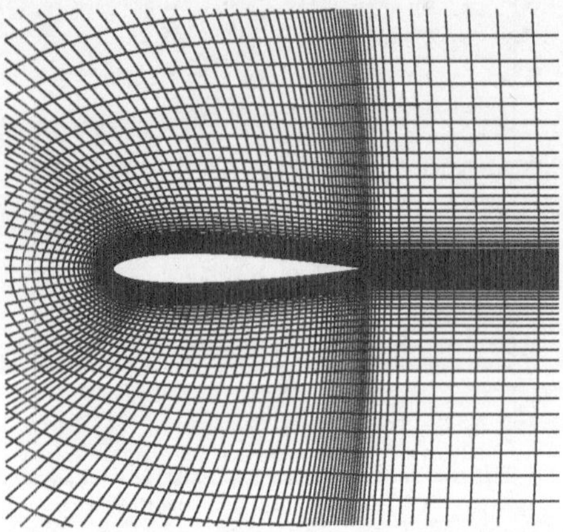

Figure 13.14 A closer view of the C-type grid (165 × 62) for NACA0012 aerofoil.

of the cell near the aerofoil $\eta_{\min}$ has been taken as 0.001 which gives sufficient number of cells inside the boundary layer for laminar flows. Outer boundary was located at about 10 chords from the aerofoil in all directions. First case considered here is a standard case (Müller and Rizzi, 1980; Müller, 1986) for the validation of the computer code for laminar flow past NACA0012 aerofoil at $M_\infty = 0.80$, $\alpha = 10°$ and $R_{e\infty} = 500$. Here, the flow separates at the middle of the aerofoil and reattaches at the trailing edge. Figure 13.15 shows the streamlines superimposed on the Mach-contours in the flow field. Separated zone is clearly visible in the figure. Chordwise pressure coefficients C_p, local skin-friction distribution and the convergence behaviour are shown in Figs. 13.16, 13.17 and 13.18 respectively. Good rate of convergence has been achieved by using several acceleration techniques like local time stepping, enthalpy damping and residual averaging discussed in Chapter 10. Typical behaviour of the separated flow can be observed from the C_p and the C_{floc} curves also since in the separated region the C_p becomes constant and the C_{floc} becomes negative.

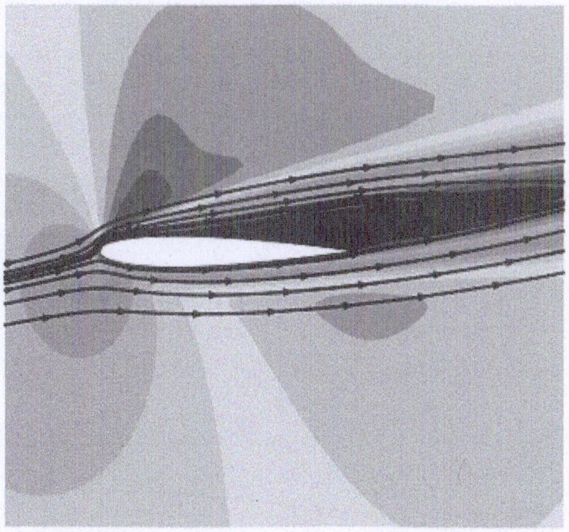

Figure 13.15 Streamlines superimposed on the Mach contours showing the separated region on the aerofoil at $M_\infty = 0.80$, $\alpha = 10°$, $Re = 500$. (The four-colour image of this figure is found at the end of this chapter.)

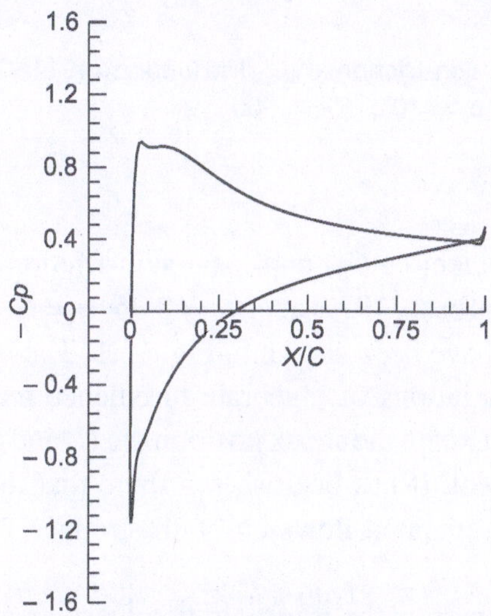

Figure 13.16 Distribution of pressure coefficient C_p on NACA0012 aerofoil at $M_\infty = 0.80$, $\alpha = 10°$, $Re = 500$.

13.9.3 Viscous Transonic Flow Past Other Aerofoils

Turbulent flow past RAE2822 aerofoil is a typical example having a strong shock with shock induced separation at the foot of the shock. A comparison of

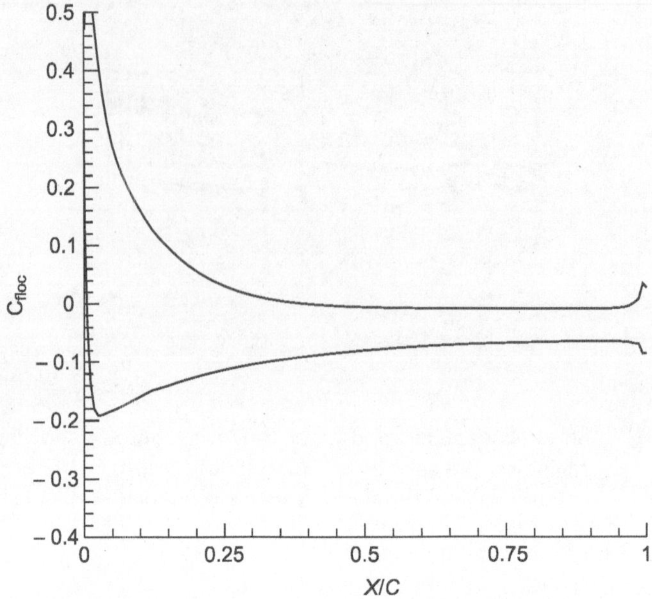

Figure 13.17 Local skin-friction C_{floc} distribution on NACA0012 aerofoil at $M_\infty = 0.80$, $\alpha = 10°$, $Re = 500$

surface pressure coefficient ($-C_p$) and local skin-friction coefficient (C_{floc}) with experimental data (Cook et al., 1979) at $M_\infty = 0.75$, $\alpha = 2.81°$, $Re_\infty = 6.2 \times 10^6$, Grid size $= 257 \times 61$ have been shown in Fig. 13.1. The results shown here were obtained using four variations of algebraic turbulence model (Chakrabartty and Dhanalakshmi, 1995). Except the shock position the C_p and C_f distribution compare well with the experiment. It has been shown there that the algebraic models are capable of predicting separated flows but with stronger shocks than observed in experiments.

Another interesting case is the transonic flow past supercritical aerofoils. This class of aerofoils was designed using transonic full potential equations (Bauer et al., 1975) to get a shock-free supercritical flow at a given Mach number M_∞ and angle of attack α (design condition). At the off-design conditions usually double shocks appear and it is a challenging task to predict the correct off-design behaviour. Inviscid computations using either full-potential or Euler equations can predict shock-free flow at design conditions (Jameson, 1987) and flow with double shocks at off-design conditions (Chakrabartty, 1992).

Viscous flow computation by solving RANS equations with Baldwin-Lomax

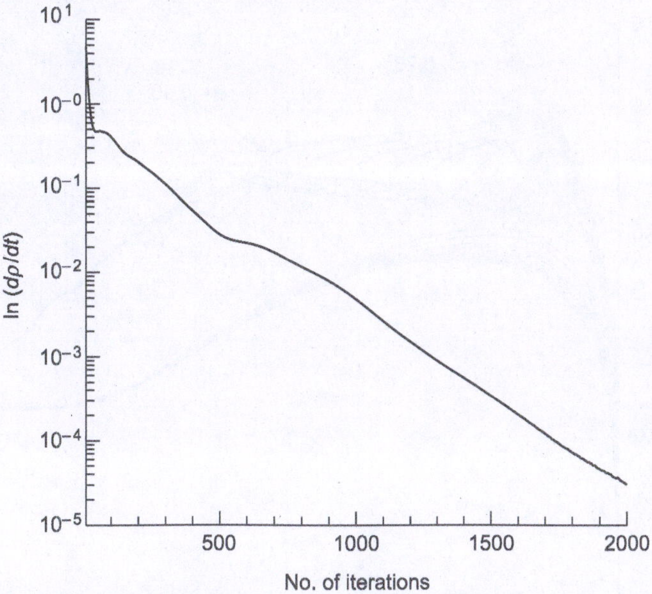

Figure 13.18 Convergence history for NACA0012 aerofoil at $M_\infty = 0.80$, $\alpha = 10°$, $Re = 500$

turbulence model and nodal point finite volume discretisation was performed by Chakrabartty and Dhanalakshmi (1996) to study the behaviour of the supercritical Korn aerofoil at design and off-design conditions. Figure 13.19 shows the C_p distributions for $M_\infty = 0.75$, $R_{e\infty} = 5.42 \times 10^6$, at angles of attack ranging from $\alpha = -2°$ to $0.90°$. The first shock appearing near the leading edge moves downstream whereas the second shock remains stationary but its strength decreases with increasing α. C_p distribution shows that the second shock completely vanishes giving a smooth transition from supersonic to subsonic flows at $\alpha \sim 0.35°$. This analysis is considered to be a good test for the accuracy of the algorithm used to solve RANS equations.

13.9.4 Internal Flow Through Nozzle

The cases considered above are for the external flow past obstacles. The same solver code can be used for internal flows also by properly defining the computational domain and the boundary conditions. For example, to compute the flow through a converging diverging nozzle, the computational domain is completely defined by the geometry of the nozzle with two open faces called inlet and outlet/exit. This is a relatively simple problem to validate the code for internal flow before we go to more difficult flows through cascades.

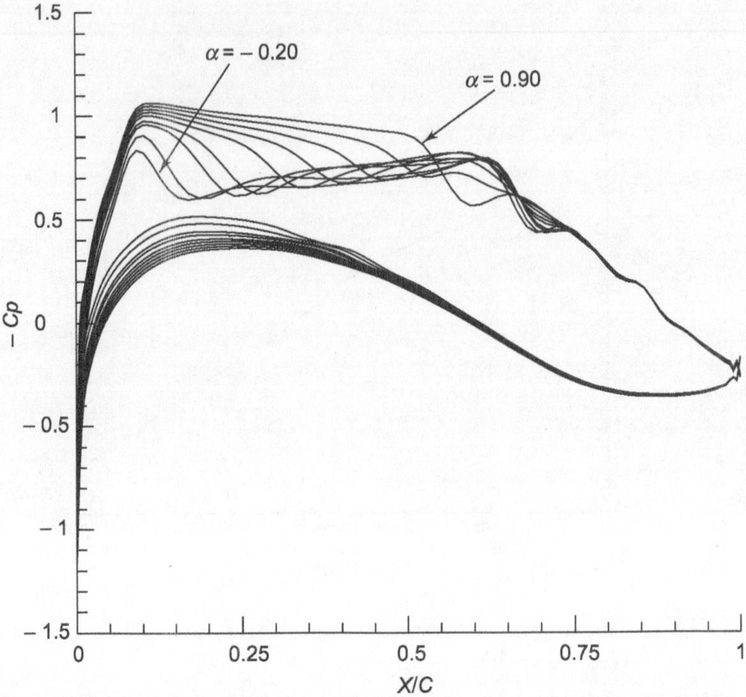

Figure 13.19 Effect of angle of attack on surface pressure coefficient on KORN aerofoil, at $M_\infty = 0.75$, $R_{e\infty} = 5.42 \times 10^6$, $\alpha = -0.20, 0.00, 0.25, 0.35, 0.50, 0.60, 0.70, 0.80, 0.90$ degrees

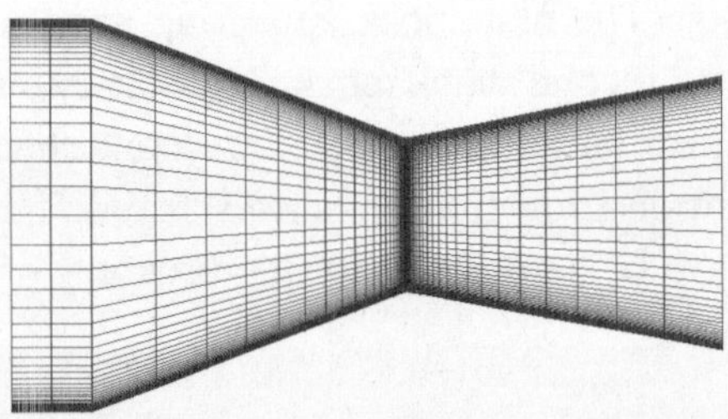

Figure 13.20 Grid (53 × 74) inside a converging diverging nozzle.

The convergent divergent nozzle considered here has an area ratio of 2.567 : 1 : 1.795. The considered flow has an inlet Mach number $M_1 = 0.2$, inlet

Reynolds number $Re = 7.5 \times 10^6$ and the ratio of the inlet stagnation pressure to exit static pressure is 5.84. Figure 13.20 shows the 53×74 H-type grid used. Reference length is the half of the height of the nozzle throat and is taken as one. Figure 13.21 shows a comparison of pressure distribution (ratio of the static pressure P and the inlet total pressure P_{T1} along the length X/X_{throat}) obtained from computation with experimental data (Mason et al., 1980). This grid has been generated by an algebraic method, with a suitable clustering of grid points near the throat and near the wall. The results show a good comparison with experiment.

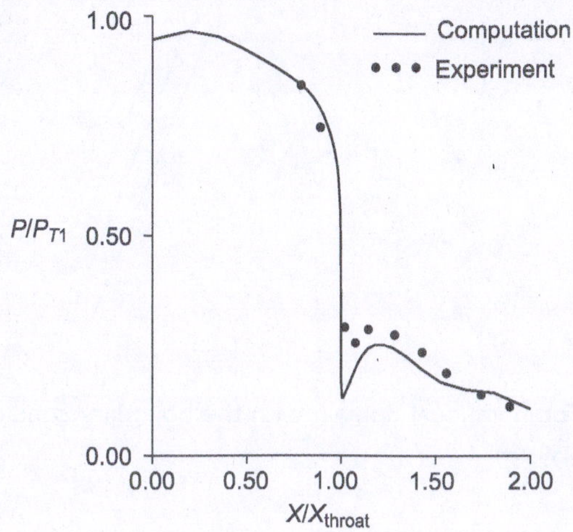

Figure 13.21 Pressure distribution on the wall of a converging diverging nozzle.

13.9.5 Turbulent Flow Through Cascades

The problem considered here is viscous turbulent flow through VKI-LS-59 turbine cascade for which experimental results are available (Sieverding, 1973). This is a two dimensional simulation of flow through turbomachinery blades. Computational domain has been shown in Fig. 13.22 along with the boundary conditions to be satisfied at different boundaries. For a subsonic inlet, the total enthalpy/total temperature, total pressure and flow angle are specified and the velocity is taken from the interior. All the variables are fixed from the prescribed inlet conditions for a supersonic inlet. For the exit, the static pressure corresponding to the desired isentropic Mach number is prescribed and the other variables are extrapolated from the interior. At the periodic boundaries the flow variables at corresponding locations on the upper and lower boundaries are set equal and the image cells are treated as shown in Fig. 13.6. A 395×41 C-type grid generated by using a commercially available code GRIDGEN (Pointwise Inc., USA) is shown in Fig. 13.23 (for clarity, only every third grid point is shown).

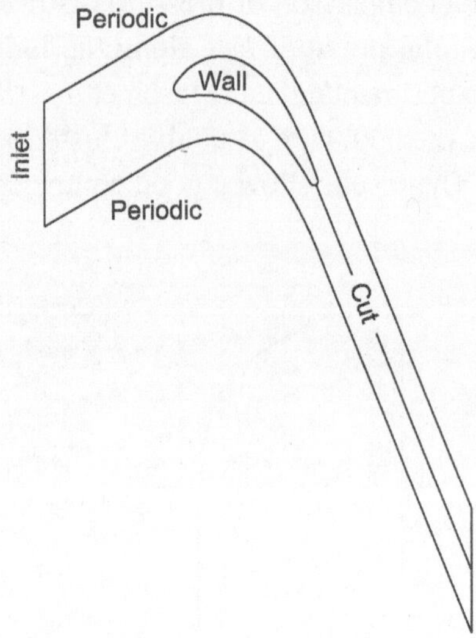

Figure 13.22 Computational domain with the boundary conditions for flow through cascade.

This grid generation code GRIDGEN generates the initial grid using transfinite interpolation and then smoothens it using an elliptic solver with control functions which cluster the grid lines near the surface of the aerofoil with the first grid line at a specified distance, and also enforce orthogonality of the grid lines close to the surface. Out of 395 points in the wraparound direction, 325 grid poins are distributed on the aerofoil. On the outer periodic boundaries the grids are also periodic so that no interpolation is necessary while satisfying the periodicity condition. The upstream distance is 0.75 and the downstream distance is 0.90 taking the chord length is unity. Non periodic grids are used by many authors (Arnone and swanson, 1993; Srinivas, 1999) but these require interpolation, which can introduce inaccuracies. To avoid these inaccuracies periodic grids have been used here, which results in a certain amount of grid skewness. In general, grid skewness reduces the accuracy of the numerical scheme. However, vertex based schemes have been shown earlier to be at least first order accurate on stretched and skewed grids and are superior to cell centred schemes. So, reasonably accurate results are obtained by using the periodic grid (Chakrabartty et al., 2001).

Two examples with exit static pressures corresponding to subsonic ($M2_{is}$ = 0.81) and supersonic ($M2_{is}$ = 1.12) isentropic Mach numbers are considered here.

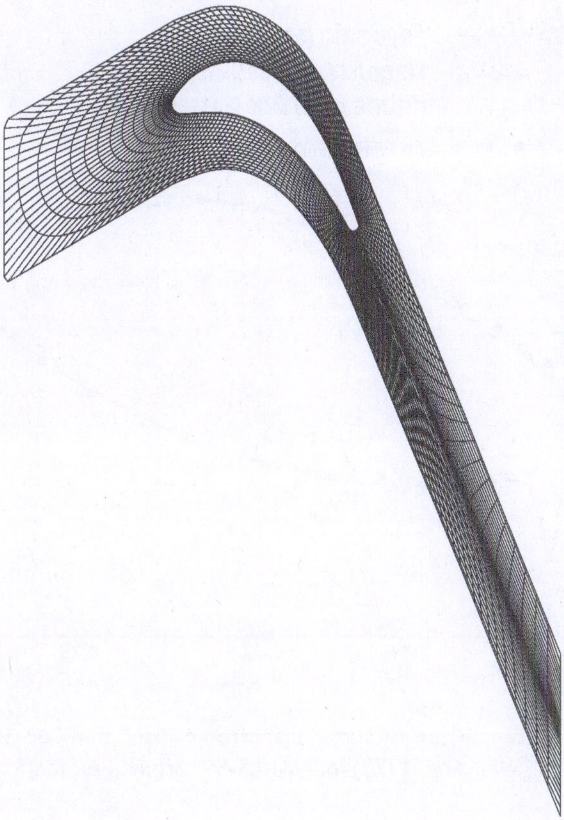

Figure 13.23 C-type grid for flow through cascade.

For both the cases inlet flow angle is 30° and the inlet total pressure and total temperature corresponding to isentropic Mach number of 0.282 is prescribed. Inlet Reynolds numbers (R_{e1}) are 0.80×10^6 and 0.65×10^6 for subsonic and supersonic exit cases respectively. Minimum height of the cell near the wall has been taken as 7.5×10^{-4}, which gives an average y^+ of about 6 at the first node point adjacent to the wall. A grid convergence study has been carried out by computing the flow using a coarser grid (195 × 25) and a finer grid (595 × 51). For the fine grid, minimum height of the cell near the wall has been taken as 5×10^{-4}, which gives an average y^+ of about 4 at the first node point adjacent to the wall. The surface isentropic Mach number M_{is} distribution computed on all the three grids is compared with experimental data in Fig. 13.24. The results show that the solutions on the medium and fine grids are very close to each other and compare well with experiments.

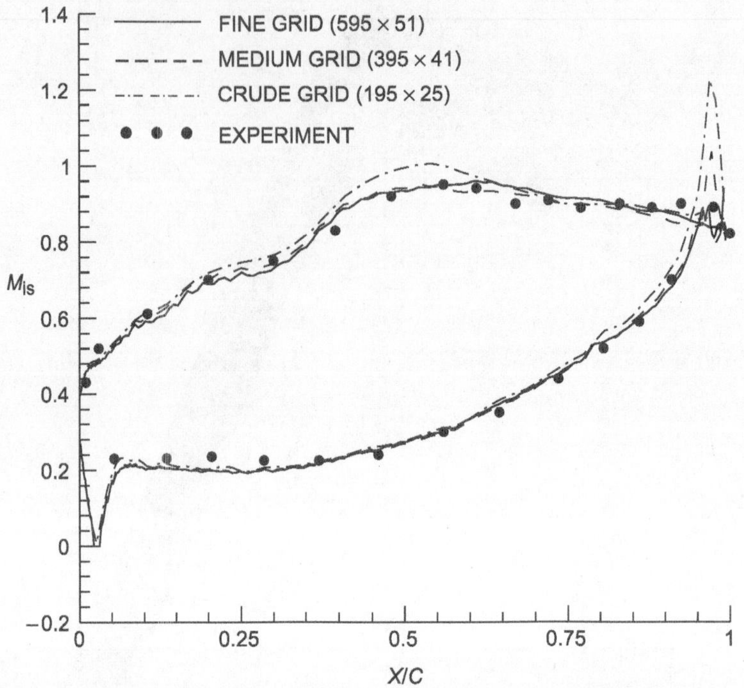

Figure 13.24 Comparison of surface isentropic Mach number M_{is} with experiment (Sieverding, 1973) for VKI-LS-59 turbine cascade for subsonic exit.

For the supersonic exit case, the computed surface isentropic Mach number distribution using medium and fine grid is compared with experiment and shown in Fig. 13.25. The results show that the two solutions are again very close to each other and compare very well with experimental data. The shock position and the separation after the shock are well predicted by computed results. The iso-Mach contours for the same case is shown in Fig. 13.26. Here also the shock and separated region near the upper side of the trailing edge are clearly visible.

13.9.6 Viscous Flow Past Aerofoil-Flap Configuration

Design-analysis studies of an aerofoil-flap configuration of a wing is an interesting problem. The study of two dimensional viscous flow past multi-element aerofoils or aerofoils with deflected flaps, prediction of maximum lift coefficient $C_{L\,max}$ attainable by varying angle of attack α and the flap deflection angle δ and the post stall behaviour of the aerofoils need a very accurate RANS solver with reasonably good and robust turbulent model. The computer code developed by using scheme-B with Baldwin–Lomax turbulence model has been used along with a suitably

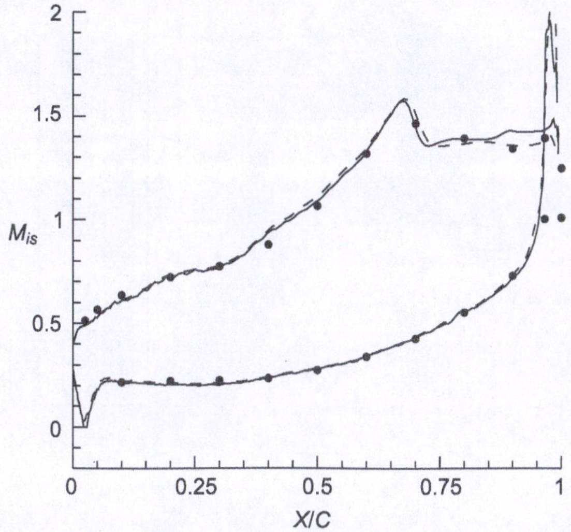

Figure 13.25 Comparison of surface isentropic Mach number M_{is} with experiment for VKI-LS-59 turbine cascade for supersonic exit, ____ Fine grid, _ _ _ Medium grid, • • • Expt.

clustered two block grid by Chakrabartty and his group (2002, 2003b) for this purpose. The code has been validated first by comparing the computed results against the experimental results for GA(W)-2 aerofoil with deflected flap and then used to design an improved aerofoil-flap configuration over an existing one to get the better performance. This has been achieved by incorporating small changes in the cove region, the flap geometry and the position of the point of rotation for the flap. After each change, the analysis was performed and the changes in flow pattern obtained were observed carefully to guide for further incremental changes. This iteration process was continued until the desirable flow pattern was achieved for various positions of the defected flap.

A typical comparison of the flow pattern is shown in Fig. 13.27 for $M_\infty = 0.3$, $R_{e\infty} = 2 \times 10^6$, $\delta = 20°$, and $\alpha = 10°$. It shows significant improvement in the flow pattern on the modified geometry of aerofoil flap configuration in comparison with the existing one. The existing configuration exhibits a large portion of separated region over the flap whereas the modified one shows a very smooth behaviour having fully attached flow throughout with a small cove vortex in the gap close to the trailing edge of the main aerofoil.

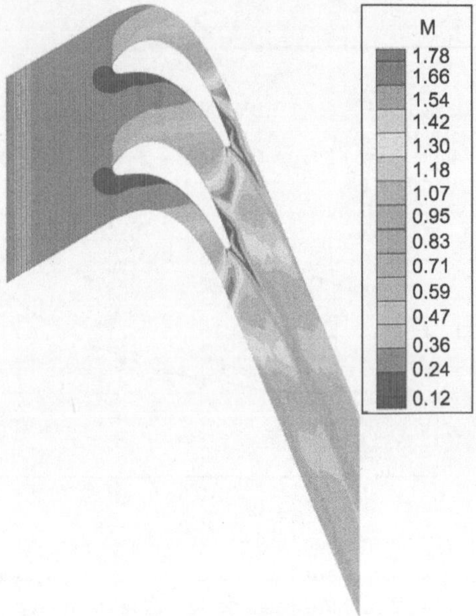

Figure 13.26 Mach contours for VKI-LS-59 turbine cascade with supersonic exit. (The four-colour image of this figure is found at the end of this chapter.)

13.10 SUMMARY

CFD has reached maturity in its various aspects. These aspects include: numerical algorithms, grid generation, advanced code development based on Euler and Navier–Stokes equations and post processing of the results for effective flow visualization. There is also a vast experience in the application of CFD for a wide range of realistic problems in design and analysis of aerospace vehicles.

After introducing the dynamic similarity parameters and the Reynolds Averaged Navier–Stokes (RANS) equations, simple algebraic turbulence modelling, different boundary conditions and their implementations have been discussed. Finite difference and cell centred finite volume discretization have been introduced in brief. Two schemes based on cell vertex or nodal point finite volume space discretization have been introduced in great detail. The scheme-B which is a novel vertex-based finite volume space discretization scheme, takes almost the same numerical effort to solve the full RANS equations as for the thin layer navier stokes (TLNS) equations and is idealy suited for computing flow past complex geometries. The codes JUEL2D and JUMBO2D developed for two dimensional flow past complex configurations solve two dimensional Euler and RANS equations

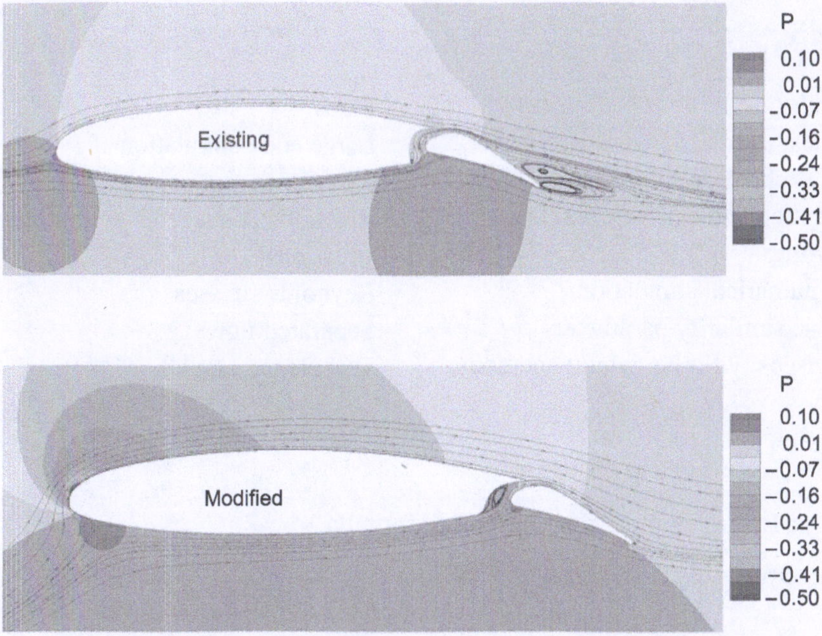

Figure 13.27 Streamlines and pressure contours of the existing and modified flap configurations, at $M_\infty = 0.3$, $R_{e\infty} = 2 \times 10^6$, $\delta = 20°$, $\alpha = 10°$. (The four-colour images of these figures are found at the end of this chapter.)

respectively using body-fitted, multi-block, structured grids. These codes along with their three-dimensional counterparts JUEL3D and JUMBO3D are being used routinely for accurate two and three dimensional inviscid and viscous flow computations with less computational cost. The algorithm uses artificial dissipation for stability, explicit fourth order Runge-Kutta time integration and algebraic turbulence model. Grid sequencing, implicit residual smoothing, enthalpy damping and local time stepping are the tools built into these codes to accelerate convergence. These codes are capable of handling single or multiple bodies with complex fluid flow structure.

These codes have been used to compute both internal and external flows mainly in the subsonic and transonic speed range. Design and analysis of flow past bodies of aerospace interest is the strength of these codes. They are well validated and documented. Transonic flow past aerofoils, aerofoils with flaps, isolated wings, aircraft configurations, satellite launch vehicles, delta wings at high angles of attack, delta wings with deflected ailerons, nose radome of fighter aircraft and internal flow through cascades are some of the problems studied using these codes.

13.11 KEY TERMS

Aerofoils
Cascades
Compressible flows
Cell-centred and cell-vertex schemes
Cove vortex
Direct numerical simulation
Dynamic similarity parameters
Eddy viscosity Finite volume methods
Flaps
Large eddy simulation
Nozzles
Reynolds Averaged Navier–Stokes equations
Reynolds stresses
Separated flows
Turbulence models.

13.12 EXERCISE 13

1. Use Euler's implicit time difference formula

$$\Delta\vec{W}^n = \Delta t\left[\frac{\partial\vec{W}^n}{\partial t} + \frac{\partial}{\partial t}(\Delta\vec{W}^n)\right] + O(\Delta t^2) \tag{13.55}$$

and Taylor's series expansion like (13.24) to get the following equation for $\Delta\vec{W}^n$

$$\left(I + \Delta t\left[\delta_x(A - M + P_x)^n - \delta_{xx}P^n + \delta_y(B - N + Q_y)^n - \delta_{yy}Q^n\right]\right)\Delta\vec{W}^n$$

$$= -\Delta t(\delta_x\overline{\overline{F}}^n + \delta_y\overline{\overline{G}}^n) \tag{13.56}$$

2. For a stretched and skewed grid as shown in Fig. 13.28, use Green's theorem to find the expression for $\frac{\partial\varphi}{\partial x}|_{i+1/2,j}$ as per scheme-A. Using Taylor's expansion show that it gives first order accurate derivatives, and also show that if the grid is either stretched or skewed it gives second order accurate first derivatives.
3. For a stretched and skewed grid as shown in Fig. 13.29, use Green's theorem to find the expression for $\frac{\partial\varphi}{\partial x}|_{i,j}$ as per derivatives at the second stage of scheme-A, using inputs $\varphi_1, \varphi_2, \cdots, \varphi_8$ (representing first order derivatives of the variables at the midpoints of the cell boundaries). Also show by using Taylors expansion that this scheme gives a first order accurate dervatives for stretched and skewed grids but for uniform Cartesian grids it gives second order accurate derivatives.
[Hints: the derivatives at the point (i, j) as per scheme-A have been approximated as the average of the fluxes across the four neighbouring cells divided by the volume V of $PQRS$]

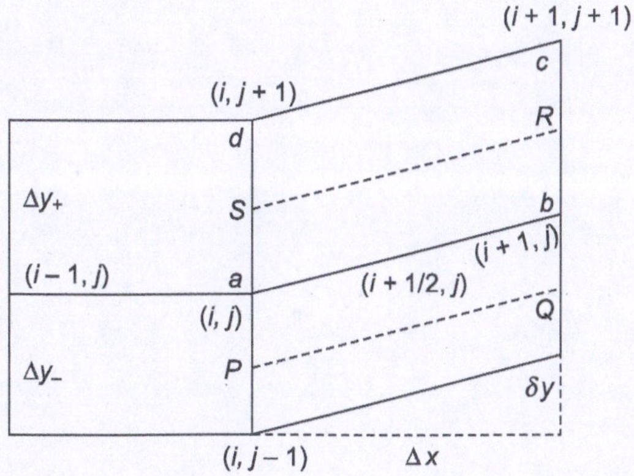

Figure 13.28 Control volume for first derivatives of scheme A.

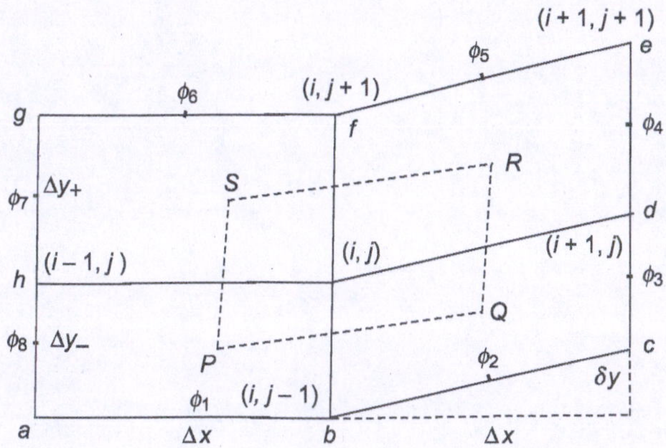

Figure 13.29 Control volume for second derivatives in scheme-A.

4. Draw a sketch of uniform grid cells and show that the of first and second derivatives of any variable (φ say) at the point (i, j) using cell centred and cell vertex finite volume schemes are of the same order of accuracy.
5. Draw stretched and skewed grid cells and find the accuracies of cell centred finite volume scheme for first and second derivatives of any variable (φ say) at the point (i, j), compare these with those of cell vertex schemes.

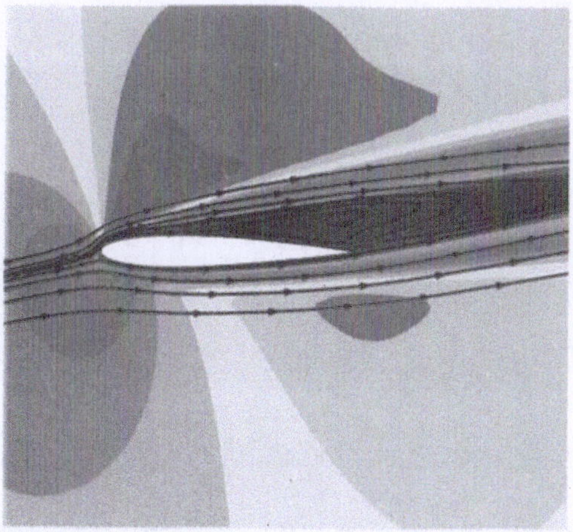

Figure 13.15 Streamlines superimposed on the Mach contours showing the separated region on the aerofoil at $M_\infty = 0.80$, $\alpha = 10°$, $Re = 500$.

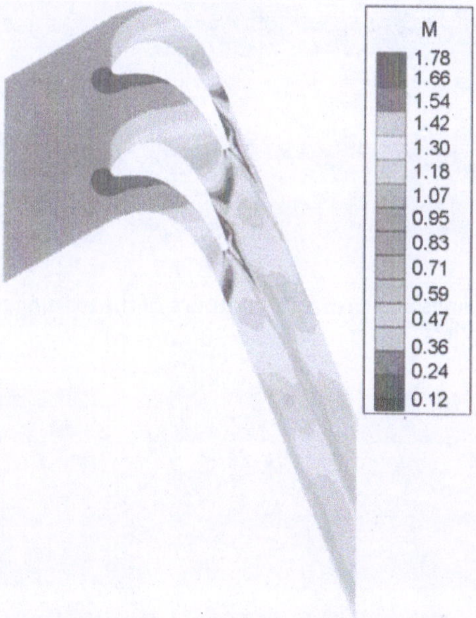

Figure 13.26 Mach contours for VKI-LS-59 turbine cascade with supersonic exit.

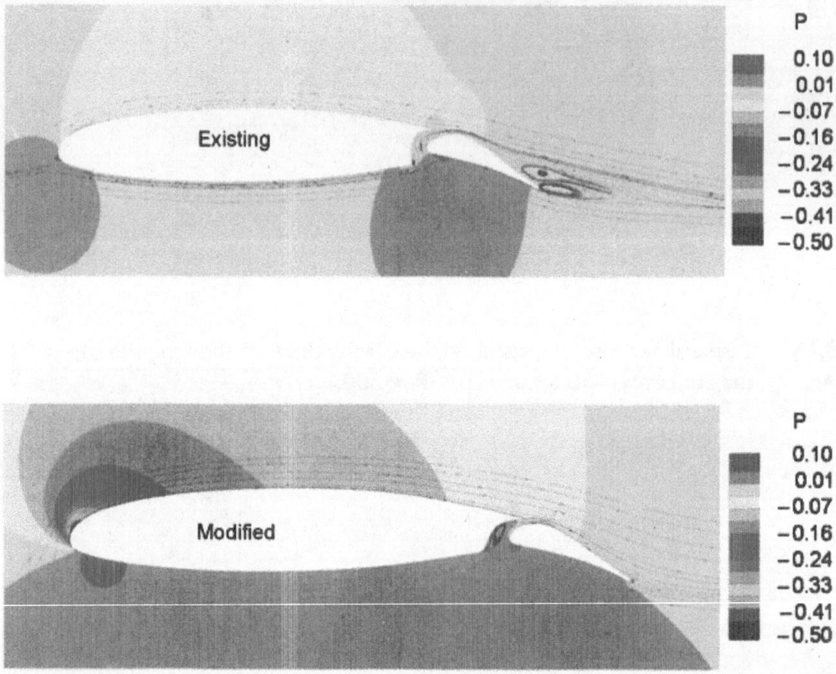

Figure 13.27 Streamlines and pressure contours of the existing and modified flap configurations, at $M_\infty = 0.3$, $R_{e\infty} = 2 \times 10^6$, $\delta = 20°$, $\alpha = 10°$.

Appendix A: Glossary

A.1 GLOSSARY

1. *Amplification factor:* In von Neumann stability analysis (or Fourier stability analysis) the effect of a small error on the scheme is studied by decomposing the error in Fourier modes and is measured by the ratio of a typical mode in two consecutive time steps.(see section 2.4).
2. *Approximate factorization:* When an implicit scheme is decomposed into factors it is known as approximate factorization.
3. *Boundary condition:* Condition that holds on the boundary of the physical domain for all time.
4. *Cauchy problem:* It is an initial value problem for a hyperbolic type partial differential equation in which continuous functions are prescribed for the value of the unknown and its first derivative on the initial line.
5. *Boussinesq hypothesis:* Boussinesq conceived that the effect of turbulence is only to increase the viscosity of the fluid and introduced the hypothesis that the eddy stress terms in the equations of mean motion may be related to the stresses of mean motion in terms of a parameter called eddy viscosity (strictly, eddy kinematic viscosity, which is not a physical quantity.)
6. *Cell Reynolds number:* The ratio $\dfrac{c\Delta x}{\alpha}$, where c denotes the convection speed and α the diffusion coefficient, Δx being the mesh spacings is known as the Cell Reynolds number.
7. *Characteristics:* These are curves in the domain of a hyperbolic (or parabolic) partial differential equation such that no inference can be drawn about the solution across them using known solution on one side of it.
8. *Closure problem:* The governing equations for the mean motion in a turbulent flow contain the unknown Reynolds stress or eddy stress terms. Requisite number of equations are not available to determine them. This difficulty is known as the closure problem.
9. *Conservative or conservation form:* A partial differential equation, in which the unknown function appears in the form of divergence of a vector.(see Section 1.8.2).
10. *Conservation law form:* A special class of conservation form of partial differential equations (see Section 1.8.2).
11. *Consistency:* A finite-difference scheme is consistent if the truncation error term approaches zero as the mesh lengths approach zero.
12. *Convergence:* If the solution of the scheme, consisting of the discretized system of equations together with associated boundary and/or initial

conditions approaches exact solution of the differential equations as the mesh spacings approach zero, then the scheme is said to be convergent.

13. *Courant number:* The ratio $c\Delta t/\Delta x$, where c denotes the convection speed and Δt and Δx denote respectively the time and space mesh spacings of the numerical scheme is known as the Courant number.

14. *Dilation:* The divergence of the velocity vector at a point of the flow field is known as the dilation of the flow at that point.

15. *Dirichlet problem:* It is a boundary value problem for an elliptic type partial differential equation in which continuous functions are prescribed as the value of the unknown on the boundary of the domain.

16. *Dispersive error:* This is a quasi-physical effect found in numerical schemes, and behaving very much like the effect of dispersion. It is indicated through the presence of odd order space derivatives in the leading term of the truncation error and acts in distorting the phase relations between the various Fourier components of the solution.

17. *Dissipative error:* This is a quasi-physical effect found in numerical schemes and behaving very much like actual viscosity. It arises through the presence of even order space derivatives in the leading term of the truncation error. This has the effect of damping out or moderating any steep gradients in the computed solution.

18. *Domain of dependence:* The interval of the intial line on which the solution at a point of the domain for an initial value problem for a hyperbolic equation depends.

19. *Eddies:* In a turbulent flow, the velocity and pressure at a fixed point experience irregular fluctuations. The fluid elements that perform such fluctuating motion are not molecules of the fluid but are small "lumps" of varying size, called eddies.

20. *Eddy stresses:* In studying the mean motion of a turbulent flow we observe that the effect of turbulence is as if certain additional stresses are introduced into the fluid when the existing flow parameters are replaced by their mean values. These additional stresses, arising on account of turbulent fluctuations are known as Reynolds stresses or eddy stresses.

21. *Error:* The error of a quantity is defined as the true value minus the approximate value of the quantity.

22. *Explicit and implicit schemes:* If a scheme be such that its solution may be obtained directly or explicitly in terms of known quantities it is said to be an explicit scheme as against an *implicit scheme* in which a system of equations has to be solved at each time step.

23. *Genuine solution:* Solution of a problem for a partial differential equation satisfying appropriate continuity and differentiability property is known as a genuine solution of the problem.
24. *Hypersonic flow:* A flow field where the local Mach number is much larger than unity.
25. *Initial condition:* Condition or conditions that hold at some initial value of time.
26. *Mach number:* The ratio of flow speed and local sound speed at a point of the flow field.
27. *Neumann problem:* It is a boundary value problem for a partial differential equation (generally elliptic) in which continuous functions are prescribed as the value of the normal derivative of the unknown on the boundary of the domain.
28. *RANS:* Abbreviation used for Reynolds averaged Navier-Stokes equations. For studying turbulent flow Reynolds (1894) conceived turbulent motion as a simple superposition of a fluctuating motion on a mean motion and derived the governing equations for the mean motion. These equations are known as RANS. Two kinds of averaging are used in practice, namely, the time averaging and the mass averaging.
29. *Reynolds number:* The dimensionless number UL/ν wher U is the flow speed, L is a representative length associated with the flow and ν denotes the kinematic viscosity of the fluid is known as the Reynolds number.
30. *Round-off error:* The error arising through the rounding of a number.
31. *Scheme:* When a problem, consisting of the differential equations together with prescribed boundary and/or initial conditions is discretized, the resulting representation is a finite-difference (or finite volume) scheme for the problem. There may be many schemes for a problem.
32. *Shock capturing and shock fitting schemes:* Schemes that detect and compute a solution with a shock discontinuity automatically, without requiring any special treatment for the shock, are known as shock capturing schemes. As against these the shock-fitting schemes require some knowledge of the location of the shock and the information is explicitly used for computing the solution with shock.
33. *Sparse systems:* A system of algebraic equations having relatively large number of zero elements in the coefficient matrix.
34. *Splitting:* When an explicit scheme is decomposed into factors it is referred to as splitting.
35. *Stability:* If the effect of a row of round-off errors on a finite-difference (or

finite volume) scheme remains bounded as we proceed stepwise in the time direction, the scheme is said to be *stepwise stable.* (see Section 2.4).

36. *Subsonic flow:* If the Mach number be less than unity at a point of the flow field, the flow is subsonic at that point. A flow field where the flow is subsonic at every point is a subsonic flow.

37. *Subcritical flow:* In subsonic flow past a body, if the flow speed is smaller than the local speed of sound everywhere in the domain the flow is said to be subcritical.

38. *Supercritical flow:* In a subsonic flow past a body, if there is an embedded supersonic region, it is said to be a supercritical flow.

39. *Supersonic flow:* If the Mach number be greater than unity at a point then it is supersonic at that point. A flow field where the flow is supersonic at every point is a supersonic flow field.

40. *Type-dependent differencing:* For numerical solution of mixed elliptic-hyperbolic equations, the choice of a central-difference operator (designated as elliptic operator) or an upwind operator (designated hyperbolic operator) at a point of the flow field according as the flow speed at the point is subsonic or supersonic respectively, is known as type-dependent differencing.

41. *Transonic flow:* A flow field where both subsonic and supersonic flows occur adjacent to each other and the local Mach number is close to unity over the flow domain is said to be a transonic flow. (see Chapter 10.)

42. *Tridiagonal matrix:* A matrix in which the only nonzero elements are those on the leading diagonal and on the two diagonals on the two sides of it.

43. *Tridiagonal systems:* A system of algebraic equations having a tridiagonal coefficient matrix.

44. *Truncation error:* The error arising through the truncation of an infinite or infinitesimal process; for example, the error arising through the truncation of an infinite series used for summing a function or replacing the derivative by a finite-difference quotient.

45. *Turbulence modeling:* For relating the Reynolds stress terms to the mean motion in a turbulent flow, one has to make ad-hoc assumptions or rely on empirical data or on empirical or semi-empirical methods called turbulence models. Development and study of assumptions and empirical results, which expresses the Reynolds stress components, comprising of correlation between fluctuating velocity components in terms of flow quantities of the mean motion is known as turbulence modeling.

46. *Unconditional stability:* If a scheme be stable, irrespective of the choice of the mesh ratios, it is said to be unconditionally stable.
47. *Weak solution:* The solution of a problem for a partial differential equation which does not satisfy requisite continuity and differentiability property is a weak solution or *generalised* solution. (see Section 1.9.)
48. *Well-posed problem:* A problem consisting of a partial differential equation (or a system of partial differential equations) together with initial and/or boundary conditions is said to be well-posed if it satisfies the three requirements: (a) existence, (b) uniqueness and (c) continuous dependence on boundary and/or initial conditions.

Appendix B: Ready-made Softwares for CFD

B.1 INTRODUCTION

This book teaches the whys and the hows of Computational Fluid Dynamics (CFD). While this knowledge is essential for anyone who plans to work in or use the results produced by CFD, it is not enough to solve completely the CFD-oriented problems that arise in typical industrial situations. Solving such problems involves additional processing—the so-called "pre-" and "post-"processing—in which a grid of some sort has to be generated and the input data described with reference to it before the CFD program can be run; and the mass of numbers produced by the CFD program has to be made accessible to the user in the form of graphs, charts and—increasingly—animations.

It is usually not possible for the CFD analyst to also be a specialist in grid generation and scientific data visualization, which is what the pre-and post-processors most often are. Hence the need for "packages" of programs or "software suites", which provide all the components necessary to completely handle a problem. Several such packages or suites are available, and their number is growing. Many research workers have made freely available the programs they have developed, and quite a few of these are very versatile. At the same time, as CFD has ceased to be merely a topic of research and become an industrial tool, software packages have become commercially available. The question naturally arises that, given a choice between free and commercial CFD tools, which should one prefer?

B.2 SOFTWARE PACKAGES FOR CFD

The answer to this question depends upon the circumstances. It must be appreciated that CFD software for solving real-life problems (as opposed to those usually posed in the classroom) typically requires new users to invest weeks or months to fully understand and use it properly. The user should have a basis for reposing confidence in a given code in order to make such a commitment.

Commercial products usually provide the complete package of CFD code and pre- and post-processors. They also provide sufficient documentation for the user to be able to use the product effectively (but rarely any that throws light on how the code itself works). In many cases the vendors of these products provide training and subsequent support to the user. However, such products are expensive, typically costing in the thousands of dollars at the low end. Quite often, these products are not sold, but "licensed", with the license having to be renewed each year on payment of a substantial fee.

Therefore, with the increasing use of CFD in industry and academia, the need for free CFD software is now greater than ever. However, free CFD tools do not often come with integrated pre- and post-processors. While excellent free tools for these purposes are available, the user has to have some knowledge about them and be willing to invest the effort to obtain and inegrate them with the CFD code. Also, most often the user is on his own when using the code, since the author likely will not provide training or support. Nevertheless, free CFD tools are being used with success, especially by those who have an understanding of how CFD programs work.

Below are given the names and brief lists of some commercial as well as free CFD tools. Most of the information is obtained from the websites `http://capella.colorado.edu/ laney/software.htm` and `http://www.cfd-online.com/Resources`, which may be consulted for more information. In addition, a search on the Internet will reveal many more relevant resources.

B.2.1 Commercial CFD Codes

FLUENT is a general purpose code that can address a wide range of problems in fluid mechanics, including chemically reacting flows. It can also handle the interaction of flows with deforming structures and calculate the stresses induced in these structures due to the interaction. It comes with integreated pre-and post-processors. PHOENICS is another general-purpose software package with capabilities similar to that of FLUENT. Its name is an acronym for "Parabolic Hyperbolic Or Elliptic Numerical Integration Code Series", which gives an idea of the range of problems it can handle. Yet another general-purpose package, AeroShape-3D, is based on the finite volume "rectangular adaptive mesh" technique for complex 3D flow simulation. It is of Russian origin.

In addition to these general-purpose packages, there are many special purpose packages dealing with particular classes of problems. FLOTHERM is designed to provide an analysis tool for all aspects of thermal problems arising in electronic equipment. WinPipeD is a program for analyzing multiphase (oil-gas-water) steady-state pipe flow hydraulics. It utilizes industry-standard equations and correlations, packaged in an easy-to-use interface (a basic version is available for download for non-commercial use). SMARTFIRE is an interactive fire field modelling CFD code that incorporates knowledge based techniques to assist in the specification of simulations. The numerical engine is based on the SIMPLE algorithm.

B.2.2 Free/public domain/shareware CFD Codes

The free codes mentioned below all have a presence on the web and the source code is available from the website mentioned in each case. Information about more such codes is available from the websites `http://capella.colorado.edu/ laney/software.htm` and `http://www.cfd-online.com/Resources` in particular.

FLUX is a numerical testbench for the linear advection and inviscid Burgers' equations. The program can be used for teaching purposes. Therefore it is written so that a student can introduce additional discretisation schemes or starting conditions. TEAMKE is a 2D finite difference/finite volume k-ϵ code using Cartesian grids. It can be applied to plane and axisymmetric flows, and laminar or turbulent flows. There is the option of using either Quadratic interpolation (QUICK) or Power Law interpolation (PLDS) to discretize the convective terms. The mesh used in the code is designed for a pre-determined grid arrangement only. Although considerable flexibility is offered through the use of non-uniform grids, there is no provision to handle unstructured grids. FLUX and TEAMKE are available from `ftp://unix.hensa.ac.uk/pub/misc/cfd/software`.

CLAWPACK is a software package designed to compute numerical solutions to hyperbolic partial differential equations using a wave propagation approach. Broadly speaking, it has the capability to handle problems related to acoustics (2D and 3D), the advection equation, the viscous Burgers' equation, Euler's equation, shallow free surface flows and even traffic congestion! See `http://www.amath.washington.edu/ claw` for more information.

NaSt3DGP is a parallel 3D flow solver for the incompressible Navier-Stokes equations. The code is developed in the division of Scientific Computing and Numerical Simulation at the Institut fur Numerische Simulation at the Rheinische Friedrich-Wilhelms-Universitat, Bonn. This is a free CFD code with source code included. It uses finite volume discretization on a Cartesian non-uniform staggered mesh, VONOS/SMART higher order upwind schemes and Adams-Bashforth time discretization. The code is parallelized with MPI and can handle complex geometries by a simple cell decomposition/enumeration technique. See the website `http://wissrech.iam.uni-bonn.de/research/projects/NaSt3DGP/index.htm`.

CHANNELFLOW is a direct numerical simulator for incompressible fluid flow on a periodic, rectangular, wall-bounded domain, written in C++. According to its homepage, CHANNELFLOW is not innovative as an algorithm. Instead, it uses modern software techniques to bring new levels of usability, flexibility, and intelligibility to a well-known algorithm. See `http://www.nongnu.org/channelflow`.

Appendix C: Programs in the 'C' Language

C.1 PROGRAM 3.1:ADI.C

```
/* ......Program 3.1: adi.c. */
/* ..... It solves by the ADI-scheme the 2-D heat-conduction eq.
 * .....  u_t =K( u_xx+ u_yy), */
/* ..... initial condition:
 * ..... u(x,y,o)= sin( pi x)sin(pi y), 0<= x,y <= 1, */
/* ......boundary conditions:
 * ..... u=0, on all the boundaries. */
/* ..... Known exact solution:
 * .....  u(x,y,t) = e^(-pi^2 t)*sin(pi x) sin(pi y). */
/* ..... Take delta x=delta y=0.2, r_1=r_2=1, KK=1. */
/* ......dth is the half time-step. nm=nmax, jm=jmax, km=kmax */
/* ............................................................. */

#include <stdio.h>
#include <stdlib.h>
#include <math.h>

static int
strid(double *, double *, double *, double *, int);

extern int
main( void )
{ int
i, j, k, n, kk, jm, km, nm, nh, jmm, kmm;

    double
u[51][51][51], ue[51][51][51], x[51], y[51];

double
aa[50], bb[50], cc[50], dd[50];

    double
  dt, pi, r1, r2, dx, dy, tt, fac, dth;

    pi = 3.1415926;
    r1 = 1.0;
    r2 = 1.0;
    kk = 1.0;
    dx = 0.125;
    dy = 0.125;
    dt = r1 * dx * dx;
    dth = dt * 0.5;

    km = 8;
```

```
    jm = 8;
    nm = 8;
    jmm = jm - 1;
    kmm = km - 1;
    for (i = 0; i <= jm; ++i)

  { x[i] = i * dx;
  y[i] = i * dy;
    }

/* ......Initial condition evaluation ........................*/
    for (j = 0; j <= jm; ++j)
for (k = 0; k <= km; ++k)
u[j][k][0] = sin(pi * x[j]) * sin(pi * y[k]);

/* .......Main iteration loop for two half  time-steps.  ..... */
/* .......Formation of the coefficient matrices................ */
    for (n = 0; n <= nm; n += 2)
{ nh = n + 1;

for (k = 1; k <= kmm; ++k)
{ for (j = 1; j <= jmm; ++j)
{ cc[j - 1] = r1 * ( -0.5 );
aa[j - 1] = r1 * ( -0.5 );
bb[j - 1] = r1 + 1.0;
dd[j - 1] = r2 * 0.5 *
( u[j][k - 1][n] + u[j][k + 1][n] ) +
(1.0 - r2) * u[j][k][n];
}
strid(aa, bb, cc, dd, jmm);
for (j = 1; j <= jmm; ++j)

u[j][k][nh] = dd[j - 1]; /* u[j - 1][k][ nh] = dd[j - 1]; */
}

for (j = 1; j <= jmm; ++j)
{ for (k = 1; k <= kmm; ++k)
{ bb[k - 1] = r2 + 1.0;
cc[k - 1] = r2 * ( -0.5 );
aa[k - 1] = r2 * ( -0.5 );
dd[k - 1] = r1 * 0.5 *
(u[j - 1][k][nh] + u[j + 1][k][nh]) +
(1.0 - r1) * u[j][k][nh];
}
strid(aa, bb, cc, dd, kmm);
for (k = 1; k <= kmm; ++k)
```

```
u[j][k][nh+1] = dd[k - 1];
}
} /* for n */

/* ......Exact solution computation................ */
    tt = nm * dt;
    fac = exp(-pi * pi * tt);

    for (j = 1; j <= jmm; ++j)
for (k = 1; k <= kmm; ++k)
ue[j][k][nm] = fac * sin(pi * x[j]) * sin(pi * y[k]);

    printf( "------------------------------------------------\n" );
    printf( "               Output of program adi.f\n" );
    printf( "------------------------------------------------\n" );

    printf( "dx=% e, dy=% e, r1=% e, r2=% e\n", dx, dy, r1, r2 );

    printf( "jm=%d, km=%d, nm=%d tt=% e, fac=% e\n",
jm, km, nm, tt, fac );

    printf( "------------------------------------------------\n" );
    printf( "          Comparison with the exact solution\n" );
    printf( "................................................\n" );
    printf("j k    u(j,k,nm)     ue(j,k,nm)\n");
    printf( "................................................\n" );

    for (j = 1; j <= jmm; j += 2)
for (k = 1; k <= kmm; k += 2)
printf("%d %d % e % e\n", j, k, u[j][k][nm], ue[j][k][nm]);

    j = 16;
    for (k = 1; k <= kmm; ++k)
{
    }

    return 0;
} /* main */

static int
strid(double a[50], double b[50], double c[50], double d[50], int m)
{ int
j, jm, jp;

double
```

```
factor, p[50];
/*  Forward elimination........................... */
    p[0] = c[0] / b[0];
    d[0] = d[0] / b[0];

    for (j = 2; j <= m; ++j)
 { jm = j - 1;
 factor = 1.0 / (b[j - 1] - a[jm - 1] * p[jm - 1]);
 p[j - 1] = c[j - 1] * factor;
 d[j - 1] = (d[j - 1] - a[jm - 1] * d[jm - 1]) * factor;
    }

/*  Back-substitution sweep....................... */
    for (j = m - 1; j >= 1; --j)
{ jp = j + 1;
d[j - 1] -= p[j - 1] * d[jp - 1];
    }

    return 0;
} /* strid */

/****/
```

C.2 PROGRAM 4.1: LXMC.C

```
      Lax-Wendroff solution of Inviscid Burger's equation, with :
      i.c. u(x,0)=x, 0<=x<=1
      b.c: u(0,t)=0, for all time.
      Exact solution: u(x,t)=.5(-t+\sqrt{t^2 + 4x}).
      Take dx=.2, r= dt/dx=0.5, 0 <=x <=2. 10 time steps.
 MacCormack soln= umc(j,n),l-w soln= ulw(j,n), exact soln= ux(j,n)
    ..............................................................
****/

#include <stdio.h>
#include <stdlib.h>
#include <math.h>

#define IM 20
#define IN 60
extern int
main( void )
```

```
{ int
ijm, imt, j, n, np;

double
dx, dt, r, t1, uj, utj, utjm, ujp, ujm, ujph, unp, umj,
ujmh, utmp;

double
x[IM + 1], ulw[IM + 1][IN + 1], ux[IM + 1][IN + 1], t[IN + 1],
utm[IM + 1][IN + 1], umc[IM + 1][IN + 1];

dx =0.1;
dt = 0.05;
r= 0.5;
imt=IM / 2;
x[0]=0.0;
ulw[0][0]=0.0;
umc[0][0]=0.0;
utm[0][0]=0.0;

/* .......Initial condition.................................. */
for ( j = 1; j <= IM; ++j )
{ x[j]= x[j-1] + dx;
ulw[j][0]= sqrt( x[j] );
umc[j][0]= ulw[j][0];
utm[j][0]= ulw[j][0];
}

/* .......Boundary condition.................................... */
for ( n = 1; n <= IN; ++n )
{ ulw[0][n]=0.0;
umc[0][n]=0.0;
utm[0][n]=0.0;
}

/* ...... Lax-Wendroff solution computation  ...................... */
for ( n = 0; n < IN; ++n )
{ np=n+1;
ijm=IM-1;

for ( j = 1; j <= ijm; ++j )
{ uj  = ulw[j][n];
ujp = ulw[j+1][n];
```

```
ujm = ulw[j-1][n];

ujph =  ujp + uj;
ujmh =  uj + ujm;
t1 = ( ujp * ujp - ujm * ujm ) / 8.0;

ulw[j][np] = uj - t1 +
( ujph * ( ujp * ujp - uj * uj ) -
ujmh * ( uj * uj - ujm * ujm )
) / 32.0;
}
}

/* ...... Computation of Exact solution uex.................... */
for ( n = 0; n <= IN; ++n )
{ t[n]= n* dt;

for ( j = 0; j <= IM; ++j )
ux[j][n] = 0.5 * ( -t[n] +
sqrt( t[n] * t[n] + 4.0 * x[j] )
);
}

/* ......Computation by MacCormack scheme ..................... */
for ( n = 0; n < IN; ++n )
{ np= n+1;
ijm=IM-1;

for ( j = 1; j <= ijm; ++j )
{ utj = utm[j][n];
utjm = utm[j-1][n];
utm[j][np] = utj - 0.5 * r * (utj * utj - utjm * utjm);
}

for ( j = 1; j <= ijm; ++j )
{ unp=utm[j][np];
utmp=utm[j+1][np];
umj=umc[j][n];
umc[j][np] = 0.5 * ( unp + umj ) -
0.25 * r * ( utmp * utmp - unp * unp ) ;
}
}

/* ............write output.................................. */
printf( "Output of L-W and Mc Sol.ofInviscid Burgers Eq\n" );
```

```
printf( " dx = % e, dt = % e\n", dx, dt );
printf( " n        x            t    ulw(j,n)  umc(j,n)  ux (j,n)\n" );

n=40;

for ( j = 1; j <= imt; ++j )
printf( "%d % e% e% e% e% e\n",
n, x[j], t[n] , ulw[j][n], umc[j][n] , ux[j][n] );

return 0;
} /* main */
```

C.3 PROGRAM 5.1: SOR.C

```
/* ......Program 5.1: sor.c */
/*      It solves 2-D Laplace eq. u_xx+u_yy=0, in the unit square */
/* ......in the 1st quadrant. B.C: u=0 on the two sides on the axes, */
/* ...... and u=x, u=y on remaining sides.  SOR itn. scheme  used. */
/* ...... SOR relaxation parameter = om, GS-scheme is om=1. */
/* .......omt= optimum relaxation parameter. */
/* ...... Known exact solution    u(x,y)= xy. */
/*  us= starting, uf= final, ux= exact solutions.... */
/*  itmax=max iteration steps performed; ep=tolerance. */
/* ...................................................... */

#include <stdlib.h>
#include <stdio.h>
#include <math.h>

int main( void )
{ int
jmax, kmax, j, k, itmax, k2, jm, km, it;

double
ep, pi, alm, omt, om, dx, dy;

double
x[306], y[306], uf[306][306], us[306][306], ux[306][306];

/* ......Set Boundary values..................................... */
    pi = 3.1415926f;
    itmax = 60000;
    ep = 5e-7f;
```

```
    kmax = 129;
    jmax = 129;}

    printf( "Input value of  relaxation parameter omega\n" );
    scanf( "%le", &om );
    printf( "SOR relaxation parameter omega = % e\n", om );
    alm = cos(pi / (double) jmax);
    omt = 2.f / (sqrt(1.f - alm * alm) + 1.f);
    jm = jmax - 1;
    km = kmax - 1;
    k2 = kmax / 2;
    dx = 1.f / jmax;
    dy = dx;
/* ........Setting boundary values.......... */
    for (j = 0; j <= jmax; ++j)
{ x[j] = j * dx;
y[j] = j * dy;
    }

    for (j = 0; j <= jmax; ++j)
{ us[j][0] = 0.0;
us[j][kmax] = x[j];
    }

    for (k = 0; k <= kmax; ++k)
{ us[0][k] = 0.0;
us[jmax][k] = y[k];
    }
/* ..... Evaluation of the exact solution................ */

    for (k = 1; k <= km; ++k)
for (j = 1; j <= jm; ++j)
ux[j][k] = j * k * dy * dx;

/* .......Testing dependence of convergence rate on omt....... */
printf( "Output of Program 5.1: sor.f\n" );
printf( "---------------------------\n" );
printf( " jmax=%d kmax = %d, itmax = %d\n", jmax, kmax, itmax );
printf( "ep = % e, omt = % e, om = % e\n", ep, omt, om );
printf( "--------------------------------------\n" );
/* .......Starting guess  us(j,k)  values  ............. */
    for (k = 1; k <= km; ++k)
for (j = 1; j <= jm; ++j)
```

```
us[j][k] = 0.1;
/* .......Main sor iteration loop............................ */
for (it = 1; it <= itmax; ++it)
{ for (k = 1; k <= km; ++k)
for (j = 1; j <= jm; ++j)
uf[j][k] = us[j][k] +
om * ( (uf[j - 1][k] + us[j + 1][k] +
uf[j][k - 1] + us[j][k + 1]
) * 0.25 - us[j][k]
);
/* .......Convergence test  ................................ */
for (k = 1; k <= km; ++k)
for (j = 1; j <= jm; ++j)
if ( fabs(uf[j][k] - us[j][k] ) > ep )
goto L65;

printf( " SOR Conv. in Steps= %d, relax. parm. = % e\n",
it, om );
printf( " Converged solution\n" );
printf( "     x(j)          y(k)          uf(j,k2)     ux(j,k2)\n" );
goto L71;

L65:
for (k = 1; k <= km; ++k)
for (j = 1; j <= jm; ++j)
 us[j][k] = uf[j][k];
    } /* for it */

    printf( "No convergence in steps=%d", itmax );
    printf( " The final values are:\n" );

L71:
    for (j = 1; j <= jm; j += 16)
printf( "% e % e % e % e\n",
x[j], y[k2], uf[j][k2], ux[j][k2] );

    return 0;
        }
```

C.4 PROGRAM 5.2: AF1.C

```
/* ..... Program 5.2: af1.c. */
/* ..... It solves 2-D Laplace equation in the unit square in the 1st */
```

```
/* ..... quadrant by approximate factorization scheme af1. */
/* ..... Boundary conditions are: u(x,y)=0.0 on the two sides on */
/* ..... coordinate axes and u= x and u= y on the remaining two sides. */
/* ..... The exact solution of this problem is  u(x,y)= x y. */
/* ..... al(i) are the acceleration parameters of AF1 scheme. */
/* ..... jmax, kmax are max no. of mesh points along x and y */
/* ..... ep=tolerance for convergence. It uses tridiagonal solver */
/* ..... subroutine trid.f */
/* ......................................................... */

#include <stdlib.h>
#include <stdio.h>
#include <math.h>

static int
trid( double *, double *, double *, double *, int );

extern int
main( void )
{ int
    jmax, kmax, i, j, k, itmax, k2, ik, kk, jm, km, it;

double
d_ _1, d_ _2;

    double
c[306][306], x[306], y[306],
f[306][306], uf[306][306], us[306][306], ux[306][306];

    double
aa[305], bb[305], cc[305], dd[305], aj, al[8];

    double
pi, ep, h4, dx, dy, hs, alh, all, alp, ddx, alu, ths;

/* ......Set Boundary values.............................. */
    itmax = 500;
    ep = 5e-7;
    pi = 3.1415927;
    kmax = 129;
    jmax = 129;
    aj = ( double )jmax;
    jm = jmax - 1;
    km = kmax - 1;
```

```
    k2 = kmax / 2;
    dx = 1.0 / aj;
    dy = dx;
    ddx = dx * dx;
    hs = 1.0 / ddx;
    ths = hs * 2.0;
    h4 = ddx / 4.0;
    alh = hs * 4.0;
    all = ths * (1.0 - cos(dx));

    for (kk = 1; kk <= 8; ++kk)
{ d__1 = (all / alh);
d__2 = (( double )(kk - 1) / 7.0);
al[kk - 1] = alh * pow(d__1, d__2);
    }

    for (j = 0; j <= jmax; ++j)
{ x[j] = j * dx;
y[j] = j * dy;
    }
/* ......Boundary values of temporary array f(j,k) set to zero. */
    for (j = 0; j <= jmax; ++j)
{ uf[j][0] = 0.0;
us[j][0] = 0.0;
f[j][0] = 0.0;
us[j][kmax] = j * dx;
uf[j][kmax] = j * dx;
f[j][kmax] = 0.0;
    }

    for (k = 0; k <= kmax; ++k)
{ uf[0][k] = 0.0;
us[0][k] = 0.0;
f[0][k] = 0.0;
us[jmax][k] = k * dy;
uf[jmax][k] = k * dy;
f[jmax][k] = 0.0;
    }

/* ..... Evaluation of the exact solution denoted by ux......... */
    for (k = 1; k <= km; ++k)
for (j = 1; j <= jm; ++j)
ux[j][k] = j * k * dy * dx;

/* ...... Approximate-Factorization scheme AF1. ........... */
```

```
    printf( "Output of af1.f: AF1 scheme\n" );
    printf( "------------------------------------\n" );
    printf( "jmax=%d, kmax=%d, itmax=%d, ep=% e\n",
jmax, kmax, itmax, ep );
    printf( "ep=% e\n", ep );
    printf( "AF1 acceleration parameters are:\n" );
    for (i = 1; i <= 8; ++i) {
printf( "% e\n", al[i - 1] );
    }
    printf( "------------------------------------\n" );

/* .......Starting values  .............................. */
    for (k = 1; k <= km; ++k)
for (j = 1; j <= jm; ++j)
      us[j][k] = 0.0;

/* .......AF1 iteration loop....................... */

    for (it = 1; it <= itmax; ++it)
{ ik = it % 8 + 1;
alp = al[ik - 1];

for (k = 1; k <= km; ++k)
{ for (j = 1; j <= jm; ++j)
{ alu = us[j+1][k] + us[j-1][k ]+ us[j][k+1]
+ us[j][k-1] - us[j][k] * 4.0;
cc[j - 1] = -hs;
aa[j - 1] = -hs;
bb[j - 1] = alp + ths;
dd[j - 1] = ths * alp * alu;

if (j == jm)
dd[j - 1] = f[jmax][k] + dd[j - 1];
}

trid(aa, bb, cc, dd, jm);

for (j = 1; j <= jm; ++j)
f[j][k] = dd[j - 1];
}

 /* ......cross-sweep............................... */
for (j = 1; j <= jm; ++j)
{ for (k = 1; k <= km; ++k)
{ cc[k - 1] = -hs;
```

```
aa[k - 1] = -hs;
bb[k - 1] = alp + ths;
dd[k - 1] = f[j][k];
}

trid(aa, bb, cc, dd, km);

for (k = 1; k <= km; ++k)
{ c[j][k] = dd[k - 1];
uf[j][k] = us[j][k] + dd[k - 1];
}

}

/* ----  Convergence test  -------------------------- */
for (j = 1; j <= jm; ++j)
{ for (k = 1; k <= km; ++k)
{ if ((fabs(c[j][k])) >= ep)
goto L65;
}

}

goto L60;

L65:
for (j = 1; j <= jm; ++j)
{ for (k = 1; k <= km; ++k)
us[j][k] = uf[j][k];
}
   } /* for it */

   printf( "No convergence in %d steps\n", itmax );
   goto L71;

L60:
   printf( "Convergence achieved in %d steps\n", it );
   printf( "Converged solution\n" );
   printf("    x(j)          y(k)          uf(j,k)       ux(j,k) \n");

L71:
   for (j = 1; j <= jm; j += 16)
printf("% e % e % e % e\n", x[j], y[k2], uf[j][k2], ux[j][k2]);

   return 0;
```

```
}/* main */

static int
trid( double a[305], double b[305], double c[305], double d[305],
   int m )
{ int
 j, jm, jp;

    double
 factor, p[205];

/*   ... forward elimination............. */
    p[0] = c[0] / b[0];
    d[0] = d[0] / b[0];

    for (j = 2; j <= m; ++j)
{ jm = j - 1;
factor = 1.0 / (b[j - 1] - a[jm - 1] * p[jm - 1]);
p[j - 1] = c[j - 1] * factor;
d[j - 1] = (d[j - 1] - a[jm] * d[jm - 1]) * factor;
    }

/*          ....... back substitution sweep ........... */
    for (j = m - 1; j >= 1; --j)
{ jp = j + 1;
d[j - 1] -= p[j - 1] * d[jp - 1];
    }

    return 0;
} /* main */
```

C.5 PROGRAM 5.3: MGC.C

```
/* .....Program 5.3: mgc.c prepared by Subhadip Niyogi.           */
/* .....It solves by multigrid method Dirichlet problem for        */
/* .....2-D Laplace equation in the unit-square in the 1st quadrant,*/
/* .....with boundary conditions: u=0 on the axes and u=4x and u=4y */
/* .....on the remaining sides. The known  exact solution is u=4xy. */
/* .....n=no. of fine grid mesh points in either direction,        */
/* .....m=grid levels.                                             */
/* .....Corrections=v and                                          */
/* .....residuals=r, point Gauss-Seidel smoother used.             */
```

```
/* ---------------------------------------------------------------        */

#include <stdlib.h>
#include <stdio.h>
#include <math.h>

/* ................................................. */
static int
residual(double us[258][258], double r[258][258], int n)
{ int
i, k;

    double
h;

    h = 1.0 / (double) (n);
    for (i = 1; i <= n -1; ++i)
for (k = 1; k <= n - 1; ++k)
r[i][k] = -( us[i - 1][k] + us[i + 1][k] +
  us[i][k - 1] + us[i][k + 1] -
  us[i][k] * 4.0 ) / ( h * h );

    for (i = 0; i <= n; ++i)
{ r[0][i] = 0.0;
r[i][0] = 0.0;
r[n][i] = 0.0;
r[i][n] = 0.0;
    }

    return 0;
} /* residual_ */

/* .............................................. */
static int
restriction(double us[258][258], double r[258][258],
double v[258][258], int n)
{ int
i, j, i2, j2;

    double
tmp1[258][258], tmp2[258][258], tmp3[258][258];

    for (i = 0; i <= n; ++i)
for (j = 0; j <= n; ++j)
```

```
{ i2 = 2 * i;
j2 = 2 * j;

tmp1[i][j] = us[i2][j2];
tmp2[i][j] = v[i2][j2];
tmp3[i][j] = r[i2][j2];
us[i][j] = tmp1[i][j];
v[i][j] = tmp2[i2][j2];
r[i][j] = tmp3[i2][j2];
}

    return 0;
} /* restriction */

/* ............................................... */
static int
solve(double us[258][258], double r[258][258], double v[258][258],
   int n)
{ int
i, j, k;

    double
h;

/* ----------------------------------------------- */
    h = 1.0 / (double)(n);
    residual(us, r, n);

    for (i = 0; i <= n; ++i)
{ v[0][i] = 0.0;
v[i][0] = 0.0;
v[n][i] = 0.0;
v[i][n] = 0.0;
    }

    residual(us, r, n);

    for (j = 1; j <= n - 1; ++j)
for (k = 1; k <= n - 1; ++k)
      v[j][k] = ( v[j - 1][k] + v[j][k - 1] - h * h * r[j][k]
   ) / 4.0;

   for (i = 1; i <= n - 1; ++i)
for (j = 1; j <= n - 1; ++j)
```

```
us[i][j] = us[i][j] + v[i][j];

    residual (us, r, n);

    return 0;
} /* solve */

/* ........................................................... */
static int
prolongation(double us[258][258], double r[258][258],
  double v[258][258], int n)
{ int
 i, j, k, i2, j2, n2, i2m, i2p;

    double
 temp1[258][258], temp2[258][258], temp3[258][258];

    for (i = 0; i <= n; ++i)
for (j = 0; j <= n; ++j)
{ temp1[i][j] = us[i][j];
temp2[i][j] = v[j][k];
temp3[i][j] = r[i][j];
}

    for (i = 0; i <= n; ++i)
for (j = 0; j <= n; ++j)
{ i2 = i * 2;
j2 = j * 2;

us[i2][j2] = temp1[i][j];
v[i2][j2] = temp2[i][j];
r[i2][j2] = temp3[i][j];
}

    for (i = 0; i <= n; ++i)
for (j = 0; j <= n; ++j)
{ i2 = i * 2;
j2 = j * 2;

us[i2][j2 + 1]     = (temp1[i][j] + temp1[i][j + 1]) / 2.0;
us[i2 + 1][j2]     = (temp1[i][j] + temp1[i + 1][j]) / 2.0;
us[i2 + 1][j2 + 1] = (temp1[i + 1][j + 1] + temp1[ i][j] +
   temp1[i + 1][j] + temp1[i][j + 1]
) / 4.0;
 v[i2][j2 + 1]     = (temp2[i][j] + temp2[i][j + 1]) / 2.0;
```

```
 v[i2 + 1][j2]      = (temp2[i][j] + temp2[i + 1][j]) / 2.0;
 v[i2 + 1][j2 + 1] = (temp2[i + 1][j + 1] + temp2[ i][j] +
  temp2[i + 1][j] + temp2[i][j + 1]
 ) / 4.0;

 r[i2][j2 + 1]      = (temp3[i][j] + temp3[i][j + 1]) / 2.0;
 r[i2 + 1][j2]      = (temp3[i][j] + temp3[i + 1][j]) / 2.0;
 r[i2 + 1][j2 + 1] = (temp3[i + 1][j + 1] + temp3[ i][j] +
  temp3[i + 1][j] + temp3[i][j + 1]
 ) / 4.0;
 }

    for (i = 1; i <= n; ++i)
 { n2 = n * 2;
 i2p = i * 2 + 1;
 i2m = i * 2 - 1;

 us[n2][i2m] = (double)( i2m * 4 ) / (double) n2;
 us[i2m][n2] = (double)( i2m * 4 ) / (double) n2;

 v[n2][i2m] = 0.0;
 v[i2m][n2] = 0.0;
 r[n2][i2m] = 0.0;
 r[i2m][n2] = 0.0;
    }

    return 0;
} /* prolongation */

extern int
main( void )
{ int
i, j, k, m, n, k2, ic, kk, it;

double
eps, rms, rmss, h;

double
    x[258], y[258];

    double
r[258][258], v[258][258], us[258][258], uex[258][258];

    n = 128;
    m = 5;
```

```
    eps = 5.0e-7;

    printf( "           Output of Program mgc.f\n" );
    printf( "-------------------------------------------\n" );
    printf( "grid pts=%d, levels=%d, eps=% e\n", n, m, eps );
    printf( "-------------------------------------------\n" );

/* .......Parameters and boundary conditions............ */
    h = 1.0 / (double) n;
    k2 = n / 2;

    for (i = 0; i <= n; ++i)
{ x[i] = ( double )i * h;
y[i] = ( double )i * h;
    }

/* .......Starting  values of u, denoted by us......... */
    for (i = 0; i <= n; ++i)
for (j = 0; j <= n; ++j)
us[i][j] = 0.0;

    for (i = 0; i <= n; ++i)
for (j = 0; j <= n; ++j)
v[i][j] = 0.0;

    for (i = 0; i <= n; ++i)
{ us[i][0] = 0.0;
us[i][n] = x[i] * 4.0;
us[0][i] = 0.0;
us[n][i] = y[i] * 4.0;

    }

/* .......Computation of exact solution, denoted by uex..... */
    for (i = 1; i <= n; ++i)
for (j = 1; j <= n; ++j)
uex[i][j] = x[i] * 4.0 * y[j];

/* ..... Start iteration..................... */
    for (i = 1; i <= n - 1; ++i)
for (j = 1; j <= n - 1; ++j)
us[i][j] = ( us[i + 1][j] + us[i - 1][j] +
  us[i][j + 1] + us[i][j - 1] ) * 0.25;

    for (ic = 1; ic <= 30; ++ic)
```

```
{ residual(us, r, n);
solve(us, r, v, n);

++it;

for (k = 1; k <= m; ++k)
{ n = n / 2;

restriction(us, r, v, n);
solve(us, r, v, n);

++it;
}

/* .......Lowest grid-level reached. Initiate prolongation.... */
for (kk = 1; kk <= m; ++kk)
{ prolongation(us, r, v, n);

n = n * 2;
solve(us, r, v, n);

++it;
}

rms = 0.0;
for (i = 1; i <= n - 1; ++i)
for (j = 1; j <= n - 1; ++j)
rms = rms + v[i][j] * v[i][j];

rmss = sqrt(rms / (double)n);
for (i = 1; i <= n - 1; ++i)
{ for (j = 1; j <= n - 1; ++j)
{ if ((fabs(v[i][j])) > eps)
goto L11;

if (rmss > eps)
goto L11;
}
}

printf( "kk=%d, it=%d, rmss=% e\n", kk, it, rmss );
printf( "Convergence reached in %d cycles, it=%d\n", ic, it );
printf( "Converged solution\n" );

goto L113;
L11:
```

```
 ;
    } /* for ic */

    printf( " No convergence in %d cycles, it=%d\n", ic, it );

L113:
    for (j = 1; j <= n - 1; j += 8)
printf( "%d % e % e\n", j, us[j][k2], uex[j][k2] );

    return 0;
} /* main */

%..................................................
```

C.6 PROGRAM 6.1: TSP.C

```
/* ................................. */
/* Program tsp.c.It computes transonic flow past a thin circular-arc */
/* profile at zero incidence under TSP model by  Murman-Cole, Murman */
/* Conservative Scheme, using formulation  of Jameson(1978).        */
/* System of Equations solved by slor. ........................     */
/* Farfield boundary conditions calculated in function farf.        */
/*  ............................................................... */

#include <stdlib.h>
#include <stdio.h>
#include <math.h>

int
m, n, l;

double
u[300][200], phi[300][200], xx[300], yy[200];

double
ah, ak, bk, gp, db, constant;

FILE
*out_14, *out_15, *out_16;

/* ............................................................. */
/* .......Function subroutine to calculate Double-Integral: ANTG(A) */
/* ...... used in far-field boundary condition evaluation......... */
```

```
/* ............................................................. */
static double
sums( int *ll )
{ int
i, k;

    double
su;

    su = u[0][*ll - 1] * u[0][*ll - 1] + u[n][*ll - 1] * u[n][*ll - 1];

    k = 2;
    for (i = 2; i <= n; ++i)
{ k = 6 - k;
su += k * (u[i - 1][*ll - 1] * u[i - 1][*ll - 1]);
    }

    return su;
} /* sums */

static double
antg( double *a )
{ int
j, k;

    double
tsum;

    j = 1;
    tsum = sums( &j );

    j = m + 1;
    tsum += sums( &j );

    k = 2;
    for (j = 2; j <= m; ++j)
{ k = 6 - k;
tsum += k * sums( &j );
    }

    return tsum * ah * ak / 9.0;
} /* antg */

/* ........................................................ */
```

```
/* .......Velocity Calculation SUBU ...................... */
/* ...................................................... */
static int
subu( void )
{ int
 i, j;

    for (i = 2; i <= n; ++i)
 for (j = 1; j <= m + 1; ++j)
 u[i - 1][j - 1] = ( phi[i][j - 1] - phi[i - 2][j - 1] ) /
   ( ah * 2.0 );

    for (j = 1; j <= m + 1; ++j)
 { u[0][j - 1] = ( phi[1][j - 1] - phi[0][j - 1]) / ah;

 i = n + 1;
 u[i - 1][j - 1] = ( phi[i - 1][j - 1] - phi[i - 2][j - 1] ) / ah;
    }

    return 0;
} /* subu */

/* ...................................................... */

/* .......Calculation of Far-field:  Farf................. */
/* ...................................................... */
static int
farf(void)
{ int
 i, j;

    double
 d, x, pi;

    pi = 3.14159260;
    d = db + gp / 2.0 * antg( &x );

 j = m + 1;

    for (j = 2; j <= m; ++j)
 { i = 1;
 phi[i - 1][j - 1] = constant * 0.5 * d * xx[i - 1] /
 ( xx[i - 1] * xx[i - 1] +
   bk * yy[i - 1] * yy[i - 1]
 );
```

```
i = n + 1;
phi[i - 1][j - 1] = constant * 0.5 * d * xx[i - 1]  /
( xx[i - 1] * xx[i - 1] +
   bk * yy[i - 1] * yy[i - 1]
);
    }

    return 0;
} /* farf */

/* ........................................................... */
/* ....................................................... */
/* .........Subroutine for solving tridiagonal system: strid ... */
/* ....................................................... */
static int
strid( double *a, double *b, double *c, double *d, int *m )
{ int
j, jm, jp;

    double
p[200], factor;

/* ........B=leading diagonal,A= left off-diagonal,C=right off-diagonal, */
/* .......D= RHS elements, solution array and used as dummy............ */
/* .......M=number of equations. D(I), I=1,M, solution vector.......... */
/* .......................................................... */
/* .......Forward elimination.................................. */
    p[0] = c[0] / b[0];
    d[0] = d[0] / b[0];

    for (j = 2; j <= *m - 1; ++j)
{ jm = j - 1;
factor = 1.0 / (b[j - 1] - a[jm - 1] * p[jm - 2]);
p[j - 1] = c[j - 1] * factor;
d[j - 1] = (d[j - 1] - a[jm - 1] * d[jm - 1]) * factor;

    }

    d[*m - 1] = ( d[*m - 1] - a[j - 2] * d[j - 2] ) /
( b[*m - 1] - p[*m - 2] * a[*m - 2] );
/* .......Back substitution Sweep................................ */
    for (j = *m - 1; j >= 1; --j)
{ jp = j + 1;
d[j - 1] = d[j - 1] - p[j - 1] * d[jp - 1];
```

```
}

return 0;
} /* strid */

extern int
main(void)
{ int
i, j, k, lb, ik, jm, nu1, nu2;

    double
beta, star, gamma, t, w, x, aminf, a1, b1, amstr, bd,
pi, xa, om, rr, ss, u1, v1, v2, x1, x2, cm, br, ah2,
al1, al2, bk1, cm1, ahh, akk, bdr, cmm, eps, tau, rsr,
cmm1, cmm2;

double
aa[300], bb[300], cc[300], dd[300];

    double
uu[300][300], san[300][300];

out_14 = fopen( "/tmp/tsc.out", "w" );
out_15 = fopen( "/tmp/tsc.u", "w" );
out_16 = fopen( "/tmp/tsc.uu", "w" );

/* .............................................................  */
/* . M,N : No. of subdivisions in y and x-directions.             */
/* . AH,AK= mesh spacings along x,y. BK= trans.similarity parameter. */
/* . EPS conv. tolerance of SLOR scheme. AL1,AL2,U1               */
/* . left, right and upper far-field boundaries respectively.     */
/* . AMINF, the free-stream Mach number.                          */
/* . STAR= reduction factor of Oswatitsch reduction,              */
/* . T=reduced thickness-ratio. OM=relax.param. of SLOR iteration.  */
/*                                                                */
/* .............................................................  */
 printf( "input values of: M,N,EPS,AL1,AL2,U1,AMINF\n" );
 scanf( "%d %d %le %le %le %le %le",
    &m, &n, &eps, &al1, &al2, &u1, &aminf );
gamma = 1.40;
gp = gamma + 1.0;
pi = atan(1.0) * 4.0;
tau = .060;
db = tau * 8.0 / 3.0;
```

```
    om = 1.850;

    beta = 1.0 - aminf * aminf;
    bk = beta / pow(tau, 2.0 /3.0 );
    amstr = sqrt(aminf * aminf / (1.0 - 0.4 / gp * beta));
    star = 1.0 / amstr - 1.0;
    t = tau / (sqrt(beta) * star);
    ah = (al2 - al1) / (double) n;
    ahh = ah * ah;
    ah2 = ah * 2.0;
    ak = ul / (double) m;
    akk = ak * ak;
    constant = 1.0 / (pi * sqrt( bk ));

fprintf( out_14, "beta = % e, bk = % e, amstr = % e\nstar = % e \
tau = % e, t = % e\nah = % e, ak = % e, const = % e\nom = % e\n",
  beta, bk, amstr, star, tau, t, ah, ak, constant, om );
/* ..............................................................*/
/* .......computation of starting values for a circular arc airfoil */
/* ....... y=TAU*(1-x**2), |x|<=1, =0, otherwise. Linearized solution */
/* ........is taken as the starting solution. */
/* ...... The far-field boundaries are: I=1, I=N+1, J=M+1. */

/* .......The far-f  values are modified through iteration......... */
/* ....... J=1 denotes body boundary. */
/* ......  PHI is  perturbation potential, new values  are SAN. */
/* .............................................................. */
    for (j = 1; j <= m + 1; ++j)
yy[j - 1] = (j - 1) * ak;

    for (i = 1; i <= n + 1; ++i)
xx[i - 1] = al1 + (i - 1) * ah;

for (i = 1; i <= n + 1; ++i)
{ x = xx[i - 1];
x1 = 1.0 - x;
x2 = x + 1.0;

phi[i - 1][0] = constant * (x * 2.0 - x1 * x2 / 2.0 *
pow( 10.0, ( x1 * x1 / (x2 * x2)) )
    );
san[i - 1][0] = phi[i - 1][0];
    }
```

```
    bk1 = sqrt(bk);
    for (i = 1; i <= n + 1; ++i)
for (j = 2; j <= m + 1; ++j)
{ a1 = bk1 * yy[j - 1];
b1 = a1 * a1;
x1 = 1.0 - xx[i - 1];
x2 = xx[i - 1] + 1.0;

phi[i - 1][j - 1] = constant *
( xx[i - 1] * 2.0 - xx[i - 1] * 2.0 * a1 *
   ( atan(x1 / a1) + atan( x2 / a1) ) -
   (x1 * x2 + b1) / 2.0 *
   pow( 10.0, ((x1 * x1 + b1) / (x2 * x2 + b1)) )
);
san[i - 1][j - 1] = phi[i - 1][j - 1];
}

/* Velocity computation subroutine SUBU.Also needed for far-f. */
/* SUBU called after each 10-iteration steps.                  */
    l = 0;
    subu();
/* .....   L  denotes iteration step.............................. */
L300:
    ik = 0;
    k = 0;
/* .......Farfield boundary values are now calculated,which involves */
/* .......double integral evaluation. ............................ */
    farf();
L200:
    ++ik;
    ++k;
    ++l;

/* Main Iteration Loop ................................        */
/* Parabolic,supersonic and subsonic and shock points tested. */
/* Velocities at points on vertical lines before and after    */
/* the profile are not tested, but taken as subsonic.         */

    for (i = 2; i <= n; ++i)
{ lb = 1;
if ( fabs( xx[i - 1] ) > 1.0 )
lb = 0;

for (j = 1; j <= m; ++j)
```

```
{ jm = j - 1;
v1 = bk - gp * (phi[i - 1][j - 1] -
phi[i - 3][j - 1]) / ah2;
v2 = bk - gp * (phi[i][j - 1] - phi[i - 2][j - 1]) / ah2;
ss = (phi[i - 1][j] - phi[i - 1][j - 1] * 2.0 +
  phi[i - 1][j - 2]) / akk;
rr = v2 * (phi[i][j - 1] - phi[i - 1][j - 1] * 2.0 +
   phi[i - 2][j - 1]) / ahh;
cm = san[i - 2][j - 1] - phi[i - 2][j - 1];

if (i == 2)
goto L51;

cmm = san[i - 3][j - 1] - phi[i - 3][j - 1];
br = v1 * (phi[i - 1][j - 1] - phi[i - 2][j - 1] * 2.0 +
   phi[i - 3][j - 1]) / ahh;
goto L52;
L51:
cmm = san[i - 2][j - 1] - phi[i - 2][j - 1];
br = v1 * (phi[i - 1][j - 1] -
   phi[i - 2][j] * 2.0) / ahh;
L52:
xa = fabs( xx[i - 1] );
if ( ( xa >= 1.0 ) || ( yy[j - 1] > 1.0 ) )
goto L440;

/* .......velocity testing  .............................. */
if (v1 > 0.0 && v2 > 0.0)
goto L440;

if (v1 > 0.0 && v2 < 0.0)
goto L450;

if (v1 < 0.0 && v2 > 0.0)
goto L460;

/* ......  HYPERBOLIC POINT  ............................ */
/* L470: */
nu1 = 1;
nu2 = 0;
goto L480;
/* ......  SUBSONIC POINT............................. */
L440:
nu1 = 0;
nu2 = 1;
```

```
goto L480;
/* ......... SONIC POINT ................................ */
L450:
nu1 = 0;
nu2 = 0;
goto L480;
/* ------------  SHOCK POINT -------------------------- */
L460:
nu1 = 1;
nu2 = 1;
/* .......case testing over........................... */
L480:
cm1 = (nu1 * 2.f * v1 - nu2 * v2) * cm / ahh;
cmm1 = v1 * cmm * nu1 / ahh;
rsr = ss + rr * nu2 + br * nu1;
if (j == 1)
goto L444;

aa[jm - 1] = 1.0 / akk;
bb[j - 1] = -2.0 * (1.0 / akk + v2 / (om * ahh) * nu2) +
nu1 * v1 / ahh;
cc[j - 1] = 1.0 / akk;
dd[j - 1] = -rsr + cm1 - cmm1;

/* ......COEFFICIENTS FOR J.NE.1  are as above............ */

goto L65;

/* .........coefficients for j=1 : Body Boundary ........... */
L444:
bb[j - 1] = -2.0 / akk - nu2 * 2.0 * v2 / (om * ahh) +
v1 * nu1 / ahh;
cc[j - 1] = 2.0 / akk;
bd = ( phi[i - 1][1] - phi[i - 1][0] ) * 2.0 /
   akk + lb * 4.0 * xx[i - 1] / ak;
bdr = nu2 * rr + bd + nu1 * br;
cmm2 = cm1 - cmm1;
dd[j - 1] = -bdr + cmm2;
L65:
;
} /* for j */

strid(aa, bb, cc, dd, &m);

for (j = 1; j <= m; ++j)
san[i - 1][j - 1] = phi[i - 1][j - 1] + dd[j - 1];
```

```
    } /* for i */
/* ......................................... */
/* ......CONVERGENCE TESTING. One cycle complete. */
/* ......................................... */
    if (ik > 10)
goto L750;

    for (i = 2; i <= n; ++i)
for (j = 1; j <= m; ++j)
{ w = fabs( san[i - 1][j - 1] - phi[i - 1][j - 1] );

if (w > eps)
goto L700;
}
/* .......Convergence achieved....................... */
    subu();
    goto L900;

L700:
/* .......Next cycle begins .......................... */
    ++l;
    for (i = 1; i <= n + 1; ++i)
for (j = 1; j <= m + 1; ++j)
      phi[i - 1][j - 1] = san[i - 1][j - 1];
L750:
    if (k >= 5)
goto L500;
else
    goto L200;

L500:
    subu();
    ik = 0;
    if (l > 10000)
goto L901;
else
    goto L300;

L900:
fprintf( out_14, "The Scheme Converges after iteration %d steps\n", l );
fprintf( out_14, " XX              U                 OSW.Red. UU \n" );

L905:
    for (i = 2; i <= n; ++i)
{ if ( fabs( xx[i - 1] ) > 1.0)
```

```
 continue;
/* ....................................................... */
uu[i - 1][0] = u[i - 1][0] * pow( tau, 2.0 / 3.0 ) / star;

fprintf( out_14, "% e     % e     % e\n",
  xx[i - 1], u[i - 1][0], uu[i - 1][0] );

fprintf( out_15, "% e     % e\n", xx[i - 1], u[i - 1][0] );

fprintf( out_16, "% e     % e\n", xx[i - 1], uu[i - 1][0] );
    }
/* ...................................................... */
    goto L1000;

L901:
fprintf(out_14, " Scheme does N O T converge in %d iteration steps\n",
1);
goto L905;

L1000:
    return 0;
 } /* main */
```

Appendix D: Answers and Hints to Solutions

D.1 CHAPTER 2

2.3(a) Let us establish the first formula (2.15).

By Taylor's expansion about the point j, assuming $u(x)$ to be sufficient number of times continuously differentiable, we have

$$\frac{1}{2\Delta x}\left[3u_j - 4u_{j-1} + u_{j-2}\right] =$$

$$\frac{1}{2\Delta x}\left[3u_j - 4\left(u_j - \Delta x u_x + \frac{1}{2}\Delta x^2 u_{xx} - \frac{\Delta x^3}{6}u_{xxx} + \ldots\right) + \left(u_j - 2\Delta x u_x + \frac{4\Delta x^2}{2}u_{xx} - \frac{8\Delta x^3}{6}u_{xxx} + \ldots\right)\right]$$

$$= \frac{1}{2\Delta x}\left[2\Delta x u_x - \frac{2}{3}\Delta x^3 u_{xxx} + \ldots\right]$$

$$= u_x - \frac{1}{3}\Delta x^3 + \ldots,$$

the derivatives being evaluated at the point j.

2.3(d) By Taylor's expansion

$$\frac{1}{12\Delta x^2}\left[-u_{j-2} + 16u_{j-1} - 30u_j + 16u_{j+1} - u_{j-2}\right] =$$

$$\frac{1}{12\Delta x^2}\left[-\left(u_j - 2\Delta x u_x + \frac{4\Delta x^2}{2}u_{xx} - \frac{8\Delta x^3}{6}u_{xxx} + \frac{16\Delta x^4}{24}u_{xxxx} - \frac{32\Delta x^5}{120}u_{xxxxx} + \frac{64\Delta x^6}{720}u_{xxxxxx} + \ldots\right) + 16\left(u_j - \Delta x u_x + \frac{\Delta x^2}{2}u_{xx} - \frac{\Delta x^3}{6}u_{xxx} + \frac{\Delta x^4}{24}u_{xxxx} - \frac{\Delta x^5}{120}u_{xxxxx} + \frac{\Delta x^6}{720}u_{xxxxxx} + \ldots\right) - 30u_j + 16\left(u_j + \Delta x u_x + \frac{\Delta x^2}{2}u_{xx} + \frac{\Delta x^3}{6}u_{xxx} + \frac{\Delta x^4}{24}u_{xxxx} + \frac{\Delta x^5}{120}u_{xxxxx} + \frac{\Delta x^6}{720}u_{xxxxxx} + \ldots\right) - \left(u_j + 2\Delta x u_x + \frac{4\Delta x^2}{2}u_{xx} + \frac{8\Delta x^3}{6}u_{xxx} + \frac{16\Delta x^4}{24}u_{xxxx} + \frac{32\Delta x^5}{120}u_{xxxxx} + \frac{64\Delta x^6}{720}u_{xxxxxx} + \ldots\right)\right].$$

The result follows on simplification.

D.2 CHAPTER 3

3.3 Exact solution, obtained by separation of variables is

$$u(x,t) = \frac{4}{\pi^2}\sum_{n=1}^{\infty}\frac{1}{n^2}\sin\frac{1}{2}n\pi\sin n\pi x e^{-n^2\pi^2 t}.$$

The values in the first square bracket in the following answers give the FTCS values while those in the second bracket give the corresponding exact solution values:

(3.3 a) [.0625, .125,.0625]; [.0896,.1309,.102]

(b) [.1983, .2855,.1983];[.1914,.2791,.2124]

(c) [.1000,.1999,.2972,.3972,.4142];[.0998,.1992,.2957,.3756,.4110], the result being symmetrical about $x = .5$.

(3.4 a) [.1414, .2000, .1414]; [.1647, .2330,.1647]

(b) [.3751, .5304,.3751]; [.3750, .5303,.3750]

(c) [.2315,.4403,.6060, .7124, ..7491]; [.2315, .4403, .6060,.7124,.7491 Result symmetrical about $x = 0.5$.

(3.5 a) [.2581, .3837, .4101, .2379];

(b) [.5179, .7341, .6682, .4077]

(3.6 a) [.4427, .6548, .7031, .4068]

(b) [.8819, 1.248, 1.146, .711]

(3.9) [.1636, .2545, .2545, .1636]; [.1074, .1752,.1752,.1074]

(3.10) [.2689, .4756, .4335,.4584]; [.2026, .3347, .4339, .4668]

D.3 CHAPTER 4

(4.1) [0.0, .05, .15, .25, .35, .45, .45, .35, .25, .15];

(0.0, .025, .10, .20, .30, .40, .45, .40, .30, .20, 0.0)

(4.2) [-.0062, .0625, .1625, .2625, .3625, .4750, .4375, .3375, .2375, .1375, 0.0];

(-.0234, .0375, .1375, .2375, .3344, .4844, .4625, .3625, .2625, .0345)

(4.6) Step I computed values: [.4478, .7245, .7245, .4478];

Exact values: [.4472, .7236, .7236, .4472]

Step II computed:[.6190, .6190]; Exact: [.6155, .6155]

Exact solution:

$$u(x, t) = 0.8 \sin(\pi x) \cos(\pi t).$$

D.4 CHAPTER 5

(5.1) From bottom to top vertical lines, left to right:
(.082, .166, .251); (.089, .180, .272); (.098, .196, .295).
The corresponding exact solutions are:
(.085, .169, .253); (.092, .183, .273); (.099, .198, .296).

(5.2) Form left to right bottom to top: (.684, .872, .686); (.872, 1.122, .874); (.686, .874, .687).

(5.3) From left to right horizontal lines, bottom to top(LRBT):
(.137, .162, .139); (.181, .216, .182); (.139, .164, .139).

The Gauss-Seidel scheme converges correct to 2 decimal places in 6 iteration steps.

(5.4) From left to right horizontal lines, bottom to top (LRBT):
(.277, .386, .279); (.355, .496,.357):(.279, .388, .280).

The Gauss-Seidel scheme converges in 7 steps correct to 2 decimals.

(5.5) LRBT: (.034, .065, .087,.079); (.065, .116, .144, .124);
(.087,.144,.173,.145); (.079, .124, .145, .124).

(5.6) LRBT: (.036, .068,.084,.069); (.068, .127, .157, .130);
(.084,.157, .197, .166); (.069, .130, .166, .147).

(5.8) LRBT: (.072, .585, 1.071); (-.406, -.111, .593); (-.929, -.416, .071).

(5.9) LRBT: (-.864, -.309, .139); (-.324, .246, .678); (.139, .6931.140).

(5.10) LRBT: (.076, .149, .210, .241); (.149, .280, .388, .446);
(.210, .388, .536, .627); (.241, .446, .627, .765).

D.5 CHAPTER 6

6.1 Hints: For preparing the modified program, the following points may be noted:

(a) Line 113, next to the statement 'LB=1', should be changed; for example, use $if(xx(i).le.0.0.or.xx(i).ge.1.0)lb = 0;$

(b) Line 37, the value of DB has to be changed, calculated from $DB = 2\tau \int_0^1 F(x)dx$, $F(x)$ denoting body shape.

(c) The starting values $phi(i, 1)$ and $phi(i, j)$ should be changed appropriately; any small value would work.

(6.3) For 6%− thickness ratio $A = 0.779$. Linearized solution is given by

$$\frac{u - u_\infty}{u_\infty} = \frac{A}{\pi\beta}\left[\frac{3}{2} + 3x - 3x^2 + (3x^2 - 1)\ln\left|\frac{1-x}{x}\right|\right]$$

(6.4) For 8% thickness ratio $A = 0.27$. Linearized solution is given by

$$\frac{u(x,0) - u_\infty}{u_\infty} = \frac{A}{\pi\beta}\left[3x - \frac{1}{2} + (3x^2 - 2x)\ln\left|\frac{1-x}{x}\right|\right]$$

D.6 CHAPTER 10

10.1 *Hints* - By superposing sources of strength $f(\xi)$ per unit length on the x-axis between 0 and 1, we get

$$\phi(x, y) = \int_0^1 f(\xi)\ln\sqrt{(x-\xi)^2 + \beta^2 y^2}d\xi,$$

where $f(xi)$ is to be determined using boundary condition Eq.(10.2). Differentiating with respect to y we get

$$v(x, y) = \beta\int_0^1 f(\xi)\frac{\beta y d\xi}{(x-\xi)^2 + \beta^2 y^2}.$$

Therefore,

$$v(x, 0) = \beta\left[\lim_{y\to 0+}\int_0^1 f(\xi i)\frac{\beta y}{(x-\xi)^2 + \beta^2 y^2}d\xi\right]$$

To evaluate the limit above, we observe that the only nonzero contribution to the integral can come from the singularity at $\xi = x$, so that

$$\lim_{y\to 0+}\int_0^1 f(\xi)\frac{\beta y\, d\xi}{(x-\xi)^2 + \beta^2 y^2} = f(x)\left[\lim_{y\to 0+}\int_0^1 \frac{\beta y\, d\xi}{(x-\xi)^2 + \beta^2 y^2}\right] = \pi f(x).$$

Form Equation (10.2) then follows that $v(x, 0) \equiv v_0(x) = \pi\beta f(x)$.

10.2 and 10.3 *Hints* - The linearized solutions of these two problems may be found in Answers to Exercise on chapter 6, problems 6.2 and 6.3. The TSP program, as modified in problem 6.1 may be used to compute the solutions for the given Mach numbers in problems 10.2 and 10.3. Note that the output of the program 6.1 tsc.f are in terms of Oswatitsch reduction. Necessary conversions may be carried out for comparison of the solutions.

D.7 CHAPTER 12

12.2 *Hints* - Use the conservative formulation for the vorticity-transport equation. It is easier to use a sequential algorithm for this problem. Appropriate under-relaxation is recommended.

12.6 For 3D incompressible flow, the continuity equation is

$$\frac{\partial u}{\partial x} + \frac{\partial v}{\partial y} + \frac{\partial w}{\partial z} = 0. \tag{D.1}$$

Now,

$$\begin{aligned}
&\frac{\partial f}{\partial t} + u\frac{\partial f}{\partial x} + v\frac{\partial f}{\partial y} + w\frac{\partial f}{\partial z} \\
&= \frac{\partial f}{\partial t} + \frac{\partial f u}{\partial x} + \frac{\partial f v}{\partial y} + \frac{\partial f w}{\partial z} - \\
&\quad f\left(\frac{\partial u}{\partial x} + \frac{\partial v}{\partial y} + \frac{\partial w}{\partial z}\right) \\
&= \frac{\partial f}{\partial t} + \frac{\partial f u}{\partial x} + \frac{\partial f v}{\partial y} + \frac{\partial f w}{\partial z},
\end{aligned}$$

by Equation (D.1).

Bibliography

J.H. AHLBERG, E.N. NILSON, and J.L. WALSH, *The Theory of Splines and Their Applications,* Academic press, New York, 1967.

A.A. AMSDEN and F.H. HARLOW, *The SMAC Method: A Numerical Technique for Calculating Incompressible Fluid Flow,* Los Alamos Scientific Laboratory Report, LA-4370, 1970.

ANAND KUMAR and K. DHANALAKSHMI, *Algebraic Grid Generation for an Airfoil*, NAL-TM-FM-8504, July 1985.

D.A. ANDERSON, J.C. TANNEHILL, and R.H. PLETCHER, *Computational Fluid Mechanics and Heat Transfer,* Hemisphere Publishing Corporation, New York, 1984.

W.K. ANDERSON, J.L. THOMAS, and D.L. WHITfiELD, Multigrid acceleration of the flux split euler equations, AIAA Paper 86-0274, AIAA 24th Aerospace Sc. Meeting, Reno, 1986.

A. ARNONE and R.C. SWANSON, A Navier–Stokes solver for turbomachinery applications, *Journal of Turbomachinery*, Vol. 115, pp. 305–313, 1993.

P.R. ASHILL, D.J. WEEKS, and J.L. FULKER, Wind tunnel experiments on aerofoil models for the assessment of computational flow methods, AGARD CP 437.

H. ASHLEY and M. LANDAHL, *Aerodynamics of Wings and Bodies,* Addison–Wesley Publishing Co., INC, Reading, U.S.A., 1965.

F.R. BAILEY, On the computation of two-and three-dimensional steady transonic flows by relaxation methods, In : *Progress in Numerical Fluid Dynamics,* Ed. H.J. Wirz, *Lecture Notes in Physics,* Vol. 41, pp. 1–77, Springer, Berlin, 1975.

T.J. BAKER, Computation of transonic potential flow, In : *Computational Methods for Tubulent, Transonic and Viscous flows,* Ed. Essers, pp. 213–289, 1984.

B.S. BALDWIN and H. LOMAX, Thin layer approximation and algebraic model for separated turbulent flows, AIAA paper 78–257, 16th AIAA Aerospace Science Meeting, Huntsville, Alabama, 16–18 January, 1978.

W.F. BALLHAUS, A. JAMESON, and J. ALBERT, Implicit approximate-factorization schemes for steady transonic flow problems, *AIAA Journal*, Vol. 16, pp. 573–579, 1978.

B.C. BASU, A mean camberline singularity method for two dimensional steady and oscillatory aerofoils and control surfaces in inviscid incompressible flow, ARC CP No. 1391, London, 1978.

G.K. BATCHELOR, *An Introduction to Fluid Dynamics,* Cambridge University Press, London, 1967.

F. BAUER, P. GARABEDIAN, D. KORN, and A. JAMESON, *Supercritical Wing Sections II, Lecture Notes in Economics and Mathematical Systems,* Springer-Verlag, Vol. 108, 1975.

H.I. BAURDOUX and J.W. BOERSTOEL, Symmetrical transonic potential flows around quasi-elliptical airfoil section, NLR-TR 69007 U, Vol. XXXVI, December, 1968.

A. BAYLISS and E. TURKEL, Radiation boundary condition for wave-like equations, *Communications On Pure and Applied Mathematics*, Vol. 33, pp. 707–723, 1980.

R.M. BEAM and R.F. WARMING, An implicit finite-difference algorithm for hyperbolic systems in conservation law form, *Journal of Computational Physics*, Vol. 22, pp. 87–110, 1976.

R.M. BEAM and R.F. WARMING, An implicit factored scheme for the compressible Navier–Stokes equations, *AIAA Journal*, Vol. 16, pp. 393–401, 1978.

L. BERS, *Mathematical Aspects of Subsonic and Transonic Gasdynamics,* Wiley, New York, 1958.

A. BETZ, *Konforme Abbildung,* 2. Auflage, Springer, Berlin, 1964.

G. BISWAS, MAC and SIMPLE algorithms, In: *Computational Fluid Flow and Heat Transfer*, Ed. K. Muralidhar and T. Sundarajan, 2nd ed., Narosa Publishing House, New Delhi, 2003.

G. BISWAS and V. ESWARAN (ED.), In: *Turbulent Flow,* Narosa Publishing House, New Delhi, 2002.

J.G. BLASCHAK and G.A. KRIEGSMANN, A comparative study of absorbing boundary conditions, Journal of Computational Physics, Vol. 77, pp. 109–139, 1988.

J.P. BORIS and D.L. BOOK, Flux corrected transport I, SHASTA, a fluid transport algorithm that works, *Journal of Computational Physics*, Vol. 31, pp. 38–69, 1973.

P. BRADSHAW, *Turbulence: The Chief Outstanding Difficulty of Our Subject Experiments in Fluids*, Vol. 16, No.2, pp 203–216, 1994.

A. BRANDT, Multilevel adaptive solutions to boundary value problems, Mathematics of Computing, Vol. 31, 1977.

M. BRAZA, P. CHASSAING, and H. HA MINH, Numerical study and physical analysis of the pressure and velocity fields in the near wake of a circular cylinder, *Journal of Fluid Mechanics*, Vol. 165, pp. 79–130, 1986.

C.A. BREBIA, *The Boundary Element Method for Engineers,* Pentech Press, London, 1978.

W.R. BRILEY and H. MCDONALD, Solution of the three dimensional compressible Navier–Stokes equations by an implicit technique, *Proc. 4th Int. Conf. Num. Methods Fluid Dyn., Boulder Colorado, Lecture Notes in Physics,* Springer, New York, Vol. 35, pp. 105–110, 1973.

J.M. BURGERS, A mathematical model illustrating the theory of turbulence, In: *Advances in Applied Mechanics,* Ed. R. von Mises and Th. von Karman, Academic Press, New York, Vol. 1, pp. 171–199, 1948.

D.A. CAUGHY, A diagonal implicit multigrid algorithm for the Euler equations, AIAA Paper, 87–453 AIAA 25th Aerospace Sciences Meeting, Reno, 1987.

T. CEBECI and P. BRADSHAW, *Physical and Computational Aspects of Convective Heat Transfer,* Springer Verlag, New York, 1988.

T. CEBECI and J. COUSTEIX, *Modeling and Computation of Boundary-Layer Flows,* Springer Verlag, 1999.

T. CEBECI and A.M.O. SMITH, *Analysis of Turbulent Boundary Layers, Academic Press,* New York, 1974.

S.K. CHAKRABARTTY, Numerical solution of two dimensional Navier–Stokes equations by finite volume method, DFVLR-IB-129-87/23 and NAL-PD-FM-8726, 1987.

S.R. CHAKRAVARTHY, High resolution upwind formulations for the Navier–Stokes equations, computational fluid dynamics, Von Kármán Inst. lecture Series, 1988-05, 1988.

S.K. CHAKRABARTTY, Numerical solution of Navier–Stokes equation for two dimensional viscous compressible flows, *AIAA Journal*, Vol. 27, pp. 843–844, 1989.

S.K. CHAKRABARTTY, Some results of compressible viscous flows past aerofoils using a modified step for stability, NAL-PD-9129, 1991.

S.K. CHAKRABARTTY, A finite volume nodal point scheme for solving two dimensional Navier–Stokes equations, *Acta–Mechanica*, Vol. 84, pp. 139–153, 1990a.

S.K. CHAKRABARTTY, Computation of transonic potential flow past wing-body configurations, *Acta Mechanica,* Vol. 81, pp. 201–209, 1990.

S.K. CHAKRABARTTY, Vertex based finite volume solution of the two dimensional Navier–Stokes equations, *AIAA Journal,* Vol. 28, pp. 1829–1831, 1990.

S.K. CHAKRABARTTY, Approximate factorization scheme for transonic potential flow computations, SADHANA, *Academy Proceedings in Engineerings Sciences*, Vol. 17, pp. 299–313, 1992.

S.K. CHAKRABARTTY and K. DHANALAKSHMI, Computation of transonic flows with shock induced separation using algebraic turbulence models, *AIAA Journal*, Vol. 33, pp. 1979–1981, 1995.

S.K. CHAKRABARTTY and K. DHANALAKSHMI, Navier–Stokes analysis of korn aerofoil, *Acta Mechanica*, Vol. 118, pp. 235–239, 1996.

S.K. CHAKRABARTTY and N.R. SUBRAMANIAN, Low frequency and small perturbation equation for transonic flow past wings, *AIAA Journal*, Vol. 18, pp. 1147–1149, 1980.

S.K. CHAKRABARTTY and N.R. SUBRAMANIAN, On the Computation of Body-Fitted Coordinates, Presented at the workshop-cum-seminar on CFD, VSSC, Trivandrum, December 19–21, 1981.

S.K. CHAKRABARTTY and N.R. SUBRAMANIAN, Numerical generation of body-fitted curvilinear coordinate system in 2D fields, *Proc. of the Workshop on Computational Methods for Flow Calculations*, Dept. Aero. Engg., Indian Institute of Technology, Kharagpur, pp. 194–199, January 6–7, 1981.

S.K. CHAKRABARTTY and N.R. SUBRAMANIAN, A second order accurate implicit scheme for conservative full potential equation in transonic flow, *Proc. AGM Aero. Soc. India,* pp. 267–270, 1985.

S.K. CHAKRABARTTY, K. DHANALAKSHMI, and J.S. MATHUR, Euler analysis of multibody launch vehicle configuration, *Proceedings of third Asian Computational Fluid Dynamics Conference,* Vol. 2, pp. 1–6, Bangalore, India, December 7–11, 1998.

S.K. Chakrabartty, K. Dhanalakshmi, and J.S. Mathur, Navier–Stokes analysis of vortex flow over a cropped delta wing, *Acta Mechanica*, Vol. 131, pp. 69–87, 1998.

S.K. Chakrabartty, K. Dhanalakshmi, and J.S. Mathur, Computation of flow past aerospace vehicles, *Current Science,* Vol. 77, pp. 1295–1302, 1999.

S.K. Chakrabartty, K. Dhanalakshmi, and J.S. Mathur, Computation of three-dimensional transonic flow using cell-vertex finite volume method for Euler equations, *Acta Mechanica*, Vol. 115, pp. 161–177, 1996a.

S.K. Chakrabartty, K. Dhanalakshmi, and J.S. Mathur, Computation of three dimensional transonic viscous flow using the JUMBO3D code, *Acta Mechanica,* Vol. 119, pp. 181–197, 1996b.

S.K. Chakrabartty, K. Dhanalakshmi, and J.S. Mathur, Navier–Stokes analysis of flow through two-dimensional cascades, *Computational Fluid Dynamics Journal,* Vol. 10, pp. 233–241, 2001.

S.K. Chakrabartty, K. Dhanalakshmi, and V. Ramesh, Navier–Stokes analysis and design of aerofoil-flap configurations for low speed aircraft, *Computational Fluid Dynamics Journal,* Vol. 11, pp. 285–289, 2002.

S.K. Chakrabartty, K. Dhanalakshmi, and V. Ramesh, Navier–Stokes analysis of GA(W)-2 aerofoil with deflected flap and redesign of HANSA flap for better performance, *Computational Fluid Dynamics Journal,* Vol. 12, pp. 89–97, 2003b.

S.K. Chakrabartty, J.S. Mathur, and K. Dhanalakshmi, Application of advanced CFD codes for aircraft design and development at NAL, *Journal of Aerospace Sciences and Technologies,* Vol. 55, pp. 74–88, 2003a.

W. Cherdron, F. Durst, and J.H. Whitelaw, Asymmetric flows and instabilities in symmetric ducts with sudden expansions, *Journal of Fluid Mechanics,* Vol. 84, pp. 13–31, 1978.

R.V. Chima, E. Turkel, and S. Schaffer, Comparison of three explicit multigrid methods for the Euler and Navier–Stokes equations, AIAA Paper 87-0602, AIAA 25th Aerospace Sc. Meeting, Reno, Nevada, 1987.

A.J. Chorin, Numerical solution of the Navier– Stokes equation, *Mathematics of Computation,* Vol. 22, pp. 745–762, 1968.

J.D. Cole, Modern developments in transonic flow, *SIAM Journal of Applied Mathematics,* Vol. 29, pp. 763–786, 1975.

P.H. COOK, M.A. MCDONALD, and M.C.P. FIRMIN, Aerofoil RAE2822—Pressure Distributions and Boundary Layer and Wake Measurements, AGARD-AR-138, 1979.

E.T. COPSON, *Theory of Functions of a Complex Variable,* Oxford, Clarendron Press, London , 1950.

R. COURANT and K.O. FRIEDRICHS, Supersonic Flow and Shock Waves, Interscience, New York, 1948.

R. COURANT, K.O. FRIEDRICHS, and M. LEWY, Über die partiellen Differenzengleichungen der Physik, *Mathematics Annals*, Vol. 100, pp. 32–76 1928; English Translation, *IBM J. Res. Dev.* Vol. 11, pp. 215–234, 1967.

R. COURANT and D. HILBERT, *Methods of Mathematical Physics,* Vol. II, Wiley, New York, 1953.

B.N. DATTA, Numerical Linear Algebra and Applications, Brooks/Cole Publishing Company, Pacific Grove, U.S.A., 1995.

R.L. DAVIS, R.H. NI, and J.E. CARTER, Cascade viscous flow analysis using the Navier–Stokes equations, AIAA 14th Aerospace Sc. Meeting, Reno, Nevada, 1986.Vol. 66, pp. 173–196, 1986.

D. DEGANI and L.B. SCHIFF, Computation of turbulent supersonic flows around pointed bodies having cross-flow separation, *Journal of Computational Physics* Vol. 66, pp. 173–196, 1986.

G.S. DEIWERT, Numerical simulation of high Reynolds number transonic flows, *AIAA Journal*, Vol. 13, pp. 1354–1360, 1975.

M.D. DESHPANDE and S.G. MILTON, Kolmogorov Scales in a driven cavity flow, *Fluid Dynamic Research*, vol. 22, pp. 359–381, 1998.

P. DUCHATEAU and D.W. ZACHMANN, *Theory and Problems of Partial Differential Equations,* Schaum's Outline Series, McGraw-Hill Book Co., New York, 1986.

E.C. DUFORT and S.P. FRANKEL, Stability conditions in the numerical treatment of parabolic differential equations, *M T A C*, Vol. 7, p. 135, 1953.

W.J. DUNCAN, A.S. THOM, and A.D. YOUNG, Chapter 6, *Boundary Layers, Wakes and Turbulence.* In *Mechanics of Fluids,* 2nd edn., The English Language Book Society and Edward Arnold (Publishers) Ltd., London, 1970.

F. DURST, A. MELLING, and J.H. WHITELAW, Low Reynolds number flow over a plane symmetric sudden expansion, *Journal Fluid Mechanics,* Vol. 64, 111–128, 1974.

F. DURST, J.C. PEREIRA, and C. TROPEA, The plane symmetric sudden expansion, *Journal Fluid Mechanics,* Vol. 248, pp. 567–581, 1993.

V. DUTTA, *Implicit 2-D Navier-Stokes solver for compressible viscous flow around aerofoils, Proc. First Asian CFD Conference*, Hong Kong, January 1995.

H.A. DWYER, R.J. KEE, and B.R. SANDERS, Adaptive grid method for problems in fluid mechanics and heat transfer, *AIAA Journal*, Vol. 18, No.10, pp. 1205–1212, 1980.

P.R. EISEMAN, Alternative direction adaptive grid generation, *AIAA Journal*, Vol. 23, No.4, pp. 551–560, 1985.

H.W. EMMONS, The numerical solution of compressible fluid flow problems, NACA TN 932, 1944.

B. ENGQUIST and A. MAJDA, Absorbing boundary conditions for the numerical simulation of waves, *Mathematical Computing*, Vol. 31, pp. 629–651, 1977.

B. ENGQUIST and S. OSHER, One-sided difference approximations for non-linear conservation laws, *Mathematical Computing*, Vol. 36, pp. 321–351, 1981.

L.E. ERIKSSON, Generation of boundary conforming grids around wing–body configurations using transfinite interpolation, *AIAA Journal*, Vol. 20, pp. 1313–1320, 1982.

A. FAVRE, Equations des Gaz Turbulents Compressibles: 1. Formes Generales, *Journal de Mechanique*, Vol. 4, pp. 361–390, 1965.

R.M. FEARN, T. MULLIN, and K.A. CLIFFE, Nonlinear flow phenomena in a symmetric sudden expansion, *Journal of fluid mechanics*, Vol. 211, pp. 595–608, 1990.

L. FERM, Open boundary conditions for stationary inviscid flow problems, *Journal of Computational Physics*, Vol. 78, pp. 94–113, 1988.

C. FERRARI and G. TRICOMI, *Transonic Aerodynamics*, English Translation, Academic Press, New York, 1968.

J.H. FERZIGER, *Numerical Methods for Engineering Applications,* Wiley Interscience Publications, New York, 1981.

C.A.J. FLETCHER, In: *Chapter 3 in Numerical Solution of Differential Equations,* Ed. J. Noye, North Holland, Amsterdam, pp. 355–475, 1983.

C.A.J. FLETCHER, *Computational Techniques for Fluid Dynamics,* Vol. I, Springer Verlag, Berlin, Heidelberg, New York, 1988a.

C.A.J. FLETCHER, *Computational Techniques for Fluid Dynamics,* Vol. II, Springer Verlag, Berlin, Heidelberg, New York, 1988b.

J.E. FROMM, *A Numerical Study of Buoyancy Driven Flows in Room Enclosures,* Lecture Notes in Physics, Vol. 8, pp. 120–126, Springer Verlag, Berlin, 1971.

P. GARABEDIAN, Estimation of the relaxation factor for small mesh size, *Mathematical Tables Aid Computing,* Vol. 10, pp. 183–185, 1956.

P. GARABEDIAN, *Partial Differential Equations,* Wiley, New York, 1964.

T.B. GATSKI, Turbulent flows: Model equations and solution methodology, In: *Hand Book of Computational Fluid Dynamics,* Ed. R. Peyret, pp. 339–415, Academic Press, 1996.

U. GHIA, K.N. GHIA, and C.T. SHIN, High-Resolutions for incompressible flow using the Navier–Stokes equations and a multigrid method, *Journal Computational Physics,* Vol. 48, pp. 387–411, 1982.

S. GHOSH, *Numerical Computation of Transonic and Supersonic Flow Past Obstacles in Two Dimensions,* Ph. D. Thesis, IIT Kharagpur, 1999.

S. GHOSH and P. NIYOGI, A study of non-oscillatory schemes based on LED Principle for inviscid flow computation past airfoils, *Acta Mechanica,* Vol. 139, pp. 73–90, 2000.

S. GHOSH, P. NIYOGI and S.K. CHAKRABARTTY, Numerical study in Transonic flow past profiles using Full-potential and Euler models, *Bulletin of the Calcutta Mathematical Society,* Vol. 92, pp. 273–294, 2000.

M.B. GILES, Nonreflecting boundary conditions for Euler equation calculations, *AIAA Journal*, Vol. 28, No. 12, pp. 2050–2058, 1990.

U.C., GOLDBERG, Separated flow treatment with a new turbulence model, *AIAA Journal*, Vol. 24, No.10, pp. 1711–1713, 1986.

G.H. GOLUB and C.F. VAN LOAN, *Matrix Computations,* 2nd Edn., John Hopkins University Press, Baltimore, 1989.

K.G. GUDERLY, *Theorie Schallnaher Stroemungen,* Springer Verlag, Berlin, 1957.

M.M. GUPTA and R.P. MANOHAR, *Journal of Computational Physics,* Vol. 31, pp. 265–288, 1979.

B. GUSTAFSSON and A.SUNDSTRÖ M, Incompletely parabolic problems in fluid dynamics, *SIAM Journal of Applied Mathematics,* Vol. 35, pp. 343–357, 1978.

W. HACKBUSCH, *Multigrid Methods and Applications,* Springer Verlag, Berlin, 1985.

W. HACKBUSCH, Introduction to multi-grid methods for the numerical solution of boundary value problems, In: *Computational Methods for Turbulent, Transonic and Viscous Flows*, Ed. J.A. Essers, pp. 45–92, Hemisphere Publishing Corp., New York, 1983.

M. HAFEZ, J.C. SOUTH, and E.M. MURMAN, Artificial compressibility method for numerical solutions of transonic full potential equation, *AIAA Journal,* Vol. 17, pp. 838–844, 1979.

M.G. HALL, Cell Vertex Multigrid Scheme for on Numerical Methods for Fluid Dynamics, University of Reading, UK, 1985, IMA Conf. Series, Oxford University Press, 1985.

M.G. HALL, and M.D. SALAS, Comparison of two multigrid methods for the two dimensional Euler equations, AIAA Paper 85-1515, 7th AIAA Computational Fluid Dynamics Conf., Cincinnati, Ohio, 1985.

G.J. HANCOCK and G. PADfiELD, Numerical solution for steady two dimensional airfoil in incompressible flow, QMC - EP 1003, Queen Mary College, London, 1972.

F.H. HARLOW and J.E. WELCH, Numerical calculation of time dependent viscous incompressible flow of fluid with free surface, *Physics of Fluids,* Vol. 8, pp. 2182–2189, 1965.

C.D. HARRIS, Two Dimensional Aerodynamic Characteristics of the NACA0012 Airfoil in the Langley 8-foot Transonic Pressure Tunnel, NASA TM 81927, 1981.

C. HäRTEL, Turbulent flows: Direct numerical simulation and large-eddy simulation, In: *Handbook of Computational Fluid Dynamics,* Ed. R. Peyret, pp. 283–338, Academic Press, New York, 1996.

A. HARTEN, High Resolution Schemes for Hyperbolic Conservation Laws, *Journal of Computational Physics*, Vol. 49, pp. 357–393, 1983.

J. HAUSER, H.G. PAAP, and M. SPEL, *Applied grid generation for complex domains in 2 and 3 dimensions, V. K. I. Lecture Series 1992-04, in Computational Fluid Dynamics,* Lecture Series Director Deconinck, H. , Von Karman Institute of Fluid Dynamics, 1992.

S.B. HAZRA, *Shock-Capturing in Computation of High-speed Inviscid and Viscous Flow past Airfoils,* Ph. D. Thesis, IIT Kharagpur, 1997.

S.B. HAZRA, Computation of high-speed flow using nonoscillatory scheme, In: *Hyperbolic Problems: Theory, Numerics, Applications, 7th International Conf., Zürich,* Ed. M. Fey and R. Jeltsch, pp. 465–474, Birkhäuser Verlag, Basel, Boston, Berlin, 1999.

S.B. HAZRA, S.K. CHAKRABARTTY, and P. NIYOGI, Compressible viscous flow past aerofoils using a nonoscillatory finite-volume scheme, *CFD Journal,* vol. 8, pp. 400–409, 1999.

S.B. HAZRA, P. NIYOGI, and S.K. CHAKRABARTTY, Study in non-oscillatory schemes for shock computation using Euler equations, *Computational Fluid Dynamics Journal,* Vol. 7, pp. 163–176, 1998.

G.W. HEDSTROM, Nonreflecting boundary conditions for nonlinear hyperbolic systems, *Journal of Computational Physics,* Vol. 30, pp. 222–237, 1979.

G. HELLWIG, *Partial Differential Equations, An Introduction* (English Translation), Blaisdell Publishing Company, U.S.A., 1964.

P.W. HENKER and S.P. SPEKREIJSE, Multigrid solution of the steady Euler equations, *Proc. Oberwohlfach Meeting on Multigrid Methods,* 1984.

J.L. HESS and R. MARTIN JR., Improved solution for potential flow about arbitrary axisymmetric bodies by the use of a higher order surface source method, NACA CR 134694, U.S.A., 1974.

J.L. HESS and A.M.O. SMITH, Calculation of potential flow about arbitrary bodies, *Progress in Aerospace Sciences*, Vol. 8, pp. 1–138, 1967.

J.O. HINZE, *Turbulence*, McGraw-Hill Publishing Co., New York, 1975.

C.W. HIRT, B.D. NICHOLS, and N.C. ROMERO, SOLA–*A numerical solution algorithm for transient fluid flows,* Los Alamos Scientific Laboratory, New Mexico, UC -34, and UC - 79 d, 1975.

K.A. HOFFMAN, S.T. CHIANG, S. SIDDIQUI, and M. PAPADAKIS, *Fundamental Equations of Fluid Mechanics,* A Publication of Engineering Education Systems™, Wichita, Kansas 67208- 1078, U.S.A., 1996.

H. HOLLANDERS and H. VIVIAND, The numerical treatment of compressible high Reynolds number flows, In: *Computational Fluid Dynamics*, Ed. W. Kollmann, Vol. 2, Hemisphere Publishing Corporations, New York, 1980.

T.L. HOLST, and W.F. BALLHAUS, Conservative implicit schemes for the full potential equation applied to transonic flows, *AIAA Journal*, Vol. 17, No. 2, pp. 145–152, 1979.

T.L. HOLST, Implicit algorithm for the conservative transonic full potential equation using an arbitrary mesh, *AIAA Journal*, Vol. 17, pp. 1038–1045, 1979.

E. HOPF, The partial differential equation $u_t + uu_x = \mu u_{xx}$, *Communication On Pure and Applied Mathematics,* Vol. 3, pp. 201–230, 1950.

H. HUANG and B.R. SEYMOUR, A finite difference method for flow in a constricted channel, *Computers and Fluids*, Vol. 24, pp. 153–160, 1995.

E. ISAACSON and H.B. KELLER, *An Analysis of Numerical Methods,* Wiley, New York, 1966.

R.K. JAIN, *Grid Generation about an Airfoil by an Algebraic Equation Method,* DFVLR Report IB-129-83/40, 1983.

M.K. JAIN, *Numerical Solution of Differential Equations,* 2nd edn., Wiley Eastern, New Delhi, 1984.

A. JAMESON, Iterative solution of transonic flows over airfoils and wings including flows at Mach 1, *Communications On Pure Applied Mathematics*, Vol. 27, pp. 283–309, 1974.

A. JAMESON, Transonic potential flow calculations in conservation form, *Proc. AIAA Second Computational Fluid Dynamics Conf.,* Hartford, pp. 148–161, 1975.

A. JAMESON, Numerical computations of transonic flows with shock waves, In: *Symposium Transsonicum II,* Eds. K. Oswatitsch and D. Rues, IUTAM Symposium Göttingen (1975), Springer Verlag, Berlin-Heidelberg-New York, pp. 384–414, 1976.

A. JAMESON, Transonic flow calculations, In : *Numerical Methods in Fluid Dynamics,* Ed. H.J. Wirz and J.J. Smolderen, Chapter I Hemisphere Pub. Corp., New York, pp. 1–87, 1978.

A. JAMESON, Accleration of transonic potential flow calculations on arbitrary meshes by the multiple grid method, AIAA Paper 79-1458, *Proc. AIAA 4th Comp. Fluid Dynamics Conf.,* Williamsburg, Va, pp. 122–144, 1979.

A. JAMESON, Solution of the Euler equations by a multigrid method, *Applied Mathematical Computations,* Vol. 13, pp. 327–356, pp. 293–302, 1983.

A. JAMESON, Multigrid algorithm for compressible flow calculations, In: *Lecture Notes in Mathematics*, Ed. W. Hackbush and U. Trottenberg, *Proc. of the 2nd European Conference on Multigrid Methods, Cologne,* Vol. 1228, pp. 166–201, 1985.

A. JAMESON, A nonoscillatory shock capturing scheme using flux limited dissipation, *Lectures in Applied Maths.,* Vol. 22, pp. 345–370, 1985a.

A. JAMESON, Successes and challenges in computational aerodynamics, AIAA Paper 87-1184, 1987.

A. JAMESON, Artificial diffusion, upwind biasing, limiters and their effect on accuracy and multigrid convergence in transonic and hypersonic flows, AIAA Paper 93 - 3359, 1993.

A. JAMESON, and BAKER, T.J., Solution of the Euler equations for complex configurations, AIAA Paper, 83-1929, *Proc. AIAA 6th Computational Fluid Dynamic Conference,* Danvers, pp. 293–302, 1983.

A. JAMESON, and D. MAVRIPLIS, Finite volume solution of the two-dimensional Euler equations on a regular triangular mesh, AIAA Paper 85-0435, AIAA 23rd Aerospace Sciences Meeting, Reno, 1985.

A. JAMESON, W. SCHMIDT, and E. TURKEL, Numerical solution of the Euler equations by finite volume methods using Runge-Kutta time stepping schemes, AIAA Paper 81-1259, *Proc. AIAA 14th Fluid and Plasma Dynamics Conference,* June 23–25, Palo Alto, California, 1981.

A. JAMESON, and YOON, S., Multigrid solution of the Euler equations using implicit schemes, *AIAA Paper* 85-0293, AIAA 23rd Aerospace Sciences Meeting, Reno, 1985.

A. JAMESON and S. YOON, LU implicit schemes with multiple grids for the Euler equations, AIAA Paper 86-0105, AIAA 24th Aerospace Sciences Meeting, Reno, 1986.

A. JAMESON, and E. TURKEL, Implicit schemes and LU decompositions, *Mathematical Computations*, Vol. 37, pp. 385–397, 1981.

M.A. JAWSON and G.T. SYMM, *Integral Equation Methods in Potential Theory and Elastostatics,* Academic Press, London, 1977.

V.G. JENSEN, Viscous flow around a sphere at low Reynolds numbers(≤ 40), *Proceedings of the Royal, Soceity of London, Series A,* Vol. 249, pp. 346–366, 1959.

D.A. JOHNSON and L.S. KING, A new turbulence closure model for boundary layer flows with strong adverse pressure gradients and separation, AIAA Paper 84-0175, 1984.

F.T. JOHNSON and P.E. RUBBERT, Advanced panel–type influence coefficient methods applied to subsonic flows, AIAA Paper No. 75–50, 1975.

W.P. JONES and B.E. LAUNDER, The prediction of relaminarization with a two-equation model of turbulence, *International Journal of Heat and Máss Transfer,* Vol. 15, pp. 301–314, 1977.

C.W. JONES and E.J. WATSON, Two dimensional boundary layers, In: *Laminar Boundary Layers,* Ed. L. Rosenhead, Oxford University Press, London, 1963.

J.C. KALITA, D.C. DALAL, and A.K. DASS, A class of higher order compact schemes for the unsteady two-dimensional convection-diffusion equation with variable convection coefficients, *International Journal for Numerical Methods in Fluids,* Vol. 38, pp. 1111–1131, 2002.

J.C. KALITA, D.C. DALAL, and A.K. DASS, Fully compact higher-order computation of steady-state natural convection in a square cavity, *Physical Review E,* Vol. 64, No. 6, p. 066703(13), 2001.

K. KARAMCHETI, *Principles of Ideal Fluid Aerodynamics,* Wiley, New York, 1966.

J.B. KELLER, and D. GIVOLI, Exact non-reflecting boundary conditions, *Journal of Computational Physics*, Vol. 82, pp. 172–192, 1989.

O.D. KELLOG, *Potential Theory,* Dover Publications, New York, 1929.

P.K. KHOSLA and S.G. RUBIN, *Computers and Fluids,* Vol. 2, pp. 207–209, 1974.

L.S. KING, A comparison of turbulence closure models about airfoils, AIAA Paper 87-0418, AIAA 25th Aerospace Sciences Meeting, Reno, 1987.

E.B. KLUNKER, Contributions to the methods for calculating the flow about thin lifting wings at transonic speeds—Analytical expressions for the far-field, NASA TND-6530, USA, 1971.

W. KORDULLA and M. VINOKUR, Efficient computation of volume in flow predictions, *AIAA Journal,* Vol. 21, No. 6, pp. 917–918, 1983.

W. KRAUS, Panel methods in aerodynamics, In: *Numerical Methods in Fluid Dynamics,* Ed. H.J. Wirz and J.J. Smolderen, Hemisphere Publications, Washington D.C., pp.237–297, 1978.

H.O. KREISS, Initial boundary value problems for hyperbolic systems, *Communications On Pure and Applied Mathematics*, Vol. 13, pp. 277–298, 1970.

N. KROLL and R.K. JAIN, *Solution of Two-Dimensional Euler Equations — Experience with a Finite-Volume Code,* (also in DFVLR-IB-129-84/19, 1984), Report DFVLR -FB 87 -41, Germany, 1987.

H. LAMB, *Hydrodynamics,* 6th edn., Dover, New York, 1945, pp. 56–61, p. 501.

L.D. LANDAU and E.M. LIFSHITZ, Fluid mechanics, In: *Course of Theoretical Physics,* Trans. J.B. Sykes and W.H. Reid, Vol. 6 Pergamon Press, Oxford, New York, 1989.

B.E. LAUNDER and D.B. SPALDING, *Lectures in Mathematical Models of Turbulence, Academic Press,* New York, 1972.

B.E. LAUNDER and D.B. SPALDING, *Computational Methods in Applied Mechanical Engineering*, Vol. 3, pp. 269–289, 1974.

P.D. LAX, Weak solutions of nonlinear hyperbolic equations and their numerical computation, *Communications On Pure and Applied Mathematics*, Vol. 7, pp. 159–193, 1954.

P.D. LAX, Hyberbolic System of Conservation Laws and the Mathematical Theory of Shock Waves, *SIAM,* Philadelphia, 1972.

P.D. LAX and B. WENDROFF, Systems of conservation laws, *Communications On Pure and Applied Mathematics,* Vol. 13, p. 217, 1960.

G.C. LAYEK, *Computational Study of Laminar Separated Flow Fields and Convective Mass Transfer,* Ph. D Thesis, Mathematics Dept., IIT Kharagpur, 1996.

G.C. LAYEK, P. NIYOGI, M.K. MAITI, and S. MUKHERJEA, Numerical study of viscous flow in a long tube with symmetrical constriction, *Computational Fluid Dynamics Journal,* Vol. 5, pp. 451–470, 1996.

S. LEE and G.S. DULIKRAVICH, Acceleration of iterative algorithms for Euler equations of gasdynamics, *AIAA Journal*, Vol. 28(5), pp. 939–942, 1990.

T.S. LEE, Numerical studies of fluid flow through tubes with double constrictions, *International Journal of Numerical Methods Fluids*, Vol. 11, pp. 1113–1126, 1990.

B.P. LEONARD, *Computational Methods of Applied Mechanical Engineering*, Vol. 19, pp. 59–98, 1979.

A. LERAT and J. SIDES, A new finite-volume method for Euler equations with applications to transonic flows, In: *Proc. Num. Methods Aeronaut. Fluid Dyn.*, Ed. P.L. Roe, Academic Press, London, 1982, pp. 245–288.

M.J. LIGHTHILL, Introduction, real and ideal fluids, In: *Laminar Boundary Layers,* Ed. L. Rosenhead, Oxford University Press, London, 1963.

R.W. MACCORMACK, *The Effect of Viscosity in Hypervelocity Impact Cratering,* AIAA Paper 69-354, Cincinnati, Ohio, 1969.

R.W. MACCORMACK, Current Status of Numerical Solutions of the Navier–Stokes Equations, AIAA Paper 85-0032, Reno, Nevada, 1985.

R.W. MACCORMACK and A.J. PAULLAY, Computational Efficiency Achieved by Time-Splitting of Finite-Difference Operators, AIAA Paper, 72–154, 1972.

R. MAGNUS and H. YOSHIHARA, *Inviscid Transonic Flow Over Airfoils,* AIAA Journal, Vol.8, pp. 2157–2162, 1970.

T.R. MAHAPATRA, G.C. LAYEK, and M.K. MAITI, Unsteady laminar separated flow through constricted channel, *International Journal of Non-linear Mechanics,* 2002.

T.K. MAIKAP, T.R. MAHAPATRA, G.C. LAYEK, P. NIYOGI, and A.K. GHOSH, Unsteady laminar separated flow through channel with double constriction, *Computational Fluid Dynamics Journal,* Vol. 12, pp. 53–64, 2003a.

T.K. MAIKAP, T.R. MAHAPATRA, G.C. LAYEK, P. NIYOGI, and A.K. GHOSH, Unsteady laminar separated flow through channel with asymmetric double constriction, *Computational Fluid Dynamics Journal,* Vol. 12, No. 4, pp. 614–621, 2003b.

A. MAJDA and S. OSHER, Numerical viscosity and the entropy condition, *Communications On Pure and Applied Mathematics*, Vol. 32, pp. 797–838, 1979.

S. MAJUMDAR, *Development of a Finite Volume Procedure for Prediction of Fluid Flow Problems with Complex Irregular Boundaries,* Technical Report, SFB 210/T/29, University of Karlsruhe, Germany, 1986.

S. MAJUMDAR, Introduction to Turbulence Models, In: *Short Course on Numerical Computation of Turbulent Flows, Lecture Notes,* Special Publication 9102, NAL Bangalore, 1991.

S. MAJUMDAR, Turbulence modelling for compressible flows, In: *Computational Aerodynamics, NAL-SP-9813,* Ed. S.K. Saxena, 1998.

C.M. MAKSYMIUK, and T.H. PULLIAM, Viscous Transonic Airfoil Workshop Results Using ARC2D, AIAA Paper 87-0415, AIAA Aerospace Sc. Meeting, Reno, 1987.

P. MALAN, *Turbulence Modelling for Industrial Computational Fluid Dynamics,* VKI lecture series 1999-06 on Industrial Computational Fluid Dynamics, 1999.

G.D. MALLINSON and G.I. VAHL DAVIS, *Journal of Computational Physics,* Vol. 12, pp. 435–461, 1973.

E. MARTENSEN, Die Berechnung der Druckverteilung an dicken Gitterprofilen mit Hilfe von Fredholmschen Integralgleichungen zweiter Art, *AVA Nr.23, Götingen 1959; also, Arch.Rat.Mech.Anlysis,* Vol. 3, p. 235, 1959.

L. MARTINELLI, and A. JAMESON, *Validation of a Multigrid Method for the Reynolds Averaged Equations,* AIAA Paper 88-0414, Reno, Nevada, 1988.

B. MASKEW, Prediction of subsonic aerodynamic characteristics : A case for low-order panel methods, *Journal of Aircraft*, Vol. 19, pp. 157–163, 1982.

M.L. MASON, L.E. PUTNAM and R.J. RE, *The Effect of Throat Contouring on Two Dimensional Converging Diverging Nozzles at Static Conditions*, NASA TP 1704, 1980.

J.S. MATHUR and S.K. CHAKRABARTTY, An approximate factorization scheme for elliptic grid generation with control functions, *Numerical Methods for Partial Differential Equations,* Vol. 10, pp. 703–713, 1994.

J.S. MATHUR, K. DHANALAKSHMI, and S.K. CHAKRABARTTY, Application of advanced CFD codes for design and development of SARAS aircraft, *Journal of Aerospace Sciences and Technologies,* Vol. 55, pp. 174–185, 2003.

S. MCKEE and A.R. MITCHELL, Alternating direction methods for two space dimension with a mixed derivative, *The Computer Journal*, Vol. 13, pp. 81–86, 1970.

R.E. MELNIK, *Turbulent Interactions on Airfoils at Transonic Speeds — Recent Developments, Proc. AGARD Conf. on Computation of Viscous–Inviscid Interactions, Colorado Springs,* AGARD CP 291, Paper No. 10, 1980.

A.R. MITCHELL and D.F. GRIFFITHS, *Finite Difference Method in Partial Differential Equations,* Wiley-Interscience, New York, 1980.

C.S. MORAWETZ, On the non-existence of continuous ansonic flow past profiles, I, II, and III, *Communications On Pure and Applied Mathematics,* Vol. 9, pp. 45–68, 1956; Vol. 10, pp. 107–131, 1957; Vol. 11, pp. 129–144, 1958.

L. MORINO, *Unsteady Compresible Potential Flow Around Lifting Bodies : General Theory,* AIAA paper 73–196, Washington D.C., 1973.

L. MORINO and C-C. KUO, Subsonic potential aeodynamics for complex configurations : A general theory, *AIAA Journal*, Vol. 12, pp. 202–208, 1974.

S.K. MUKHERJEA, *Study of Incompressible Separated Flow Problems,* Ph.D Thesis, Dept. of Aerospace Sc., IIT Kharagpur, 1990.

B. MüLLER, *Navier–Stokes Solution of Laminar Transonic Flow over a NACA0012 Airfoil,* FFA Report-140, 1986.

B. MüLLER and A. RIZZI, *Runge-Kutta Finite Volume Simulation of Laminar Transonic Flow over a NACA0012 Airfoil using the Navier–Stokes Equations,* FFA-TN-1980-60, 1980.

E.M. MURMAN, Analysis of embedded shock waves calculated by relaxation methods, *AIAA Journal* Vol. 12, pp. 626–632, 1974.

E.M. MURMAN and J.D. COLE, *Calculation of plane steady transonic flows, AIAA Journal*, Vol. 9, pp. 114–121, 1971.

R.H. NI, A multiple grid scheme for solving the Euler equations, *AIAA Journal*, Vol. 20, pp. 1565–1571, 1982.

G.Y. NIEUWLAND, and B.M. SPEE, Transonic shock-free flow, fact or fiction ?, AGARD Conf. Proc. No. 35, NLR MP 68004U, Netherlands, 1968.

P. NIYOGI, *Inviscid Gasdynamics,* The MacMillan Co. of India Ltd., New Delhi, 1977.

P. NIYOGI, Recent developments in integral equation method in transonic flow, *Proceedings of Indian Academy of Science (Engineering Science),* Vol. 3, pp. 143–167, 1980.

P. NIYOGI, *Transonic flow past thin wings, Proceedings of Indian Academy of Science (Engineering Science),* Vol. 4, pp. 347–361, 1981.

P. NIYOGI, *Integral Equation Method in Transonic Flow,* Vol. 157, Lecture Notes in Physics, Springer Verlag, Berlin Heidelberg, New York, 1982a.

P. NIYOGI, Existence and uniqueness of shock-free transonic flow past symmetrical thin wings at zero incidence, *Acta Mechanica,* Vol. 45, pp. 177–184, 1982b.

P. NIYOGI, Transonic flow past thin wings and profiles, In: *Problems of Nonlinear Continuum Mechanics,* Ed. C.N. Kaul and M. Maiti, Proc. International Symp. on Nonlinear Continuum Mechanics, Indian Society of Theoretical and Applied Mechanics (ISTAM) Kharagpur, 1982c.

P. NIYOGI, *Numerical Analysis and Algorithms,* Tata McGraw-Hiill Pub Company, New Delhi, 2003.

J. NORDSTRöM, The influence of open boundary conditions on the convergence to steady state for the Navier–Stokes equations, *Journal of Computational Physics*, Vol. 85, pp. 210–244, 1989.

J. OLIGER, and A. SUNDSTRöM, Theoretical and practical aspects of some initial boundary value problems in fluid dynamics , *SIAM Journal of Applied Mathematics*, Vol. 35, No. 3, pp. 419–446, 1978.

J.M. ORTEGA and W.C. RHEINBOLDT, *Iterative Solution of Nonlinear Equations in Several Variables,* Academic Press, New York, 1970.

S. OSHER and F. SOLOMON, Upwind difference schemes for hyperbolic system of conservation laws, *Mathematics and Computing,* Vol. 38, pp. 339–374, 1982.

K. OSWATITSCH, *Gasdynamics,* English Trans., G. Kuerti, Academic Press, New York, 1956.

K. OSWATITSCH, *Spezialgebiete der Gasdynamik,* Springer Verlag, Wien, 1977.

K. OSWATITSCH, *Die Geschwindigkeitsverteilung an symmetrischen Profilen beim Auftreten lokaler Überschallgebiete. Acta Physika Austriaca,* Vol. 4, pp. 228–271, 1950; also, ZAMM, Vol. 30, pp. 17–24, 1950; English translation, In: *Contributions to the Developments of Gasdynamics,* Eds., W. Schneider and M. Platzer, Fiedr. Vieveg and Sohn, Braunschweig, pp. 150–187, 1980.

K. OSWATITSCH and J. ZIEREP, *Das Problem des senkrechten Stösses an einer gekrümmten Wand, ZAMM*, Vol. 40, pp. 143–144, 1960.

S.I. PAI, *Introduction to the Theory of Compressible Flow,* Affiliated East-West Press Pvt. Ltd., New Delhi, 1959.

S.V. PATANKAR, *Numerical Heat Transfer and Fluid Flow,* McGraw-Hill, New York, 1980.

S.V. PATANKAR and D.B. SPALDING, A calculation procedure for heat mass and momentum transfer in three dimensional parabolic flows, *International Journal of Heat and Mass Transfer,* Vol. 15, pp. 1787–1806, 1972.

D.W. PEACEMAN and H.H. RACHFORD, Numerical solution of parabolic and elliptic differential equations, *Journal of society for Indusrial and applied Mathematics,* Vol. 3, pp. 28–41, 1955.

H.H. PEARCY, The aerodynamic design of section shapes for swept wings, *Advances in Aerospace Sciences*, Vol. 3, pp. 277–322, 1962.

R. PEYRET and T.D. TAYLOR, *Computational Methods for Fluid Flow,* Springer Verlag, Berlin, Heidelberg, New York, 1983.

W. PRAGER, Die Drukverteilung and Körpern in ebener Potential Strömung, *Physik. Zeitschrift,* XXIX, 865, 1928.

L. PRANDTL, ‘UBER FLÜSSIGKEITSBEWEGUNG SEHR KLEINER REIBUNG. *Verhandl. III. Int. Math. Kongress, Heidelberg (1904), Teubner, Leipzig (1905),* pp. 484–491.

P. PRASAD and R. RAVINDRAN, *Partial Differential Equations,* Wiley Eastern Ltd., New Delhi, 1985.

T.H. PULLIAM and J.T. BARTON, *Euler Computations of AGARD Working Group 07 Airfoil Test Cases*, AIAA Paper 85-0018, 1985.

T.H. PULLIAM and J.L. STEGER, Implicit finite difference simulations of three-dimensional compressible flow, *AIAA Journal*, Vol. 18, pp. 159–167, 1980.

R. RADESPIEL, *A Cell-Vertex Multigrid Method for the Navier–Stokes Equations, NASA TM* 101557, 1989.

R. RADESPIEL and N. KROLL, *Progress in the Development of an Efficient Finite Volume Code for the Three Dimensional Euler Equations,* DFVLR FB 85-31, Institut für Entwurfsaerodynamik, Braunschweig, 1985.

M.M. RAI and P. MOIN, Direct simulations of turbulent flow using finite-difference schemes, *Journal of Computational Physics,* Vol. 96, p. 15, 1991.

J.V. RAKICH, E. VENKATAPATHY, J.C. TANNEHILL, and D. PRABHU, Numerical solution of space shuttle orbiter flowfield, *J. Spacecraft Rockets,* Vol. 21, pp. 9–15, 1984.

V. RAMESH, *Least-Squares Grid-Free Kinetic Upwind Method,* Ph. D. Thesis, Dept. of Aerospace Engineering, Indian Institute of Science, Bangalore, 2002.

A. RESTIVO and J.H. WHITELAW, Turbulence characteristics of the flow downstream of a symmetric plane sudden expansion, *Journal of Fluids Engineering, Transactions of ASME,* Vol. 100, p. 308, 1978.

W.C. REYNOLDS, The potential and limitations of direct and large eddy simulation. In: *J.L. Lumley, Whither turbulence? Turbulence in Crossroads,* Lecture Notes in Physics, (Ed.), Vol. 357, Springer, New York, 1990.

L.F. RICHARDSON, The approximate arithmetical solution by finite differences of physical problems involving differential equations, with an application to the stresses in a masonry dam, *Philosophical Transactions of the Royal Society of London, Series A,* Vol. 210, pp. 310–357, 1910.

R.D. RICHTMEYER and K.W. MORTON, *Difference Methods for Initial-Value Problems,* Interscience, New York, 1967.

A. RIZZI, Computational mesh for transonic aerofoils, In: *Numerical Methods for the Computation of Inviscid Transonic Flow with Shocks*, Eds., A. Rizzi and H. Viviand, Friedr. Vieweg & Sohn, Braunschweig/Wiesbaden, 1981.

A. RIZZI and M. INOUYE, Time split finite volume method for three dimensional blunt body flow, *AIAA Journal,* Vol. 11, pp. 1478–1485, 1973.

P.J. ROACHE, *Computational Fluid Dynamics,* Hermosa, Albuquerque N.M., 1972.

W. RODI, *Turbulence Models and Their Application in Hydraulics,* State of Art Paper, I. A. H. R., Delft, Netherlands, 1980.

W. RODI, S. MAJUMDAR and B. SCHOENUNG, Finite volume methods for two dimensional incompressible flows with complex boundaries, *Computer Methods in Applied Mechanics and Engineering,* Vol. 75, pp. 369–392, 1989.

C. ROSSOW, Comparison of cell centered and cell vertex finite volume schemes, In: *Proc. of the 7th GAMM Conf. on Numerical Methods in Fluid Mechanics*, 1987; Also appeared in Notes on Numerical Fluid Mechanics, Vol. 20, pp. 327–334, Vieweg: Braunschweig, 1988.

P.E. RUBBERT and G.R. SAARIS, *Review and evaluation of a lifting three dimensional lifting potential flow analysis method for arbitrary configurations,* AIAA Paper No. 72-188, 1972.

S.G. RUBIN and P.K. KHOSLA, Navier–Stokes calculations with a coupled strongly implicit method-1. Finite-difference solutions, *Computers in Fluids,* Vol. 9, pp. 163–180, 1981.

M. SALAS, A. JAMESON, and R. MELNIK, *A Comparative Study of the Nonuniqueness Problem of the Potential Equation,* AIAA Paper 83-1888, 1983.

H. SCHLICHTING, and K. GERSTEN, Boundary-Layer Theory, 8th edn., Springer, Berlin, Heidelberg, 2000.

W. SCHNEIDER, *Mathematische Methoden der Strömungs- mechanik,* Vieweg Braunschweig, 1978.

L.J. SEGERLIND, *Applied Finite Element Analysis,* John Wiley & Sons, Inc., New York, London, Sydney, Toronto, 1976.

C.C.L. SELLS, Iterative Method for Thick Cambered Wings in Subcritical Flow, ARC R & M No.3786, London, 1974.

C.C.L. SELLS, Plane subcritical flow past a lifting aerofoil, *Proceedings of the Royal Society A,* Vol. 308, pp. 377–401, 1968.

P.N. SHANKAR and M.D. DESHPANDE, Fluid mechanics in the driven cavity, *Annual Reviews of Fluid Mechanics* Vol. 32, pp. 93–136, 2000; Ann. Rev. U.S.A., 2000.

W. SHYY and C.S. SUN, *Computers in Fluids,* Vol. 22, p. 51, 1993.

C.H. SIEVERDING, Experimental data on two transonic turbine blade sections and comparison with various theoretical methods, In: *Transonic Flows in Turbomachinery,* VKI-LS- 59, 1973.

J.P. SINGH and D. SCHWAMBORN, Navier–Stokes solution for supersonic flow past airfoils, IMACS International Symposium, Bangalore, 1992.

R.E. SMITH (Ed.), *Numerical Grid Generation Techniques,* NASA CP 2166, Proceedings, Workshop held at NASA Langley Research Centre, Hampton, Virginia, October 6–7, 1980.

G.D. SMITH, *Numerical Solution of Partial Differential Equations,* Oxford University Press, Oxford, 1965.

I.J. SOBEY, Observation of waves during oscillatory channel flow, *Journal of Fluid Mechanics,* Vol. 151, p. 395, 1985.

J.C. SOUTH and A. BRANDT, Application of a multi-level grid method to transonic flow calculations, In: *Proc. of Workshop on Transonic Flow Problems in Turbomachinery, Monterey, 1976,* Ed. T.C. Adamson and M.F. Platzer, Hemisphere, pp. 180–206, 1977.

P. SPALART and S. ALLMARAS, A One Equation Turbulence Model for Aerodynamic Flows, Technical Report AIAA-92-0439, 1992.

B.M. SPEE, and R. UIJLENHOET, *Experimental Verification of Shock-Free Transonic Flow around Quasi-Elliptical Aerofoil Sections,* NLR Report MP 68003U, Amsterdam, 1968.

K. SRINIVAS, Computation of cascade flows by a modified CUSP scheme, *Computational Fluid Dynamics J.*, Vol. 8, No. 2, pp. 285–295, 1999.

E. STANEWSKY, W. PUFFET, R. MüLLER, and T.E.B. BATEMAN, Super-critical Aerofoil CAST-7 Surface Pressure, Wake and Boundary Layer Measurements, AGARD-AR-138, 1979.

J.L. STEGER and B.S. BALDWIN, *Shock waves central difference and upwind schemes, Journal of Computational Physics*, Vol. 101, pp. 292–306, 1972.

J.L. STEGER and B.S. BALDWIN, Shock Waves and Drag in the Numerical Calculation of Isentropic Transonic Flow, *NASA TN D*-6997, 1972.

J. STEINHOFF, and A. JAMESON, Multiple solutions of the transonic potential flow equation, *AIAA Journal*, Vol. 20, 1982.

R.C. SWANSON and E. TURKEL, A Multistage Time Stepping scheme for the Navier–Stokes Equations, AIAA-85-0035. AIAA 23rd Aerospace Sc. Meeting, Reno, Nevada, 1985.

R.C. SWANSON, and E. TURKEL, On central difference and upwind schemes, *Journal of Computational Physics*, Vol. 101, pp. 292–306, 1992.

P.K. SWEBY, *High resolution TVD schemes using flux limiters, Lectures in Applied Mathematics,* Vol. 22, pp. 289–209, 1985.

T. TANIGUCHI, K.-P. HOLZ, and C. OHTA, *Grid generation for 2D flow problems, International Journal for Numerical Methods in Fluids*, Vol. 15, pp. 985–997, 1992.

T. THEODORSEN, *Theory of Wing Sections of Arbitrary Shape,* N.A.C.A Report 411, U.S.A., 1931.

P.D. THOMAS and J.F. MIDDLECOFF, Direct control of the grid point distribution in meshes generated by elliptic equations, *AIAA Journal*, Vol. 18, p. 652, 1980.

J.F. THOMPSON, B.K. SONI, and N.P. WEATHERILL (Eds.), *Handbook of Grid Generation,* CRC Press, Boca Raton, London, New York, Washington D. C., 1999.

J.F. THOMPSON, F.C. THAMES, and C.W. MASTIN, "TOMCAT" - A code for numerical generation of boundary–fitted curvilinear coordinate systems on fields containing any number of arbitrary two dimensional bodies, *Journal of Computational Physics,* Vol. 24, p. 274, 1977.

J.F. THOMPSON, Z.U.A. WARSI, and C.W. MARTIN, *Numerical grid generation,* Foundations and Applications, North Holland, New York, 1985.

B. THWAITES (Ed.), *Incompressible Aerodynamics,* Oxford University Press, London, 1960.

A.I. VAN DE VOOREN and L.S. JONG, *Calculation of Incompressible Flow About Airfoils Using Source, Vortex or Doublet Distribution,* Report TW - 86, Delft University, Netherlands, 1970.

J. VAN DER VOOREN, J.W. SLOOF, G.H. HUIZING, and A. VAN ESSEN, Remarks on the suitability of various transonic small perturbation equations to describe three-dimensional transonic flow, Examples of computations using fully conservative rotated difference scheme, In: *Symposium Transsonicum II,* Eds. K. Oswatitsch and D. Rues, IUTAM Symposium Göttingen (1975), Springer Verlag, Berlin-Heidelberg-New York, pp. 557–566, 1976.

J.P. VAN DOORMAL and G.D. RAITHBY, Enhancement of the SIMPLE method for prediction of incompressible fluid flow, *Numerical Heat Transfer,* Vol. 7, pp. 147–163, 1984.

M.D. VAN DYKE, *Perturbation Methods in Fluid Mechanics,* Academic Press, New York, 1964.

B. VAN LEER, *Flux vector splitting for the Euler equations, Proc. 8th Int. Conf. Num. Methods Fluid Dyn., Aachen,* Ed. E. Krause, *Lecture Notes in Physics*, Vol. 170, Springer-Verlag, pp. 507–512, 1982.

R. VARGA, *Matrix Iterative Analysis,* Prentice-Hall, Englewood Cliffs, N.J., 1962.

V. VENKATAKRISHNAN and A. JAMESON, Computation of unsteady transonic flows by the solution of Euler equations, *AIAA Journal*, Vol. 26, pp. 974–981, 1988.

M. VINOKUR, *An Analysis of Finite-Difference and Finite-Volume Formulations of Conservation Laws,* NASA Contractor Report 177416, 1986.

H. VIVIAND, In: *7th Int. Conf. Num. Methods Fluid Dyn.*, Ed. W.C. Reynolds and R.W. MacCormack, Lecture Notes in Physics, Berlin, New York, Vol. 141, pp. 44–54, 1981.

TH. VON KARMAN, Engineer grapples with nonlinear problems, *Von karman, Collected Works,* Pergamon Press, New York, 1941.

J. VON NEUMANN and R.D. RICHTMYER, A Method for the numerical calculation of hydrodynamical shocks, *Journal of Applied Physics,* Vol. 21, pp. 232–257, 1950.

R.F. WARMING and R.M. BEAM, Upwind second-order difference schemes and applications in unsteady aerodynamic flows, *Proc. AIAA 2nd Computational Fluid Dynamics Conf.*, Hartford, Connecticut, pp. 17–28, 1975.

R.F. WARMING and B.J. HYETT, The modified equation approach to the stability and accuracy analysis of finite-difference methods, *Journal of Computational Physics,* Vol. 14, pp. 159–179, 1974.

N.P. WEATHERILL, Grid generation, VKI Lecture series 1990-10, Von Karman Institute of Fluid Dynamics, Belgium, 1990.

G.B. WHITHAM, *Linear and Nonlinear Waves,* Wiley, New York, 1974.

F.A. WOODWARD, Analysis and design of wing-body combinations at subsonic and supersonic speeds, *Journal of Aircraft*, Vol. 5, pp. 528–534, 1968.

N.N. YANENKO, *The Method of Fractional Steps,* M. Holt, (trans.), Springer Verlag, Berlin, 1971.

J.Y. YANG, C.A. HSU, and W.H. HUI, Solution of the steady Euler equations in a generalized lagrangian formulation, *AIAA Journal*, Vol. 31, pp. 266–271, 1993.

H.C. YEE, On the implementation of a class of upwind schemes for systems of hyperbolic conservation laws, NASA TM 86839, 1985.

H.C. YEE, A class of high-resolution explicit and implicit shock-capturing methods, NASA TM 101088, February, 1989.

Index